Bo Kågström Jack Dongarra
Erik Elmroth Jerzy Waśniewski (Eds.)

Applied Parallel Computing

Large Scale Scientific and Industrial Problems

4th International Workshop, PARA'98
Umeå, Sweden, June 14-17, 1998
Proceedings

Volume Editors

Bo Kågström
Erik Elmroth
Umeå University, Dept. of Computing Science and HPC2N
S-901 87 Sweden
E-mail: {bo.kagstrom,elmroth}@cs.umu.se

Jack Dongarra
University of Tennessee, 107 Ayres Hall
Knoxville, TN 37996-1301, USA
E-mail: dongarra@cs.utk.edu

Jerzy Waśniewski
Danish Computing Centre for Research and Education
DTU, UNI C, Bldg. 304, DK-2800 Lyngby, Denmark
Jerzy.Wasniewski@uni-c.dk

Cataloging-in-Publication data applied for

Die Deutsche Bibliothek - CIP-Einheitsaufnahme

Applied parallel computing : large scale scientific and industrial problem ; 4th international workshop ; proceedings / PARA '98, Umeå, Sweden, June 14 - 17, 1998. Bo Kågström ... (ed.). - Berlin ; Heidelberg ; New York ; Barcelona ; Hong Kong ; London ; Milan ; Paris ; Singapore ; Tokyo : Springer, 1998
(Lecture notes in computer science ; Vol. 1541)
ISBN 3-540-65414-3

CR Subject Classification (1998): G.1-2, G.4, F.1-2, D.1-3, J.1

ISSN 0302-9743
ISBN 3-540-65414-3 Springer-Verlag Berlin Heidelberg New York

Typesetting: Camera-ready by author
SPIN 10693025 06/3142 – 5 4 3 2 1 0 Printed on acid-free paper

Lecture Notes in Computer Science 1541
Edited by G. Goos, J. Hartmanis and J. van Leeuwen

Springer

Berlin
Heidelberg
New York
Barcelona
Hong Kong
London
Milan
Paris
Singapore
Tokyo

Preface

The Fourth International Workshop on Applied Parallel Computing (PARA'98) was held in Umeå, Sweden, June 14-17, 1998. The workshop was organized by the High Performance Computing Center North (HPC2N) and the Department of Computing Science at Umeå University. The general theme for PARA'98 was Large Scale Scientific and Industrial Problems, focusing on:

- High-performance computing applications in academia and industry,
- Tools, languages and environments for high-performance computing,
- Scientific visualization and virtual reality applications in academia and industry,
- Future directions in high-performance computing and networking.

The workshop attracted over 140 people representing 18 different countries. PARA'98 was an international forum for idea and competence exchange for specialists in high performance and parallel computing, visualization, and scientists from industry and academia involved in solving large scale computational problems. The workshop program included 20 invited presentations and 64 contributed presentations that were selected by the PARA'98 steering committee. These proceedings reflect the results of this meeting.

The PARA'98 meeting began with a one-day tutorial followed by a three-day workshop. The tutorial program included three topics: Tools and Languages for High Performance Computing (Dennis Gannon, Indiana University), Projection Based Virtual Environments for Collaboration in Scientific Visualization, Industrial Design and Art (Dan Sandin, University of Illinois, Chicago), and Scientific Visualization and Computational Steering, (Chris Johnson, University of Utah). Over 80 people attended the tutorials.

The first three PARA workshops were held at the Technical University of Denmark (DTU), Lyngby (1994, 1995, and 1996). Following PARA'96 an international steering committee for the PARA meetings was appointed and the committee decided that a workshop should take place every second year in one of the Nordic countries. One important aim of these workshops is to strengthen the ties between HPC centers, academia, and industry in the Nordic countries as well as worldwide. Sweden and Umeå University organized the 1998 workshop and the next workshop in the year 2000 will take place at the University of Bergen in Norway.

September 1998

Bo Kågström
Jack Dongarra
Erik Elmroth
Jerzy Waśniewski

Organization

PARA'98 was organized by the High Performance Computing Center North (HPC2N) and the Department of Computing Science at Umeå University.

Organizing Committee

Conference Chair:	Bo Kågström (Umeå University, Sweden)
Conference Coordinator:	Erik Elmroth (Umeå University, Sweden)
Local Organization:	Krister Dackland (Umeå University, Sweden)
	Lena Hellman (Umeå University, Sweden)
	Per Ling (Umeå University, Sweden)
	Mats Nylén (Umeå University, Sweden)
	Peter Poromaa (Umeå University, Sweden)

Steering Committee

Petter Bjørstad	University of Bergen (Norway)
Jack Dongarra	University of Tennessee and Oak Ridge National Laboratory (USA)
Björn Engquist	PDC, Royal Institute of Technology, Stockholm (Sweden)
Kristjan Jonasson	University of Iceland, Reykjavik (Iceland)
Bo Kågström	Umeå University and HPC2N (Sweden), *PARA'98 Chairman*
Risto Nieminen	Helsinki University of Technology, Espoo (Finland)
Karstein Sørli	SINTEF, Dept of Industrial Mathematics, Trondheim (Norway)
Olle Teleman	Center for Scientific Computing (CSC), Espoo (Finland)
Jerzy Waśniewski	Danish Computing Centre for Research and Education (UNI•C), Lyngby (Denmark), *PARA'94–96 Chairman*

Sponsoring Institutions

Swedish Council for High Performance Computing (HPDR)
Swedish Natural Science Research Council (NFR)
Swedish Research Council for Engineering Sciences (TFR)
Umeå University (Rector, HPC2N, Department of Computing Science)
IBM Sweden

Acknowledgments

We acknowledge the enthusiastic work of the steering committee and the local organizing committee. We also acknowledge the following people for the assistance and support in the organization of PARA'98: Inga Bohman, Lena Carneland, Anne-Lie Persson, Inger Sandgren; Anders Backman, Erik Bågfors, Torbjörn Johansson, Mikael Rännar, Åke Sandgren, and Björn Torkelsson. PARA'98 would not have been possible without the personal involvement of all these people. We thank Krister Dackland, Isak Jonsson, and Per Ling for their professional assistance in editing the proceedings. Finally, we also would like to thank the sponsoring institutions for their generous financial support.

Table of Contents

Speaker information.[1]

[1] Bold style indicates the invited speaker. Underline indicates the speaker.

Communications Latency Hiding Techniques for a Reconfigurable Optical Interconnect: Benchmark Studies[1]

Ahmad Afsahi Nikitas J. Dimopoulos
Department of Electrical and Computer Engineering, University of Victoria
P.O. Box 3055, Victoria, B.C., Canada, V8W 3P6
{aafsahi, nikitas}@ece.uvic.ca

Abstract. Communication overhead adversely affects the performance of multicomputers. In this work, we present evidence (through the analysis of several parallel benchmarks) that there exists communications locality, and that it is "structured". We have used this in a number of heuristics that "predict" the target of subsequent communications. This technique, can be applied directly to reconfigurable interconnects (optical or conventional) to hide the communications latency by reconfiguring the interconnect concurrently to the computation.

1.0 Introduction

Message-passing multicomputers are composed of a large number of processor/memory modules that communicate with each other by exchanging messages through their interconnection networks. Optics is ideally suited for implementing interconnection networks because of its superior characteristics over electronics [13,15]; optical signals do not interact with each other, and an optical network can be reconfigured on demand. We have introduced [1] a *reconfigurable optical network*, OK_N, consisting of N computing nodes. A node is capable of connecting directly to any other node. Connections are established dynamically by reconfiguring the interconnect, and remain established until they are explicitly destroyed. A block diagram of the network is shown in Figure 1.

Circuit-switching with k-port or *single*-port models with full-duplex communication is assumed.

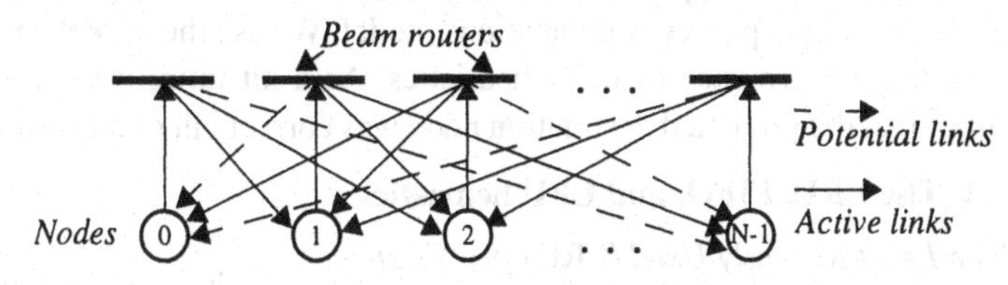

FIGURE 1. OKN, a massively parallel computer interconnected by a complete free-space optical interconnection network

The node-to-node communication delay is modeled as $T = d + t_s + l_m\tau$ with d being the reconfiguration delay, t_s the setup time, l_m the length of the message, τ the per unit transmission time. The setup time t_s [6] and reconfiguration delay d, are the major contributors of the communication delay T, both being of the order of several μs . Several researchers are working to minimize this communication delay by using *active* [7] or *fast* messages [14]. In this work, we are particularly interested in techniques that hide the reconfiguration delay, d.

It is obvious that if a link is already in place, then the configuration phase is not needed with a commensurate savings in the message transmission time. This can be

1. This research was supported through grants from the Natural Sciences and Engineering Research Council of Canada.

accomplished, if the target of the communication operation can be "predicted" before the message itself is available. If the communication operation is regular and known, it is possible to determine the destinations and the instances that these are to be used [1]. However, if the algorithm is not known, the above approach cannot be used.

Many parallel algorithms are built from loops consisting of computation and communication phases. Hence, communication patterns may be repetitive. This has motivated researchers to find *communications locality* properties for parallel applications [9,10]. By communications locality, we mean that if a certain source-destination pair has been used it will be reused with high probability by a portion of code that is "near" the place that was used earlier, and that it will be reused in the near future. If communications locality exists in parallel applications, then it is possible to *cache* the configuration that a previous communication request has made and reuse it at a later stage. Caching in the context of this discussion will mean that when a communication channel is established it will remain established until it is explicitly destroyed.

The main focus of our work is to explore whether the target of a communication request can be "predicted". For these studies, we used the MPI [11] implementation of five parallel applications (BT, SP, LU, MG, CG) of the NAS parallel benchmark suite (NPB) [3] on a network of SUN, and HP workstations using MPICH [8]. The target prediction heuristics developed adapt based on the past history and can be used in circuit switched interconnects including [5,16]. Section 2 analyzes the proposed heuristics, while we conclude with Section 3.

2.0 Latency hiding heuristics

The heuristics proposed in this section predict the destination node of a subsequent communication request based on a past history of communication patterns. Although our heuristics are applicable to any-port model, we shall present most of our results under the single port communications model. We use the *hit ratio* to establish and compare the performance of these heuristics. As a hit ratio, we define the percentage of times that the predicted destination node was correct out of all communication requests.

2.1 The LRU, FIFO, and LFU heuristics

The *Least Recently Used* (LRU) [9], *First-In-First-Out* (FIFO) and *Least Frequently Used* (LFU) heuristics, all maintain a set of k (k is the *window size*) message destinations. If the next message destination is already in the set, then a hit is recorded. Otherwise, a miss is recorded and the new destination replaces one of the destinations in the set according to which of the LRU, FIFO or LFU strategies is adopted.

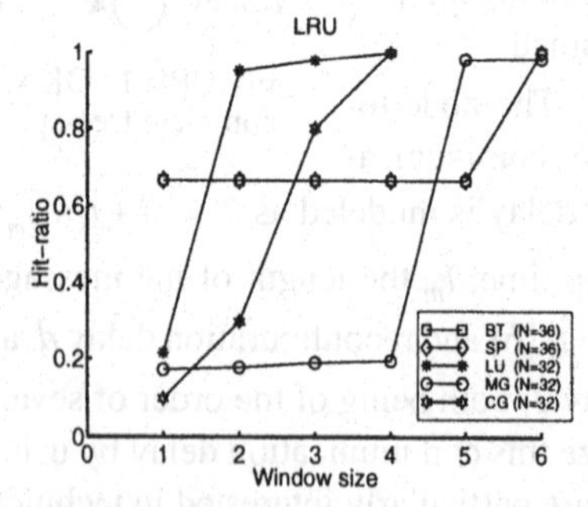

FIGURE 2. Effects of the LRU heuristic on the NAS benchmarks

The window size, k, corresponds to the number of ports used. Figure 2, shows the result of the LRU heuristic on the benchmarks. It can be seen that the hit-ratios in all benchmarks approach 1 as the window size increases. The performance of the FIFO algorithm is almost the same with LRU for all benchmarks. Additionally, the performance of the LFU algorithm [2] is better than LRU and FIFO, the exception being the LU benchmarks for $k = 2$.

2.2 The Single-cycle heuristic

The *Single-cycle* heuristic is based on the fact that if a group of destinations are requested cyclically, then a single port can accommodate these requests by ensuring that the connection to the subsequent node in the cycle can be established as soon as the current request ends. This heuristic implements a simple cycle discovery algorithm. Starting with a *cycle-head* node (this is the first node that is requested at start-up, or the node that causes a miss), we log the sequence of requests until the cycle-head node is requested again. This stored sequence constitutes a cycle and is used to predict the subsequent requests. If the predicted node coincides with the subsequent requested node, then we record a hit. Otherwise, we record a miss and the cycle formation stage commences with the cycle-head being the node that caused the miss.

Request sequence	1	3	5	6	1	3	5	6	7	7	1	3	5	6
Predicted	-	-	-	-	3	5	6	1	-	7	-	-	-	-
	Cycle formation										*Cycle formation*			

The example to the right illustrates the heuristic used. The top trace represents the sequence of requested destination nodes, while the bottom trace represents the predicted nodes according to the Single-cycle heuristic. The arrows with the cross represent misses, while the ones with the circle represent hits. The "dash" in place of a predicted node indicates that a cycle is being formed; thus no prediction is offered.

Figure 3, shows the behavior of this algorithm. This algorithm behaves much better than the LRU, FIFO and LFU heuristics for the LU, MG, and CG benchmarks. However its performance deteriorates for the BT and SP benchmarks. One of the reasons is that in these benchmarks there exist cycles of length one (such as the one composed of node 7 in the example above) always resulting into two misses. The Single-cycle2 heuristic presented next improves the performance ..

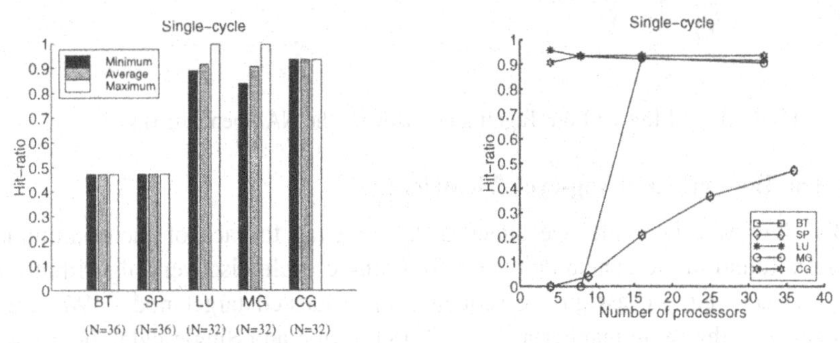

FIGURE 3. Effects of the Single-cycle heuristic on the NAS benchmarks

2.3 The Single-cycle2 heuristic

The *Single-cycle2* heuristic is identical to the single-cycle heuristic with the addition that during cycle formation, the previously requested node is offered as the predicted node. This heuristic performs much better than the LRU, FIFO, and LFU algorithms for the LU, MG, and CG benchmarks under single-port modeling and almost identically for the BT, and SP benchmarks [2].

2.4 The Tagging heuristic

The *Tagging* heuristic assumes a static communications environment in the sense that a particular communications request (send) in a section of code, will be to the same target node with a large probability. Therefore, as the execution trace nears the section of code in question, it can cause the communications environment to establish the connection to the target node before the actual communications request is issued[1].

For our experiments, we attach a different tag to each of the communication requests found in the benchmarks. To this tag, we assign the requested target node. A hit is recorded if in subsequent encounters of the tag, the requested communications node is the same as the target already associated with the tag. Otherwise, a miss is recorded and the tag is assigned the newly requested target node. This tagging technique is similar to the technique used in the branch cache of the MC68060 [4].

As it can be seen in Figure 5 the tagging heuristic results in an excellent performance (hit ratios in the upper 90%) for all the benchmarks except CG. The reason is that the CG benchmark includes send operations with a target address calculated based on a loop variable. Thus, the same section of code cycles through a number of different target addresses. As we have seen before, the Single-cycle and the Single-cycle2 heuristics are excellent in discovering such cyclic occurrences. In the following, we propose combined tag and cycle heuristics which are able to alleviate this problem.

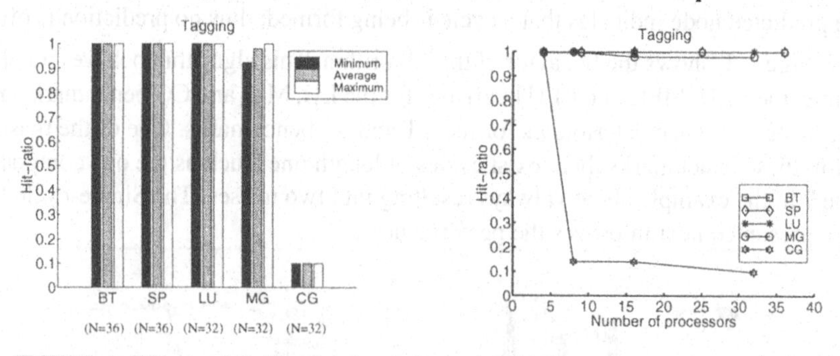

FIGURE 4. Effects of the Tagging heuristic on the NAS benchmarks

2.5 The Tag-cycle and Tag-cycle2 heuristics

In the *Tag-cycle* heuristic, we attach a different tag to each of the communication requests found in the benchmarks and do a Single-cycle discovery algorithm on each tag. To each tag, we assign the sequence of requested target nodes. We use these sequences in the same manner as in the Single-cycle and Single-cyle2 heuristics (c.f.

1. This can be implemented with the help of the compiler through a *pre-connect(tag)* operation which will force the communications system to establish the communications connection before the actual communications request is issued. This technique is similar to the prefetch operation advocated by T. Mowry and A. Gupta[12].

Section 2.2 and Section 2.3). The Tag-cycle heuristic performs exceptionally well across all the benchmarks, as shown in Figure 5.

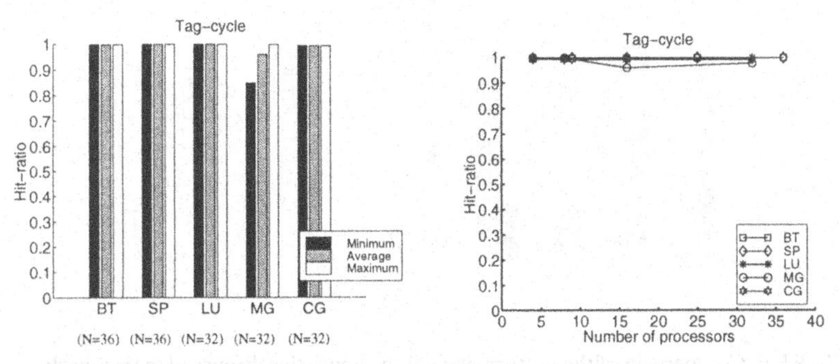

FIGURE 5. Effects of the Tag-cycle heuristic on the NAS benchmarks

The *Tag-cycle2* heuristic is identical to the Tag-cycle heuristic with the addition that during cycle formation, similar to the Single-cycle2 heuristic, the previously requested node is offered as the predicted node. The performance of Tag-cycle2 heuristic is slightly better than Tag-cycle.

3.0 Discussion and Conclusions

In this work, we presented a number of heuristics which can be used to "predict" the target of a communications request before the actual request is issued. These heuristics use the pattern of communications and are designed to extract dependencies which are embedded in these patterns. For these studies, we used the publicly available NAS parallel benchmarks. The results of our studies give strong evidence for the existence of "communications locality" at least in the benchmark programs studied.

The heuristics proposed are only possible because of the existence of *communications locality*. This is a very desirable property since it allows us to effectively hide the cost of establishing such communications links, providing thus the application with the raw power of the underlying hardware (e.g. a reconfigurable optical interconnect).

Figure 6, presents a comparison of the performance of the heuristics presented in this work only under single-port assumption. That is only one communications channel is available at any time, and this is reconfigured on demand. As it can be seen, the tagging+cycle2 heuristic is the best for all benchmarks and its hit-ratio is consistently very high (approaching 100%) for all the benchmarks considered here.

We are confident that the existence of "communications locality" and the resulting latency hiding techniques will usher a new era in interconnection technologies by allowing the use of reconfiguarbility and fast optical fabrics.

References

1. A. Afsahi and N. J. Dimopoulos, "Collective Communications on a Reconfigurable Optical Interconnect," Proceedings of the International Conference on Principles of Distributed Systems, Dec., 1997, pp. 167-181
2. A. Afsahi and N. J. Dimopoulos "Communications Latency Hidinng Techniques for a Reconfigurable Optical Interconnect: Benchmark Studies" *Technical Report* ECE-98-2, Department of Electrical and Computer Engineering, University of Victoria, June 1998.

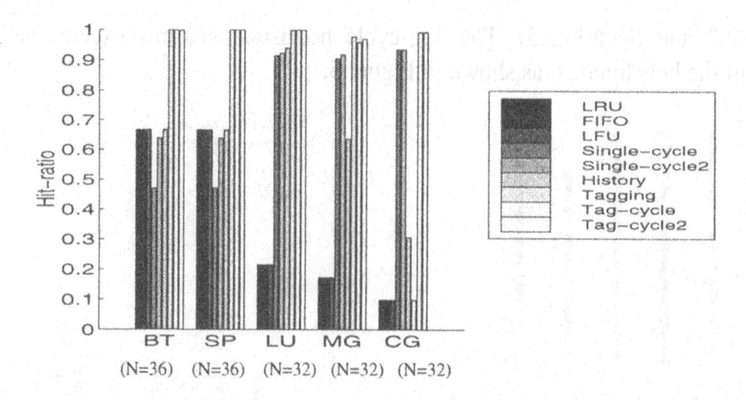

FIGURE 6. Comparison of the performance of the heuristics discussed in this work for the five NAS benchmarks under single-port modelling

3. D. H. Bailey, et al., "NAS Parallel Benchmark Result 3-94," Proceedings of Scalable High-Performance Computing Conference, 1994, pp. 111- 120
4. J. Circello, et al., "The Superscalar Architecture of the MC68060," IEEE Micro, Volume 15, Number 2, April 1995, pp. 10-21
5. B. V. Dao, S. Yalamanchili, and J. Duato, "Architectural Support for Reducing Communication Overhead in Multiprocessor Interconnection Networks" Proceedings, Third International Symposium on High Performance Computer Architecture, 1997, pp. 343-352
6. J. J. Dongarra and T. Dunigan, "Message-Passing Performance of Various Computers," Concurrency, Vol. 9, No. 10, Dec. 1997, pp. 915-926
7. T. V. Eicken, et al., "Active Messages: A Mechanism for Integrated Communication and Computation," Proceedings of the 19th Annual International Symposium on Computer Architecture, May 1992, pp. 256-265
8. W. Gropp and E. Lusk, "User's Guide for MPICH, a Portable Implementation of MPI," Argonne National Laboratory, Mathematics and Computer Science Division, ANL/MCS-TM-ANL-96/6
9. J. Kim and D. J. Lilja, "Characterization of Communication Patterns in Message-Passing Parallel Scientific Application Programs, "Workshop on Communication, Architecture, and Applications for Network-based Parallel Computing, International Symposium on High Performance Computer Architecture, February 1998, pp. 202-216
10. D. G. de Lahaut and C. Germain, "Static Communications in Parallel Scientific Programs" Proceedings of PARLE'94, Parallel Architecture and Languages, Athen, Greece, July 1994
11. Message Passing Interface Forum: MPI: A Message-Passing Interface Standard. Version 1.1 (June 1995)
12. T. Mowry and A. Gupta, "Tolerating Latency Through Software-Controlled Prefetching in Shared-Memory Multiprocessors," Journal of Parallel and Distributed Computing, 12(2), 1991, pp. 87-106
13. R . A. Nordin, et al., "A System Perspective on Digital Interconnection Technology," IEEE Journal of Lightwave Technology, Vol. 10, June 1992, pp. 801-827
14. S. Pakin, M. Lauria, and A. Chien, " High Performance Messaging on Workstation: Illinois Fast Messages (FM) for Myrinet," Proceedings of Supercomputing'95, Nov., 1995
15. G. I. Yayla, P.J. Marchand and S.C. Esener "Speed and Energy Analysis of Digital Interconnections: Comparison of On-chip, Off-chip and Free-Space Technologies" Applied Optics, Vol. 37, No.2, Jan. 1988, pp. 205-227.
16. X. Yuan, R. Melhem, and R. Gupta, "Compiled Communication for All-Optical TDM Networks," Proceeding's of Supercomputing'96, 1996

Multifrontal Solvers Within the PARASOL Environment*

Patrick Amestoy[1], Iain Duff[2], and Jean-Yves L'Excellent[3]

[1] ENSEEIHT-IRIT, Toulouse, France.
amestoy@enseeiht.fr
[2] RAL, Oxfordshire, England and CERFACS, Toulouse, France.
isd@rl.ac.uk
[3] CERFACS, Toulouse, France.
excelle@cerfacs.fr

Abstract. PARASOL is an ESPRIT IV Long Term Research Project whose main goal is to build and test and portable library for solving large sparse systems of equations on distributed memory systems. There are twelve partners in five countries, five of whom are code developers and five end users. The software is written in Fortran 90 and uses MPI for message passing. There are routines for both direct and iterative solution of symmetric and unsymmetric systems. The final library will be in the public domain. We discuss the PARASOL Project with particular emphasis on the algorithms and software for direct solution that are being developed by RAL and CERFACS. The underlying algorithm is a multifrontal one with a switch to ScaLAPACK processing later in the factorization (and solution).

1 Introduction

We consider the solution of both symmetric and unsymmetric systems of sparse linear equations. A new parallel distributed memory multifrontal approach is described. A parallel asynchronous algorithm with dynamic scheduling of the computational tasks has been developed to handle numerical pivoting efficiently. We discuss some of the main algorithmic choices and compare both implementation issues and the performance of the LDL^T and LU factorizations. Performance results on an IBM SP2 and an SGI Origin 2000 shows the efficiency and the potential of the method.

This work has been performed as a Work Package within the PARASOL Project. PARASOL is an ESPRIT IV Long Term Research Project (No 20160) for "An Integrated Environment for Parallel Sparse Matrix Solvers". The main goal of this Project, which started on January 1 1996, is to build and test a portable library for solving large sparse systems of equations on distributed memory systems. There are twelve partners in five countries, five of whom are code developers and five end users. The software is written in Fortran 90 and

* This work was supported in part by the EU ESPRIT IV LTR Project Number 20160

uses MPI for message passing. There are routines for both the direct and the iterative solution of symmetric and unsymmetric systems. The final library will be in the public domain.

CERFACS and RAL with the collaboration of ENSEEIHT-IRIT are developing the direct solver based on a multifrontal approach originally developed by Duff and Reid (1983,1984) and extended to shared memory computers by Duff (1986) and Amestoy and Duff (1989,1993) and subsequently to a prototype version using PVM by Espirat (1996). The integration of this direct code into the PARASOL Library and comments on the performance of earlier versions of the code can be found in Amestoy, Duff, L'Excellent and Plecháč (1998*b*) and Amestoy, Duff and L'Excellent (1998*a*). In this paper, we give an extended abstract but also include some new results in Section 3.

2 Main implementation issues

The current version of our sparse direct solver MUMPS ("MUltifrontal Massively Parallel Solver") solves the linear system of equations $\mathbf{Ax} = \mathbf{b}$, where $\mathbf{A}$ is either unsymmetric or symmetric positive definite. In addition, there are various options that include the solution of the transposed system, error analysis, iterative refinement, scaling of the original matrix, and the input of a given ordering for the factorization.

In the current version of MUMPS (Version 2.1.3), parallelism arising both from sparsity and from dense factorizations kernels are exploited, and we distribute the pool of work among the processors. Our model, however, still requires an identified host node to perform the analysis phase, distribute the incoming matrix, collect the solution, and generally oversee the computation. The code is organized with a designated host node and other processors as follows:

1. Analysis. The host constructs an ordering using an approximate minimum degree algorithm based on the symmetrized pattern $\mathbf{A} + \mathbf{A}^T$, and then performs symbolic factorization to generate an assembly (or computational) tree. A mapping of the multifrontal computational tree is then computed, and symbolic information is transferred from the host to the other processors. Using this information, the processors estimate the memory necessary for factorization and solution.
2. Factorization. The host sends appropriate entries of the original matrix to the other processors that are responsible for the numerical factorization. The numerical factorization on each frontal matrix is conducted by a *master* processor (determined by the analysis phase) and one or more *slave* processors (determined dynamically). Each processor allocates an array for contribution blocks and factors; the factors must be kept for the solution phase.
3. Solution. The right-hand side is broadcast from the host to the other processors. These processors compute the solution using the (distributed) factors computed during Step 2, and the solution is assembled on the host.

For both the symmetric and the unsymmetric algorithms used in the code, we have chosen a fully asynchronous approach with dynamic scheduling of the computational tasks. In fact, we combine the main features of static and dynamic approaches. We use the estimation done during analysis to map some of the main computational tasks; the other tasks are dynamically scheduled at execution time. Asynchronous communication was chosen to enable overlapping between communication and computation. Dynamic scheduling was initially used to accommodate numerical pivoting in the factorization. The other important reason for this choice is that, with dynamic scheduling, the algorithm has the potential to adapt itself at execution time, and can remap work and data to a more appropriate processor. The main data structures (original matrix and matrix of the factors) are partially mapped according to the analysis phase. Part of the initial matrix is replicated to enable rapid task migration without data redistribution.

3 Performance

The Approximate Minimum Degree (AMD) ordering of Amestoy, Davis and Duff (1996) is used as the ordering algorithm to permute the matrix and all timings are given in seconds.

Statistics for two test problems from the PARASOL set are given in Table 1. Both of the test problems QUER and CRANKSEG2 are symmetric positive definite. Advantage is taken of symmetry by the "SYMMETRIC CODE", but the matrix is treated as a general unsymmetric sparse matrix by the "UNSYMMETRIC CODE". Characteristics to note, which are common with nearly all PARASOL test problems, are that there are very many more entries in the factors than in the original matrix, that there is a significant number of floating-point operations in the factorization, and that the root node is large enough for computations within it to be efficiently performed in parallel. We also note that, on the matrix QUER, the factors generated by the symmetric version of the code have more than half the number of entries than those generated by the unsymmetric version. This is because, in the symmetric version of the code, we store the L-factors in rectangular rather than lower trapezoidal form and also do not use packed storage at the root node.

We report, in Table 2, results of the symmetric and unsymmetric versions of our code for the test problem QUER on an IBM SP2 located at GMD (Bonn, Germany), where each node is a 66 MHertz processor with 128 MBytes of physical memory and 512 MBytes of virtual memory.

We observe a problem with memory paging due to the use of virtual memory on the SP2. However, speed-ups are good when comparing times with uniprocessor CPU times. Furthermore, comparison of symmetric and unsymmetric performance shows a good performance of the symmetric code, which is almost twice as fast as the unsymmetric code. Note that, even when several processors are being used, the analysis times are still less than the factorization times thus supporting our decision not to parallelize the analysis phase.

Matrix	Order	Nonzeros in matrix ($\times 10^6$)	Nonzeros in factors ($\times 10^6$)	Flops ($\times 10^9$)	Size of root	Time for analysis (on SP2)
SYMMETRIC CODE						
QUER	59122	1.5	12	4.0	1043	3.0
CRANKSEG2	63838	7.1	73	101.9	3127	14.5
UNSYMMETRIC CODE						
QUER	59122	3.0	20	8.0	1043	5.5

Table 1. Statistics for two PARASOL test problems (ordering based on AMD).

Working processors	Time for factorization unsymmetric	symmetric
1 (CPU)	64.9	41.1
1 (elapsed)	299.2	150.4
2 (elapsed)	109.1	21.0
4 (elapsed)	19.1	12.9
8 (elapsed)	15.2	9.3
16 (elapsed)	11.5	6.4
24 (elapsed)	10.1	6.6
32 (elapsed)	10.5	5.8

Table 2. Performance of the unsymmetric and symmetric versions of the code on test problem QUER on an IBM SP2.

Table 3 presents results for the larger test problem CRANKSEG2 both on the IBM SP2 and an Origin 2000 located at Parallab (Bergen).

Machine	Working processors	Time for factorization
SP2	16	1045.34
	24	457.26
	32	139.66
Origin	1	635.4
	2	411.0
	3	275.7
	4	220.4
	5	175.1
	6	158.3
	7	142.9
	8	135.7

Table 3. Results for the symmetric version of the code on CRANKSEG2.

The speed-ups on the SP2 are both due to algorithmic and memory effects. This is not the case on the Origin 2000 where the speed-ups only come from the algorithm.

References

Amestoy, P. R. and Duff, I. S. (1989), 'Vectorization of a multiprocessor multifrontal code', *Int. J. of Supercomputer Applics.* **3**, 41–59.

Amestoy, P. R. and Duff, I. S. (1993), 'Memory management issues in sparse multifrontal methods on multiprocessors', *Int. J. Supercomputer Applics* **7**, 64–82.

Amestoy, P. R., Davis, T. A. and Duff, I. S. (1996), 'An approximate minimum degree ordering algorithm', *SIAM J. Matrix Analysis and Applications* **17**(4), 886–905.

Amestoy, P. R., Duff, I. S. and L'Excellent, J.-Y. (1998*a*), Multifrontal parallel distributed symmetric and unsymmetric solvers, Technical Report RAL-TR-1998-051, Rutherford Appleton Laboratory.

Amestoy, P. R., Duff, I. S., L'Excellent, J.-Y. and Plecháč, P. (1998*b*), PARASOL. An integrated programming environment for parallel sparse matrix solvers, Technical Report RAL-TR-98-039, Rutherford Appleton Laboratory. To appear in Proceedings of Conference HPCI 1998.

Duff, I. S. (1986), 'Parallel implementation of multifrontal schemes', *Parallel Computing* **3**, 193–204.

Duff, I. S. and Reid, J. K. (1983), 'The multifrontal solution of indefinite sparse symmetric linear systems', *ACM Trans. Math. Softw.* **9**, 302–325.

Duff, I. S. and Reid, J. K. (1984), 'The multifrontal solution of unsymmetric sets of linear systems', *SIAM J. Scientific and Statistical Computing* **5**, 633–641.

Espirat, V. (1996), Développement d'une approche multifrontale pour machines à mémoire distribuée et réseau hétérogène de stations de travail, Rapport de stage de 3ieme année, ENSEEIHT-IRIT, Toulouse, France.

Parallelization of a 3D FD-TD Code for the Maxwell Equations Using MPI

Ulf Andersson

PSCI/NADA, Royal Institute of Technology, S-100 44 Stockholm, Sweden.
ulfa@nada.kth.se

Abstract. We have parallelized an existing code that solves the Maxwell equations in the time domain. The code uses finite differences in 3D for discretization. The method is based on a leap frog scheme introduced by Yee in 1966. The first order Mur scheme is used as outer (absorbing) boundary condition. Wave excitation is done with dipoles, i.e. using point sources. All parts of this scheme are local in space. It is therefore suitable to parallelize the code using domain decomposition. This is done with MPI. It is possible to achieve negligible communication time on an IBM SP-2 when each node houses a large enough problem. This is demonstrated with a problem size of $100 \times p \cdot 100 \times 100$ where p is the number of processors. We also give speed-up results for the IBM SP-2 and a Cray J932.

1 Introduction

For linear, isotropic and non dispersive materials in a region of space where there are no electric or magnetic current sources, the Maxwell equations are,

$$\begin{aligned} \nabla \cdot (\epsilon \mathbf{E}) &= \mathbf{0} \,, \\ \nabla \cdot (\mu \mathbf{H}) &= \mathbf{0} \,, \\ \frac{\partial}{\partial t}\mathbf{H} &= -\frac{1}{\mu}\nabla \times \mathbf{E} \,, \\ \frac{\partial}{\partial t}\mathbf{E} &= \frac{1}{\epsilon}\nabla \times \mathbf{H} - \frac{\sigma}{\epsilon}\mathbf{E} \,, \end{aligned} \tag{1}$$

where $\mathbf{E}$ is the electric field vector, $\mathbf{H}$ is the magnetic field vector, σ is the electric conductivity, μ is the magnetic permeability and ϵ is the electric permittivity. The first two equations in (1) are Gauss's Laws for the electric and the magnetic fields. The third equation is Faraday's Law and the fourth equation is Ampere's Law.

The Maxwell equations describe electromagnetic waves such as micro, radio and radar waves. There is thus a substantial need to solve these equations. They can only be solved analytically for a few very simple shapes such as a sphere or a circular cylinder. Hence one has to rely on a mix of experiments and approximative and/or numerical methods. Numerical methods for the Maxwell equations are usually referred to as Computational ElectroMagnetics (CEM).

The Maxwell equations can be solved either in the time domain or the frequency domain. Time domain methods have the advantage that they in on calculation can solve a problem for several frequencies. Frequency domain methods, such as the Methods of Moments (MoM) [3], reduce the volumetric equations to surface equations and thus reduces the problem with one dimension. Another advantage is that after solving a particular problem for one angle of incidence it is relatively easy to solve the same problem for another angle of incidence. It is however difficult to handle cases with varying material properties. This is easier to treat with time domain methods. Another advantages of the time domain techniques is the possibility to follow the pulse evolution in time.

In time domain methods one needs at least ten mesh points per wavelength for practical engineering accuracy. It follows that for moderately high frequencies a large number of mesh points is required to be able to resolve the problem. This implies a need for large memory capacity and computing power, i.e. parallel computers. For really high frequencies it becomes impossible to resolve the problem using time domain methods. Here one has to use high frequency methods such as the Geometrical Theory of Diffraction [1].

In the time domain there are for intermediate frequencies several possible techniques, e.g. finite difference (FD) [6], finite volumes (FV) [4], finite elements (FE) [5] and Transmission Line Modeling (TLM) [2]. Of these methods, FD is the most commonly used.

2 The Yee scheme

Traditionally, the finite difference time domain scheme introduced by Yee in 1966 [8] has been the dominating method within CEM. This explicit FD scheme uses central differences in both space and time on staggered Cartesian grids, i.e. it is a leap frog scheme. Only Ampere's and Faraday's Laws are explicitly discretized. The two Gauss's Laws are implicitly enforced. The Yee scheme is usually referred to by the acronym FD-TD. This acronym was coined by Taflove who gives an excellent description on this scheme in his book [6].

Ampere's and Faraday's Laws constitute a first order hyperbolic system with six unknowns. All six equations in this system have the same structure (assuming $\sigma = 0$). As example on the discretization process, we consider one of the six equations. We take

$$\frac{\partial H_x}{\partial t} = \frac{1}{\mu}\left(\frac{\partial E_y}{\partial z} - \frac{\partial E_z}{\partial y}\right) . \tag{2}$$

The FD-TD updating stencil for (2) is

$$H_x\Big|_{i,j+\frac{1}{2},k+\frac{1}{2}}^{n+\frac{1}{2}} = H_x\Big|_{i,j+\frac{1}{2},k+\frac{1}{2}}^{n-\frac{1}{2}} + \frac{\Delta t}{\mu}\left[\begin{array}{c} \frac{E_y|_{i,j+\frac{1}{2},k+1}^{n} - E_y|_{i,j+\frac{1}{2},k}^{n}}{\Delta z} \\ -\frac{E_z|_{i,j+1,k+\frac{1}{2}}^{n} - E_z|_{i,j,k+\frac{1}{2}}^{n}}{\Delta y} \end{array}\right] , \tag{3}$$

where we have assumed that μ is constant. The updating stencils for the other five field components are similar. Implemented using Fortran (3) becomes

```
Hx(i,j,k) = Hx(i,j,k) +                                          &
          (  (Ey(i,j,k+1)-Ey(i,j  ,k))*Cbdz +                    &
             (Ez(i,j,k  )-Ez(i,j+1,k))*Cbdy  ) .
```

3 Parallelization issues

A 3D code, pscyee, based on the Yee scheme has been developed within the center of excellence PSCI [7]. pscyee is written in Fortran 90. The parallel version of pscyee have been implemented using MPI. The major part of the implementation have been performed on the IBM SP-2 at PDC, but pscyee have also been tested on several other parallel computers.

The parallel version of pscyee uses first order Mur as absorbing boundary condition. Wave excitation is done with point sources or with Huygens' surfaces which are used to model incident electromagnetic fields. For a detailed description of the Mur absorbing boundary condition and Huygens' surfaces, see Taflove's book [6].

Since all involved operations are local in space it is convenient to perform parallelization using domain decomposition. During one time step we need to perform 36 arithmetic operations per cell (24 additions and 12 multiplications).

The domain decomposition makes it necessary for each node to send four and receive four field variables in each space direction. Two magnetic field components are sent upwards and two electric field components are sent downwards. If every node houses a $100 \times 100 \times 100$ block, these messages will contain 10 000 floating point values if the two field variables are sent in separate messages.

We have implemented four different parallelization strategies using MPI. We denote them by:

- SSEND: where all nodes send first using synchronized send and then receive.
- ISEND: where all nodes send first using non blocking send and then receive.
- SENRECV: uses `MPI_SENDRECV` and thus lets the MPI implementation handle the order in which messages are sent and received.
- red-black: where every second node send first using synchronized send and then receive and vice versa for the other half of the nodes.

4 Performance results

The performance of pscyee on a 160 MHz node of the IBM SP-2 is 179 MFlop/s for a problem size of $100 \times 100 \times 100$. One iteration takes approximately 0.2 seconds. The peak performance of this node is 640 MFlop/s. A substantial effort has been put into optimizing the sequential code.

4.1 Scale-up

Figure 1 illustrates the performance of **pscyee** when the problem size is scaled up with the number of processors, p, so that it is $100 \times p \cdot 100 \times 100$. It displays the result for four different MPI implementations. We can see that all but one of them achieve negligible communication time, i.e. the execution time is independent of the number of processors.

The performance model for the SSEND implementation is based on the assumption that the time needed to take a timestep is given by:

$$t = m(p-1)t_{40000} + t_{calc} \tag{4}$$

where p is the number of processors, m is the number of messages sent, t_{40000} is the time it takes to send a message of 40000 bytes and t_{calc} is the time it takes to perform the calculations on a $100 \times 100 \times 100$ block. In this case we have $m = 4$, $t_{40000} \approx 0.6$ms and $t_{calc} \approx 0.2$s.

The performance model for the red-black implementation (NOT drawn in Figure 1) is based on the assumption that the time for each timestep is given by

$$t = 2mt_{40000} + t_{calc} \approx t_{calc} \approx 0.2\text{s} . \tag{5}$$

Notice that time is independent of the number of processors. This is in agreement with the measured performance displayed in Figure 1.

4.2 Speed-up

Figure 2 displays performance results on a problem size of $100 \times 100 \times 100$. It displays the best results achieved for a given implementation and a given number of processors. All possible domain decompositions with the constraint that all nodes should have equal sized blocks have been tested.

It is more difficult to create a performance model for this case. There are two major reasons for this:

1. The computing speed of the codes core, the leap frog updates, changes with the problem size. The speed varies between 160 Mflop/s and 190 Mflop/s on one node.
2. The time to copy to and from the temporary arrays that are sent and received depends on the axis along which the message is sent. Temporary arrays are needed since we want to send 2D slices of the 3D fields that do not lie consequently in memory. It is more efficient to handle this copying explicitly than letting MPI take care of it. The time to perform the copying depends on the stride of the field values in the 2D slice. When the computational domain on each node becomes small, the time to perform this copying will no longer be negligible.

An estimate for the maximum performance for the red-black implementation can be given by comparing calculation time and communication time. For $p = 64$, each node houses a $25 \times 25 \times 25$ block. Measurements show that a timestep takes about $t_{calc} = 3.09$ms for such a block. A node which has a block with no outer boundaries need to send two messages and receive two messages per space dimension. (Two field variables are sent in each message.) The size of these messages is $25^2 * 4 * 2 = 5000$ bytes. Measurement show that such a message takes roughly 0.164ms meaning that the total communication time is $t_{com} = 12 * 0.164 = 1.97$ms. Maximum possible speedup is then given by ($t_1 = 192$ms)

$$\frac{t_{calc} + t_{com}}{t_1} \approx 38. \tag{6}$$

This estimate neglects the copying time mentioned above. As seen in Figure 2 the actual speed-up for 64 nodes is about 30 for the red-black implementation. This difference is due to the copying mentioned above. This conjecture has been proved by removing all copying from the code, which of course will make the results erroneous. This procedure gave a measured speed-up of 36 which is near enough the estimated maximum possible speed-up calculated in (6).

4.3 Shared Memory Processors

The main parallelizing effort has been put into the MPI implementation but the code has been tried on several shared memory parallel computers using automatic parallelization. This includes an SMP node of the IBM SP-2 and a Cray J932.

The SMP nodes of the IBM SP-2 have four 332MHz processors capable of two floating point operations per clock cycle. This gives a peak performance of 2656 Mflop/s. The performance of `pscyee` is 74 Mflop/s using one processor and 220 Mflop/s using all four processors, i.e. a speedup of 3.0. The poor relation between peak performance and the performance of `pscyee` is most likely due to the low memory bandwidth of the SMP node. The compiler option `-qsmp=schedule=dynamic` was used. However, it gave only fractionally better performance than the other schedules.

Figure 3 shows the performance on a Cray J932. Parallelization was performed using autotasking with the compiler option `-Otask3`. The problem size is now $128 \times 128 \times 128$ which fits well with the vector length of the Cray J932 which is 64. The one node performance on the Cray J932 is 100 Mflop/s for this problem size, which is exactly half the peak performance.

5 Conclusions

We have demonstrated that it is possible to achieve negligible communication time on the IBM SP-2 when the problem size is increased with the number of processor if the problem size on each node is large enough. On the other hand, ideal speed-up can not be achieved for a fixed problem size since the communication time will not be negligible when the problem size on each node decreases.

In all results presented here we have used point sources for excitation. In an industrial FD-TD code it will be necessary to efficiently parallelize Huygens' surfaces, PML (a more advanced absorbing boundary condition), subcell models for thin wires etc., models for curved boundaries (such as local unstructured grid) etc. These issues will be addressed in the near future within the Large Scale FD-TD project [7].

6 Acknowledgments

This work was supported with computing resources by the Swedish Council for Planning and Coordination of Research (FRN) and Parallelldatorcentrum (PDC), Royal Institute of Technology.

Part of this work was performed during a research visit at Department of Mathematics at UCLA. The visit was financed by Svenska Institutet and Telefonaktiebolaget L M Ericssons stiftelse för främjandet av elektroteknisk forskning.

References

1. V. A. Borovikov and B. Y. Kinber. *Geometrical Theory of Diffraction.* IEE, UK, 1994.
2. C. Christopoulos. *The Transmission-Line Modeling Method: TLM.* IEEE Press and Oxford University Press, 1995.
3. R. F. Harrington. *Field Computations by Moment Methods.* The Macmillan Co., New York, 1968.
4. V. Shankar, W. Hall, and H. Mohammadian. A CFD-based finite-volume procedure for computational electromagnetics - interdisciplinary applications of CFD methods. *AIAA*, pages 551–564, 1989.
5. P. P. Silvester and R. L. Ferrari. *Finite Elements for Electrical Engineers 2nd Ed.* Cambridge University Press, Cambridge, 1990.
6. A. Taflove. *Computational Electrodynamics: The Finite-Difference Time-Domain Method.* Artech House, Boston, MA, 1995.
7. Large Scale FDTD. `http://www.nada.kth.se/~ulfa/CEM.html`.
8. K. S. Yee. Numerical solution of initial boundary value problems involving Maxwell's equations in isotropic media. *IEEE Trans. Ant. Prop.*, AP-14(3):302–307, May 1966.

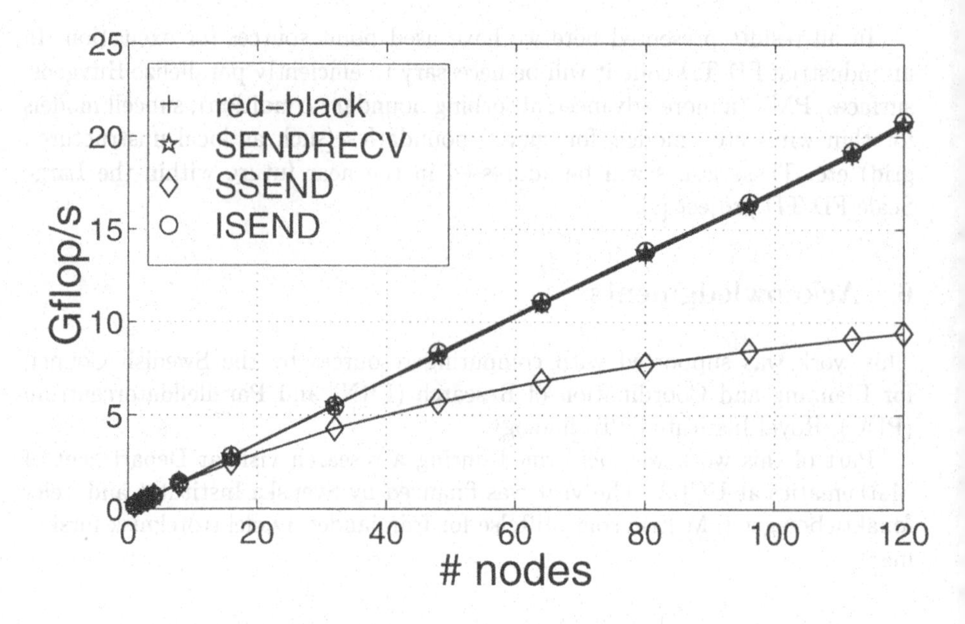

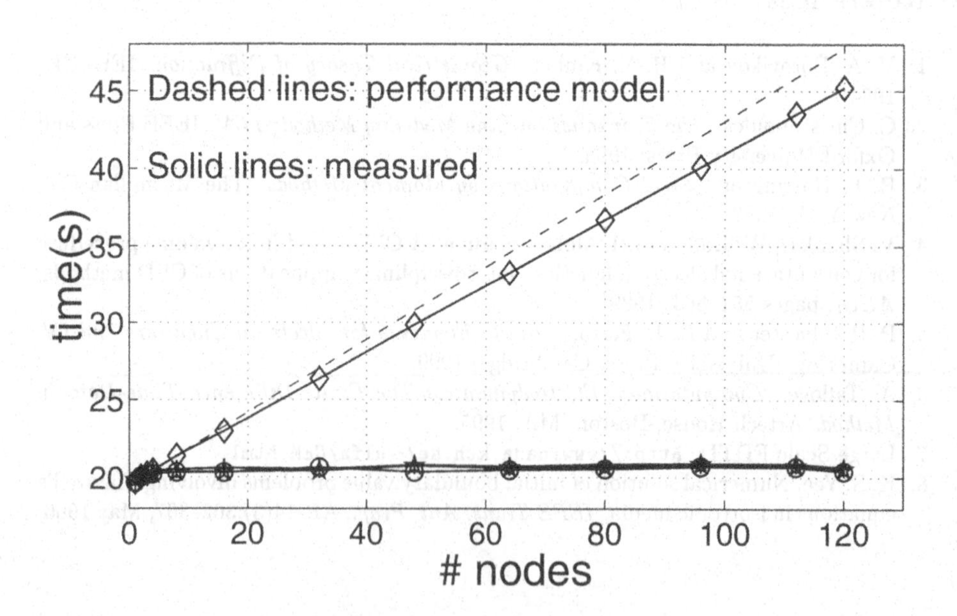

Fig. 1. Code performance on the IBM SP-2 when the problem size is scaled with the number of processors (p) so that it is $100 \times p \cdot 100 \times 100$.

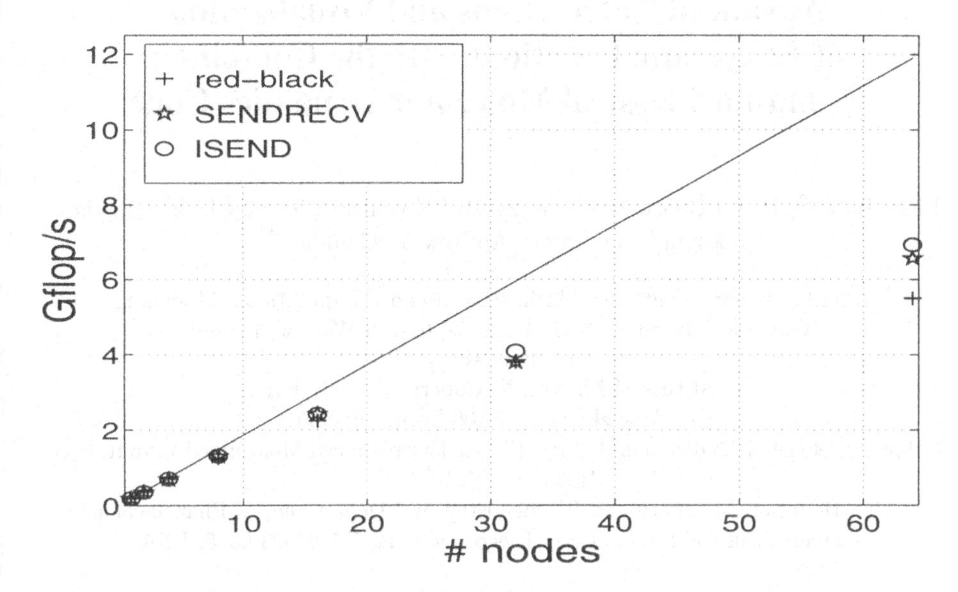

Fig. 2. Code performance on the IBM SP-2 for a fixed problem size of $100 \times 100 \times 100$. The solid line represents ideal speed-up.

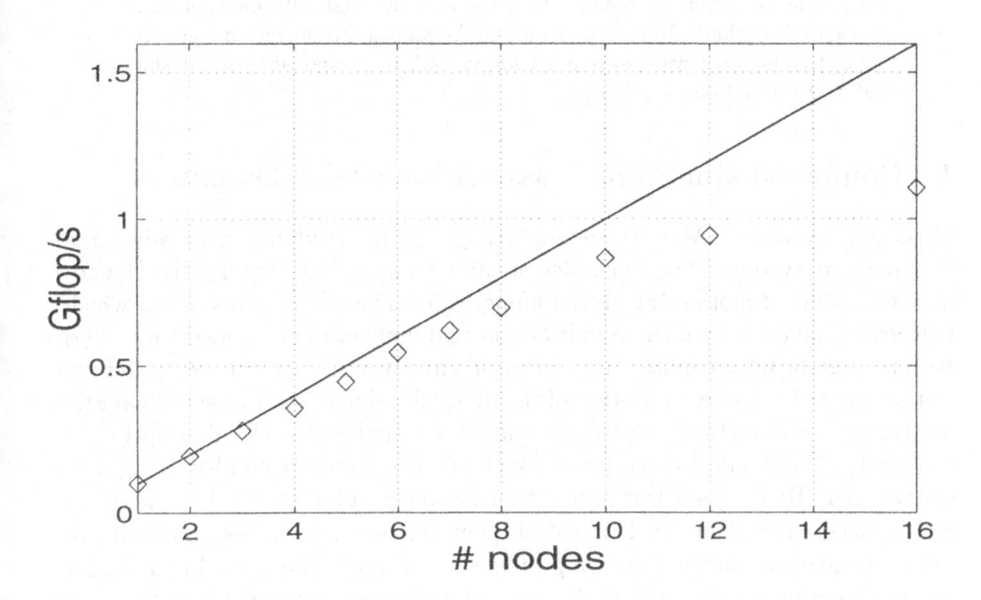

Fig. 3. Code performance on the Cray J932 for a fixed problem size of $128 \times 128 \times 128$. The solid line represents ideal speed-up.

Advanced Calculations and Visualization of Enzymatic Reactions with the Combined Quantum Classical Molecular Dynamics Code

Piotr Bała[1,2], Paweł Grochowski[1], Krzysztof Nowiński[1], Terry Clark[3], Bogdan Lesyng[1] and James Andrew McCammon[4]

[1] Interdisciplinary Centre for Mathematical and Computational Modelling, Warsaw University, Pawińskiego 5a, 02-106 Warsaw, Poland
bala@icm.edu.pl
[2] Institute of Physics, N. Copernicus University, Grudziądzka 5/7, 87-100 Toruń, Poland
[3] Baylor College of Medicine and Texas Center for Advanced Molecular Computation Houston TX, USA
[4] Department of Chemistry and Biochemistry, and Department of Pharmacology, University of California at San Diego, La Jolla, CA 92093-0365, USA

Abstract. The parallel version of the Quantum Classical Molecular Dynamics code is presented. The execution time scales almost linearly with the number of processors. The measured overhead of the parallelization paradigm is extremely small which ensures the high efficiency of the presented method. Tools based on the Advanced Visualization System (AVS) framework were developed for visualization and analysis of the QCMD simulations.

1 Combined Quantum Classical Molecular Dynamics

Molecular dynamics (MD) is an important tool for studying relatively large biomolecular systems. The dynamics of the system is described by the Newtonian equations of motion integrated numerically. The size of the system, which reaches nowadays several thousands atoms, and number of performed time steps requires significant computer time for such simulations. However, computer resources are not the only limitation of the molecular dynamics. Conventional MD methods with analytical force fields cannot be applied for the description of chemical or biochemical reactions in which bond-breaking or bond-forming processes occur. To overcome this problem during the past few years, there has been an increased interest in development of quantum-classical models and their applications to chemical and physical problems at the molecular level. In particular one can mention a wide range of the molecular dynamics simulations with forces evaluated with *ab initio* calculations [1, 2]. Within such models electronic degrees of freedom are treated as quantum variables, whereas positions and momenta of heavy atoms are treated as classical ones.

We have developed a Quantum-Classical Molecular Dynamics model (QCMD), which allows one to simulate dynamical processes in quantum-classical systems

[3–5]. This model can be used to simulate the motion of quantum particles, protons or electrons, in molecular systems and in the solid state. In the QCMD model, the proton dynamics in an enzyme active site is described by the time-dependent Schroedinger equation with the dynamics of the remaining atoms is performed using classical molecular dynamics. The potential energy function appearing in the time-dependent Schroedinger equation is numerically computed using an approximate valence bond (AVB) method [6]. The method was designed and parametrized by us for proton transfer processes [4, 13] but the model can be used to simulate the motion of quantum particles, protons or electrons, in molecular systems and in the solid state. The ultimate targets of our studies are enzymatic reactions, and the AVB matrix elements are parameterized to reproduce *ab initio* MP2/6-31G* and/or density functional calculations.

1.1 QCMD model

In the QCMD model, selected degrees of freedom are treated quantum dynamically, and the others classically or stochastically. The model consists of a system of coupled equations which govern the dynamics of the quantum and classical subsystems.

The dynamics of the quantum subsystem is described by the time-dependent Schroedinger equation, while the rest of the system is described by the Newtonian equations of motion. The coupling between the quantum and classical domains is described by the time-dependent potential function $V = V(x, \{R_\alpha(t)\})$ in the Schroedinger equation, where V is a function of the instantaneous atomic positions R_α, and by extended Hellmann-Feynman forces [7] modifying the classical forces in the Newtonian equations of motion (Eq. (2) and (3) below). These coupling terms govern the energy transfer between the quantum and classical domains, which influences the time evolution of the whole system [7, 2].

Assuming that the positions of the classical nuclei change slowly in time, one can treat them as parameters in the quantum Hamiltonian, and in the simplest representation the QCMD equations are the following:

$$i\hbar \frac{\partial \Psi(x,t;\{R_\alpha\})}{\partial t} = \left(-\frac{\hbar^2}{2m}\Delta_x + V(x;\{R_\alpha\}) \right) \Psi(x,t;\{R_\alpha\}), \tag{1}$$

$$M_\alpha \ddot{R_\alpha} = F_\alpha + F_\alpha^{HF} \tag{2}$$

with:

$$F_\alpha^{HF} = -\sum_k \left(b_k^2 \langle \Phi_k | \frac{\partial V(x;\{R_\alpha\})}{\partial R_\alpha} | \Phi_k \rangle + E_k \frac{\partial b_k^2}{\partial R_\alpha} \right), \tag{3}$$

where Φ_k and E_k are instantaneous eigenfunctions and eigenvalues of the Hamiltonian $H = -\frac{\hbar^2}{2m}\Delta_x + V(x;\{R_\alpha\})$. b_k are the expansion coefficients of the wavefunction in the basis of the eigenfunctions of the Hamiltonian H: $\Psi(x,t;\{R_\alpha\}) = \sum_k b_k e^{-\frac{i}{\hbar}E_k t}\Phi_k(x;\{R_\alpha\})$.

In the limit of a small time step the extended Hellmann-Feynman forces are equal to the conventional Hellmann-Feynman forces [5].

The forces between classical atoms, F_α are computed using standard potential energy functions. The forces between quantum particles and classical atoms located sufficiently far from the quantum subsystem are calculated in the classical limit; in practice, when the distance to a remote atom is much larger than the hydrogen bond length, the classical forces can be used instead of the rigorous extended Hellmann-Feynman forces.

1.2 Potential energy surface

In the case of proton transfer, the potential energy function $V(x;\{R_\alpha\})$ is the Born-Oppenheimer potential energy surface which is obtained numerically by solving the time-independent electronic Schroedinger equation, varying the proton position and computing the energy of the system. For this purpose conventional *ab initio*, semi-empirical molecular orbital methods or density functional methods could be applied. In practical applications for proton transfer processes in molecular complexes or in the active site of phospholipase A_2 the fast AVB method is used [6]. The influence of remote classical atoms on the enzyme active site is included by explicit inclusion of electrostatic interactions with remote atomic charges in the AVB Hamiltonian's diagonal elements. The discretized energy surface changes its shape in time according to the current atomic positions and the current value of the electrostatic field generated in the active site region by the remote atoms. Changes of the shape of the potential energy surface play a key role in the quantum proton dynamics which leads to the proton transfer.

1.3 Algorithm

The QCMD code, because of complexity of the model, consists of different parts characterized by a variety of numerical methods. The classical MD dynamics is performed using a *leap frog* algorithm. The quantum dynamics described by the time-dependent Schroedinger equation is performed based on the discrete representation of the proton's wavefunction on a uniform Cartesian grid. A Chebychev polynomial method is used for the propagation of the wavefunction [8–10]. For this purpose the potential $V(x;\{R_\alpha\})$ must be calculated at each grid point. Moreover, because of the evaluation of the Hellmann-Feynman forces, which is performed using numerical derivatives of the potential with respect to the classical atoms positions R_α, additional evaluations of the potential at all grid points must be performed. The calculation of the potential at each grid point involves three main steps:

- evaluation of the electrostatic field generated by the charged atoms,
- evaluation of the valence band matrix elements as a sum of several empirical potentials described by analytical functions,
- diagonalization of the AVB matrix.

Since all above steps must be performed at each time-step, several times for each point of the 3D grid, this part of the code is the most time consuming (see

Table 1. The time spent in different parts of the QCMD code in the case of sequential execution.

Subroutine	CPU time
Evaluation of the electrostatic potential	7.20 %
Calculation of the AVB matrix elements	13.60 %
Diagonalization of AVB matrix	73.60 %
Propagation of the proton wavefunction	0.81 %
Molecular Dynamics step	0.10 %
Other	4.96 %

Table 1) and together consumes about 99% of the total CPU time. One should note that the potential energy surface could, in principle, be calculated on the fly with any *ab initio* method, but since the fastest quantum mechanical code is at least 100 times more time consuming than the developed AVB method, such a task would exceed any available computer power.

2 Parallelization paradigm

Since the most time-consuming part of the code is performed on a 3 dimensional spatial grid and evaluations of the potential at each grid point require only local information, all variables evaluated on the grid, e.g. the potential $V(x; \{R_\alpha\})$ and the wavefunction $\Psi(x,t)$, can be distributed over the processor set and evaluated in parallel.

The potential $V(x; \{R_\alpha\})$ defined on the discrete grid $N_x \times N_y \times N_z$ is distributed on the processor grid $np_x \times np_y \times np_z$. Uniform mapping between the Cartesian grid and the processor grid is performed.

Distribution of the calculations of the potential function results in parallelization of the majority of the code. However, the potential is evaluated very often during simulations and time spent in a single potential evaluation is relatively small. All operations performed independently at the grid points are performed in parallel, and in consequence all variables represented on the grid (potential, wavefunction, gradient of the potential) are distributed in the same way as the potential.

Evaluation of the global properties of the wavefunction such as position, energy or Hellmann-Feynman forces requires calculation of the sum over the grid. Partial sums on the distributed grids are performed at each processor and than partial results are communicated and summed over the processor set. Because of the identical operations performed at each grid point during evaluation of the potential an equal workload can be easily obtained with uniform distribution of the Cartesian grid over the processor set.

Unfortunately not all operations can be easily distributed. The propagation of the protons wavefunction requires evaluation of the Hamiltonian acting on the wavefunction. The evaluation of the kinetic part of the Hamiltonian $\frac{1}{2m}\Delta_x\Psi(x)$ is performed using a Fast Fourier Transform (FFT) [14]. In the QCMD the Chebychev polynomial method is used to propagate the wavefunction and Hamiltonian must be evaluated $2N_{CZ}$ times for each time-step, where N_{CZ} is length of the Chebychev polynomial expansion.

Since parallelization of the FFT algorithm is not straightforward and has been of long interest we have applied a 3-dimensional parallel Fast Fourier Transform, PCFFT3D, from the Cray T3E Scientific libraries. This approach requires proper distribution of the matrix over the processor set, eg. $np_x = 1$.

2.1 Code parallelization

The QCMD code consists of about 20,000 lines of Fortran 77 code [3]. The parallelization of the code is performed based on the PFortran package [12] together with the Cray MPI package. PFortran, used as preprocessor, allows for the parallelization of the code together with its easy portability. The developed PFortran code can be preprocessed for use with different message passing packages, such as MPI, PVM or TCGMSG. Because all variables are replicated at the different processors, the scalar parts of the code were performed on each processor. The operations that involved grids were performed in parallel with different parts of the grid distributed across processors. In this case at each processor only memory required for local part of the grid is allocated.

The I/O is performed from the designated processor. All files are opened and read by this processor and then communicated to all processors. The outputted variables, if not stored at every processor, are first exchanged and then outputted from the designated processor. Since disk operations take insignificant amount of the CPU time, this approach does not affect the code performance. The control data can be printed out by each processor independently which was helpful during code development.

One should note that in this case large portions of the code must be performed in parallel, which requires careful distribution of variables and prohibits parallelization scheme based on the distribution of the innermost loops. This task was easy to achieve with PFortran as well as particular matrix distribution described above.

The QCMD performance measured on Cray T3D and T3E machines with the MPI library exhibits almost linear scaling with the number of processors (Fig. 1). which ensures a high level of the achieved parallelism. The overhead of the performed parallelization was measured by the comparison of the execution time of the sequential code without usage of PFortran and the parallel version executed at one processor. The overhead time is equal to 0.22 % of the execution time.

3 Simulations and visualization

The QCMD code was applied for the investigation of the enzymatic reaction in the active site of Phospholipase A_2. The detailed system preparation and equili-

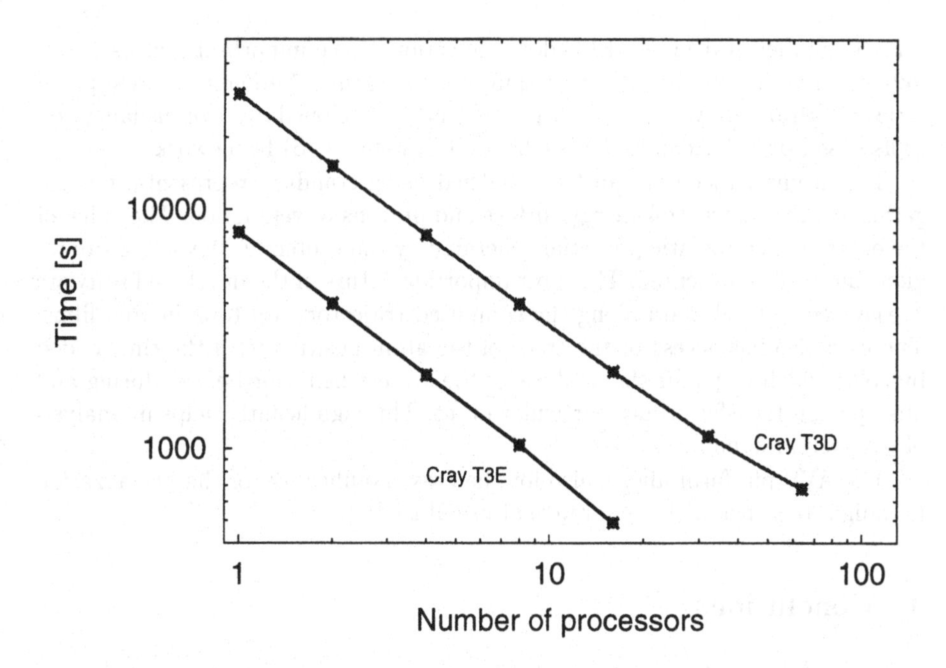

Fig. 1. Performance of the QCMD code on the Cray T3D and T3E computers.

bration can be found elsewhere [13]. A 0.5 ps QCMD simulation of the enzyme dynamics was performed. At the initial part of the simulation the proton is part of a water molecule located in the active site of the enzyme. The proton wavefunction is therefore located in a deep potential well close to the water oxygen atom. During next 0.2 ps the proton wavefunction spreads out and starts to occupy a metastable state near histidine nitrogen. In the same time the potential changes significantly its shape and the minimum near nitrogen becomes deeper. The proton wavefunction follows this change and localizes near histidine forming a covalent bond. In the same time the nucleophilic attack of the formed OH^- ion takes place. Almost at the same time cleavage of the substrate bond is observed.

The enzymatic reaction can be therefore investigated in details without any *a priori* assumptions about reaction path or about sequence of the events taking place in the active site of the enzyme. The QCMD simulations take into account all physical processes which can take place during the enzymatic reaction, together with bond breaking and proton tunneling.

The analysis of the MD trajectories requires advanced techniques either statistical or visual. The QCMD model introduces new challenges to the standard post-processing methods. The classical representation of the atoms as points or spheres should be extended in the case of *quantum* particles to the continuous cloud of the proton density. At the same time the detailed information of the potential energy surface for the proton motion is available with no additional computational cost. Because of the fact that the investigated system contains a large number of degrees of freedom, detection of the important factors which influence reaction is difficult and requires visualization of different system properties simultaneously. For this purpose we have developed a set of visualization tools based on the Advanced Visualization System (AVS) framework.

The atoms trajectory can be visualized using standard representations together with the potential energy surface and protons wavefunction. The value of the external electrostatic potential generated by the protein in the active site region can be also presented. The most important future of the developed software is easy access to all data along the computed trajectory, up to animated films. The user also has access to the trace of the atom position with the time coded by color which helps for the analysis of the system behavior before, during and after proton transfer or any particular event. This significantly helps in analysis of the system dynamics.

The AVS platform allows also for the easy modification of the visualization paradigm together with wide range of useful tools.

4 Conclusions

The Quantum Classical Molecular Dynamics code has been successfully parallelized using the PFortran preprocessor, which allows wide portability of the code. The execution time scales almost linearly with the number of processors. The measured overhead of the parallelization paradigm is extremely small which ensures the high efficiency of the presented method. The developed code opens new fields for practical applications of the QCMD model. In particular it allows for studies of enzymatic reactions such as electron and proton transfer and proton transfer phenomena in the condensed phase. The parallel QCMD code allows for an economic usage of scalable parallel architectures.

Acknowledgements The computations were performed using the Cray T3D computer in SDSC, San Diego (USA) and Cray T3E at ICM, Warsaw University. The research was supported by the Polish State Committee for Scientific Research and N. Copernicus University, and in the US, by NPACI and NFS.

References

1. Bertran J.: Some fundamental questions in chemical reactivity. Theor. Chem. Acc. **99** (1998) 143-150
2. Bala P., Grochowski P., Lesyng B., McCammon J. A.: Quantum–classical molecular dynamics. models and applications. In M. Field, editor, *Quantum Mechanical Simulations Methods for Studying Biological Systems*, pages 115–196. Springer-Verlag Berlin Heidelberg and Les Editions de Physique Les Ulis, 1996.
3. Bala P., Grochowski P., Lesyng B., McCammon J. A.: Quantum – Classical Molecular Dynamics and its computer implementation. Computers & Chemistry **19** (1995) 155–160

4. Bala P., Grochowski P., Lesyng B., McCammon J. A.: Quantum-classical molecular dynamics simulations of proton transfer processes in molecular complexes and in enzymes. J. Phys. Chem. **100** (1996) 2535–2545
5. Bala P., Clark T., Grochowski P., Lesyng B., McCammon J. A.: Parallel Version of the Combined Quantum Classical Molecular Dynamics Code for Complex Molecular and Biomolecular Systems. *This Proceedings*, Springer Verlag, 1998.
6. Grochowski P., Bala P., Lesyng B., McCammon J. A.: Density functional based parametrization of a valence bond method and its applications in quantum-classical molecular dynamics simulations of enzymatic reactions. Int. J. Quant. Chemistry **60** (1996) 1143–1164
7. Bala P., Lesyng B., McCammon J. A.: Extended Hellmann-Feynman theorem for non–stationary states and its application in Quantum-Classical Molecular Dynamics simulations. Chem. Phys. Lett. **219** (1994) 259–266
8. Tel-Ezer H. Kosloff R.: An accurate and efficient scheme for propagating the time dependent Schroedinger equation. J. Chem. Phys. **81** (1984) 3967–3971
9. Leforestier C., Bisseling R. H., Cerjan C., Feit M. D., Freisner R., Guldberg A., Hammerich A., Jolicard G., Karrlein W., Meyer H.-D., Lipkin N., Roncero O., Kosloff R.: A comparison of different propagation schemes for the time dependent Schrodinger equation. J. Comput. Phys. **94** (1991) 59–80
10. Truong T. N., Tanner J. J., Bala P., McCammon J. A., Kouri D. J., Lesyng B., Hoffman D.. A comparative study of time dependent quantum mechanical wavepacket evolution methods. J. Chem. Phys. **96** (1992) 2077–2084
11. Van Gunsteren W.: GROMOS (Groningen Molecular Simulation Computer Program Package). Biomos, Laboratory of Physical Chemistry, University of Groningen, 1987.
12. Bagheri B., Clark T. W., Scott L. R.: PFortran: a parallel dialect of Fortran. ACM Fortran Forum **11** (1992) 20–31
13. Bala P., Grochowski P., Lesyng B., McCammon J. A.: Quantum Dynamics of Proton Transfer Process in Enzymatic Reactions. Simulations of Phospholipsae A_2. Ber. Bunsenges. Phys. Chem. **102** (1998) 580–586
14. Kosloff D., Kosloff R.: A fourier method solution for the time dependent schroedinger equation as a tool in molecular dynamics. J. Comput. Phys. **52** (1983) 35–53

Memory Access Profiling Tools for Alpha-based Architectures

Susanne M. Balle and Simon C. Steely, Jr.

Compaq Computer Corporation,
Maynard, MA 01754, USA
{susanne.balle, simon.steely}@digital.com

Abstract. The development of efficient algorithms on today's high performance computers is far from straight-forward. Applications need to take full advantage of the deep memory hierarchy which implies that the user has to know exactly how his/her implementation gets executed. With today's compilers it can be very difficult to understand or predict the execution path without having to look at the machine code. We present a set of tools designed to help us better understand programs and their memory access pattern. These tools enable the programmer to compare memory access patterns of different algorithms as well as provide insight into the algorithm's behavior e.g. potential bottlenecks resulting from memory accesses.

1 Introduction

The development of efficient algorithms on today's high performance computers can be a challenge. One of the major issues in implementing high performing algorithms is to take full advantage of the deep memory hierarchy. In order to better understand a program's performance two things need to be considered: computational intensiveness as well as the amount of memory traffic involved. In addition to the latter it is important to consider the pattern of the memory references since the success of hierarchy is attributed to locality of reference and reuse of data in the user's program.

In this paper, we investigate the memory access pattern of FORTRAN programs and we present a set of experimental Atom [4, 6, 7] tools which help us understand how the program is executed. These tools are very useful to understand how different compiler switches impact the algorithm implemented or if the algorithm is actually doing what it is intended to do. In addition, they help the process of translating an algorithm into an efficient implementation on a specific machine. Related work for Basic Linear Algebra Subroutine implementations is described in [3]. In most scientific programs the data elements are matrix-elements which are usually stored in 2-D arrays (column-major in Fortran). Knowing the order of array referencing is important in determining the amount of memory traffic.

Section 2 describes an experimental memory pattern analysis tool we developed to better understand a program's consecutive memory reference patterns.

We illustrate its usefulness with a case study. Section 3 presents additional memory access profiling tools as well as gives examples of outputs resulting from these tools. Guidelines on how to use the tools are given as well as comments about conclusions to be derived from the histograms generated by the different tools.

2 Memory Access Pattern Profiling Tool

The tool presented in this Section collects a set of histograms taking into account each memory reference in the program. The first histogram measures strides from the previous reference, the second histogram gives the stride from the second-to-last reference, and so on, for a total of MAXEL histograms for each memory reference in the program. By stride we mean the distance between two memory references (load or store) in bytes. In the investigation described below we chose a MAXEL of 5 but MAXEL can be given any value.

The tool only takes into account for the next histogram the memory references which stride is more than 128 bytes. It doesn't consider in the $(i+1)^{th}$ histogram strides which are less than 128 bytes in the ith histogram. The reason behind that choice is that it highlights memory accesses that are stride one for a while and then have a stride greater than 128 bytes. 128 bytes was chosen arbitrarily and could be changed to any value.

The output histograms presented in our case study are on a per-array basis instead of over all references in the program. This gives us a clearer picture of the memory access pattern involved in the piece of the program we want to investigate. We present separate histograms for the loads and the stores of each of the arrays involved in the memory traffic of the subroutine we considered.

It is important when looking at memory access patterns to have the ability not to include load instructions that perform prefetching. Even though it adds to the memory traffic they can pollute the memory access pattern picture.

we now investigate the FORTRAN loop given in Figure 1 using the profiling tool described above. We show that the code doesn't get executed the way we originally expected it to be. Often the developer would have to dig in to the assembler of the given loop in order to understand how and when the different instructions get executed. The use of our experimental tool ease that process and the resulting histograms give a clear picture of the memory reference patterns. The understanding of the memory references' access pattern are much easier and less time consuming for the developers.

```
1    Q(i)=0, i=1, n
2    do k1= 1, 4
3      do j=1,n
4        p1=COLSTR(j,k1)
5        p2=COLSTR(j+1,k1)-1
6        p3= [snip]
7        sum0=0.d0
8        sum1=0.d0
9        sum2=0.d0
10       sum3=0.d0
11       x1 = P(index+ROWIDX(p1,k1))
12       x2 = P(index+ROWIDX(p1+1,k1))
13       x3 = P(index+ROWIDX(p1+2,k1))
14       x4 = P(index+ROWIDX(p1+3,k1))
15       do k = p1, p3, 4
16         sum0 = sum0 + AA(k,k1) * x1
17         sum1 = sum1 + AA(k+1,k1) * x2
18         sum2 = sum2 + AA(k+2,k1) * x3
19         sum3 = sum3 + AA(k+3,k1) * x4
20         x1 = P(index+ROWIDX(k+4,k1))
21         x2 = P(index+ROWIDX(k+5,k1))
22         x3 = P(index+ROWIDX(k+6,k1))
23         x4 = P(index+ROWIDX(k+7,k1))
24       enddo
25       do k = p3+1, p2
26         x1=P(index+ROWIDX(k,k1))
27         sum0 = sum0 + AA(k,k1)*x1
28       enddo
29       YTEMP(j,k1)=sum0+sum1+sum2+sum3
30     enddo
31     do i = 1, n, 4
32       Q(i) = Q(i) + YTEMP(i,k1)
33       Q(i+1) = Q(i+1) + YTEMP(i+1,k1)
34       Q(i+2) = Q(i+2) + YTEMP(i+2,k1)
35       Q(i+3) = Q(i+3) + YTEMP(i+3,k1)
36     enddo
37   enddo
where n = 14000,
real*8 AA(511350,4), YTEMP(n,4)
real*8 Q(n), P(n)
integer*4 ROWIDX(511350,4), COLSTR(n,4)
```

Fig. 1. Fortran loop

Considering the loop in Figure 1 we would expect from line 4 and 5 the array COLSTR to be read stride one 100% of the time. Line 29 indicates that YTEMP is accessed stride one through the whole j loop. From line 32 to 35, YTEMP's stride should be one all the way through the i loop but should be equal the number of columns in the array when YTEMP gets accessed first time in the i loop. Q should be accessed stride one for both the loads and the stores. From lines 11 to 14, 20 to 23, and 26 we expect ROWIDX to be accessed stride one between the p1 and p2 bounds of the k loop. We expect array P to have non adjacent memory references since this algorithm deliberately sacrifices the array P's access patterns in order to improve the memory references of Q and AA.

We investigate the memory access patterns achieved by the loop Figure 1 when compiled with the following switches: f77 -g3 -fast -O5. The *-g3* switch is needed to extract the addresses of the arrays from the symbol table. For more information on Digital fortran compiler switches see [5].

Figure 2 through Figure 4 show histograms of the strides between the loads for arrays Q, COLSTR, and YTEMP. Figure 5 and Figure 6 show the strides between consecutive stores for Q and YTEMP. The remaining arrays ROWIDX, AA, and P's strides are not shown since their behavior correspond to our expectations.

From Figure 2 and Figure 5 we see that array Q is accessed as we expected, 100% stride one for the loads and the stores. Since Q is accessed contiguously 100% of its memory references we will not have any entries in the next four histograms. As described earlier we only record in the next histogram the strides that are greater than 128 bytes in the current histogram.

On the other hand as illustrated by Figure 3 COLSTR gets accessed 50% stride zero and 50% stride one. This is unexpected since Figure 1 suggests that array to be accessed stride one 100% of the time. The fact that we only have entries for the strides between the current and the previous loads indicates that the elements of COLSTR are accessed in a nearly contiguous way. A closer look at Figure 1 tells us that the compiler is loading COLSTR twice even though it already has a copy of the earlier used COLSTR(j) in one of its registers. The work-around is to perform a scalar replacement as described by Blickstein et al [2]. We put $p2 = COLSTR(1, k1) - 1$ outside the j loop and substitute inside the j loop $p1 = COLSTR(j, k_1)$ with $p1 = p2 + 1$. p2 inside the j loop remains the same. Getting rid of the extra loads didn't enhance performance and a possible assumption is that the analysis done by the compiler concluded that no gain would result from that optimization.

Figure 4 and Figure 6 illustrate the strides for the loads and the strides for the stores for the array YTEMP. One more time, the implementation is not being executed the way we thought it would. We expected YTEMP's loads to be referenced stride one approximatively 99% of the time (Figure 1 lines 17 to 20) since for every 14000 iterations we have one stride greater than 32KB. By considering Figure 4 along with Figure 1 we conclude that YTEMP gets unrolled by four in the k1-direction in the i loop. The fact that all the strides between the current load and the load two loads back/three loads back/four loads back and five loads back are greater than 32KB is consistent with traversing the matrix along the rows. Figure 6 tells us that the j loop doesn't get unrolled by four in the k1-direction since all the loads of YTEMP are 100% stride one.

Going through these arrays access patterns provide us with an idea of how these arrays are accessed as well as if the implementation is doing what it is suppose to do. An important conclusion from this investigation is that unless the developer looks at the assembler generated from his code he/she cannot assume anything about how his/hers program is executed. Our easy-to-use tool provide a very useful interface to understand the memory access patterns of a program as well as to hint about where possible memory bottlenecks might occur.

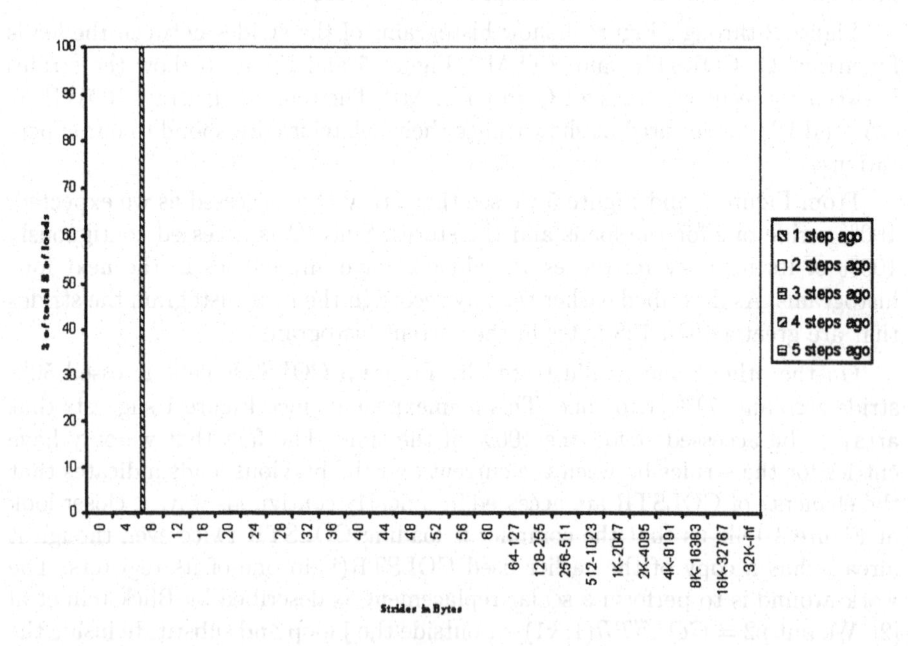

Fig. 2. Strides for array Q between the current load and the load 1–5 steps ago.

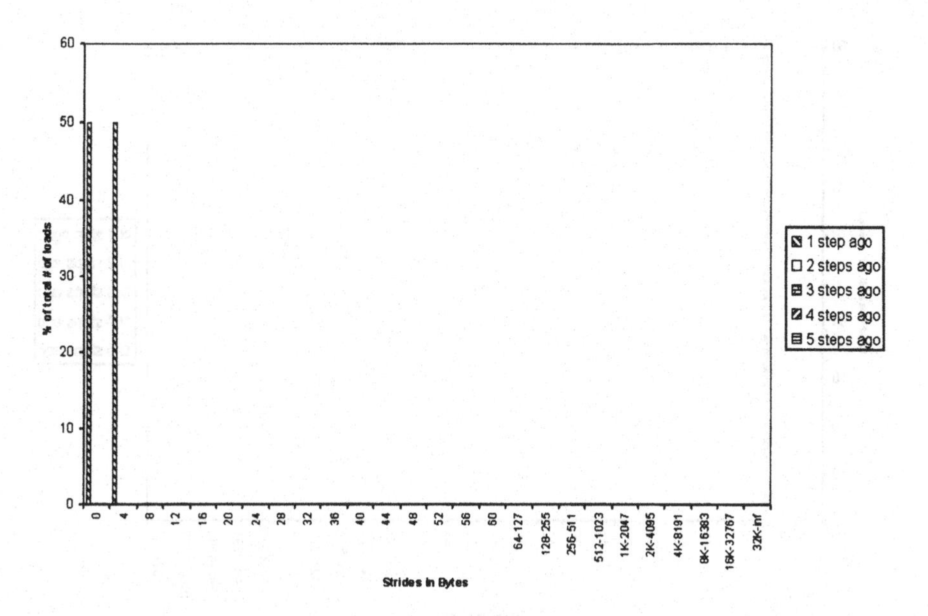

Fig. 3. Strides for array COLSTR between the current load and the load 1–5 steps ago.

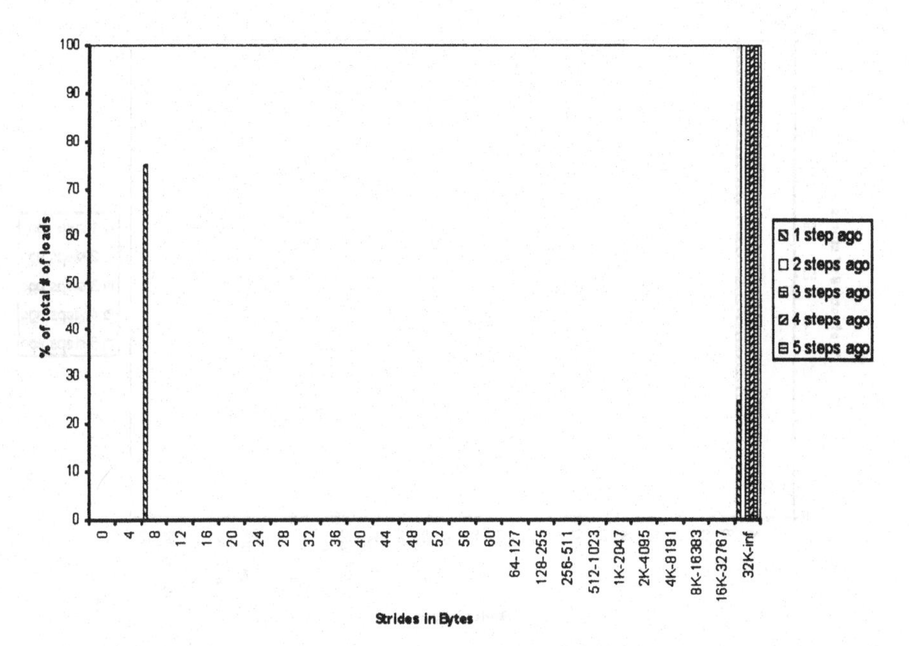

Fig. 4. Strides for array YTEMP between the current load and the load 1–5 steps ago.

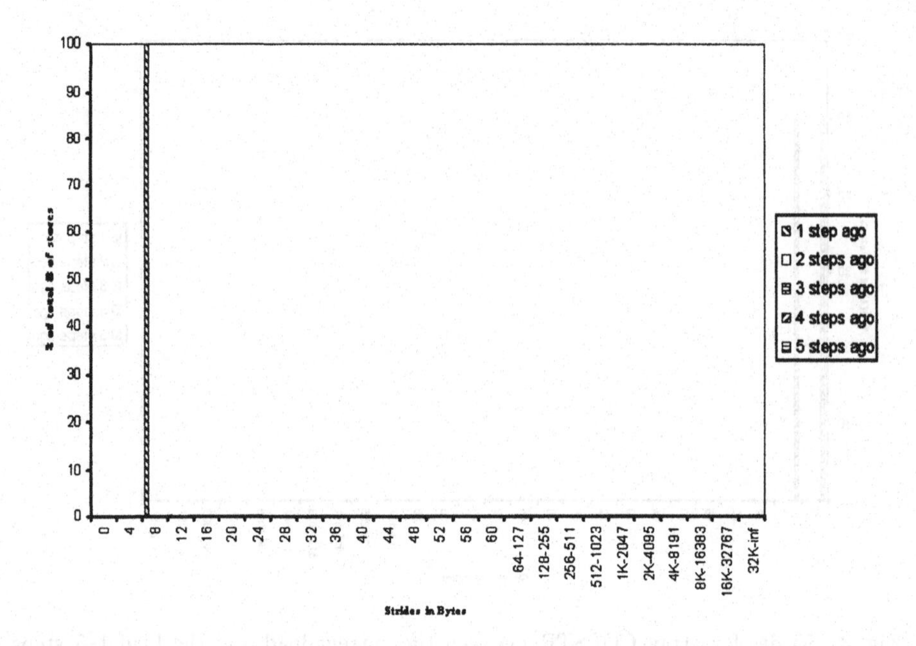

Fig. 5. Strides for array Q between the current store and the store 1–5 steps ago.

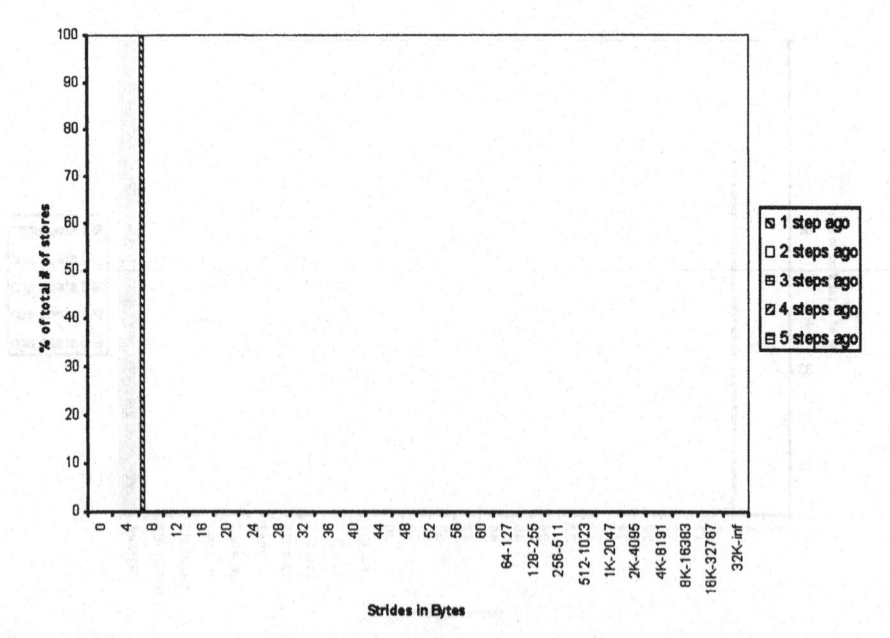

Fig. 6. Strides for array $YTEMP$ between the current store and the store 1–5 steps ago.

3 Other Memory Access Pattern Analysis Tools

In this Section we list additional tools we developed which have been useful in determining and helping us better understand memory references and their access patterns.

3.1 Array Interaction Tool

The Array Interaction Tool (AIT) was primarily developed to help us better understand the interaction between different arrays involved in a computation. The output histograms from this tool include the following information:

- # of references per array (stores and loads)
- # of instructions between a store and its first load
- # of loads per stores

Figure 7 shows the number of stores and the number of loads per array for the loop described in Figure 1.

Array names	# of Stores	# of Loads
P	0	46327600
Q	1400000	1050000
COLSTR	0	1456104
ROWIDX	0	48180704
AA	0	48180704
Extra loads and stores out-of-memory-region-range		
(null)	494	754

Fig. 7. Output from AIT.

3.2 Per-page Memory Reference Investigation Tool

This tool gives the developer insight on his/her program's behavior based on how the different pages are accessed. The page size on Alpha-based architectures is 8KB. The output histograms from this tool are the following:

- # of loads per page
- # of stores per page
- Runlength for the loads per page
- Runlength for the stores per page

We define runlength as the number of consecutive references in the same page. These histograms are useful in order to fully understand the memory access picture. For example the histograms from the tool described in Section 2 could look the same for two different types of access patterns since we only consider strides between different consecutive memory references. The histograms created by this tool would tell us if effectively their runlength is the same and thereby their behavior.

4 Conclusion

The memory pattern case study we presented in Section 2 shows that given the right program analysis tools a program developer can take better advantage of the computer he/she is working on. The experimental tools we designed were very useful in giving us insight in the algorithm's behavior.

Balle and Steely [1] describe a casestudy performed using some of these tools to better understand a program's execution. The investigation was definitely a worthwhile exercise since a big reduction in execution time was obtained using straight-forward and easy guidelines. Using an approach close to the one described in Section 2 they achieved an improvement in performance of 8 % (or 139 seconds).

Given such a set of tools, the data collected helps the user better understand a given program as well as its memory access patterns. It is also an easier and faster way to get insight in a program than to look at the listing and the assembler generated by the compiler. These tools enable the programmer to compare memory access patterns of different algorithms and thereby are very useful when optimizing codes. Probably the most important thing of all is that it tells the developer if his/her implementation is doing what he/she thinks the algorithm is doing as well as highlight potential bottlenecks resulting from memory accesses. Optimizing an application is an iterative process and being able to use relatively easy to use tools like atom is a very important part of the process.

Acknowledgments

The authors wish to thank David Lafrance-Linden, Dwight Manley, Jean-Marie Verdun, and Dick Foster from Digital Equipment Corporation for fruitful discussions.

References

1. Balle, S. M. and Steely, S. C., Jr., Analyzing Memory Access Patterns of Programs on Alpha-based Architectures, Digital Technical Journal, Vol 9, No 4: 21-32, 1998.
2. Blickstein, D. S., Craig, P. W., Davidson, C. S., Fairman, R. N., Glossop, K. D., Grove, R. B., Hobbs, S. O., and Noyce, W. N., The GEM Optimizing compiler system, Digital Technical Journal, Vol 4, No 4, 1992.
3. Brewer, 0., Dongarra, J., and Sorensen, D., Tools to aid in the analysis of memory access patterns for fortran programs, Technical report, Argonne national Laboratory, June 1988.
4. Digital Equipment Corporation, Programmer's Guide, Digital UNIX Version 4.0, chapter 9, Maynard, MA, March 1996.
5. Digital Equipment Corporation, DEC Fortran Language Reference Manual, Maynard, MA, 1997.
6. Srivastava, A. and Eustace, A., ATOM: A System for Building Customized Program Analysis Tools, Proceedings of SIGPLAN'94, Conference on Programming Language Design and Implementation, Orlando, June 1994.

7. Wilson, L. S., Neth, C. A., and Rickabaugh, M. J., Delivering Binary Object Modification Tools for Program Analysis and Optimization, Digital Technical Journal, Vol 8, No 1: 18-32, 1992.

Parallelized Block–Structured Newton–Type Methods in Dynamic Process Simulation *

Jürgen Borchardt

Weierstrass Institute for Applied Analysis and Stochastics
Mohrenstrasse 39, D - 10117 Berlin, Germany
borchardt@wias-berlin.de

Abstract. The parallelization of numerical methods for the solution of initial value problems for large systems of differential algebraic equations (DAE's) arising from the dynamic process simulation of chemical plants is considered. Due to a unit-oriented modeling the systems of DAE's are structured into subsystems. By merging subsystems to blocks, block–structured Newton–type methods are used for their solution. These methods enable a coarse grained parallelization. Results from dynamic simulation runs for industrial distillation plants on parallel computers are given.

1 Introduction

Complex chemical production plants are networks of coupled process units as pumps, reboilers, condensers, or trays of distillation columns. A mathematical model is assigned to each unit and the units are linked by streams in a flow-sheet. For the time domain simulation, this leads to initial value problems for large nonlinear systems of differential–algebraic equations (DAE's), which are structured corresponding to the units into m coupled subsystems

$$F_i(t, y(t), \dot{y}(t), u(t)) = 0, \quad i = 1(1)m, \tag{1}$$

$$F_i : \mathbb{R} \times \mathbb{R}^n \times \mathbb{R}^n \times \mathbb{R}^q \to \mathbb{R}^{n_i}, \quad \sum_{i=1}^{m} n_i = n, \quad t \in [t_0, t_{END}],$$

with given vector valued parameter function $u(t)$ and unknown function $y(t)$.

In today's applications, the system of DAE's can involve tens of thousands of equations or more. For solving such large scale problems, a two level hierarchical structure of the system is considered. The first level of the structure is build by the subsystems of the DAE–system, and the second level of the hierarchy is obtained by merging subsystems to blocks

$$\tilde{\mathcal{F}}_j = (F_{j_1}, F_{j_2}, \ldots, F_{j_{m_j}})^T, \quad j = 1(1)p, \quad \sum_{j=1}^{p} m_j = m. \tag{2}$$

* This work was supported by the Federal Ministry of Education, Science, Research and Technology, Germany under grants GA7FVB-3.0M370 and GR7FV1.

The algorithm for generating such a block partitioning (2) from (1) exploits only topological informations. It merges subsystems to nearly equal sized blocks while trying to minimize the number of coupling variables between blocks and to "optimize" the number of blocks in relation to the number of equations as well as the number of available processors. The implemented algorithm is reliable for general problems, it is of low complexity and therefore it is fast.

In Section 2, it is described how the hierarchical structure of the DAE–system can be used to construct effectively parallelizable block–structured Newton–type methods. These methods require to solve many linear systems with the same pattern structure of the sparse nonsymmetric matrices but with different right hand sides. For this a direct solver [2] is used, which generates a pseudo code for an efficient LU-refactorization and solution. Finally, in Section 3, results for large scale real life applications of the Bayer AG Leverkusen are given.

2 Block–structured Newton–type methods

Based on a block partitioning (2), the system of nonlinear equations arising from backward differentiation formulas (BDF) [3] is extended to use block–structured Newton–type methods for its solution on parallel computers. The extension is done by determining the internal variables $x = (x_1, \ldots, x_p)^T$ of the blocks, duplicating of external couple variables $z = (z_1, \ldots, z_p)^T$, and appending linear identification equations $\mathcal{G}(z) = 0$, yielding the extended system

$$\begin{aligned} \mathcal{F}_j(x_j, z_j) &= 0, \qquad j = 1(1)p, \\ \mathcal{G}(z) &= 0, \end{aligned} \tag{3}$$

where the nonlinear functions $\mathcal{F}_j$, corresponding to the blocks $\tilde{\mathcal{F}}_j$, have disjunctive arguments. Each block function $\mathcal{F}_j : \mathbb{R}^{r_j} \times \mathbb{R}^{s_j} \longrightarrow \mathbb{R}^{q_j}, \quad r_j + s_j \geq q_j$, is splitted into $\mathcal{F}_j = (\mathcal{F}_j^1, \mathcal{F}_j^2)^T$ by determining r_j pivot elements in the $q_j \times r_j$ dimensional matrix $\partial_{x_j}\mathcal{F}_j := \partial\mathcal{F}_j / \partial x_j$, so that the r_j pivot rows determine $\mathcal{F}_j^1$, and the regularity of $\partial_{x_j}\mathcal{F}_j^1$ is ensured. With that, a basic parallel algorithm for the evaluation of the the corrections $\Delta x^{k+1} = x^{k+1} - x^k$ and $\Delta z^{k+1} = z^{k+1} - z^k$ in the k'th iteration step of a modified Newton method can be obtained:

(1) **do parallel** for all $j \in \{1, \ldots, p\}$:

(a) for new Jacobian: (i) compute the Jacobians $\partial_{x_j}\mathcal{F}_j$, $\partial_{z_j}\mathcal{F}_j$ and do LU–factorization of the coefficient matrix in (4)

(ii) solve:

$$\begin{bmatrix} \partial_{x_j}\mathcal{F}_j^1 & \emptyset \\ \partial_{x_j}\mathcal{F}_j^2 & I \end{bmatrix} \begin{bmatrix} B_j \\ C_j \end{bmatrix} = \begin{bmatrix} \partial_{z_j}\mathcal{F}_j^1 \\ \partial_{z_j}\mathcal{F}_j^2 \end{bmatrix} \tag{4}$$

(b) compute the function $\mathcal{F}_j(x_j^k, z_j^k)$ and solve:

$$\begin{bmatrix} \partial_{x_j}\mathcal{F}_j^1 & \emptyset \\ \partial_{x_j}\mathcal{F}_j^2 & I \end{bmatrix} \begin{pmatrix} \Delta\hat{x}_j^{k+1} \\ \hat{\mathcal{F}}_j^{k+1} \end{pmatrix} = c \begin{pmatrix} \mathcal{F}_j^1(x_j^k, z_j^k) \\ \mathcal{F}_j^2(x_j^k, z_j^k) \end{pmatrix} \tag{5}$$

enddo

(2) **do sequential**
(a) for new Jacobian: do LU–factorization of the main system matrix in (6)
(b) solve:
$$\begin{bmatrix} C \\ \partial_z \mathcal{G} \end{bmatrix} \Delta z^{k+1} = -\begin{pmatrix} \hat{\mathcal{F}}^{k+1} \\ c\,\mathcal{G}(z^k) \end{pmatrix} \qquad (6)$$
enddo
(3) **do parallel** for all $j \in \{1,\ldots,p\}$:
$$\Delta x_j^{k+1} = -\Delta \hat{x}_j^{k+1} - B_j \Delta z_j^{k+1}$$
enddo

with scalar constant c and the notations:
$\Delta\hat{x}_j := (\partial_{x_j}\mathcal{F}_j^1)^{-1}\mathcal{F}_j^1$, $\hat{\mathcal{F}}_j := \mathcal{F}_j^2 - \partial_{x_j}\mathcal{F}_j^2 \Delta\hat{x}_j$, $B_j := (\partial_{x_j}\mathcal{F}_j^1)^{-1}\partial_{z_j}\mathcal{F}_j^1$,
$C_j := \partial_{z_j}\mathcal{F}_j^2 - \partial_{x_j}\mathcal{F}_j^2 B_j$, $C :=$diag(C_j) and $\hat{\mathcal{F}} := (\hat{\mathcal{F}}_1, \hat{\mathcal{F}}_2, \ldots, \hat{\mathcal{F}}_p)^T$.

It should be remarked, that for solving the equations (4) and (6) the same LU–factorizations are used several times.

Both main parts of the computational amount, namely all the calculation of functions and Jacobians as well as most of the amount for the solution of the linear systems, are included together in one parallel loop (step 1) resulting into a coarse grain parallelism. The bottleneck is the sequential part (step 2), which is dominated by the LU–factorization of the main system matrix. To reduce this sequential amount of the algorithm and to increase the efficiency of the implementation on parallel computers, various modifications of the method can be considered. Here, only some of them can be mentioned.

At first, the main system factorization can be speeded up by eliminating zero elements (*sparsing*) from the dense computed submatrix blocks C_j.

At second, by using *multilevel Newton iteration* techniques [6] it is possible to shift computational costs from the main system solution (outer iteration) to the solution of the blocks by substituting step (1)(b) by an inner iteration loop:

set $\hat{x}_j^{k+1,0} := x_j^k$
do $l = 0(1)l_j$: compute $\mathcal{F}_j(\hat{x}_j^{k+1,l}, z_j^k)$, solve:
$$\begin{bmatrix} \partial_{x_j}\mathcal{F}_j^1 & \emptyset \\ \partial_{x_j}\mathcal{F}_j^2 & I \end{bmatrix} \begin{pmatrix} \Delta\hat{x}_j^{k+1,l+1} \\ \hat{\mathcal{F}}_j^{k+1,l+1} \end{pmatrix} = c \begin{pmatrix} \mathcal{F}_j^1(\hat{x}_j^{k+1,l}, z_j^k) \\ \mathcal{F}_j^2(\hat{x}_j^{k+1,l}, z_j^k) \end{pmatrix} \qquad (7)$$
and set $\qquad \hat{x}_j^{k+1,l+1} := \hat{x}_j^{k+1,l} - \Delta\hat{x}_j^{k+1,l+1}$

enddo
set $\Delta\hat{x}_j^{k+1} := \Delta\hat{x}_j^{k+1,l_j+1}$ and $\hat{\mathcal{F}}_j^{k+1} := \hat{\mathcal{F}}_j^{k+1,l_j+1}$

At third, by formally extending the main system (6) with the last factorized block–diagonal matrix $\hat{C}$, a new factorization of the main system matrix can be avoided, in some cases, by using instead of step (2)(b) the following *iterative scheme for solving the main system*:

set $z_j^{k+1,0} := z_j^k,\ j = 1(1)p$
do $l = 0(1)l_0$ **:**
do parallel for all $j \in \{1,\ldots,p\}$

$$\begin{bmatrix} \partial_{x_j}\mathcal{F}_j^1 & \emptyset \\ \partial_{x_j}\mathcal{F}_j^2 & I \end{bmatrix} \begin{pmatrix} B_j \Delta z_j^{k+1,l} \\ (C_j - \hat{C}_j)\Delta z_j^{k+1,l} \end{pmatrix} = \begin{pmatrix} \partial_{z_j}\mathcal{F}_j^1 \Delta z_j^{k+1,l} \\ (\partial_{z_j}\mathcal{F}_j^2 - \hat{C}_j)\Delta z_j^{k+1,l} \end{pmatrix} \tag{8}$$

enddo

$$\begin{bmatrix} \hat{C} \\ \partial_z \mathcal{G} \end{bmatrix} \Delta z^{k+1,l+1} = -\begin{pmatrix} \hat{\mathcal{F}}_2^{k+1} + (C - \hat{C})\Delta z^{k+1,l} \\ c\, \mathcal{G}(z^{k,l}) \end{pmatrix} \tag{9}$$

enddo
set $\Delta z^{k+1} := \Delta z^{k+1,l_0+1}$

Step (3) can then be replaced by:

do parallel for all $j \in \{1,\ldots,p\}$: $\Delta x_j^{k+1} = -\Delta\hat{x}_j^{k+1} - B_j \Delta z_j^{k+1,l_0}$ **enddo**

In this modification the explicit evaluation of the matrices B_j is avoided by computing only the vectors $(B_j \Delta z_j^{k+1,l})$ from (8). Because in (8) as well as in (9) only existing factorizations are used, and because the corrections in (8) can be computed in parallel for all blocks $j = 1(1)p$, this approach tends to be efficient for very large problems.

It should be remarked, that the integer constants $l_j, j = 0(1)p$ depend on the convergence behavior.

3 Applications

The block–structured Newton–type methods have been proven successfully for the dynamic process simulation of real life chemical plants on parallel computers with shared memory. They are now included in a block oriented process simulation code BOP using topological block partitioning algorithms. To use numerical integration with BDF methods, the DASSL code has been modified with respect to the nonlinear and linear solver, consistent initialization, and handling of discontinuities. BOP uses a hierarchically structured data interface, which is currently generated out of the data supplied by the commercial process simulator SPEEDUP[1] [1]. It is implemented on computers Cray J90 and SGI Origin 2000 using multiprocessing compiler directives for parallelization.

Carrying out dynamic process simulations for various large distillation plants of the Bayer AG Leverkusen have shown the potential parallelism of the methods realized in BOP. The times given for the examples in Table 1 and 2 are measured for whole simulation runs on non dedicated machines Cray J90 and include the times for sequential pre– and post–processing, which are usually about 5% of the overall CPU time.

[1] Used under licence 95122131717 for free academic use from AspenTech, Inc.

Table 1. Dynamic simulation of plant *bayer13* (296 subsystems, 18 350 equations)

Processors	1	1	8	24
Blocks	1	24	24	24
$\sum$ CPU time (sec.)	606.40	636.19	717.01	676.24
Wall clock time (sec.)	621.56	653.86	146.66	84.75
Speedup factor	1	0.95	4.24	7.33

Table 2. Dynamic simulation of plant *bayer12* (170 subsystems, 19 558 equations)

Processors	1	1	7	21
Blocks	1	21	21	21
$\sum$ CPU time (sec.)	1 250.10	1 124.01	1 160.59	1 142.34
Wall clock time (sec.)	1 284.98	1 147.56	244.72	142.29
Speedup factor	1	1.12	5.25	9.03

From a performance analysis using the Cray tool ATExpert, a speedup factor of 11.5 was predicted, for example, for the dynamic simulation part (without pre- and post-processing) of plant *bayer12*, when 16 processors and a 16 block partitioning are used. Due to the coarse grain parallelism only a very low overhead caused by the parallelization was found.

Acknowledgements. The author thanks his colleagues F. Grund and D. Horn for useful discussions. The valuable assistance and the technical support from the Bayer AG Leverkusen are gratefully acknowledged.

References

1. AspenTech: SPEEDUP, User Manual, Library Manual, Aspen Technology, Inc., Cambridge, Massachusetts, USA, (1995)
2. Borchardt, J., Grund, F., Horn, D., Michael, T.: Parallelized Numerical Methods for Large-Scale Dynamic Process Simulation. In: Sydow, A. (ed.): Proceedings of the 15th IMACS World Congress on Scientific Computation, Wissenschaft & Technik Verlag, Berlin, Volume I, (1997) 547–552
3. Brenan, K.E., Campbell, S.L., Petzold, L.R.: Numerical Solution of Initial–Value Problems in Differential–Algebraic Equations, North–Holland, New York, (1989)
4. Brüll, L., Pallaske U.: On Consistent Initialization of Differential–Algebraic Equations with Discontinuities. In: Wacker, Hj., Zulehner W. (eds.): Proceedings of the Fourth Conference on Mathematics in Industry, (1991) 213–217
5. Gräb, R., Günther, M., Wever, U., Zheng, Q.: Optimization of Parallel Multilevel-Newton Algorithms on Workstation Clusters. In: L. Bouge et al. (Eds.): Euro–Par96 Parallel Processing, Berlin, Lecture Notes in Computer Science 1124, Springer-Verlag, Berlin Heidelberg, (1996) 91–96
6. Hoyer, W., Schmidt, J.W.: Newton–Type Decomposition Methods for Equations Arising in Network Analysis, ZAMM 64, No. 9, (1984) 397–405

Tuning the Performance of Parallel Programs on NOW's Using Performance Analysis Tool

Marian Bubak[1,2], Włodzimierz Funika[1], Jacek Mościński[1,2]

[1] Institute of Computer Science, AGH, al. Mickiewicza 30, 30-059, Kraków, Poland
[2] Academic Computer Centre - CYFRONET, Nawojki 11, 30-950 Kraków, Poland
{bubak,funika,jmosc}@uci.agh.edu.pl

Abstract. This paper presents the functionality and use of an off-line tool for performance analysis of parallel programs running on networks of workstations. The tool consists of a set of facilities used for instrumentation, monitoring, trace data processing and visualisation. It was applied to tune parallel programs implementing a lattice gas automata method and a travelling salesman problem solution with genetic algorithms. A study of how coupling or splitting messages could influence application's performance in terms of communication overhead and load imbalance is presented.

1 Introduction

In recent years parallel distributed systems and parallel programming gave evidence to ever growing use of message passing paradigm for solving complex computational problems [1–3]. Networks of workstations (NOWs), which now constitute a popular class of distributed computer systems, can be effectively used to realise high performance computing, thus competing with massively parallel systems (MP). Investigation of the reasons for which a parallel program's execution on NOWs may be inefficient is a more complicated task than for parallel programs on MP due to an additional number of factors affecting performance. An overview of tools for performance analysis of parallel programs can be found in [4]. In order to be accepted by the user, tools should meet a large set of requirements.

The paper considers the functionality and use of an off-line tool for performance analysis and tuning parallel programs running on NOWs. The tool has been applied to tune parallel programs implementing a lattice gas automata method and a travelling salesman problem solution using genetic algorithms. A study of the influence of coupling or splitting messages on application's performance is presented.

2 Functionality of the Tool

Although now a large number of performance tools exists, the user of NOWs still face the lack of adequate tools for their work [5]. In part, this is due to the implementation complexity of monitoring systems and the diversity of actions and data to be handled. This forced us to design a set of facilities for performance analysis using the best of the experiences gained by other research teams [6]. The tool is aimed at trace-based evaluation of a parallel application. The functional structure of the tool is shown in Fig. 1.

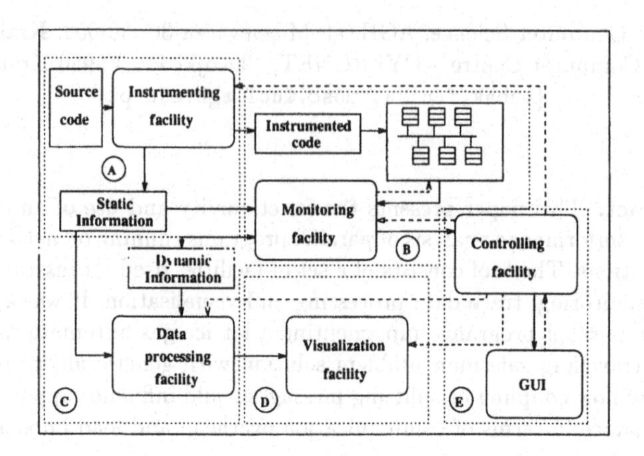

Fig. 1. Functional structure of the tool

Instrumentation phase provides a GUI to allow for a graphical specification of instrumentation actions: tracing – generating events, and extended profiling – counting and timing. When parsing the application source code, each instrumentable object entry is described by a set of static information which are used in further stages of performance analysis. Extended profiling allows to lower significantly perturbations compared to tracing.

The goal of trace-oriented *monitoring* consists in generating particular events of the program's run and cumulative performance data relevant to the selected level of detail. Trace data are buffered in local memories. A clock synchronisation mechanism enables to produce globally consistent time-stamps. Performance data are transformed into SDDF format [7].

Trace processing is aimed at obtaining a set of performance metrics and visualizations to help the user find out those performance phenomena which affect the efficiency of application execution. *Visualisation* modules are designed to generate statistics as well as profiling and dynamic views. Mechanisms for filtering, reducing data, and for correlating performance data with source code of application are provided.

3 Samples of Performance Analysis and Tuning

3.1 Description of LGA and TSP–GA Parallel Programs

Below we present examples of performance tuning for two typical distributed applications running on NOW under PVM which employ master/worker model:

a lattice gas simulation program (LGA) [8] and a travelling salesman problem solving program exploiting genetic algorithms (TSP-GA) [9].

In LGA program, a lattice is divided into domains along y axis. Communication features a twofold topology: that between master and workers is star-oriented, while the inter-slave one is a ring-like. A lattice of 480×2880 sites was used.

In TSP-GA program, which employs the island model, the whole population of chromosomes is divided into subpopulations which are sent to slaves by the master process. The results of simulation are sent back to master to be mixed, selected and re-divided among slaves. The solution was searched for 100 points.

Experiments were done on a network of RS6000/320 stations. In the figures below the *val* alias stands for master node.

3.2 Tuning LGA Program

The main contributors to communication and idling overhead are *output* and *trans_fhp* procedures which perform communication with master and inter-slave communication, respectively. Whereas in the case of *trans_fhp* procedure, the idle time is cumulative over the evolution of the system which includes hundreds of steps, the idle time of *output* procedure is due to few data exchanges with master process. A fragment of Gantt diagram (Fig. 2) illustrates this situation.

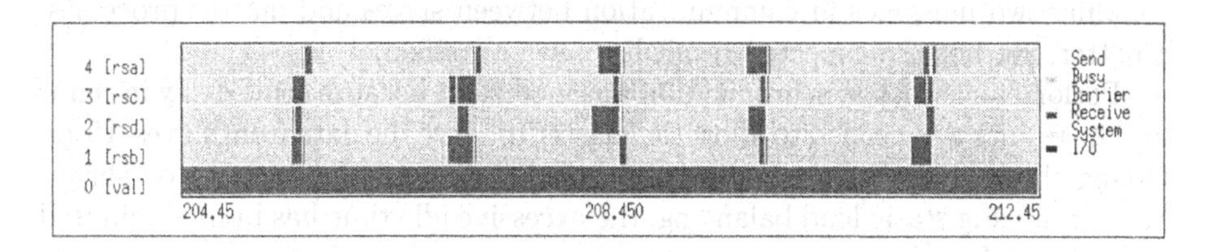

Fig. 2. Gantt diagram of LGA before tuning

In order to reduce the idle time in *output* procedure, the communication scheme was rearranged through splitting slave-master communication pattern into two ones. This allowed to cut the idle time more than by half.

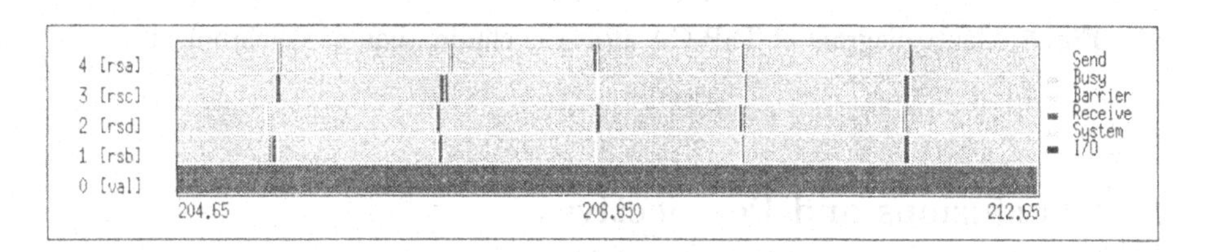

Fig. 3. Gantt diagram of LGA after tuning

When running master and one of slave processes on the same computing node, multiprogramming affected *output* procedure. This was eliminated by moving the master process to another machine. The final level of communication overhead does not exceed 5% of execution time. The Gantt diagram illustrating the performance obtained after the reorganization is shown in Fig. 3.

3.3 Tuning TSP-GA Program

The efficiency of the program implementation can be evaluated indirectly by investigating the distance in function of the number of evaluation steps done.

Gantt diagram (Fig. 4) illustrates that the original version of the program features a significant imbalance between idle and communication time.

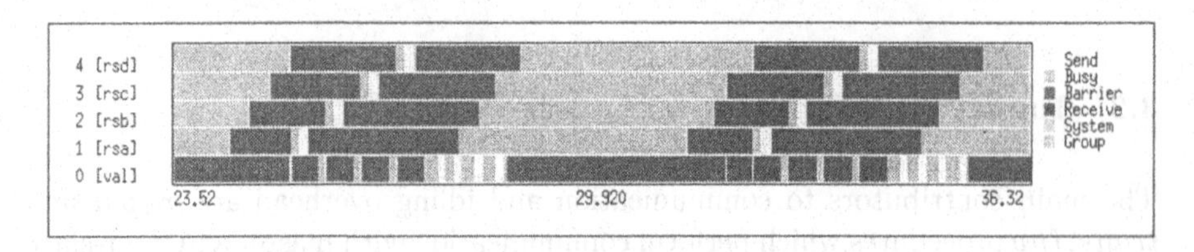

Fig. 4. Gantt diagram of TSP-GA before tuning

After having introduced synchronisation points into slave computations and coupling two messages in communication between slaves and master processes, a better, yet not the expected result has been obtained.

Performance with synchronization was inefficient because some delay in sending the results was needed to allow the master to serve the data from slaves. Upon having eliminated synchronization and tuned the population number on slaves, i.e. introducing static load balancing, the excessive idle time has been eliminated as shown in Fig. 5.

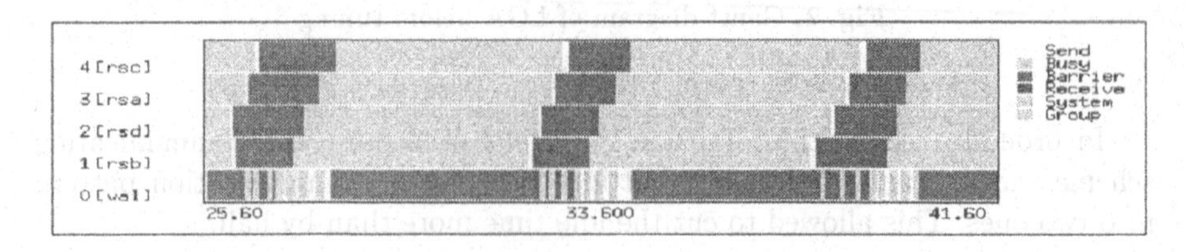

Fig. 5. Gantt diagram of TSP-GA after the eliminatiion of synchronization

4 Conclusions and Perspectives

In the paper the off-line performance analysis tool for applications employing message passing paradigm on networked workstations was described. The tool consists of a set of facilities which enable instrumentation, trace-oriented monitoring, performance data processing, and visualisation. A set of flexible views enables to locate and isolate the phenomena affecting performance.

In the examples presented, to achieve a desirable efficiency of parallel program, in one case we needed to split a single communication pattern into two ones to eliminate the dependence between two different types of communication (LGA program), whereas in the other case the way out was found with the help of merging two communication patterns into a single one (TSP-GA program).

Acknowledgements. We would like to acknowledge the assistance of Dr. Renata Słota and the contribution of Grzegorz Kazior and Radosław Gemrarowski. This research was partially supported the AGH grant.

References

1. Viersbach, U., Hesse, M., Zellermann, G., Schlanke J.: Parallelization: Readiness of Workstation Clustering for ComputerAided Engineering. In: Bubak, M., Mościński, J., (eds.): Conf. Proc. High Performance Computing and Networking on Hewlett-Packard Systems HiPer'97, Kraków, November 5-8, 1997, ACC CYFRONET-KRAKÓW (1997) 95
2. Geist, A., et al.: PVM: Parallel Virtual Machine. A Users' Guide and Tutorial for Networked Parallel Computing. MIT Press, Cambridge, Massachusetts (1994)
3. MPI: A Message Passing Interface Standard. In: Int. Journal of Supercomputer Applications, **8** (1994); Message Passing Interface Forum: MPI-2: Extensions to the Message Passing Interface, July 12, (1997); `http://www.mpi-forum.org/docs/`
4. Pancake, C.M., Simmons, M.L., and Yan., J. C., (eds.): Special Issue on Performance Evaluation Tools for Parallel and Distributed Systems. In: IEEE Computer, Vol. **28**, (1995) No 11
5. Ludwig, T., Wismueller, R., Sunderam, V., and Bode, A.: OMIS – On-line Monitoring Interface Specification (Version 2.0). Shaker Verlag, Aachen, Vol. 9, LRR-TUM Research Report Series (1997)
6. Bubak M., Funika, W., Mościński J.: Evaluation of Parallel Application's Behavior in Message Passing Environment. In: Marian Bubak, Jack Dongarra, Jerzy Wasniewski (eds.): Recent Advances in Parallel Virtual Machine and Message Passing Interface, Proceedings of 4th European PVM/MPI Users' Group Meeting, Cracow, Poland, November 1997. Lecture Notes in Computer Science, Vol. 1332, Springer (1997) 234-241
7. http://www-pablo.cs.uiuc.edu/Projects/Pablo/sddf.html
8. Bubak, M., Mościński, J., Słota, R.: FHP lattice gas on networked workstations. In: Gruber, R., Tomassini, M., (eds.): The 6th Joint EPS-APS International Conference on Physics Computing, Lugano, Switzerland, August 22-26, (1994) 427-430
9. Bryt, A., Funika, W., Gąciarz, T.: Parallel Solution of Travelling Salesman Problem using Genetic Algorithms. In: Szmuc, T., Szymkat, M., Tadeusiewicz, R., Uhl, T.(eds.): The 1st Polish Conference on Computer Methods and Systems in Scientific Research and Engineering, Cracow, Poland, AGH, November 25-26, (1997) 21-26 (In Polish)

Numerical Simulation of 3D Fully Nonlinear Water Waves on Parallel Computers

Xing Cai

Department of Informatics, University of Oslo, P.O. Box 1080 Blindern, N-0316 Oslo, Norway
xingca@ifi.uio.no

Abstract. A new programming model is used to parallelize a sequential simulator for 3D fully nonlinear water waves. This so-called simulator-parallel model has its numerical foundation in additive Schwarz methods, so it possesses inherent parallelism and inherits the intrinsic numerical efficiency of domain decomposition algorithms. Within an object-oriented framework implementing this model, the actual parallelization work for the sequential wave simulator becomes straightforward. Numerical experiments show that the resulting parallel simulator achieves good parallel efficiency.

1 Introduction

In this paper, we are interested in numerical simulations of 3D fully nonlinear water waves on parallel computers. Mathematically, this reduces to solving a system of nonlinear partial different equations (PDEs). The issues on devising an efficient numerical method for running sequential simulations were discussed in an earlier work [3]. The resulting sequential wave simulator is a very good starting point for parallel computing. However, to achieve good parallel efficiency, a good parallelization strategy is necessary. For this purpose, we use a so-called *simulator-parallel* programming model that has its numerical foundation in additive Schwarz methods. In this way, the intrinsic numerical efficiency of domain decomposition (DD) algorithms is added upon the efficiency of the original sequential wave simulator, ensuring good overall computational performance of the resulting parallel wave simulator. The actual parallelization will be done in a generic framework implementing the simulator-parallel model. This process is efficient and systematic. Numerical experiments will show that the overall computational efficiency is good for the parallel wave simulator.

The rest of the paper is organized as follows. Section 2 gives a brief description of the mathematical model and an efficient sequential numerical method. An introduction to the resulting sequential wave simulator is also included. Section 3 discusses the simulator-parallel programming model, which is used as the parallelization strategy. Thereafter, Section 4 sketches how a parallel wave simulator is implemented based on the existing sequential simulator, inside a generic implementation framework. One numerical experiment is presented in Section 5 where different issues on efficiency are studied. Finally, Section 6 concludes the paper with some remarks.

2 Mathematical Model and Numerical Method

We begin with a brief description of the mathematical model. Let $(\bar{x},\bar{y},\bar{z})$ be the spatial coordinates and let t denote time. By the assumption that the wave induced velocity field is divergence free and irrotational, we arrive at the following system of PDEs (see e.g. [9]),

$$-\nabla^2\varphi = 0 \quad \text{in the water volume,} \tag{1}$$

$$\eta_t + \varphi_{\bar{x}}\eta_{\bar{x}} + \varphi_{\bar{y}}\eta_{\bar{y}} - \varphi_{\bar{z}} = 0 \quad \text{on the free surface,} \tag{2}$$

$$\varphi_t + \frac{1}{2}(\varphi_{\bar{x}}^2 + \varphi_{\bar{y}}^2 + \varphi_{\bar{z}}^2) + g\eta = 0 \quad \text{on the free surface,} \tag{3}$$

$$\frac{\partial\varphi}{\partial n} = 0 \quad \text{on solid boundaries.} \tag{4}$$

In the above system, the velocity potential $\varphi(\bar{x},\bar{y},\bar{z},t)$ and the water surface elevation $\eta(\bar{x},\bar{y},t)$ are the two primary unknowns. Here, subscripts denote derivatives and g is the gravitational acceleration in the negative $\bar{z}$-direction. Equation (1) is the Laplace equation and has to be solved in the 3D water volume. The nonlinear PDEs (2) and (3) govern the water motion on the free surface and are referred to as the kinematic and dynamic boundary condition, respectively. On the rest of the boundary, which is assumed to be solid in this paper, equation (4) guarantees that no water flows through. Furthermore, we restrict the water volume to be of the form:

$$\overline{\Omega}(t) = \{(\bar{x},\bar{y},\bar{z}) \,|\, (\bar{x},\bar{y}) \in \Omega_{\bar{x}\bar{y}},\ -H \le \bar{z} \le \eta(\bar{x},\bar{y},t)\}\,. \tag{5}$$

A typical example of $\overline{\Omega}(t)$ is a water tank. Noticing that the 2D cross section $\Omega_{\bar{x}\bar{y}}$ is time independent, we introduce a time dependent transformation ρ : $(\bar{x},\bar{y},\bar{z}) \to (x,y,z)$:

$$x = \bar{x}, \quad y = \bar{y}, \quad z = \left(\frac{\bar{z}+H}{\eta+H} - 1\right)H\,, \tag{6}$$

which maps the dynamic 3D physical domain $\overline{\Omega}(t)$ onto a stationary computational domain Ω of the form:

$$\Omega = \{(x,y,z) \,|\, (x,y) \in \Omega_{xy},\ -H \le z \le 0\}\,.$$

The transformation is well-defined as long as $\eta(x,y,t) + H > 0$. We thus avoid the expensive regridding of $\overline{\Omega}(t)$ at each time step. Instead of the original Laplace equation (1) in $\overline{\Omega}(t)$, an elliptic boundary value problem

$$-\nabla\cdot(\boldsymbol{K}(x,y,z,t)\nabla\varphi) = 0\,, \tag{7}$$

which has a time dependent variable coefficient needs to be solved in Ω. More precisely, we have

$$\boldsymbol{K}(x,y,z,t) = \frac{1}{H}\begin{bmatrix} \eta+H & 0 & -(z+H)\eta_x \\ 0 & \eta+H & -(z+H)\eta_y \\ -(z+H)\eta_x & -(z+H)\eta_y & \dfrac{H^2+(z+H)^2(\eta_x^2+\eta_y^2)}{\eta+H} \end{bmatrix}.$$

We refer to [3] for details of the transformation and the resulting elliptic boundary value problem.

We devised an efficient sequential numerical method in [3]. Our numerical method consists of two substeps at each time level: 1. Solution of the elliptic boundary value problem (7) by the finite element method (FEM); 2. Explicit time marching of (2) and (3) to generate η and the values of φ on the surface to be used at the next time level. Obviously, Substep 1 is the more CPU intensive one. The reason for choosing the FEM for (7) is that the resulting linear system of equations (at each time level) can be solved efficiently by a preconditioned conjugate gradient (PCG) method. The computational efficiency is determined by the preconditioner that can e.g. be the modified incomplete LU-factorization (MILU, see [7]). However, to achieve optimal convergence of the PCG method, i.e., independent of the mesh size, better preconditioners should be chosen. A fast Poisson solver based on the fast Fourier transform was used in [3] and we will consider the more general multilevel methods in this paper. More specifically, we will use the combination of DD and multigrid methods, i.e., DD as the underlying structure for parallelization and multigrid V-cycles as the efficient subdomain solver.

A sequential water wave simulator has been implemented in Diffpack [2, 6], which is an object-oriented (O-O) scientific computing environment containing, among other things, a large collection of efficient numerical methods and preconditioners for solving linear systems of equations. It is worth mentioning that the design of Diffpack has taken numerical efficiency into consideration by confining O-O techniques to high-level administrative tasks, while using low-level C codes and carefully constructed for-loops in CPU intensive numerics. Therefore, the sequential wave simulator has reached a high level of computational efficiency and maintained good flexibility due to O-O programming techniques. Consequently, we are interested in a parallelization strategy that maintains the computational efficiency and, at the same time, promotes extensive reuse of sequential codes. This will be addressed in the next section.

3 Parallelization

We proposed in [1] a so-called simulator-parallel programming model for parallelizing existing sequential simulators. The programming model has its numerical foundation in *additive Schwarz methods* that fall under the category of overlapping DD methods. Roughly speaking, DD methods search for the solution of the original large problem by iteratively solving smaller subproblems associated with the subdomains. For an overview of DD methods, we refer to [5, 8]. Nowadays DD methods have been widely applied in the solution of large systems of equations arising from the discretization of PDEs. They also serve as a standard parallelization strategy because of their inherent parallelism and numerical efficiency. However, the tradition has been to use DD at the level of linear algebra in the parallelization, and the implementation is often done from scratch using procedural programming languages.

Our simulator-parallel programming model, on the other hand, applies DD at the level of subdomain simulators by making extensive use of O-O programming techniques. More specifically, the simulator-parallel model puts one or several O-O sequential simulators on each processor, coordinated by some global administration and communication. The basic idea roots in the observation that the main ingredient of any overlapping DD method is the solution of subproblems, which are the restriction of the original problem to the subdomains. So the subdomain solves can essentially be carried out by the original sequential simulator. Meanwhile, O-O programming techniques make it very easy to extend/modify the original sequential simulator to fit in a DD setting. An immediate advantage of a parallelization strategy using the simulator-parallel model is that different types of grid, linear system solvers, preconditioners, convergence monitors etc. are allowed for different subdomains. Moreover, the global data distribution is implied by the local data structure of the subdomain simulators, so the data distribution issue can be hidden from the users.

For the elliptic boundary value problem (7), overlapping DD methods are well known for their superior numerical efficiency. Incorporating coarse grid corrections, the DD methods are asymptotically optimal, even on sequential computers. The most robust and efficient approach is to apply one DD iteration as the preconditioner in the PCG method. Another property of the DD methods is that they allow inexact solves of the subproblems. This gives the possibility of using iterative methods in subdomain solves and thus reducing the work amount considerably. An optimal combination is to use DD as the underlying parallel structure and multigrid V-cycles as the subdomain solver. Consequently, the overall computational work for solving (7) is (nearly) $\mathcal{O}(1)$ per degree of freedom because: 1. The work per multigrid V-cycle is proportional to the number of unknowns in a subdomain; 2. $\mathcal{O}(1)$ CG-iterations are sufficient for the global convergence. It is now clear that a parallel wave simulator based on the simulator-parallel model will at least maintain the numerical efficiency from the sequential case. Meanwhile, experiences indicate that overheads of communication and coarse grid corrections are relatively small in comparison with the subdomain solves. Therefore, good overall parallel efficiency can be expected under the assumption that subdomains have approximately the same size.

4 Implementation

In this section, we first give a brief presentation of a general-purpose O-O implementation framework for the simulator-parallel programming model. Then we sketch the actual parallelization work for the wave simulator in form of C++ terminologies.

We consider a general-purpose implementation framework consisting of three main parts: The sequential subdomain simulator, a communication part and a global administrator. The function of the framework is to simplify the parallelization process and, at the same time, offer a systematic approach suitable for a wide range of sequential simulators. *Flexibility*, *extensibility* and *portability* are three key words describing the framework. Using O-O programming techniques, two C++ class hierarchies are built to represent the sequential subdomain simulator and the communication part, i.e., the `SubdomainSimulator` hierarchy and the `Communicator` hierarchy. So users can plug-and-play different ready-made classes or derive new classes to handle specific situations. Low-level communication codes are encapsulated in `Communicator` that provides users with high-level communication commands. Using the MPI standard, the codes can be run on any distributed computing system supporting MPI.

The global administrator implements the mathematical algorithm of DD at a high abstraction level. Its actions include invoking subdomain solves, initiating communication and coordinating operations between different processors. The design of the global administrator allows the user to choose, among other things, the specific Krylov subspace method at run-time. Readers are referred to [4] for a detailed description of the implementation framework. We mention that a prototype C++ implementation of the framework has been made as an add-on library to the standard Diffpack libraries.

The above framework promotes reusing sequential codes to a large extent. It thus greatly simplifies the parallelization work of a concrete sequential PDE simulator. The essential work consists of extending the existing sequential simulator to inherit the generic interface of `SubdomainSimulator` and making a connection between the extended sequential simulator and the global administrator. During the process of extending the existing sequential simulator, users usually bind the *virtual* member functions of `SubdomainSimulator` to the concrete member functions of the target sequential simulator. Meanwhile, users are also free to make necessary modifications inside the extended sequential simulator. We refer to [4] for the details. In the case of the existing sequential wave simulator `WaveSimulator`, the C++ coding can be of the following form:

```
class ParaWaveSimulator : public WaveSimulator,
                          public SubdomainFEMSolver
{
  // ....
  virtual void createLocalMatrix () {
    WaveSimulator::makeSystem ();
  }
};
```

The above code segment shows a simplified definition of `ParaWaveSimulator` that is the extended sequential simulator to be used in the framework. We see that class `ParaWaveSimulator` is derived from both the existing sequential simulator `WaveSimulator` and `SubdomainFEMSolver`, which is a class from the `SubdomainSimulator` hierarchy and offers a generic interface for simulators solving a scalar/vector elliptic PDE discretized by FEMs. We also see an example of binding a virtual member function of `SubdomainFEMSolver` to a concrete member function of `ParaWaveSimulator`.

For the explicit time marching of the free surface conditions (2) and (3), the parallelization is straightforward and uses the communication facilities of the framework for exchanging information. The parallelization of our water wave simulator is thus done in an efficient and systematical way.

5 One Numerical Experiment

In this section, we report parallel simulations of a specific numerical experiment. The purpose is twofold: 1. Examine the parallel efficiency of the resulting parallel wave simulator; 2. Demonstrate the superior efficiency of the multigrid V-cycles in the subdomain solves.

More specifically, We study the motion of water waves in a 3D water tank where $\Omega = [0, L_1] \times [0, L_2] \times [-H, 0]$ with $L_1 = L_2 = 80$ and $H = 50$. The initial condition is that the wave is at rest and the surface elevation is of the form:

$$\eta(x, y, 0) = \left(-0.9 \cos\left(\frac{\pi x}{L_1}\right) + \cos\left(\frac{2\pi x}{L_1}\right)\right)\left(1 - 0.9 \cos\left(\frac{\pi y}{L_2}\right) + \cos\left(\frac{2\pi y}{L_2}\right)\right).$$

We run simulations in the time interval $0 < t \leq T = 4$ with a time step of 0.125. The fixed 3D global grid is $41 \times 41 \times 41$ and we make 2D partitions of it for desired numbers of subdomains. We denote by M the number of subdomains and by P the number of processors in use. Two case studies have been carried out. In case study number one, we make a fixed partition with $M = 16$ and use various P for the *same* simulation. (That is, when $P = 1$, all the 16 subdomains are placed on a single processor, whereas $P = 16$ means that 16 processors are in work, each responsible for one subdomain.) In Table 1, we list the CPU-times associated with different choices of P for the whole simulation. The measurements are obtained on a SGI Cray Origin 2000 parallel computer with R10000 processors. As stated before, one DD iteration is used as the preconditioner for a global parallel PCG method. For the global convergence at each time level, we require that the discrete L_2-norm of the global residual is smaller than 10^{-8}. We have tried two types of subdomain solver, one is a local PCG method using a MILU preconditioner and the other is one local multigrid V-cycle. The convergence requirement for the local PCG method is that the local residual's discrete L_2-norm is reduced by a factor of 100. In the local multigrid V-cycle we use one pre and one post SSOR smoothing on each grid level. Some snapshots from a parallel simulation are available at `http://www.ifi.uio.no/~xingca/pwave/`.

Form Table 1 we can see that the parallel efficiency of our parallel wave simulator is good. This is clearly reflected by the measurements obtained from using one multigrid V-cycle as the subdomain solver. Although there are three sources of overhead: communication, coarse grid correction and synchronization (due to a slightly difference in the size of the subdomains), we are able to achieve almost perfect speedups. The comparatively poorer performance associated with the local PCG subdomain solver is due to the fact that different number of local CG iterations are needed to obtain local convergence on different subdomains, thus destroying the balance of work load.

In case study number two, we have $P = M$ which is the most common situation of parallel computing. This time we only consider the case of using one

Table 1. CPU-times (in seconds) for the whole simulation for different choices of P and the subdomain solver, where the partition is fixed with $M = 16$.

	Multigrid V-cycle		PCG(MILU)	
P	CPU-time	Speedup	CPU-time	Speedup
1	1404.40	~	4936.62	~
2	715.32	1.96	2567.19	1.92
4	372.79	3.77	1432.90	3.45
8	183.99	7.63	764.50	6.46
16	90.89	15.45	394.57	12.51

multigrid V-cycle as the subdomain solver. For $P = 1$ we actually run global sequential PCG iterations preconditioned by one global multigrid V-cycle. For $P > 1$, on the other hand, one DD iteration works as the preconditioner for the parallel CG method as before. The convergence requirement for the global PCG method is the same as in case study number one. Some CPU measurements are shown in Table 2, in which we also list the maximum number of degrees of freedom per subdomain and the average number of global PCG iterations I needed at each time level.

Table 2. CPU-times (in seconds) for the whole simulation ($P = M$). *The average number of global iterations for $P = 2$ is for a parallel Bi-CG-Stabilized method.

P	CPU-time	Max Subgrid	I
1	642.14	68,921	7.69
2	597.47	38,663	9.00*
4	265.62	21,689	13.59
8	172.23	12,259	17.25
16	90.89	6,929	16.56

6 Concluding Remarks

We have discussed the parallelization of a sequential wave simulator based on the simulator-parallel model. The work is done in a generic implementation framework where O-O programming techniques are essential. We feel that more numerical experiments need to be carried out in the future and simulations on different parallel platforms are also necessary to verify the good overall performance obtained.

Acknowledgement. The author is indebted to Professor Aslak Tveito and Dr. Klas Samuelsson for numerous discussions on various subjects. The work has been supported by the Research Council of Norway (NFR) under Grant 110673/420 (*Numerical Computations in Applied Mathematics*) and through *Programme for Supercomputing* in form of a grant of computing time.

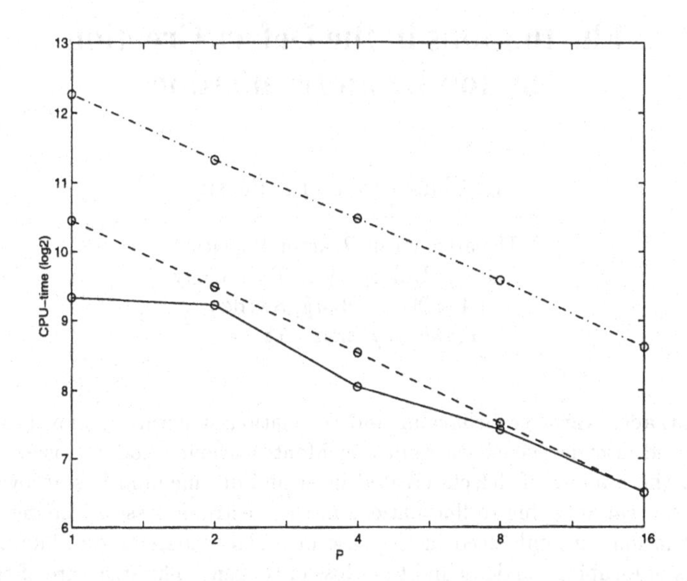

Fig. 1. Comparison of CPU-times for the whole simulation in different situations. The solid line indicates the case where $P = M$ and one multigrid V-cycle is used as the subdomain solver. The dashed line and the dash dotted line correspond to different P but fixed $M = 16$, where the subdomain solver used is one multigrid V-cycle and a local PCG method with MILU preconditioning, respectively.

References

1. A. M. Bruaset, X. Cai, H. P. Langtangen, A. Tveito, *Numerical solution of PDEs on parallel computers utilizing sequential simulators.* In Y. Ishikawa et al. (eds): Scientific Computing in Object-Oriented Parallel Environment, Springer-Verlag Lecture Notes in Computer Science 1343, 1997, pp. 161–168.
2. A. M. Bruaset, H. P. Langtangen, *A comprehensive set of tools for solving partial differential equations; Diffpack.* In M. Dæhlen and A. Tveito (eds): Numerical Methods and Software Tools in Industrial Mathematics, Birkhäuser, Boston, 1997, pp. 61–90.
3. X. Cai, B. F. Nielsen, H. P. Langtangen, A. Tveito, *A finite element method for fully nonlinear water waves.* J. Comput. Phys., 143 (1998), pp. 1–25
4. X. Cai, *An object-oriented model for developing parallel PDE software.* Preprint 1998-4, Department of Informatics, University of Oslo.
5. T. F. Chan, T. P. Mathew, *Domain decomposition algorithms.* Acta Numerica (1994), pp. 61–143.
6. Diffpack Home Page, *http://www.nobjects.com/Products/Diffpack.*
7. I. Gustafsson, *A class of first order factorization methods*, BIT, 12 (1978), pp. 142–156.
8. B. F. Smith, P. E. Bjørstad, W. D. Gropp, *Domain Decomposition, Parallel Multilevel Methods for Elliptic Partial Differential Equations.* Cambridge University Press, 1996.
9. G. B. Whitham, *Linear and Nonlinear Waves.* John Wiley & Sons, Inc., New York, 1974.

Fluctuations in the Defect Creation by Ion Beam Irradiation

R. Chakarova and I. Pázsit

Department of Reactor Physics,
Chalmers University of Technology,
S-412 96 Göteborg, Sweden
rum@nephy.chalmers.se

Abstract. Variance behaviour and correlation structures in ion beam induced cascades have been studied by Monte Carlo method. It was found that the number of defects created in an infinite medium has an over-Poisson variance due to fluctuations in the electronic losses. The defect fluctuations are enhanced in the case of a slab geometry due fluctuations of escaping particles and their loss of progeny. The structure of the vacancy depth correlations supports the variance results and indicates clustering effects in an individual cascade.
The algorithm is suitable for parallelization by particle decomposition method. Linear speed-up is easily achieved. The usage of MPI library makes the code portable to parallel computers with different architectures (Cray T3E, IBM SP, SGI Origin 2000). However optimization and code modifications are needed to obtain comparable performance results.

1 Introduction

Ion irradiation of a solid initiates developing of collision cascades in the material, resulting in creation of defects (vacancies and interstitials). The process is stochastic and the defect yield fluctuates around a certain mean value. In many applications only mean values are needed and variance is calculated to know how reliable the mean is. Another reason why study of fluctuations is of constant interest, [1], is the fact that the cascade statistics is non-trivial. Due to the branching, (recoil production), there are correlations between the energetic atoms and thus between the defects produced in each individual cascade.

It can be shown, [2], that the variance contains integrated information on the correlations. Let R^6 denote the cascade phase space. The two-point correlation function is defined as

$$c(R_1, R_2) = n(R_1, R_2) - n(R_1)n(R_2) \tag{1}$$

where $n(R)$ is the one point density, i.e. $n(R)dR$ is the expected number of defects in a volume dR around R, and $n(R_1, R_2)$ is two-point density, involving the joint probability of having defects at two different phase points. The variance of the number of defects in a finite volume R, σ_N^2, is expressed as

$$\sigma_N^2 = \langle N \rangle + \int_R c(R_1, R_2)dR_1dR_2 \tag{2}$$

Here, $\langle N \rangle$ is the mean number of defects in the volume. It is seen that a negative correlation integral causes a sub-Poisson behaviour of the variance and a positive one leads to over-Poisson variance.

Conceptual studies, based on a hard-sphere and on a power law scattering with a constant power, when neglecting the electronic losses,([3] and [4]), show that the variance of the number of the defects in a half-space is over-Poisson and increasing with the ion energy but deeply sub-Poisson and constant in an infinite medium. The objectives of our present work are to study the variance behaviour as well as correlation structures for different medium geometries in a more realistic interaction model.

2 Transport model

Analytical and numerical calculations of the higher moments are very complicated even in simple model cases. Therefore here a Monte Carlo method has been used. Classical binary collisions are assumed and the collision kinematics is derived from cross sections. The scattering kernel has a power law dependence on both the initial energy, E, and the transferred energy, T:

$$d\sigma(E,T) = C_m E^{-m} T^{-1-m} dT \tag{3}$$

where C_m is a constant depending on the projectile/target masses and nuclear charges. m is a parameter related to the steepness of the repulsive particle/particle interaction potential. In general, it varies from $m = 1$ at high energies (Rutherford scattering) down to $m \approx 0$ at very low energies. Here the energy dependence of the power exponent is obtained so that the nuclear stopping power corresponding to the Kr-C potential is reproduced. A finite minimum energy transfer is assumed to avoid divergence of the total cross section.

Non-local electronic losses are included by implementing the model of Lindhard and Scharff. The energy loss, ΔE_e, over a distance, L, travelled between two successive collision events is given by

$$\Delta E_e = L.N.k.E^{0.5} \tag{4}$$

where N is the target density and k is a function of the nuclear charges and projectile mass. A threshold energy is introduced for both recoil production (vacancy creation) and projectile trapping (interstitial). The binding energy is accounted in the damage model as well. The surface is treated as a planar potential barrier.

3 Programming model

The Monte Carlo algorithm is based on the statistical simulation of the history of each particle (ion or host atom recoil) moving from one interaction point to another until its energy drops below a certain cut-off energy or until it escapes from the medium. The ion energy may be several hundred keV, whereas the cut off energy is several eV. Soft collisions and small angle scattering dominate so that each particle loses energy in small portions. Since the threshold energy is of the order of eV, a thousand recoils might be produced in each cascade initiated. The outcomes for each ion are recorded. The final values of the output quantities, (sputtering and defect yield, energy and depth distributions, variance and correlation functions), are obtained by using data from 10 to 20 statistical batches with 10000 ions in one batch.

Thus the calculations are very demanding on CPU time and the usage of high performance computer resources is necessary. Fortunately, the algorithm is suitable for parallelization by particle decomposition due to the independence of the cascades initiated by each ion (linear transport theory assumptions). A parallel FORTRAN90 code has been written using the MPI library. The total number of ion histories is divided among the processors. The results obtained are then collected by one node for further statistical evaluation. An important point is the generation of independent random number sequences for each processor.

4 Variance and correlation results

Calculations have been performed for the case of Si medium and a monodirectional beam of Si or Ge ions with energies up to 300 keV. The mean and the relative variance of the number of vacancies vs Si ion energy for semi-infinite and infinite medium are shown in fig. 1. The difference between the results for the two geometries is so small that it can not be distinguished in the figure. This is due to the fact that the small angle scattering dominates. The cascade develops in highly anisotropic, forward biased manner and the effect of the half-space free surface becomes negligible. It is seen in the figure that the variance is over-Poisson and increasing with the ion energy. Detailed estimations prove that the large fluctuations are related to the fluctuations in the total amount of electronic losses per cascade.

The structure of the vacancy depth correlations confirms the over-Poisson behaviour of the variance, see fig. 2. Large positive correlations are concentrated around the line $z_1 \approx z_2$, whereas for large $|z_1 - z_2|$, they are negative. This is an expression of the fact that defect pairs are very likely to be created in the vicinity of each other, but the probability decreases very fast with increasing distance (the depth in this calculations). The negative correlations between pairs at large depth differences are actually an expression of exclusion for creation of such pairs. Thus the correlation structure indicates a large probability of vacancy clustering. This feature in the cascade development completely dissappears when considering the first moments only. For example, when plotting a histogram of the vacancy depth distribution, the clusters of many cascades are overlapping and the averaged effect is only registered.

The half-space medium is not suitable for investigation of boundary effects, especially when the probability of particle transmitted through the boundary is not negligible. Therefore here calculations have been performed for a medium with a slab geometry. In the case of 100 keV Si ions and a 2500 Å Si slab the parti-

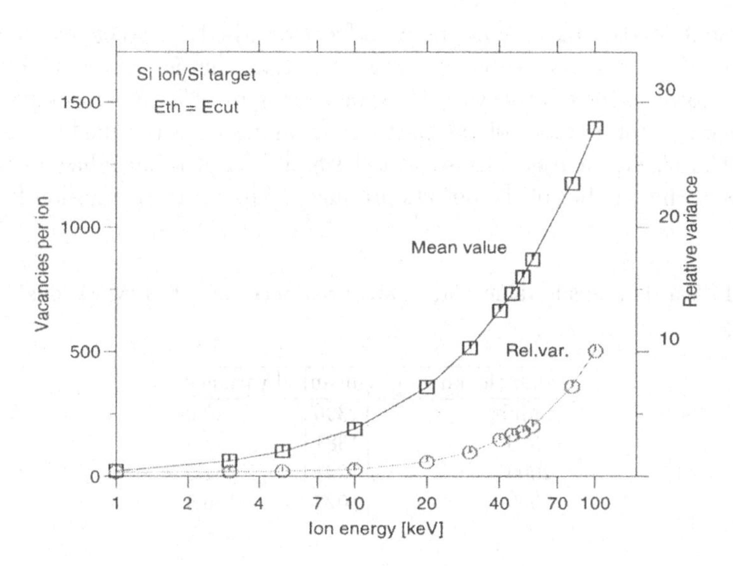

Fig. 1. Vacancy mean value and relative variance vs energy, symbols - semi-infinite medium; line - infinite medium.

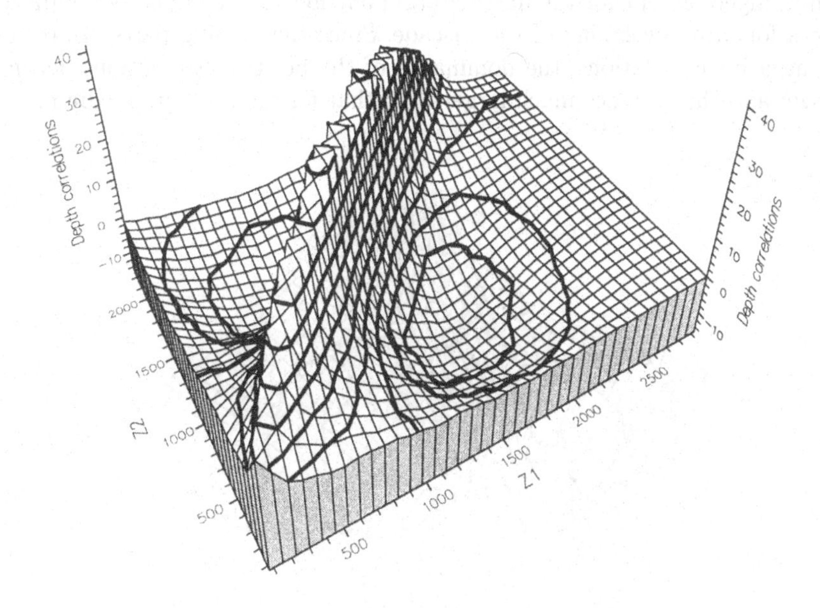

Fig. 2. Depth correlations for vacancies created by 100 keV ions in Si semi-infinite medium

cle transmission through the boundary affects very little the vacancy mean. The number of the vacancies created is only 1% less than that for infinite medium, see Table 1. However the relative variance is increasing by 53%. When decreasing the slab thickness the number of the particles escaping from the medium becomes larger. The stochastic loss of those particles and their progeny enhances the fluctuations in the number of the defects produced. The depth correlation functions

Table 1. Vacancy mean value and relative variance for 100 keV Si ions and slab geometry

slab thickness [Å]	mean	rel.variance
infinite	1396	10.3
2500	1381	15.8
2000	1291	62.5
1500	1022	176.5

of the vacancies in case of 100 keV Si ions and 1500 Å slab are shown in figs. 3 and 4. The structure compared to that in case of semi-infinite medium is not much changed, except for the missing parts outside the slab. The fact indicates again a forward developing of the cascade. Since the missing parts are regions with negative correlations, the dominance of the positive correlations becomes stronger and the variance much larger than that for semi-infinite medium.

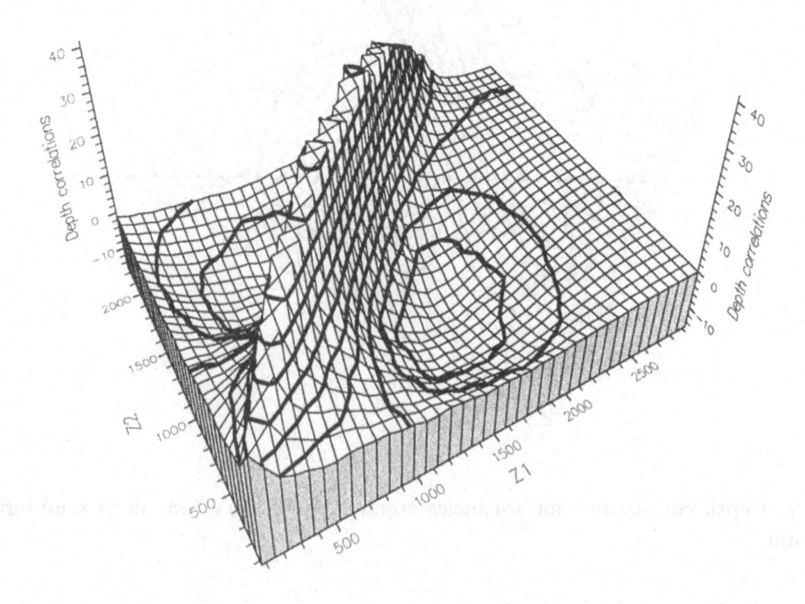

Fig. 3. Vacancy depth correlations for 100 keV Si ions and 1500 Å slab

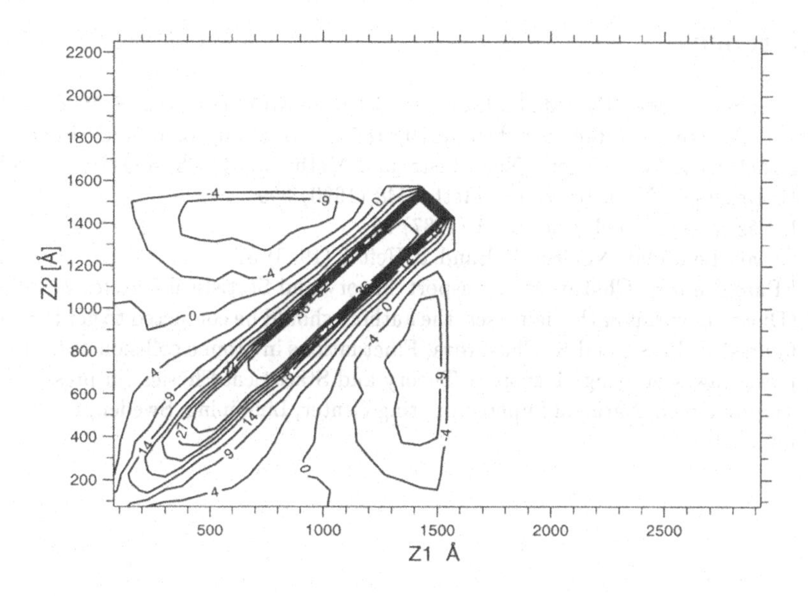

Fig. 4. 2D projection of the correlation structure in fig. 3

5 Computer implementation

Due to availability of the message passing standard, MPI, the code is highly portable to parallel computers with different architectures. It can be used as an example to practice and test User-Agent relations, where the User sends its computational task to the Agent who coordinates the usage of certain computer resources.

The results, described above, have been obtained on Cray T3E and IBM SP. Cray T3E at the National Supercomputing Center, Linköping, is shared memory machine with 216 DEC Alpha EV5, 300 MHz. IBM SP at the Center for Parallel Computers, Stockholm, is distributed memory system with 146 POWER2 RS/6000 160 MHz. Test calculations have been performed on SGI Origin 2000 (64 MIPS R1000, 195 MHz) and SUN Enterprise 10000 (64 Ultra SPARC, 250 MHz). Both are shared memory resources at Chalmers University of Technology, Gothenburg.

The code has been optimized at each particular system by using different compiler flags. Linear speedup has been obtained, reciprocal to the number of processors involved, since the algorithm requires minimum interprocessor communication. However, large difference in the performance has been registered, namely, 120 MFlops per SGI Origin processor vs 40 MFlops on Cray T3E. After modification of the scalar math library calls in the code, 90 MFlops performance on Cray T3E has been achieved [5]. It can be concluded that simple User-Agent relations can not guarantee high performance results.

References

1. P. Sigmund, Mat. Fys. Medd. Dan. Vid. Selsk. 36 No 14 (1978); J. E. Westmoreland and P. Sigmund, Radiat. Eff. 6 (1970) 187; K. B. Winterbon, Nucl. Instr, Meth. B18(1986)1; W. Eckstein, Nucl. Instr. and Meth. B33 (1988) 489; U. Conrad and H. Urbassek, Nucl. Instr. and Meth. B48 (1990) 399
2. I. Pázsit, Ann. nucl. Energy 14 (1987) 25
3. G. von Leibfried, Nukleonik, Band 1, Heft 1 (1958) 57
4. I.Pázsit and R. Chakarova, Transport Theory and Statistical Physics 27 (1997) 1. (Due to misprint of the figure set, the authors should be contacted to get the correct figures), I. Pázsit and R. Chakarova, Fluctuations in atomic collision cascades with power law scattering, Transport Theory and Statistical Physics (in press)
5. Torgny Faxen, National Supercomputing Center, Linköping, Sweden, Private communication

Parallelisation of an Industrial Hydrodynamics Application Using the PINEAPL Library

Thomas Christensen[1], Arnold R. Krommer[2], Jesper Larsen[1], and Lars Sørensen[3]

[1] Math-Tech, Admiralgade 22, DK-1066 Copenhagen K, Denmark
{thomas,jesper}@math-tech.dk
http://www.math-tech.dk

[2] Numerical Algorithms Group Ltd, Wilkinson House, Jordan Hill Road, Oxford, OX2 8DR, UK
arnoldk@nag.co.uk
http://www.nag.co.uk

[3] Danish Hydraulic Institute, Agern Alle 5, 2970 Hørsholm, Denmark
lss@dhi.dk
http://www.dhi.dk

Abstract. The present paper reports the parallelisation of an industrial hydrodynamics application, MIKE3, using the PINEAPL parallel, numerical library. Some background information and a profiling of the application provides the basis for a parallelisation strategy. A two-stranded approach with both a functional parallelisation of part of the application and the use of library routines is reported. Excellent scalability properties of the two approaches are reported.

1 Introduction

The authors are currently participating in the European HPCN Fourth Framework project on Parallel Industrial NumErical Applications and Portable Libraries (PINEAPL). The main goals of this project are (i) to increase the suitability of the existing NAG Parallel Library for dealing with computationally intensive industrial applications by appropriately extending the range of library routines, and (ii) to port several industrial applications onto parallel computers by replacing sequential code sections with calls to appropriate parallel library routines. This article focuses on the parallelisation of one of the applications, a 3D hydrodynamic modelling system called MIKE3.

MIKE3 consists of a basic hydrodynamic module, based on a pressure correction scheme, and a number of add-on modules. These modules include an environmental module dealing with advection-dispersion, water quality and eutrophication, and a sediment processes module dealing with mud transport and particle tracking. The application area of MIKE3 thus spans from estuarine and coastal hydraulics and oceanography to environmental hydraulics and sediment processes modelling.

The code is very time consuming on large problems, resulting, for instance, from modelling the water flow through Øresund and the effect of the planned connection between Denmark and Sweden. The program is less demanding in its memory requirements. The application holders are interested in a parallel version of the program in order to bring down compute time (some of the Øresund cases took several weeks to run on high-end workstations) and to enable them to solve even bigger problems in reasonable time.

The ambition with the parallelisation of MIKE3 within the PINEAPL framework is to achieve a speedup of at least two on as few processors as possible, the goal is not to obtain a massively parallel program. This is important to have in mind, because it has implications for the design decisions made.

2 The MIKE3 application

This section will describe the initial analysis and profiling of the application, and then the chosen parallelisation strategy will be explained.

2.1 Code structure

In order to understand the profiles and the analysis, we need to take a closer look at the code structure.

The hydrodynamic module is the fundamental engine of the modelling system. This module can be subdivided into a basic flow module, a density module and a turbulence module. In case of stratified flows with variable density, the density module is applied in order to calculate the local density as a function of temperature and/or salinity. In case of turbulent flows the turbulence module is applied in order to calculate the eddy viscosity based on one of the three available turbulence models.

The environmental and sediment processes modules all depend on the three-dimensional velocity fields resulting from the hydrodynamic module, but do not couple back to the hydrodynamic module. The advection dispersion module simulates simple advective-dispersive transport of a number of passive pollutants. The water quality module and the eutrophication module simulate both the hydrodynamic transport and the biological processes and chemical interactions of a number of environmental parameters. The water quality module simulates the oxygen conditions (biological oxygen demand and dissolved oxygen) controlling the bacterial concentrations, while in addition the eutrophication module simulates the nutrient cycle, the algae growth and the distribution of bentic vegetation. The total number of state variables in the eutrophication module is 12.

2.2 Profiling

Profiling purely hydrodynamic benchmark cases revealed that the majority of the CPU time (approximately 90%) is used when solving the momentum equations and the pressure correction equation. These equations are all reduced to large sparse systems of linear equations which have to be solved in each timestep, and this is what makes the solution of the momentum equations and the pressure correction equation computationally intensive.

When profiling an environmental benchmark case, a different pattern emerges. In this case the solution of the transport equations turns out to be the most time consuming. A transport equation has to be solved for each substance being treated in the eutrophication module. This amounts to 11, since the bentic vegetation is not carried around with the water flow. The percentage of the CPU time used to solve the transport equations is typically around 60, so although the solution of these equations dominate, the time used to solve the momentum equations and the pressure correction equation is still significant.

In other cases, even more transport equations may have to be solved, since other modules also solve such equations. Thus even more than 60% of the total CPU time may be used to solve transport equations.

2.3 Parallelisation strategy

Having identified the three most computationally intensive parts of the code, the solution of the momentum equations, the pressure correction equation and the transport equations, a parallelisation strategy remains to be chosen. The most appropriate solution turns out to be a two-stranded approach.

As stated earlier, it is the solution of large sparse systems that makes the solution of the momentum and pressure correction equations time consuming, thus these equations are well suited for being parallelised by using parallel library routines. This approach is discussed in further detail in a later section.

The solution of the transport equations are not well suited for being parallelised in this way, but the solution of the equation for one substance is independent of the solution for another, and they can thus take place simultaneously on separate processors. This functional parallelisation scheme is implemented by using MPI.

3 The functional parallelisation

The easiest way to solve the transport equations on individual processors is to run instances of the entire program on every processor, doing all calculations redundantly except the solution of the transport equations. This strategy is CPU scalable but not memory scalable, but since the application is CPU bound and not memory bound for large problems, this has been considered a feasible approach.

A more detailed discussion on the choice of this parallelisation approach can be found in [1].

3.1 Implementation

The implementation of this parallelisation strategy was very simple. Three issues had to be dealt with: File i/o, logical control of the parallelism and communication.

The application reads very little input from files, so in the parallel case, each program instance simply reads all input files. Simple modifications was made to the program in order to make the different program instances write separate output files. The output files was simply given the programs process number as an extension.

The flow control necessary to implement the functional parallelism is simply to skip the solution of some of the transport equations. The processors are assigned an approximately equal number of equation in order to achieve a good load balance.

When a transport equation has been solved for a given substance on one processor, the solution (the concentration of the substance in the entire domain) has to be send to all other processors. The equations are solved at each timestep, so this communication must also take place at each timestep.

The communication of the concentrations could be implemented as a broadcast for each concentration, but in order to optimise performance, the communication scheme depends on the number of processors used. If we look at the eutrophication module, where 11 transport equations are solved, then all communication can be performed in a singe call to MPI_SENDRECV if two processors are used, which is very efficient. If from 3 to 10 processors are used, each concentration must be send with a MPI_BCAST, because there wont be the same number of concentrations on all processors (11 being a prime number). On 11 processors all communication can be performed in a single call to MPI_ALLGATHERV, which again is very efficient.

3.2 Results

Considering that the parallelised part only takes up around 60% of the total CPU time, the result is quite good. On eleven processors the speedup is actually more than two (though we cannot reasonably claim to have achieved our goal yet, since a speedup of two on eleven processors hardly is cost efficient!).

All tests reported in this article have been conducted on the IBM SP2 machine at UNI-C in Denmark.

On Fig. 1 only the parallel part, the solution of the transport equations, is shown. The bars are spilt into time used on calculations, synchronisation and communication. The significant synchronisation time is due to the fact that 11 transport equations are solved (because it uses the eutrophication module), and this work cannot be distributed evenly on anything but a single, or 11 processors. Thus, the synchronisation time would be close to zero on 11 processors because of a perfect load balance.

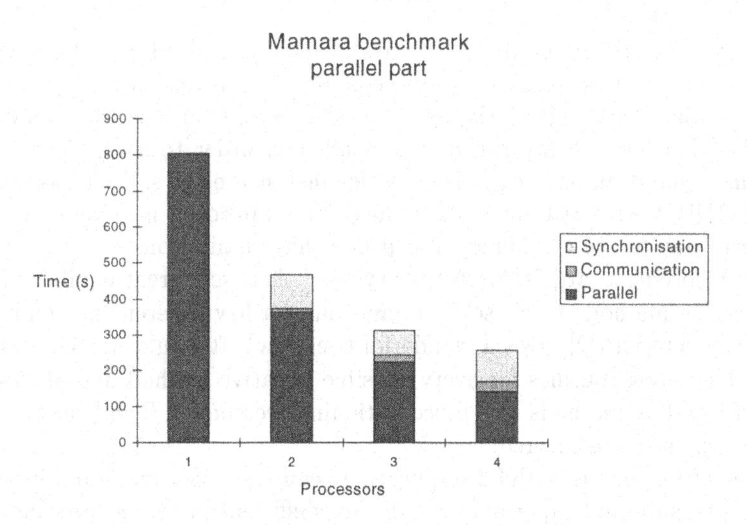

Fig. 1. The parallel part of the program running an environmental benchmark case.

4 Library parallelisation

Having discussed the explicit functional parallelisation of the solution of the transport equations, we now turn to the implicit parallelisation using parallel library routines, of the momentum and pressure correction equations.

These mathematical equations have different physical origin which influences their nature. The momentum equations deal with the velocity of the water, which is a relatively local physical phenomena, compared to the global nature of pressure, which is dealt with in the pressure correction equation. This fact is reflected in the numerical methods needed to solve the linear systems resulting from the discretisation of these differential equations. The linear system resulting from the momentum equations are thus solved using a local method, line Jacobi, whereas the linear system resulting from the pressure correction equation is solved using conjugate gradient (CG).

To assure and improve convergence both methods use preconditioning, and the choice of preconditioner is also governed by physics. The grid cells in the domain are numbered in the vertical direction first, which means that interactions between cell neighbours in the vertical direction will appear as non-zero elements in the sub- and super diagonal of the coefficient matrix, i.e. the elements $A_{i,i-1}$ and $A_{i,i+1}$. These elements will usually be much larger than other elements in the matrix, since the vertical dimension of the grid cells from this type of application nearly always are much smaller than the horizontal dimensions, making the physical interactions in the vertical direction strongest. Having large elements on the sub- and super diagonal of the coefficient matrix reveals that the inverse of this part of the matrix will constitute a good preconditioner for the system.

At this point in the PINEAPL project only the PCG method has been parallelised, and the remaining of this section will therefore focus on this method.

4.1 Integration of library routines

There are two principally different ways of using the PINEAPL library; using black-box routines or by using reverse communication routines.

Using black-box routines is far the easiest way to utilise the library. Here a single, or a few, library routines are called in order to solve the numerical problem at hand, in one go. This could for instance be to solve a linear system using GMRES with ILU preconditioning. This approach is a very convenient high-level way to use the library, but it does have limitations.

When solving a large sparse linear system, there is a great number of different iterative methods to chose from, and quite a few different preconditioning techniques as well ([2] gives a wonderful overview). It would not be feasible to supply black-box routines for every possible iterative method and all combinations of iterative methods and preconditioning techniques. Thus, only the most popular schemes are covered.

As mentioned, the MIKE3 application requires a CG method with the inverse of the sub- and super diagonal of the coefficient matrix as preconditioner. This is not a main stream combination, but rather very application specific, and therefore there is no black-box routine provided. In order to implement this functionality by means of PINEAPL routines, one must resort to reverse communication.

Reverse communication is a low-level approach, which is more difficult to use, but which provides a much larger degree of flexibility. Here, the functionality is implemented by calling a larger (~15) number of library routines that each perform a simpler task, like a matrix-vector multiplication. The term 'reverse communication' is due to the fact that certain library routines communicates back to the user program, requesting for some specific action to take place.

For an in depth discussion on the sparse linear algebra routines in the PINEAPL library see [3] and [4]. A related article also appear elsewhere in this volume of LNCS.

4.2 Data distribution

When integrating the library routines into the application, the most straight forward approach is to split the coefficient matrix into a number of equally sized row-blocks corresponding to the number of processors used. This was indeed the initial approach used when parallelising MIKE3, but it turned out to be rather inefficient due to the close couplings between row blocks. This can be improved by exploiting the structure of the matrix.

The matrix has seven bands of nonzero element due to the seven point stencil used in the discretisation of the differential equation. As explained earlier, the non-zero elements in the innermost bands (in the sub- and super diagonal of the matrix) corresponds the interaction between cell neighbours in the vertical direction. Thus, the tridiagonal will decouple at the surface and the bottom of the domain. Likewise, the middle bands will decouple at the boarders of the domain in one of the horizontal directions. Only the outermost bands will not decouple at all. The bands narrows in at both ends of the matrix, because the number of cells in the vertical columns becomes fewer towards the shores of the domain, as the water becomes more shallow.

When exploiting the sparsity pattern of the matrix by partitioning the matrix where the innermost bands decouple, the tridiagonal will not couple between different processors. This is a big advantage, because the preconditioning step of the PCG method then becomes embarrassingly parallel, such that a sequential tridiagonal solver can be applied on each processors instead of using a parallel tridiagonal solver.

4.3 Results

Fig. 2 shows the scalability of the parallel PCG method compared to the original sequential version. The figure shows that there is a performance penalty to pay, when going from the sequential version to the parallel version on one processor, but since the parallel version outperforms the sequential version already on two processors, this is not regarded as a severe drawback. The other important note to make is that the parallel version scales wonderfully. It is certain that it also will scale beyond 8 processors, provided the problem is sufficiently large.

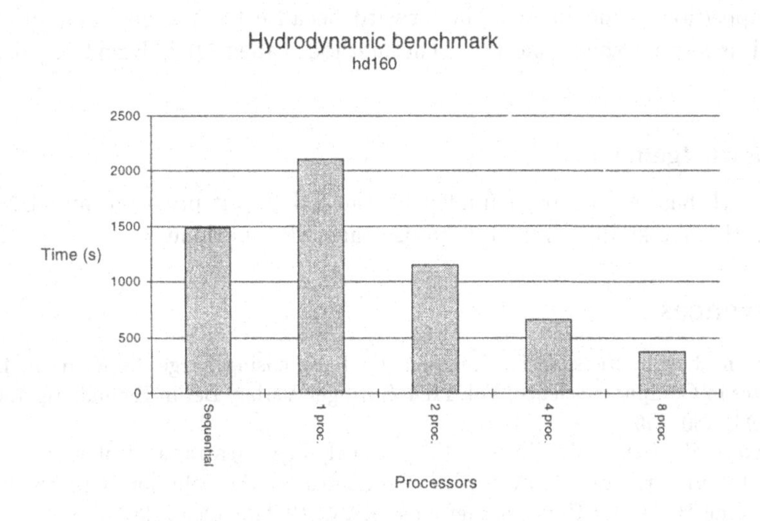

Fig. 2. The scalability of the PCG method.

5 Conclusion

This article has described a two-stranded approach to the parallelisation of the MIKE3 application, and given results for those parts of the code parallelised at this point in the PINEAPL project. No results have been reported for the entire application utilising both types of parallelism, since such results would be misleading until the remaining part of the application has been parallelised. However, based on the very satisfying results of those parts that have been parallelised, we are confident that at the end of the project we will achieve a speedup well above the factor of two on a few processors, which was set to be a minimum at the beginning of the project.

An application like MIKE3 is often parallelised using domain decompositioning. This approach could also have been chosen for MIKE3, but this would have required much more resources than have been allocated in the PINEAPL framework, since modifications to a much larger part of the code would have been necessary. The chosen strategy has the great advantage that the code only has to be modified in a very few places, and still gives a significant speedup. Naturally this requires much fewer resources than a domain decompositioning.

If a massively parallel solution was required, a domain decompositioning would have had to be chosen, since the chosen strategy does not possess the necessary degree of parallelism. It is however possible to extend the current parallelisation using the library routines into a domain decompositioning, since the partitioning of the matrices of the linear equations corresponds to a domain decompositioning. The functional parallelisation would have to be removed, since it is incompatible with a domain decompositioning, but the transition to a domain decomposition would be straight forward because the transport equations are solved using an explicit method. The method, called QUICKEST, is discussed in [5].

Acknowledgements

This work has in part been funded by the EU Esprit program, and UNI-C in Denmark have kindly allowed us to use their SP2 machine.

References

1. Larsen, J., Christensen, T., Frellesen, L.: Parallelising Large Application. Lecture Notes in Computer Science, Vol. 1184. Springer-Verlag, Berlin Heidelberg New York (1996) 450–456
2. Barrett, R., Berry, M., Chan, T.F., Demmel, J., Dongarra, J., Eijkhout, V., Pozo, R., Romine, C., van der Vorst, H.: Templates for the Solution of Linear System: Building Blocks for Iterative Methods. SIAM, Philadelphia (1994)
3. Krommer, A.: Parallel Sparse Matrix Computations Using the PINEAPL Library: A Performance Study. Technical Report TR1/98, The Numerical Algorithm Group Ltd, Oxford, UK (1998)
4. Krommer, A., Derakhshan, M., Hammarling, S.: Solving PDE Problems on Parallel and Distributed Computer Systems Using the NAG Parallel Library. High-Performance Computing and Networking (B. Hertzberger, P. Sloot, Eds.), Vol. 1225 of LNCS, Springer-Verlag, Berlin (1997) 440–451
5. Vested, H.J., Justesen, P., Ekebjærg, L.: Advection-dispersion modelling in three dimensions. Appl. Math. Modelling, Vol. 16 (1992) 506–519

Hyper-Rectangle Selection Strategy for Parallel Adaptive Numerical Integration

Raimondas Čiegis[1], Ramūnas Šablinskas[2], and Jerzy Waśniewski[3]

[1] Institute of Mathematics and Informatics, Akademijos 4, LT-26000 Vilnius, Lithuania
rc@fm.vtu.lt
[2] Vytautas Magnus University, Vileikos 8, LT-3035 Kaunas, Lithuania
ramas@omnitel.net
[3] The Danish Computing Centre for Research and Education
UNI-C, Bldg. 305, DK-2800 Lyngby, Denmark
Jerzy.Wasniewski@uni-c.dk

Abstract. In this paper we consider the problem of numerical integration of multidimensional integrals. A new hyper-rectangle selection strategy is proposed for the implementation of globally adaptive parallel quadrature algorithms. The well known master-slave parallel algorithm prototype is used for the realization of the algorithm. Numerical results on the $SP2$ computer and on a cluster of workstations are reported. A test problem where the integrand function has a strong corner singularity is investigated.

1 Introduction

The problem considered in this paper is the numerical approximation of the multi-dimensional integral

$$I(f,\Omega) = \int_{\Omega} f(x)\,dx \tag{1}$$

to some given accuracy ε, where $\Omega = [a_1,b_1] \times [a_2,b_2] \cdots [a_n,b_n]$ is the range of integration and $f(x)$ is the integrand function. We are interested in implementing globally adaptive integration algorithms on distributed memory parallel computers. The algorithms of this paper are targeted at clusters of workstations with standard message passing interface. We use PVM in our numerical experiments. The important features of such parallel computers are heterogenity of the cluster and unfavourable ratio between computation and communication rates.

Numerical integration algorithms attempt to approximate $I(f,\Omega)$ by the sum

$$S_N(f) = \sum_{i=1}^{N} w_i f(x_i). \tag{2}$$

The numbers w_i are called *weigts*, and the x_i are called *knots*. Efficient numerical algorithms are adaptive and a strategy for selection of hyper-rectangles for further subdivision is included into the algorithm. We use the cubature rule pair, which was proposed by Genz and Malik [3]. Parallel adaptive algorithms were investigated in many papers, e.g. [1], [4], [5]. Mostly these algorithms are targeted at shared memory computers.

In our previous paper [2] we have investigated variuos algorithms for static subdivision of the whole task. These smaller subtasks are solved in parallel by different processors. The main result of [2] is that parallel static subdivision algorithms are efficient if numbers of subdivisions required within the different subregions vary only moderately. The situation is different when the integrand function f has strong singularities. The load balancing of the static subdivision – dynamic distribution algorithm becomes very poor and the quality of the parallel adaptive algorithm also degrades seriuosly.

In this paper we develop a new hyper-rectangle selection algorithm. The quality of the parallel adaptive cubature method is improved considerably, but additional costs of communications are also increased since a more fine-grain subdivision of the integration region Ω is implemented.

We summarize the remainder of this paper. In Section 2 we review the Static Subdivision – Dynamic Distribution (SSDD) algorithm. The analysis is given for the case when the integrand function f has a strong corner singularity. Section 3 describes the new hyper-rectangle selection algorithm. Numerical results for the $SP2$ computer and for a cluster of workstations are presented in Section 4.

2 Static Subdivision – Dynamic Distribution algorithm

We describe very briefly the parallel algorithm for multidimensional cubature that is based on a static initial subdivision of the integration region Ω. Such algorithms are investigated in [2].

We use the well known parallel algorithm prototype: master - slave algorithm. The pool of tasks $P_j, j = 1, 2, \dots, M$ is formed from subproblems $I(f, \Omega_j)$, where we make the initial partitioning of the region $\Omega = \Omega_1 \bigcup \Omega_2 \bigcup \cdots \bigcup \Omega_M$. The error tolerance ε_j for each subregion is chosen proportional to the volume of the subregion. The master process dynamically redistributes unsolved subtasks to slaves and collects the results. Slave processes calculate the approximation of the integral over the given subregion and return the result to the master process.

Remark 1. At the end of computations the same unsolved subproblem can be send to several different slaves. This modification is important for heterogeneous clusters of workstations.

Numerical results given in [2] lead to a conclusion that the $SSDD$ algorithm is efficient for integrals with the integrand function f which has no strong singularities. It is important to note that the $SSDD$ method is well adapted to the variations of workload on workstations during computations.

Now we investigate the $SSDD$ algorithm in the case when the integrand function f has strong singularities. Let consider the following test problem (see [1])

$$\int_0^1 \int_0^1 \int_0^1 (x_1 x_2 x_3)^{-0.9} \, dx_1 \, dx_2 \, dx_3, \tag{3}$$

where the integrand function has a strong corner singularity.

We compute a numerical approximation of the integral (3) to the relative accuracy $\varepsilon = 0.003$. Table 1 summarizes the performance of the *SSDD* method. Here M denotes the number of subregions Ω_j, N_M is the total number of required subdivisions

$$N_M = n_1 + n_2 + \cdots + n_M, \tag{4}$$

where n_j is the number of subdivisions required for the numerical approximation of the integral $I(f, \Omega_j)$, Q is the quality of the algorithm, which is defined as

$$Q = \frac{N_M}{N_1}, \tag{5}$$

In order to investigate the load balancing property of the *SSDD* algorithm we define the parameter L

$$L = \frac{max_j \, n_j}{N_M}. \tag{6}$$

Table 1. Numerical results for the *SSDD* algorithm

M	Number of subdivisions N_M	Quality Q	Load balancing L
2	163249	1.524	0.98
4	213123	1.990	0.94
8	255345	2.384	0.91

From the Table 1 we see that the quality of the parallel adaptive qubature algorithm decreases when M becomes larger. The load balancing of this algorithm is also bad since only one process is involved in useful computations.

3 Hyper-Rectangle Selection Strategy

The numerical integration problem can be considered as a dynamic resource problem. If our parallel computer is a cluster of multi-user workstations then the available processing capacity per node may change during computations, hence we have a dynamic resource computer. The load balancing problem can be considered as the mapping of a dynamic resource problem onto a dynamic resource computer. The algorithm must include some mechanism for the migration of subproblems from over-loaded processes to under-loaded processes at run-time.

We propose the following hyper-rectangle selection algorithm.

Hyper-rectangle selection algorithm

1. The *master process* keeps the list of hyper-rectangles and this list is ranked by error estimates so that $\varepsilon_1 < \varepsilon_2 < \ldots < \varepsilon_s$.
2. The *master process* receives from a slave the ranked list of R_{recv} hyper-rectangles and puts these results to the main ranked list of subregions. It also updates the integral approximation and error estimates.
3. If the accuracy is not achieved the *master process* sends to the slave a new list of R_{send} hyper-rectangles with the largest errors.
4. A *slave process* applies the adaptive numerical integration algorithm and makes $R_{recv} - R_{send}$ subdivisions. We use $R_{recv} = kR_{send}$, $k > 2$. The *slave process* also ranks the obtained list of hyper-rectangles according their error estimates and sends the result to the master process.

The algorithm contains two free parameters R_{send} and R_{recv}. The optimal values of these parameters depend on the dimension of the integral and on the ratio between compution and communication rates of a parallel computer.

Remark 2. The *quick-sort* method is used by slave processes to rank the list of subproblems.

4 Numerical Results

In this section we show the performance of the proposed hyper-rectangle selection algorithm. We calculate the multidimensional integral (3) to the relative accuracy $\varepsilon = 0.002$. The $SP2$ computer and a cluster of $RS6000$ workstations were used in computations. In all numerical experiments we set the following values of parameters $R_{send} = 200$, $R_{recv} = 700$.

We define the speed-up S_p as

$$S_p = \frac{T_s}{T_p}, \tag{7}$$

where T_p is the time used to solve the given problem on p processors, T_s represents the time for the sequential globally adaptive algorithm.

Table 2 presents the numbers of subdivisions N_p, timings T_p in seconds, speed-ups S_p and the efficiency E_p on the $SP2$ computer.

Table 3 gives analogous information for the cluster of workstations $RS6000$. We wish to test the quality of the proposed algorithm, therefore we choose to add appropiate delays in the evaluation of integrand function. A dummy loop is incorporated into the code and it is repeated 5 times.

From the tables we see that the quality of the proposed algorithm degrades only slightly when the number of processors is increased. Numerical experiments show that the algorithm is not very sensitive to the selection of parameters R_{send} and R_{recv} and the time of computations varies by about ten percents for a large range of values of these parameters.

Table 2. Numerical results on the $SP2$ computer

p	Number of subdiv. N_p	Time T_p	Speed-up S_p	Efficiency E_p
1	217698	77.9	0.957	0.957
2	217897	39.8	1.874	0.937
4	219295	20.8	3.586	0.896
7	220392	12.7	5.873	0.839
15	226984	7.4	10.080	0.672

Table 3. Numerical results on the cluster of workstations

p	Number of subdiv. N_p	Time T_p	Speed-up S_p	Efficiency E_p
1	217698	313.3	0.945	0.945
2	217897	167.4	1.768	0.884
3	218596	117.7	2.515	0.838
4	219295	92.0	3.217	0.804
5	219494	74.5	3.973	0.795
6	220193	64.0	4.625	0.771
7	220392	57.8	5.121	0.732

References

1. Bull, J.M., Freeman, T.L.: Parallel Gloabally Adaptive Algorithms for Multidimensional Integration. Appl. Numer. Mathem. **19** (1995) 3–16
2. Čiegis, R., Šablinskas, R., Waśniewski, J.: Numerical Integration on Distributed Memory Parallel Systems. In: Bubak, M., Dongarra, J., Waśniewski, J. (eds.): Recent Advances in PVM and MPI. Lecture Notes in Computer Science, Vol. 1332. Springer-Verlag, Berlin Heidelberg New York (1997) 329–336
3. Genz, A.C., Malik, A.A.: Remarks on Algorithm 006: an Adaptive Algorithm for Numerical Integration Over an N-dimensional Rectangular Region. J. Comput. Appl. Math. **6** (1980) 295–302
4. Keister, B.D.: Multidimensional Quadrature Algorithms. Computers in Physics **10** (1996) 119–122
5. Miller, V.A., Davis, G.J.: Adaptive Quadrature on a Message Passing Multiprocessor. J. Parallel Distributed Comput. **14** (1992) 417–425

Parallilising Fuzzy Queries for Spatial Data Modelling on a Cray T3D

A.Clematis[1], A.Coda[1], M. Spagnuolo[1], S.Spinello[1]

T. Sloan[2]

[1] Istituto per la Matematica Applicata, CNR, Via De Marini 6, 16149- Genova, Italy
[2]Edinburgh Parallel Computing Centre, King's Buildings, Edinburgh, UK

Abstract. In this paper we present some results about the use of parallel computing for fuzzy modelling in Geographic Information Systems (GIS) applications. Fuzzy modelling is going to gain an increasing popularity in the GIS community, but its application find an obstacle in the high computational cost. It is possible to design efficient parallel algorithms for fuzzy modelling: here we present an approach to parallelisation of fuzzy queries for digital terrain models. Load balancing is shown to be the key point to obtain good performances, and suitable load distribution strategy is discussed. Experimental results obtained on a Cray-T3D using MPI are presented as well.

1. Introduction

Parallel computing is becoming attractive for an increasing number of applications which requires an high computational power. Spatial Information Systems (SIS) play an interesting role in this sense, because the complexity of the data to be handled and of the functions to be supported in SIS offer plenty of examples where a distributed environment of computation and storage could surely enhance the performance of such systems [1].

In this context, a method will be described for the interrogation of terrain models based on fuzzy arithmetic, which are very interesting for the SIS community as they are able to code the uncertainty usually related to the terrain samples. When data are acquired from the real world, indeed, the capability offered to explicitly represent the uncertainty related to the measures is a parameter which should be taken into account for assessing the quality of a model [2]. Beside the loss of information due to the discretisation procedure itself, there exists also an imprecision associated to the samples, which can be sometimes quantified, and is generally caused by the imprecision of the measurement instrument. A model able to effectively represent this uncertainty has been presented in [3] where the *Fuzzy-B-Splines* (FBS) have been proposed to model terrain data. The FBS approach is based on the idea of replacing the concept of crisp number, typical of traditional approaches, with the concept of *fuzzy* number which can be roughly defined making use of the interval analysis and relating two important concepts: confidence intervals and presumption levels. A confidence interval is an interval of real numbers that provides a representation for an imprecise numerical value by means of its sharpest enclosing range. A presumption level may be regarded as an estimate of the truth value attributed to some knowledge about the value. Presumption levels belong to the [0,1] interval: the maximum estimated truth to be at level 1, while the minimum is at level 0. See [4,5] for further details on this topic.

2. Evaluating fuzzy queries

The FBS approach and related interrogation processes have then been proved very interesting and promising from the application point of view, however, they are computationally intensive with respect to traditional approaches [6,7]. As an example, consider that to produce solution of a simple case test it takes around 30 minutes on an unloaded Sparc-Ultra1 station with 128 Mbyte of ram, while computing solution of a more complex one takes around 4 hours on the same machine. Thus, the need of a distributed environment of computation appears a necessary condition for developing an implementation of the approach for practical use.

The interrogation is of crucial importance for the usefulness of the proposed model. We focused our attention on the following typical query: "given a crisp level k, where the spline, or the surface, modelling the phenomenon, takes value k?". Such typical query is translated into a system of non-linear equations and for this reason insights into the investigated phenomena can be obtained only solving efficiently such systems [8]. Since our model is made with families of B-splines, an efficient technique to interrogate B-splines must be used. The main task of the algorithm is to compute, for a very huge set of points, the solutions of a non linear differential system with interval coefficients: we used a quadrisection method with a Monte Carlo approach.

We start with a box containing all the domain: this box is initially partitioned in four sub boxes, equal in size, and solutions are looked for in each of them. In order to see if a box contains solutions, the value of the equations is computed in a fixed number of points inside the box itself. There are, so, three possibilities:

1. If the box contains only solutions, then it is taken as a whole as a solution;
2. If the box contains no solutions it is thrown away;
3. If the box contains some solutions and its dimension is smaller then the a priori fixed precision, it is accepted as a solution, otherwise it is subdivided again in four sub boxes and the whole process starts again

The boxes are managed in a LIFO way, using a stack (note that the order of box processing is completely indifferent from the correctness point of view).

The main parameters which determine the cost and quality of this algorithm are:

- Number of points used to determine if a box is a solution: too few points lead to wrong results, too many points lead to very high execution times;

- Maximum dimension of a box to be accepted as a solution (precision): this value influences the "goodness" of the result on the borders (smoothness of the curves). Lower the value, higher is the number of box generated, hence the execution time.

3. The parallelisation strategy

It is possible to take a data parallel approach, since the data domain can be partitioned in independent sub-domains. The key idea is to use the same box subdivision employed by the sequential program: e.g., using four processor we may partition the initial domain in four boxes of equal size and apply the sequential algorithm independently to each one. A Master-Worker paradigm can be used for the parallel program, the master spawns initial data among workers and each worker applies the sequential algorithm on its own portion of data.

Unfortunately, using this structure it can occur very often that one or more workers become idle, typically because either the portion of data assigned to them contains no solutions or all the solutions for that part have been found. It is not possible to know, a priori, which is the "best" domain partitioning to use, such as it is not possible to find a static domain that prevents unbalancing to occur (doing so requires knowing how solutions are distributed across the solution's space). Even assuming that solutions are equally distributed in space (hence equal work load for every process), load unbalancing will occur because of the "shape" of the local solution. For this reason it is necessary to employ a dynamic load balancing policies.

We notice that the load of a worker corresponds to the number of boxes queued in its local stack. An empty stack identifies an idle worker, thus load redistribution is achieved by equalising the number of elements in local stacks. This can be done without side effects, since each box can be handled in a complete independent way.

The outline of the redistribution strategy may be summarised in the following way:

- Each time a worker becomes idle, it notifies the master this situation by sending it a message
- As soon as a threshold value *tr* of idle workers is reached, the master force a synchronisation step. During this step:
  - Each worker sends a message to the master, communicating the total number of boxes actually queued (0, if idle)
  - When the master has collected all the results, it determines if a redistribution load has to be performed (are there at least *n* boxes?, where *n* is a threshold value), and in that case how jobs can be exchanged between workers in order to minimise the unbalancing, then
  - It sends a message to each worker, communicating the partner to which it has to send (receive) data, if any
  - Then, after no more than *nw*/2 (*nw* = number of workers) communications between workers, the workload is equally redistributed
- The computation can start again.

There are several parameters that can be used to optimise load redistribution to the target machine:

1. Threshold value of idle workers *tr* (master): This value determines a compromise between communication time and computation time (in other words, between the total number of messages that have to be exchanged between processes and the maximum number of processes that can be idle at a certain time). On computers with a very fast communication network this value can be set to 1, avoiding workers to remain idle, on other machines it can be increased to minimise communication overhead;
2. Minimum number of total boxes to let a load redistribution occur (*n*) (master). When the computation is near to its end, the global number of boxes will remain low, going to 0. With this parameter it is possible to avoid an excessive and useless communication overhead near the program termination;
3. How often, in number of iteration, to check redistribution need (worker): again, a low value will raise communication but minimise "response times", a high value will introduce latency in worker behaviour.

4. Experimental Results

In the program development we have paid great attention to portability, performance optimisation on different architectures is achieved by a proper tuning of the above mentioned parameters. For the sake of portability the MPI library [9] has been selected to develop the parallel code. Hereafter we report results obtained on Cray T3D: another interesting architecture for GIS is workstation network where a different method has to be pursued.

On the Cray T3D we got very good results both in terms of speed-up and scalability: thanks to the fast interconnection network between processors it was possible to optimise the code by taking the following decisions:

1. Imposing the workers to check for a redistribution need at every iteration, minimising the latency in worker response to synchronisation requests from master, hence minimising also the overall workers' idle time;
2. Setting the threshold value for redistribution at the minimum value (1), assuring that in every moment all of the workers are working or engaged in a load redistribution;
3. Setting the minimum number of total boxes to let a load redistribution occur to the minimum value allowed (number of workers).

The obtained results are summarised in Figure 1 to Figure 3, which provide different types of information. All the figures refer to the same test case, in some of them different levels of precision are considered.

The first graph shows the ratio between computation and communication time for a fixed precision and for different processors number (Figure 1): as it is possibile to see, the communication time remains in most cases under 20% of the whole execution time.

The graph of Figure 2 shows the good effect of redistribution comparing the maximum and minimum computation times for the different workers at a fixed precision: the unbalancing index so provided is very close to zero. The results presented in Figure 1 and 2 show that the cost benefit ratio of redistribution is very good.

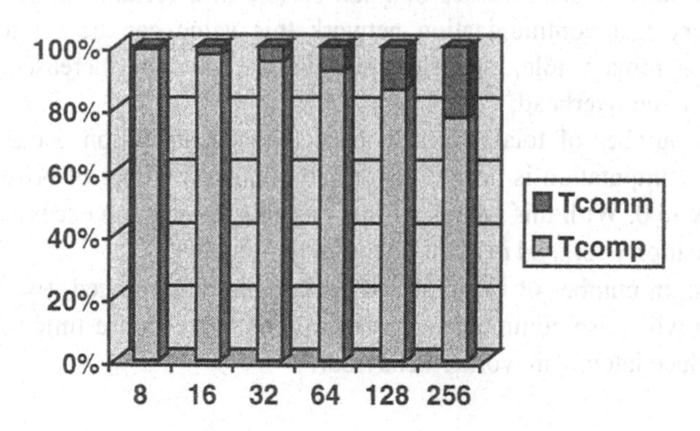

Fig. 1. Percentage of communication with respect to computation time for a precision of 0.0005 and for up to 256 processors

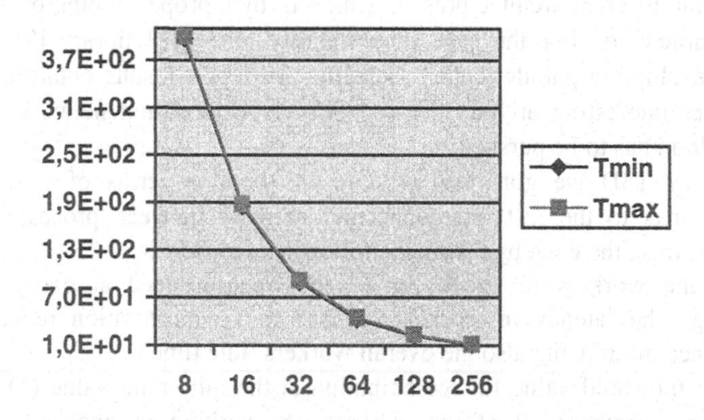

Fig. 2. Maximum versus minimum computation times (in seconds) for a precision of 0.0005 and for up to 256 processors (the difference between the plot provides the unbalancing index)

Finally, the graph of Figure 3 represents the measured efficiency for up to 256 processors and for different precision values compared to the optimal one: it is interesting to notice that when a high precision is required it is possible to exploit a high degree of parallelism, and the algorithm scalability is very satisfactory.

Future works will include the experimentation of the proposed algorithm on workstation network, since this computing platform is very interesting for most users of GIS. The first results we have obtained on this kind of architectures are quite promising, even if it is necessary to properly evaluate the redistribution parameters in order to take in consideration the use of nodes with different computing power.

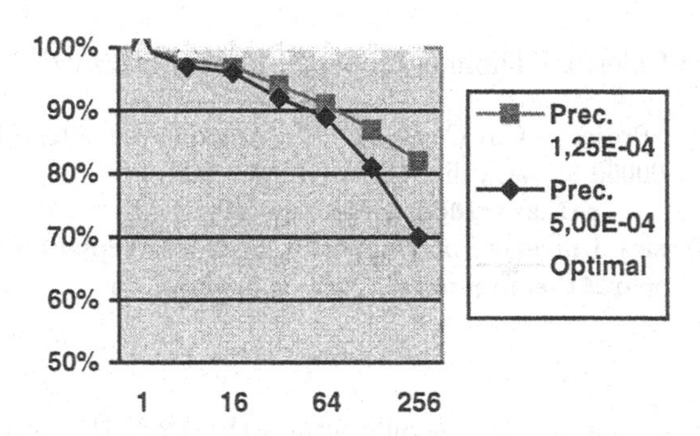

Fig. 3. Measured vs. optimal efficiency for up to 256 processors, and different precision values

Acknowledgements

The support of the TRACS (Training and Research on Advanced Computing Systems) programme funded by the European Commission, and of the EPCC (Edinburgh Parallel Computing Centre) is kindly acknowledged.

References

1. R.G. Healey (Editor) Special Issue on parallel processing in GIS, Int. J. Geographic Information Systems 1996, vol. 10 No. 6.
2. K. Beard. Representation of data quality. In M. Craglia and H. Couclelis, editors, Geographic Information Research: Bridging the Atlantic, chapter 18. Taylor and Francis, London, 1996.
3. G. Gallo, and M. Spagnuolo. Uncertainty Coding and Controlled Data Reduction Using Fuzzy-B-Splines, In Computer Graphics International '98, to appear.
4. A. Kauffman, and M.M. Gupta. Introduction to Fuzzy Arithmetic: Theory and Applications. New York : Van Nostrand Reinhold, 1991.
5. H.J. Zimmermann. Fuzzy Set Theory and its application. Dordrecht : Kluwer, 1991.
6. R.H. Bartels, J.C. Beatty and B.A. Barsky. Introduction to splines for Use in Computer Graphics and Modeling. Morgan Kauffman, 1987.
7. C. DeBoor. On calculating with B-splines. Journal of Approximation Theory , 6, 50-62, 1972.
8. C. Floreno and G. Novelli. Implementing Fuzzy Polynomial interpolation. Le Matematiche vol.LI , 59-76, 1996.
9. W. Gropp, E. Lusk, A. Skjellum, Using MPI, MIT Press, 1994.

Hyper-Systolic Implementation of BLAS-3 Routines on the APE100/Quadrics Machine

Marco Coletta[1], Thomas Lippert[2], Paolo Palazzari[1]

[1] ENEA - HPCN Project - C.R.Casaccia - Via Anguillarese, 301, S.P. 100, 00060 S.Maria di Galeria (Rome - Italy)
palazzari@casaccia.enea.it

[2] Department of Physics, University of Wuppertal, D-42097 Wuppertal (Germany)
lippert@theorie.physik.uni-wuppertal.de

Abstract. Basic Linear Algebra Subroutines (BLAS-3) [1] are building blocks to solve a lot of numerical problems (Cholesky factorization, Gram-Schmidt ortonormalization, LU decomposition,...). Their efficient implementation on a given parallel machine is a key issue for the maximal exploitation of the system's computational power. In this work we refer to a massively parallel processing SIMD machine (the APE100/Quadrics [2]) and to the adoption of the hyper-systolic method [3, 6, 4] to efficiently implement BLAS-3 on such a machine. The results we achieved (nearly 60-70% of the peak performances for large matrices) demonstrate the validity of the proposed approach. The work is structured as follows: section 1 is devoted to review BLAS-3, in section 2 we recall the hyper-systolic method, subsequently (section 3), the target machine is described and (section 4) the HS implementation is shown. Finally (section 5), some experimental results are given.

1 Introduction

Basic Linear Algebra Subroutines (BLAS-3) [1] deal with matrix-matrix multiply and add operations. They can be grouped according to the following categories:

a) Matrix matrix products	b) multiplying a matrix by a triangular matrix
$C = aAB + bC$	$B = aTB$
$C = aA^TB + bC$	$B = aBT$
$C = aAB^T + bC$	$B = aT^TB$
$C = aA^TB^T + bC$	$B = aBT^T$

c) rank k and 2k updates of symmetric matrices	d) solving triangular systems of equations with multiple right-hand sides
$C = aA^TA + bC$	$B = aT^{-1}B$
$C = aAA^T + bC$	$B = aT^{-T}B$
$C = aAB^T + aAB^T + bC$	$B = aBT^{-1}$
$C = aA^TB + aA^TB + bC$	$B = aBT^{-T}$

A, B and C are rectangular matrices, a and b are scalar values, T is a (upper or lower) triangular matrix. In this work we address the efficient implementation of operation a) on a massively parallel processing SIMD machine through the Hyper-Systolic (HS) algorithm [5].

2 Hyper-systolic computing

HS [3, 6, 4, 5] is based on the embedding of a virtual loop topology onto a physical machine with larger connection capabilities. The computation, referred to the systolic model, is carried out by reducing communication overhead through the adoption of few additional registers (namely k) and reorganizing the algorithm in three phases:

- the first phase performs a distribution of input data on the k additional registers, creating shifted replicas (with different and appropriate shifting steps) of the data on the virtual loop;
- in the second phase the computations are locally executed, accumulating k partial results on the k registers;
- the third phase performs k back shift operations (corresponding to the ones of phase one) and the partial results are accumulated to obtain the final result.

In the works cited above it is demonstrated, along with details on HS implementation, that the communication overhead is reduced from $O(p)$ to $O(\sqrt{p})$ time complexity, with p being the number of processors.

3 The APE100/QUADRICS machine

Developed by the Italian Institution for Nuclear Physics (INFN) in order to study QCD problems, this machine [2] has been successfully used in a broad range of applications (material science, electromagnetism, neural networks, fluidodynamic, image analysis and compression, atmospheric modeling,...). The APE100/QUADRICS machine is based on the SIMD architectural model, using VLIW pipelined custom processors with 50 Mflops of peak power and connected according to a 3-dimensional torus topology. Each machine, with a number of processors ranging from 8 to 512, is hosted by a SUN Sparc machine which can be connected to the parallel system through an HPPI channel (sustained I/O bandwidth 20 Mbyte/sec). In its maximal configuration, the APE100/QUADRICS machine offers a peak power of 25.6 Gflops. By the end of the year (1998) this machine will be upgraded to the APEmille system, a SIMD machine with a peak performance of 1 Tflop.

4 HS implementation of BLAS-3

In a first step we present our systolic version of the matrix-matrix multiplication on a 1-dimensional ring of processors. In the result section, we are going to benchmark this algorithm together in comparison to the HS algorithm. The systolic algorithm is given by:

foreach processor $i = 1 : p$ ($\in$ systolic array)
  for $j = 1 : p$
    for $l = 1 : j - 1$
!! Mult. of submat $[p - (j-1) + l - 1]$ of A in proc i with submat j of B in proc i
      for $t = 1 : n/p$
        for $k = 1 : n/p$
          for $h = 1 : n/p$
            $c_{k+\frac{n}{p}(l-1),t+\frac{n}{p}(i-1)} +=$
            $a_{k+\frac{n}{p}(l-1)-\frac{n}{p}(j-1)+n,h+\frac{n}{p}(i-1)} \cdot b_{h+\frac{n}{p}(j-1),t+\frac{n}{p}(i-1)}$
      end for h, k, t
!! End of multiplication submatrix-submatrix
    end for l
    for $l = j : p$
!! Mult. of submat $[l - j - 1]$ of A in proc i with submat j of B in proc i
      for $t = 1 : n/p$
        for $k = 1 : n/p$
          for $h = 1 : n/p$
            $c_{k+\frac{n}{p}(l-1),t+\frac{n}{p}(i-1)} +=$
            $a_{k+\frac{n}{p}(l-1)-\frac{n}{p}(j-1),h+\frac{n}{p}(i-1)} \cdot b_{h+\frac{n}{p}(j-1),t+\frac{n}{p}(i-1)}$
      end for h, k, t
!! End of multiplication submatrix-submatrix
    end for l
!! Shift of the matrix A by 1 to the left
    for $l = 1 : n$
      for $k = 1 : n/p$
        $a_{l,k+\frac{n}{p}(i-1)} = a_{l,[k+\frac{n}{p}i]\mathrm{mod}p+1}$
  end for l, k, j
end foreach i

The matrices are partitioned into $(n/p \times n/p)$ blocks distributed in the skewed way discussed in [4].

Note that this simple scheme shows a communication overhead of $O(p)$. With the hyper-systolic method, we can reduce this communication overhead to $O(\sqrt{p})$. Thus, as to communication, the HS scheme scales as well as standard methods like Cannon's algorithm!

The HS version of the matrix multiplication is given in the following algorithm:

foreach processor $i = 1 : p$ ($\in$ systolic array)
for $j = 1 : n$ step $k\frac{n}{p}$
for $l = 2 : k$
for $t = 1 : n/p$
for $s = 1 : n/p$
$b_{j-1+t+\frac{n}{p}(l-1),[s+\frac{n}{p}(i-1)]\mathrm{mod}p+1} = b_{j-1+t+\frac{n}{p}(l-1),s+\frac{n}{p}(i-1)}$
end for s, t, l, j
for $j = 1 : K$ step $k\frac{n}{p}$
for $l = 1 : k$
for $d = 1 : (j-1)k + l - 1$
for $h = 1 : n/p$
for $t = 1 : n/p$
for $s = 1 : n/p$
$c^{l}_{t+\frac{n}{p}(d-1),h+\frac{n}{p}(i-1)} + =$
$a_{t+\frac{n}{p}[d-(j-1)k-l]+n,s+\frac{n}{p}(i-1)} \cdot b_{s+\frac{n}{p}[(j-1)k+l-1],h+\frac{n}{p}(i-1)}$
end for s, t, h, d
for $d = (j-1)k + l : p$
for $h = 1 : n/p$
for $t = 1 : n/p$
for $s = 1 : n/p$
$c^{l}_{t+\frac{n}{p}(d-1),h+\frac{n}{p}(i-1)} + =$
$a_{t+\frac{n}{p}[d-(j-1)k-l],s+\frac{n}{p}(i-1)} \cdot b_{s+\frac{n}{p}[(j-1)k+l-1],h+\frac{n}{p}(i-1)}$
end for s, t, h, d
end for l
for $l = 1 : n$
for $h = 1 : n/p$
$a_{l,h+\frac{n}{p}(i-1)} = a_{l,[h+\frac{n}{p}(i-1)+k-1]\mathrm{mod}p+1}$
end for h, l, j
for $j = l : k-1$
for $l = 1 : p$
for $h = 1 : n/p$
for $t = 1 : n/p$
$c^{k-j}_{t+\frac{n}{p}(l-1),h+\frac{n}{p}(i-1)} + = c^{k-j+1}_{t+\frac{n}{p}(l-1),[h+\frac{n}{p}(i-1)]\mathrm{mod}p+1}$
end for t, h, l, j
end foreach i

In an analogous way, we have developed the HS formulation for $C = aA^TB + bC$ and $C = aAB^T + bC$; we underline that we have structured the algorithms without explicitly performing the very time-consuming transposition operation.

5 Results

In order to demonstrate the validity of the proposed approach, we have implemented either the systolic or the HS version of $aAB + bC$, $aA^TB + bC$ and $aAB^T + bC$ for the real and complex cases. Particular effort has been devoted to the implementation of the sub-matrix products within the single processors, trying to exploit as much computational power as possible (optimization of the memory to internal register data transfer, maximal attention to avoid jump instructions interrupting the processor pipeline,...).

In the following figures 1 and 2 we show, for the best and worst real and complex cases, the performances (referred to the peak of the used 128 processor machine, i.e. 6.4 Gflops) achieved as a function of matrix dimensions (for the sake of simplicity we consider square matrices).

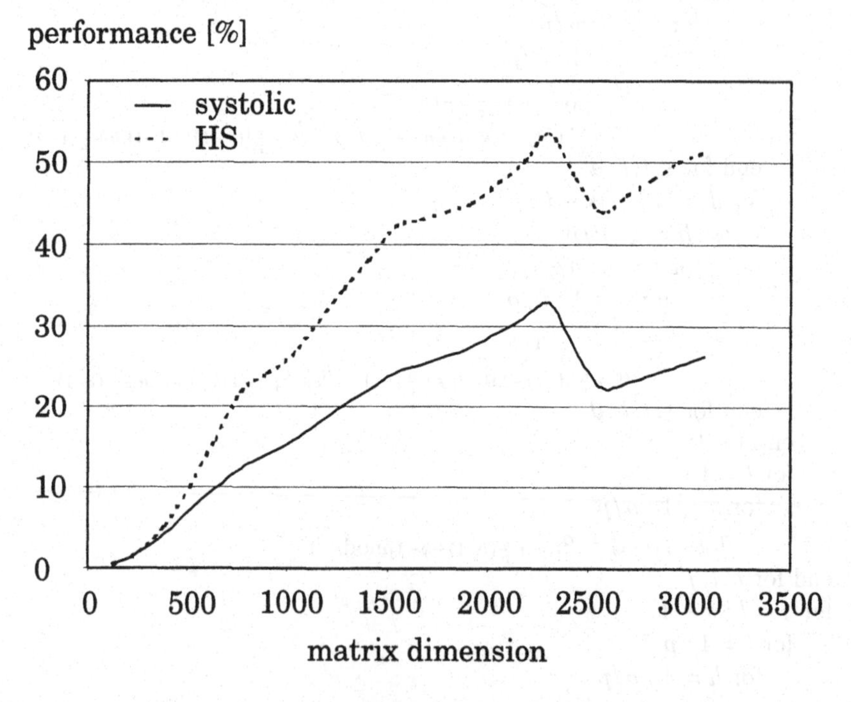

Fig. 1. Performances for the computation of $C = aAB + bC$, real case, 128 processor machine.

As we can see, the HS method allows the achievement of performances varying (in the best cases) from nearly 53% to 70% of the peak performance of the machine.

Acknowledgments T.L. thanks Klaus Schilling for many illuminating discussions.

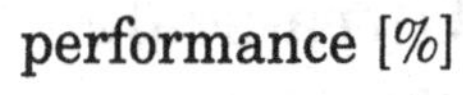

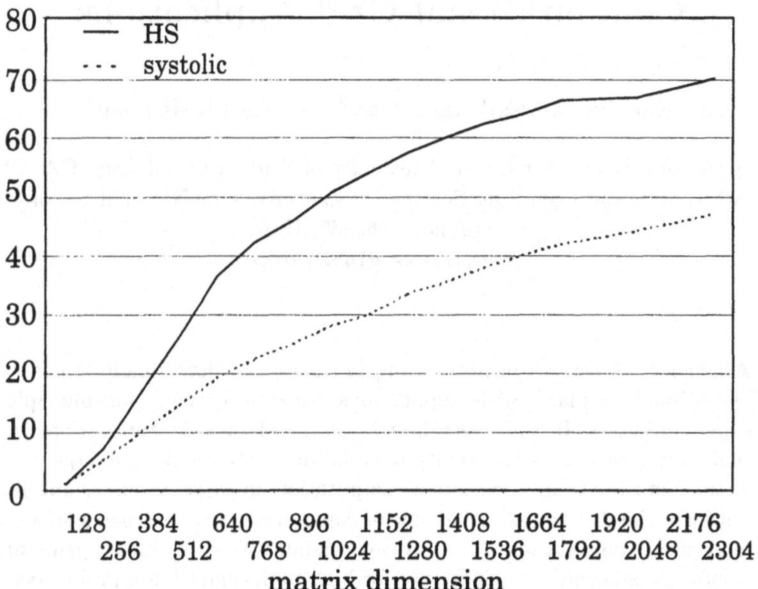

Fig. 2. Performances for the computation of $C = aA^TB + bC$, complex case, 128 processor machine.

References

1. J. Choi, J.J. Dongarra, D.W. Walker: 'The design of scalable software libraries for distributed memory concurrent computers'. J.J. Dongarra and B. Tourancheau editors. Environments and Tools for Parallel Scientific Computing. Elsevier 1982.
2. A. Bartoloni et al: 'A hardware implementation of the APE100 architecture'. International Journal of Modern Physics **C4** 1993.
3. T. Lippert, P. Palazzari, K. Schilling: 'Automatic template generation for solving n^2 problems on parallel systems with arbitrary topology'. Proceedings of the IEEE Workshop on Parallel and Distributed Software Engineering, Boston (MA) May 1997.
4. T.Lippert et al.: 'Hyper-systolic matrix multiplication'. Proceedings of the Proceedings of PDPTA '97, CSREA 1997.
5. T. Lippert, A. Seyfried, A. Bode, and K. Schilling: 'Hyper-Systolic Parallel Computing '. IEEE Trans. On Parallel and Distributed Systems, Vol. 9, No. 2, February 1998.
6. P.Palazzari, T. Lippert, K. Schilling: 'Simulated Annealing Techniques for communication-efficient Hyper-Systolic parallel computing on Quadrics'. Nato advance research whorkshop on High Performance Computing - Technology and Applications, June 24-25 1996, Cetraro (Italy).

Resource Management for Ultra-scale Computational Grid Applications

Karl Czajkowski[1], Ian Foster[2], and Carl Kesselman[1]

[1] Information Sciences Institute, University of Southern California, CA, USA.
[2] Mathematics and Computer Science Division, Argonne National Laboratory, Argonne, IL 60439, USA.
http://www.globus.org/

Abstract. Advances in networking infrastructure have made it possible to build very large scale applications whose execution spans multiple supercomputers. In such very large scale or ultra-scale applications, a central requirement is the ability to simultaneously co-allocate large collections of resources, to initiate a computation on those resources and to initialize the distributed collection of components to construct a single, integrated computation. In a previous paper [3], we defined a general resource management architecture for high-performance distributed systems in which resource co-allocation was an integral component. In this extended abstract, we examine co-allocation in more detail and describe the implementation of a specific resource co-allocator called the Dynamically Updated Request Online Co-allocator, or DUROC. DUROC has been implemented as part of the Globus grid toolkit. We briefly describe the design of DUROC and discuss how is has been used to support a range of large-scale grid applications.

1 Introduction

Recent trends in high-performance networking are resulting in an significantly increased availability of high-bandwidth network connections. With these advances in networking infrastructure, it is now possible to construct large-scale distributed computing environments, or "computational grids" as they are sometimes termed [6]. Computational grids provide an application with predictable, consistent and uniform access to a wide range of remote resources, including compute resources, data-repositories, scientific instruments, and advanced display devices. This access makes it possible to construct whole new classes of applications, such as supercomputer enhanced instruments, desktop supercomputing, tele-immersive environments, and distributed supercomputing [2].

While the grid is clearly about more then just computing resources, one compelling use of the grid is for distributed supercomputing [7] —- coupling together multiple supercomputers to construct a single "meta-computer" that can solve problems that are too large to be solved on any one supercomputer alone. With the deployment of persistent grid environments, such as the Globus Supercomputing Testbed Organization (GUSTO) [5], it is now possible to make significant numbers of the largest supercomputers available and apply them in a uniformed way to solve extremely large, or *ultra-scale* problems, allowing metacomputing to transition from one-off stunts, to a more production oriented environment.

One characteristic of ultra-scale computations that distinguish them from other classes of grid applications is the need to simultaneously initiate a computation on a large number of resources. Thus, it is important for a grid to not only support the management of resources one at a time, but to also provide an infrastructure for initialization and coordination across sets of resources. In a previous paper [3], we identified the simultaneous coordinated management of a collection of resources as the co-allocation problem. In this extended abstract, we examine the issues of co-allocation in more detail and describe a specific co-allocation service developed as part of the Globus metacomputing toolkit.

2 The Globus Resource Management Architecture

We now present a brief overview of the resource management architecture implemented by the Globus grid toolkit. A more complete discussion of the architecture can be found in [3].

The Globus resource management architecture, illustrated in Figure 1, consists of four main components:

- a resource specification language,
- resource brokers,
- local resource managers, and
- resource co-allocators.

All resource requirements within Globus are specified in terms of the Globus resource specification language, or RSL. An RSL specification consists of a set of attribute/value pairs combined by a small set of operators. The attributes in an RSL expression can be used to express both resource requirements and job parameters.

While an application may construct an RSL expression and direct it towards a specific resource, it is also likely that applications will want to express their requirements in terms of high level specifications, such as: "I need 25 GFLOPS", not "I need a 512 node T3E". It is the job of the resource broker to take these high-level specifications and transform them into detailed specifications indicating the use of specific resources.

Once the RSL has been refined to the point that it identifies a specific resource, it can be directed towards a local resource manager, called a Globus Resource Allocation Manager, or GRAM. The GRAM takes an RSL expression as input, and returns a job handle as output, if the request was successful. The GRAM also enables the requesting application to monitor the state of the job running on the allocated resource as well as control the job (i.e. kill it).

The final component of the our resource management architecture is the co-allocator. A co-allocator is responsible for coordinating the submission of a resource request requiring several different resources by dispatching components

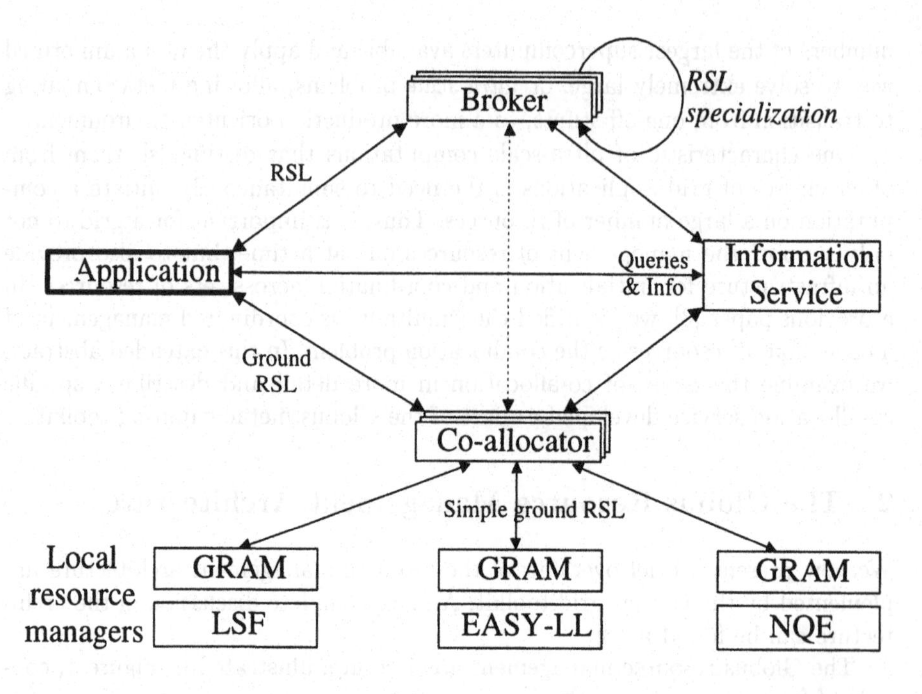

Fig. 1. The Globus resource management architecture, showing how RSL specifications pass between application, resource brokers, resource co-allocators, and local managers (GRAMs).

of that request to the appropriate GRAMs and synthesizing a single job abstraction from the individual components.

We will now discuss co-allocation in more detail.

3 Resource Co-allocation

Resource co-allocation provides the mechanisms needed to bridge the gap between applications and resource brokers that require collections of resources, and local resource managers that only know how to provide access to a single resource. It is the job of the co-allocator to provide an abstraction by which a collection of resources may be treated as a unit, stitching together a distributed collection of resources into an integrated application.

When composing a co-allocation request, the application or resource broker makes the underlying assumption that there exists a (non-zero) interval of time during which the resource specifications for the all the individual components can all be met and the resources can be allocated. A co-allocator is not required to provide any mechanism to ensure that this is the case and it is possible for a co-allocation request to fail because the specified resources could not be acquired simultaneously. In general, the success of a co-allocation request can not be guaranteed, as it is always possible for a requested resource to fail. However, we have developed a range of techniques by which it is possible to construct an allocation request so that it is *likely* to succeed. Space limitations prevent our discussing these techniques further.

4 Globus Co-allocation Services

As part of the Globus toolkit, we have designed and implemented a specific co-allocation service. A distinguishing feature of this service is the ability to dynamically edit the details of a pending co-allocation request, hence the its name: Dynamically Updated Request Online Co-allocator, or DUROC.

The general DUROC model is that a co-allocation request expressed as a single RSL expression which is submitted via the DUROC API. In response to a request, DUROC returns a single resource handle, which can be used to monitor and control the status of a whole job. In addition to these core capabilities, DUROC provides an API that allows the user to alter the contents of a pending request, adding and deleting components up until the time an explicit commit operation is made.

4.1 DUROC Resource Specifications

The resource specifications processed by DUROC are standard RSL expressions augmented with a small number of DUROC specific attributes. Within an RSL expression, the + operator is used to conjoin two independent requests to form a single *multi-request*. For example:

```
+ (nodes=2) (memory=64)
```

actually specifies two different resources, one with two nodes and the other with sixty four megabytes of memory.

Input to DUROC is a single multi-request, which DUROC uses to make GRAM request for each component of this RSL expression. Recall that a co-allocator does not make any decisions about where to place a request, thus, in order to fulfill a co-allocation request, DUROC must be able to determine which resource manger to contact for each element of the multi-request. This information is provided by augmenting each element of a multi-request with a:

```
resourceManagerContact
```

attribute, whose value specifies the name of the GRAM to which the request should be sent. This attribute must be provided for every element of a multi-request that is submitted to DUROC.

4.2 Support for Atomic Co-allocation in DUROC

One of the goals of DUROC is to support atomic co-allocation: prohibiting a program from proceeding until all of the requested resources are available and failing if one or more elements of a mult-request cannot be satisfied. Because the GRAM interface does not provide support for inter-manager atomicity, the user code must be augmented to implement a job-start barrier. As distributed components of the job become active, they rendezvous with the allocating agent to be sure all components were successfully started prior to performing any non-restartable user operations.

The start barrier allows the processes to synchronize before performing any non-restartable operations. In the absence of a start barrier, there is no way to guarantee that all job components are successfully created prior to executing user code.

4.3 Dynamic configuration in DUROC

Experience has shown that it is difficult to determine the exact set of resources that will be used by an application far in advance of that application actually being ready to execute. Failures can occur during application startup, subjobs can fail, or we may have requested more resources then required to improve our chances of co-allocation success, pruning off unneeded elements of the job once adequate resources have been allocated. In all of these cases, discarding a partially successful co-allocation can be expensive, yet the atomic allocation model provided by DUROC requires us completely identify the resources participating in a computation prior to starting actual application execution.

To enable flexibility in the actual configuration of a job while still supporting atomic allocation, DUROC distinguishes between submitting a resource request and committing to a resource request: the elements of a co-allocation request can be edited up until the time the controlling application issues an explicit commit operation. At that time, the elements of the request are frozen, and the runtime applications are allowed to proceed past the job start barrier once all the components of the frozen request have been allocated. Prior to committing to a configuration, the pending request can have elements added or deleted from it via the DUROC editing API. Note that a deleted component of a DUROC request may already have been allocated resources. As the program should be waiting at the job start barrier, no unreversable operations should have been performed, so the application is killed using standard GRAM process control functions.

4.4 Job Bootstrapping

While co-allocation is complete once the resources have been obtained in the requested processes have been started, DUROC also provides basic communication methods that can be use by the application to integrate the separate components into an integrated whole. The communications component of DUROC provides two simple mechanisms to send start-up and bootstrapping information between processes: an inter-subjob mechanism to communicate between “node 0” of each subjob, and an intra-subjob mechanism to communicate between all the nodes of a single subjob.

DUROC also provides a library of common bootstrapping operations built on top of the basic inter-subjob and intra-subjob communication primitives. These operations include an inter-subjob broadcast routine and the ability to construct an all-to-all communication network between all nodes in a job.

5 Uses of DUROC

DUROC has been used in a wide range of settings, and has proven to provide an effective co-allocation service. In one example, DUROC was used to start the largest distributed interactive simulation ever performed, starting a computation on over 1300 processors distributed across 13 different parallel supercomputers [1].

DUROC has also been used to construct of other Globus components. For example, we have produced a grid enabled version of the Message Passing Interface [4] which uses DUROC to start the elements of an MPI job. In this case, all DUROC calls are hidden in the MPI library, and an application does not have to make any modifications to benefit from DUROC co-allocation.

6 Conclusions and Future Work

We are currently investigating a number of ways in which the functionality of DUROC can be extended. We are looking at how the DUROC API can be extended to enable a tighter coupling between resource brokers and the co-allocator. Such a coupling would enable a broker to adapt to co-allocation failures in a sensible manner, perhaps via the existing DUROC editing API. We are also considering ways in which advanced reservation can be incorporated into the co-allocation process. Advanced reservation can greatly improve the chances that a co-allocation request will succeed.

In this abstract, we have provided a brief overview of the role of co-allocation in the Globus resource management architecture and we have shown how co-allocation has been implemented in Globus via the DUROC co-allocation service. DUROC has been used extensively as part of the Globus toolkit and has proven to be an effective and flexible means for coordinating resource allocation activities across multiple platforms.

References

1. Sharon Brunett, Karl Czajkowski, Ian Foster, Steven Fitzgerald, Andy Johnson, Carl Kesselman, Jason Leigh, and Steven Tuecke. Application experiences with the globus toolkit. IEEE Computer Society Press, 1998.
2. C. Catlett and L. Smarr. Metacomputing. *Communications of the ACM*, 35(6):44–52, 1992.
3. K. Czajkowski, I. Foster, N. Karonis, C. Kesselman, S. Martin, W. Smith, and S. Tuecke. A resource management architecture for metacomputing systems. In *The 4th Workshop on Job Scheduling Strategies for Parallel Processing*, 1998.

4. I. Foster, J. Geisler, W. Gropp, N. Karonis, E. Lusk, G. Thiruvathukal, and S. Tuecke. A wide-area implementation of the Message Passing Interface. *Parallel Computing*, 1998. to appear.
5. I. Foster and C. Kesselman. The Globus project: A status report. In *Proceedings of the Heterogeneous Computing Workshop*, pages 4–18. IEEE Computer Society Press, 1998.
6. I. Foster and C. Kesselman, editors. *The Grid: Blueprint for a Future Computing Infrastructure.* Morgan Kaufmann Publishers, 1998.
7. Paul Messina. Distributed supercomputing applications, 1998.

A ScaLAPACK-Style Algorithm for Reducing a Regular Matrix Pair to Block Hessenberg-Triangular Form

Krister Dackland and Bo Kågström

Department of Computing Science and HPC2N, Umeå University, S-901 87 Umeå, Sweden.
{dacke, bokg}@cs.umu.se

Abstract. A parallel algorithm for reduction of a regular matrix pair (A, B) to block Hessenberg-triangular form is presented. It is shown how a sequential elementwise algorithm can be reorganized in terms of blocked factorizations and matrix-matrix operations. Moreover, this LAPACK-style algorithm is straightforwardly extended to a parallel algorithm for a rectangular 2D processor grid using parallel kernels from ScaLAPACK. A hierarchical performance model is derived and used for algorithm analysis and selection of optimal blocking parameters and grid sizes.

1 Introduction

The problem under consideration is to reduce a regular matrix pair (A, B) to condensed form (H, T) using orthogonal transformations $Q \in \mathcal{R}^{n\times n}$ and $Z \in \mathcal{R}^{n\times n}$ such that

$$Q^T AZ = H, \quad Q^T BZ = T, \tag{1}$$

where H is upper r-Hessenberg with r subdiagonals and T is upper triangular. If $r = 1$, then H is an upper Hessenberg matrix. Without loss of generality we assume that r is a multiple of the matrix size n. Then by an appropriate partitioning H can be viewed as a block upper-Hessenberg matrix

$$\begin{bmatrix} H_{11} & H_{12} & H_{13} & \dots & H_{1N} \\ H_{21} & H_{22} & H_{23} & \dots & H_{2N} \\ 0 & H_{32} & H_{33} & \dots & H_{3N} \\ 0 & \ddots & \ddots & \ddots & \vdots \\ 0 & \dots & 0 & H_{NN-1} & H_{NN} \end{bmatrix}, \tag{2}$$

where each H_{ij} $(i \le j)$ is a dense $r \times r$ matrix and the subdiagonal blocks H_{i+1i} are square upper triangular.

The condensed form (1) may serve as a first step in the solution of the generalized eigenvalue problem $Ax = \lambda Bx$ [8]. It is also important in its own right, for example, when solving matrix pencil systems of the type $(A - sB)x = b$ for many different values of the scalar s and possibly different right hand sides [6].

In this contribution we present a parallel version and implementation of a generalization of one of the blocked algorithms presented in [3] based on Householder reflections and the compact WY representation of the Householder matrices [9]. The target architecture is a scalable MIMD computer system that can embed any rectangular $P_r \times P_c$ processor grid and that can emulate the corresponding communication topology efficiently. This may include distributed memory as well as shared memory systems, and hybrid architectures that combine shared and distributed memory. A parallel algorithm to reduce just one matrix to block upper-Hessenberg from is presented in [1].

The rest of this paper is organized as follows. Section 2 describes the transformation of an elementwise algorithm to blocked algorithms. The costs in operations are quantified and analyzed. Section 3 presents a parallel ScaLAPACK-style implementation and a hierarchical performance model of the generalized block-reduction algorithm. Finally, in Section 4 measured and modeled parallel performance results are discussed.

2 Reduction to Block Hessenberg-Triangular Form

The reduction of a regular matrix pair (A, B) to block Hessenberg-triangular form (1) is performed in two major steps:

- Determine an orthogonal matrix $U \in \mathcal{R}^{n \times n}$ such that $U^T B$ is upper triangular. To preserve the generalized eigenvalues of the matrix pair, U is also applied to A: $A \leftarrow U^T A$. Now A is dense and B is triangular.
- To further reduce A to (block) upper Hessenberg form while preserving B upper triangular, orthogonal transformations are applied to A and B from the left and right hand sides.

In the following we assume that B is already on triangular form, so we can focus on the second step in the reduction procedure.

2.1 From elementwise to blocked algorithms

The elementwise algorithm for reducing A to upper Hessenberg form while preserving B upper triangular can be expressed in terms of Givens rotations or 2×2 Householder reflections [7]. Orthogonal transformations applied from the left annihilate elements in A but destroy the structure of B. The triangular form of B is restored by right hand side orthogonal transformations. We illustrate one step of the reduction using Givens rotations (A and B are of size 4×4):

$$Q_{34}^T A = \begin{bmatrix} a_{11} & a_{12} & a_{13} & a_{14} \\ a_{21} & a_{22} & a_{23} & a_{24} \\ a_{31} & a_{32} & a_{33} & a_{34} \\ & a_{42} & a_{43} & a_{44} \end{bmatrix}, \quad Q_{34}^T B = \begin{bmatrix} b_{11} & b_{12} & b_{13} & b_{14} \\ & b_{22} & b_{23} & b_{24} \\ & & b_{33} & b_{34} \\ & & \mathbf{b_{43}} & b_{44} \end{bmatrix},$$

$$BZ_{34} = \begin{bmatrix} a_{11} & a_{12} & a_{13} & a_{14} \\ a_{21} & a_{22} & a_{23} & a_{24} \\ a_{31} & a_{32} & a_{33} & a_{34} \\ & a_{42} & a_{43} & a_{44} \end{bmatrix}, \quad AZ_{34} = \begin{bmatrix} b_{11} & b_{12} & b_{13} & b_{14} \\ & b_{22} & b_{23} & b_{24} \\ & & b_{33} & b_{34} \\ & & & b_{44} \end{bmatrix}.$$

The rotation Q_{34} eliminates a_{41} and when applied to B introduces a non-zero element b_{43}. This fill-in is set to zero by a right rotation Z_{34}. Since Z_{34} only operates on columns 3 and 4, no new non-zero elements are introduced in A. To annihilate the remaining elements below the subdiagonal, similar two-sided rotations are used.

Our blocked variants of the elementwise reduction algorithm perform the operations on square blocks of size $r \times r$ and produce a condensed form where A is block upper-Hessenberg and B is triangular [3]. In the annihilation phase of the elementwise algorithm two consecutive elements in a column are referenced and the lower of the two is zeroed while the other is updated. In the blocked variants elements are substituted for blocks and the lower block is annihilated while the upper is triangularized. This is performed by a QR factorization of a rectangular $2r \times r$ block in the current block column. When the transformations are applied to B, an $r \times r$ block instead of a single element will be filled in. This fill-in is annihilated by an RQ factorization of a square block of size $2r \times 2r$.

The procedure starts in the first block column of A and two blocks are reduced in each iteration until the first block column is on block upper-Hessenberg form. Then the annihilations in the second block column are performed and so on.

2.2 A generalized block Hessenberg-triangular reduction algorithm

In the blocked variants discussed so far, the number of reduced A-blocks is limited to two per block iteration. This choice minimizes the fill-in in each block iteration. A possible generalization is to let $p \geq 2$ blocks be involved in each step; $p-1$ are annihilated and one is triangularized. This will cause a fill-in in B of size $pr \times pr$ in each iteration, which in turn impose more computations for restoring B to triangular form. The upper bound for p is the number of blocks below the diagonal in the current block column of H. We propose the use of a fixed $p \geq 2$ for the complete reduction.

In Figure 1 we present this generalized blocked algorithm, where on input $A, B \in \mathcal{R}^{n \times n}$, B is upper triangular, r is the block size and $p \geq 2$ is the number of $r \times r$ blocks to be reduced in each block iteration. On output A is on block upper-Hessenberg form, B is upper triangular, and Q, Z are the orthogonal transformation matrices.

The A matrix is reduced by QR factorizations of rectangular $pr \times r$ blocks and B is restored by RQ factorizations of square $pr \times pr$ blocks.

Since the fill-in overlap for consecutive iterations, it is possible to apply the RQ factorization to blocks of size $(p-1)r \times pr$ in all iterations except the last one in each block column. This fact is exploited in the parallel algorithm.

```
function [A, B, Q, Z] = blockHT (A, B, r, p)
  k = n/r;                    # blocks in the first block column.
  for j = 1:r:n-r
    k = max(k - 1, 2);        # blocks to reduce in current block column j.
    l = ceil((k-1)/(p-1));# steps required for the reduction.
    i = n;
    for step = 1:l
      nb = min(p*r, i-j-r+1);
      Phase 1: Annihilation of p r x r blocks in block column j of A.
      [q, A(i-nb+1:i,j:j+r-1)] = qr(A(i-nb+1:i,j:j+r-1));
      A(i-nb+1:i,j+r:n) = q'*A(i-nb+1:i,j+r:n);
      B(i-nb+1:i,i-nb+1:n) = q'*B(i-nb+1:i,i-nb+1:n);
      Q(:,i-nb+1:i) = Q(:,i-nb+1:i)*q; Q = I_n initially.
      Phase 2: Restore B - annihilation of fill-in.
      [z, B(i-nb+1:i,i-nb+1:i)] = rq(B(i-nb+1:i,i-nb+1:i));
      A(1:n,i-nb+1:i) = A(1:n,i-nb+1:i)*z;
      B(1:i-nb,i-nb+1:i) = B(1:i-nb,i-nb+1:i)*z;
      Z(:,i-nb+1:i) = Z(:,i-nb+1:i)*z; Z = I_n initially.
      i = i - nb + r;         Pointer for next block annihilation.
    end
  end
```

Fig. 1. Matlab-style algorithm for block Hessenberg-triangular reduction

2.3 Cost in flops versus performance

Minimizing the number of floating point operations (*flops*) is not always equivalent to minimizing the execution time. It might, for example, be profitable to perform extra flops in order to get a more favorable memory reference pattern.

Algorithm blockHT has two tuning parameters, the block size r and the number of blocks reduced per block iteration p. Notice that the resulting condensed form is different for different block sizes but is the same for different values of p and a fixed r. The total number of flops performed on A and B in blockHT is

$$\frac{(20p^2+54p-44)}{12(p-1)}n^3 + \frac{(-4p^3r-30p^2r-123pr+60r)}{12(p-1)}n^2 + \frac{(-2p^4r^2+6p^3r^2-15p^2r^2+11pr^2+28r^2)}{12(p-1)}n + \frac{4p^4r^3+4p^3r^3-10p^2r^3+38pr^3+56r^3}{12(p-1)}. \quad (3)$$

The cost for accumulation of the transformations in Q and Z is not included in this expression.

It is also possible to use one of the two elementwise algorithms and stop the reduction of A r rows below the diagonal. The cost in flops when using 2×2

Householder reflections is

$$\frac{32}{3}n^3 + (-20r - 2)n^2 + (8r^2 - 8r - \frac{26}{3})n + \frac{4}{3}r^3 + 10r^2 + \frac{26}{3}r. \quad (4)$$

See [5] for details about the cost in flops of the different suboperations. In Table 1 we show the cost-ratios in flops between `blockHT` and the elementwise algorithm for $r = 32$ and $p = 2, 4, 8, 16$ and 32. The last column shows the ratios between the leading terms in the denominator and nominator, respectively.

p	Flop ratio `blockHT`/elementwise	$\approx$ Ratio
2	$\frac{18n^2-776n-24064}{16n^2-451n-2541}$	1.13
4	$\frac{123n^3-9344n^2-75776n+10878976}{96n^3-5778n^2+71346n+487872}$	1.28
8	$\frac{417n^3-39136n^2-1526784n+148701184}{224n^3-13482n^2+166474n+1138368}$	1.85
16	$\frac{1485n^3-207776n^2-28193792n+2266169344}{480n^3-28890n^2+356730n+2439360}$	3.09
32	$\frac{5541n^3-1325344n^2-490374144n+35360014336}{992n^3-59706n^2+737242n+504184}$	5.56

Table 1. Extra flops for generalized block algorithm, block size $r = 32$

In the left diagram in Figure 2 we show the number of flops required when reducing a matrix pair of size 1088 for different block sizes r and $p = 2$ or $p = 4$. Although `blockHT` requires about 13% more flops when $p = 4$ compared to $p = 2$, the execution time for $p = 4$ is less than the execution time for $p = 2$ up to a certain block size. The break even point in this specific case is around $r = 32$.

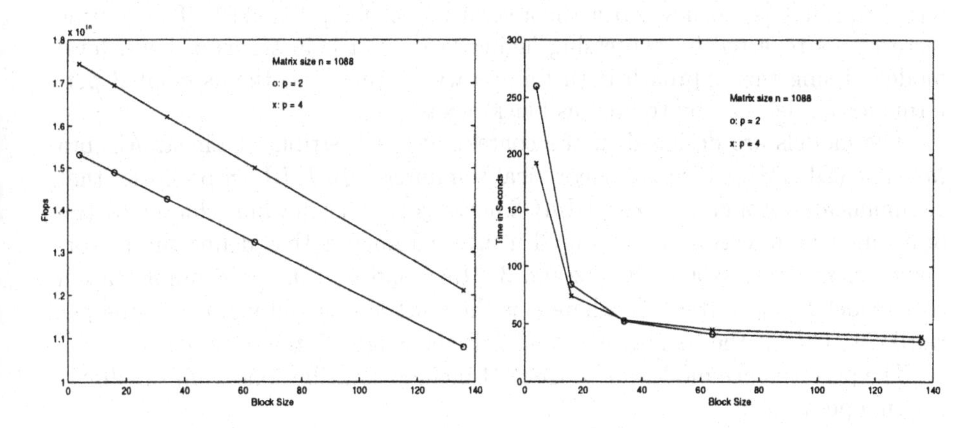

Fig. 2. Cost in flops and execution time on one processor for `blockHT` using different block sizes (r) and number of blocks annihilated per block iteration (p)

3 Parallel Reduction to Block Hessenberg-Triangular Form

The benefit of algorithm blockHT is enlarged in a parallel setting. In the following we assume a parallel environment that can embed any rectangular $P_r \times P_c$ processor grid and that can emulate the corresponding communication topology efficiently. Moreover, the matrices, A and B, are partitioned and distributed over the processor grid using a square block scattered (cyclic) decomposition.

3.1 A ScaLAPACK-style implementation

Our parallel implementation of blockHT, named p_blockHT, is built on top of standard building blocks from the ScaLAPACK library [2], which guarantees portability as well as high performance. All routines used are from ScaLAPACK or its underlying libraries.

As before, $B \in R^{n \times n}$ is assumed to be in upper triangular form. If this is not the case, the ScaLAPACK routines PDGEQRF and PDLARFB are used to perform a QR factorization of B and to update the matrix A accordingly.

Except for some minor differences in the reduction of fill-in elements in B, the implementation of p_blockHT follows the implementation of blockHT (see Figure 3). Notice, the standard blocked algorithms that only reduce two blocks per iteration limit the parallel degree to two processor rows in each block iteration. With the generalized blocked algorithm we can choose p between 2 and the number of blocks in current block column and thereby enlarge the parallel degree.

3.2 Hierarchical performance modeling

We use our hierarchical approach for performance modeling of ScaLAPACK-based routines [4] to develop a theoretical model for p_blockHT. The routine hierarchy is reflected by expressing higher level models in terms of lower level models. Using this approach both the library routines and the associated performance models can be treated as black boxes.

The models are designed in the context of the Distributed Linear Algebra Machine (DLAM), which is a theoretical computer with P BLAS processes that communicate through a $P_r \times P_c$ BLACS network [2]. The machine characteristics of a target architecture are captured in five parameters that define the performance of uni-processor level 1, 2 and 3 BLAS and α, β in the communication cost model $T_{\text{comm}} = \alpha + \beta L$, where α is the startup cost (latency), β is the per word cost ($1/\beta$ is the bandwidth), and L is the length of the message.

The performance model for p_blockHT include contributions from the different suboperations:

$$T_{\text{p\_blockHT}}(n, r, p, P_r, P_c) = T_{QR} + T_{Q^TA} + T_{Q^TB} + T_{RQ} + T_{AZ} + T_{BZ}. \quad (5)$$

The first three components in (5) correspond to the reduction of column panels of A and the update from left of A and B with respect to all QR factorizations.

n = matrix size, r = block size, p = # blocks reduced per block iteration

```
k = n/r, # blocks in the first block column.
for j = 1 to n - r step r
  k = max(k - 1, 2), # blocks to reduce in current block column j
  l = ⌈(k - 1)/(p - 1)⌉, # steps required for the reduction
  i = n, pointer to block to be factorized
  for step = 1 to l
      nb = min(p · r, i - j - r + 1)
      PDGEQR2, QR-factorize A_{i-nb+1:i, j:j+r-1}
      PDLARFT, compute triangular factor T of block reflector
      PDLARFB, apply block reflector to A_{i-nb+1:i, j+r:n}
      PDLARFB, apply block reflector to B_{i-nb+1:i, i-nb+1:n}
      if step = l then top = 0 else top = r
      for ib = i - r to i - nb + top step -r
          PDGERQ2, RQ-factorize B_{ib+1:ib+r, i-nb+1:ib+r}
          PDLARFT, compute triangular factor T of block reflector
          PDLARFB, apply block reflector to A_{1:n, i-nb+1:ib+r}
          PDLARFB, apply block reflector to B_{1:ib, i-nb+1:ib+r}
      end
      i = i - nb + r, pointer for next block annihilation.
  end
end
```

Fig. 3. p_blockHT expressed in terms of ScaLAPACK routines

Similarly, the three last components in (5) correspond to the annihilation of fill-in in B and the update from right of A and B with respect to all RQ factorizations of row panels. Hierarchical models for each of these six components are derived and presented in [5].

The model for p_blockHT is used as an algorithm analysis tool and for selection of optimal block sizes (r and p) and grid sizes ($P_r \times P_c$). The predicted results are verified by real experiments.

4 Measured and Predicted Performance Results

The measured performance results of p_blockHT are obtained by using up to 64 Thin Nodes (P2SC, 120 MHz) on the IBM Scalable POWERParallel System at High Performance Computing Center North (HPC2N).

To be able to evaluate our hierarchical performance model (5) we compare the measured performance results to predicted results. The five parameters that specify DLAM have been empirically determined from experiments on the SP system and represent estimates of the average performance of computational and communication routines used in p_blockHT; BLAS 1, 2 and 3: 50, 150 and 320 Mflops/s; α = 40 microsecs; β^{-1} = 80 Mbytes/s.

In Table 2 we report a selection of performance results. A larger set of results is presented in [5]. In all these tests we use $p = \max(2, P_r)$. Columns five to seven show the real time (SP), the predicted time ($Model$) in seconds used to compute the block Hessenberg-triangular form of (A, B), and their ratios ($SP/Model$). These performance results verify the accuracy of the performance model.

Problem Configuration				*Time in secs*			*Performance*	
n	r	P_r	P_c	*SP*	*Model*	*SP/Model*	*Mflops/s*	*Speedup*
2048	32	1	1	353	312	1.13	251	1.00
2048	64	1	1	291	302	0.96	295	1.00
2048	128	1	1	265	281	0.94	306	1.00
5760	32	8	1	2070	2072	1.00	974	3.88
5760	64	4	2	1663	1668	1.00	1200	4.06
5760	128	4	2	1425	1487	0.96	1371	4.48
11520	128	2	16	13040	12632	1.03	1225	4.0
11520	128	4	8	7244	7234	1.00	2204	7.0
11520	128	8	4	5382	5623	0.96	2967	9.7
11520	128	16	2	6897	7027	0.98	2315	7.6
16384	128	32	2	19214	17682	1.09	2406	7.9
16384	128	16	4	11275	11659	0.97	4100	13.4
16384	128	8	8	11768	11446	1.03	3928	12.8

Table 2. Performance results for `p_blockHT` running on 1, 8, 32, and 64 IBM SP nodes

The last two columns in Table 2 show the performance measured in Million floating point operations per second (*Mflops/s*) and the parallel speedup delivered by the SP system for the problem setup. The performance in Mflops/s is computed using the operation count (4) for the standard elementwise algorithm divided by the measured execution time (SP). The actual performance is higher since `p_blockHT` requires much more flops than the elementwise algorithm.

While r and p affect the number of required flops, the processor grid configuration affects the data distribution and thereby the communication overhead and execution rate for the underlying computational kernels. For example, using 64 nodes the performance is almost doubled when going from a 32×2 to a 16×4 grid (see Table 2). The results also show that that choosing $P_r > P_c$ gives a better performance.

Table 3 illustrates the impact on the performance from the choice of the number of reduced blocks per iteration, p, and $P_r \times P_c$. In this specific case ($n = 10008$, $r = 128$) the best configuration is $p = 4$ and an 8×4 grid.

The last six columns in Table 3 show the predicted execution time of the different suboperations in the reduction. The total cost is the modeled time (*Model*). Since the updates from the right hand side are the most expensive suboperations it is most profitable to reduce the time for these updates. An update of A from the right hand side with respect to an RQ factorization of a block of B of size $pr \times pr$ is executed as a sequence of updates of size $n \times ir, i =$

p	#flops	P_r	P_c	SP	Model	Ratio	T_{QR}	T_{QT_A}	T_{QT_B}	T_{RQ}	T_{AZ}	T_{BZ}
4	1.77e+13	4	8	6275	6329	0.99	99	401	231	249	3185	2162
4	1.77e+13	8	4	4446	4296	1.03	99	763	423	249	1616	1143
8	2.55e+13	4	8	6170	6590	0.94	59	277	164	310	3430	2348
8	2.55e+13	8	4	4640	4829	0.96	58	374	223	326	2245	1601

Table 3. Performance of p_blockHT and predicted cost in seconds for different suboperations ($n = 11008$, $r = 128$)

$2, \ldots, p$. This implies that some processors do not participate in the update when $i < P_c$. Specifically, $P_c - i, i = 2, \ldots, p$ processor columns are idle in each update. This is of course also true for the corresponding update of B. This observation explains why grids where $P_r > P_c$ show better performance than grids where $P_r \leq P_c$.

A more detailed analysis of p_blockHT, including a scalability analysis based on the hierarchical model and real experiments are presented in [5].

References

1. M. Berry, J. Dongarra and Y. Kim. A Highly Parallel Algorithm for the Reduction of a Nonsymmetric Matrix to Block Upper-Hessenberg Form. *Parallel Computing*, Vol. 21, No. 8, pp 1189-1212, 1995.
2. S. Blackford, J. Choi, A. Clearly, E. D'Azevedo, J. Demmel, I. Dhillon, J. Dongarra, S. Hammarling, G. Henry, A. Petit, K. Stanley, D. Walker, and R.C. Whaley. *ScaLAPACK Users' Guide.* SIAM Publications, Philadelphia, 1997.
3. K. Dackland and B. Kågström. Reduction of a Regular Matrix Pair (A, B) to Block Hessenberg-Triangular Form. In Dongarra et.al., editors, *Applied Parallel Computing: Computations in Physics, Chemistry and Engineering Science*, pages 125–133, Berlin, 1995. Springer-Verlag. Lecture Notes in Computer Science, Vol. 1041, Proceedings, Lyngby, Denmark.
4. K. Dackland and B. Kågström. An Hierarchical Approach for Performance Analysis of ScaLAPACK-based Routines Using the Distributed Linear Algebra Machine. In Dongarra et.al., editors, *Applied Parallel Computing: Computations in Physics, Chemistry and Engineering Science*, pages 187–195, Berlin, 1996. Springer-Verlag. Lecture Notes in Computer Science, Vol. 1184, Proceedings, Lyngby, Denmark.
5. K. Dackland. Parallel Reduction of a Regular Matrix Pair to Block Hessenberg-Triangular Form - Algorithm Design and Performance Modeling. Report UMINF-98.09, Department of Computing Science, Umeå University, S-901 87 Umeå, 1998.
6. W. Enright and S. Serbin. A Note on the Efficient Solution of Matrix Pencil Systems. *BIT*, 18:276–281, 1978.
7. G. H. Golub and C. F. Van Loan. *Matrix Computations*, Second Edition. The John Hopkins University Press, Baltimore, Maryland, 1989.
8. C. B. Moler and G. W. Stewart. An Algorithm for Generalized Matrix Eigenvalue Problems. *SIAM J. Num. Anal.*, 10:241–256, 1973.
9. R. Schreiber and C. Van Loan. A Storage Efficient WY Representation for Products of Householder Transformations. *SIAM J. Sci. and Stat. Comp.*, 10:53-57, 1989.

Parallel Tight-Binding Molecular Dynamics Code Based on Integration of HPF and Optimized Parallel Libraries

B. Di Martino[1,2], M. Celino[3],
M. Briscolini[5], L. Colombo[4], S. Filippone[5], and V. Rosato[3]

[1] Second University of Naples, Dip. di Ingegneria dell' Informazione (Italy)
dimartin@grid.unina.it or dimartin@par.univie.ac.at
[2] University of Vienna, Inst. for Software Technology and Parallel Systems (Austria)
[3] ENEA - HPCN Project, C.R. Casaccia, CP 2400, 00100 Roma AD (Italy)
celino@casaccia.enea.it
[4] INFM and Dip. di Scienza dei Materiali, Università di Milano (Italy)
[5] IBM Italia, Roma (Italy)

Abstract. In this paper we describe the parallelization of a molecular dynamics code, based on a Tight-Binding Hamiltonian, on a DMMP parallel platform. The data-parallel implementation has been carried out within the HPF framework, and tested on IBM SP2 architectures. The integration of an optimized parallel library routine (PESSL) for the full diagonalization of large symmetric matrices is also described. The scalability of this approach, the performance of the parallel code and that of the PESSL routine are shown.

1 Introduction

Computer simulations represent one of the major theoretical tool used to investigate the properties of matter, from complex molecules to semiconductors. Materials Science has largely benefited of the enormous progress achieved in the last twenty years in this area which now allows for the availability of a large set of techniques ranging from classical [1] to fully *ab-initio* [2] representations. It is now possible to select the computational scheme which offers the best compromise between the accuracy of the involved physical representation and the resulting computational cost. The Tight Binding (TB) representation has recently called the attention as being able to combine accuracy and computational cost in the field of simulations of semiconductors systems. TB implements an empirical parametrization of the atomic bonding interactions based on the expansion of the electronic wave functions on a very simple basis set [3, 10].

Although being much simpler than so-called *ab-initio* approaches, the computational complexity of TB algorithms is still considerable when used in combination to a Molecular Dynamics technique (TBMD): as it requires a large matrix diagonalization (O(N^3) scaling) matrix at each time step. As a consequence, the practical size of the simulated systems cannot exceed the limit of 3-400 atoms on a workstation. While such a figure already allows the study of a number of relevant problems in semiconductor physics, it still leaves a large gap towards the true domain of Materials Science simulations, *e.g.* grain boundaries, dislocations, interfaces, nanostructures.

This limit can be overcome by using parallel computers to perform the diagonalization task. In this way, the study of systems in the range of about 10^3 atoms could become accessible.

In this paper we report the results of the parallelization and porting of a TBMD code on a Distributed Memory MultiProcessor (DMMP) architecture (IBM SP2), using the High Performance Fortran (HPF) language and the IBM parallel mathematical libraries PESSL (Parallel Engineering and Scientific Subroutine Libraries).

The paper will proceeds as follows: in the next section, the sequential TBMD code will be analyzed from an algorithmic point of view. The main issues involved in the parallelization will be underlined. In the last section, the implementation of the code and its perfomances on the IBM SP2 will be discussed.

2 The sequential TBMD code

The goal of a MD code is to generate time trajectories of a system of n atoms by solving the classical equations of motion driven by forces, acting on each atom, which can be calculated in a classical or in a quantum mechanics representation. Force calculation is, thus, the main computational task of a generic MD code in both classical or quantum regime. Whereas in the classical representation force evaluation is essentially related to the calculation of the interparticles distance, in the TB representations there is a further computational complexity given by diagonalization of the interaction matrix whose construction is related to the computation of the interparticle distances.

The TB formulation is based on the adiabatic approximation of the Hamiltonian H_{tot} of a system of atoms and electrons in a solid [3]

$$H_{tot} = T_i + T_e + U_{ee} + U_{ei} + U_{ii} \tag{1}$$

where T_i and T_e is the kinetic energy of ions and electrons, U_{ee}, U_{ei}, U_{ii} are the electron-electron, electron-ion and ion-ion interactions, respectively.

Referring to the theory of one electron moving in the presence of the average field due to the other valence electrons and ions, the reduced one-electron Hamiltonian can be written

$$h = T_e + U_{ee} + U_{ei} \tag{2}$$

giving the eigenvalues (energy levels) ϵ_n and the eigenfunctions $|\Psi_n>$. In a TB scheme, the eigenfunctions are represented as a linear combination of atomic orbitals $|\phi_{l\alpha}>$

$$|\Psi_n> = \sum_{l\alpha} c^n_{l\alpha} |\phi_{l\alpha}> \tag{3}$$

where l is the quantum number index and α labels the ions. The expansion coefficients $c^n_{l\alpha}$ represent the occupancy of the l-th orbital located at the α-th site.

In the present TB approach, the elements of the h matrix, $< \phi_{l'\beta}|h|\phi_{l\alpha} >$, connecting α and β nearest neighbours sites are constituted by two contributions

$$< \phi_{l'\beta}|h|\phi_{l\alpha} >= a_{\beta\alpha} f(r_{\beta\alpha}) \tag{4}$$

where a_{ij} represent the distance independent part of the matrix elements which are fitted on *ab-initio* results. As a further approximation, a minimal basis set is usually adopted: four basis functions (s, p_x, p_y, p_z) per atom are known to be sufficient for a satisfactory description of the valence bands in the case of elemental semiconductors. In order to obtain the single-particle energies ϵ_n and the eigenvectors $c^n_{l\alpha}$ it is necessary to solve the secular problem at each MD time-step. This implies repeated diagonalization of the matrix h, which introduces the $O(N^3)$ scaling of this TBMD formulation, where N is the rank of the matrix. The rank of the matrix is determined by $N = n * n_b$ where n is the number of atoms in the simulated systems and n_b is the dimension of the basis set ($n_b = 4$ in the simplest case, of elemental semiconductors). Once the eigenvalues and eigenvectors are known, the attrattive potential energy can be computed and summed to the repulsive part derived from a many-body approach [8].

The Hamiltonian of eq.(1) refers only to the atomic system. In order to simulate the interaction with the surrounding, suitable MD schemes have been devised to account for a coupling between the internal degrees and the external degrees of freedom (thermal bath and possibility of imposing to the system an external stress) which are simulated by introducing further degrees of freedom. The MD simulation in the isothermal-isobaric ensemble, for instance requires the presence of 10 extra variables which are coupled to the variables of the atomic system via external parameters which can be adjusted to ensure the fastest convergence to equilibrium [9]. After force calculation, the equations of motion can be integrated by using some finite difference scheme. In the implemented code, a fifth-order predictor-corrector scheme [5] (Gear algorithm) has been used in view of its good accuracy.

The layout of the sequential TBMD code consists in the following steps:

1. Predictor for the integration of the dynamical equations for the evolution of the $10 + 3n$ degrees of freedom;
2. Calculations of the interparticle distance $r_{\alpha\beta}$. Once the atomic coordinates are given as input, the nearest neighbour array must be computed;
3. Computation of the matrix elements $< \phi_{l'\beta}|h|\phi_{l\alpha} >$;
4. Diagonalization of the real skew matrix h for the computation of the half spectrum of eigenvectors and eigenvalues. This is the most computational intensive part of the code;
5. Computation of the attractive part of the atomic forces (Feynman routine): the attractive part of the forces acting on each atom are computed using both the eigenvalues and the eigenvectors computed in the previous step;

6. Computation of the repulsive part of the atomic forces: this part of the forces depends only by the distances among the atoms: the nearest neighbour array is used again;
7. Corrector for the integration of equations of motions;
8. All the physical quantities are evaluated. Atoms coordinates and atoms velocities are stored for restarting purposes.

The predictor-corrector scheme requires the knowledge of the atomic coordinates till the fifth derivative: the whole atomic system is thus described by 3*5 arrays of length n for the coordinates and by further 6 arrays of the same length to store the forces. A $n*n$ matrix is requested to storage the interparticle distances and the h matrix of dimensions at least $4n*4n$ must be allocated.

The main difficulties in the parallelization of a TBMD code are:

- the system cannot be mapped on a regular grid (atoms can move inside the simulation box);
- diagonalization of a large sparse matrix is involved.

3 The parallelization strategy and its implementation in HPF

The several phases in which the computation is decomposed, described in sec. 2, are all amenable to be parallelized by following the well known *data parallel* execution model.

HPF [6,7] is a programming language designed to support the data parallel paradigm. HPF eliminates the complex, error prone task of explicitly programming how, where and when to pass messages between processors on distributed memory machines. The underlying HPF compiler is responsible for producing code sections which automatically distribute the array elements on the available processors. The HPF framework is broadly recognized as being the tool of choice for porting huge legacy sequential codes to parallel architectures, when the computation exhibits regular data-parallel behavior, and it can be effective even when the computation presents characteristics of irregularity (see, e.g., [4]).

The main issue which arises in depicting the parallelization strategy is the selection of the layout for the data structures, most of them involved in more than one (or all) computational tasks.

Most of the schemes for data distribution which allow reducing data locality (and, thus, minimize communications) or maximize the iterations in the loop distribution (and, thus, maximize the degree of parallelization), are often conflicting.

Step (1) and (7) (the predictor and corrector steps) involve sweeps over the monodimensional arrays: `x,y,z` (particles coordinates), `x1, y1, z1, x2, y2, z2, x3, y3, z3, x4, y4, z4, x5, y5, z5` (up to the fifth order derivatives of the particles coordinates), `fx,fy,fz` (the total forces acting on each particle).

The do loops implementing those sweeps do not present loop-carried dependences; thus their iterations can be distributed among the processors, by following the optimal data layout, which corresponds to an alignment, and cyclic distribution, of the structures above:

```
distribute (cyclic) :: x,y,z
align with x :: x1,y1,z1,x2,y2,z2,x3,y3,z3,x4,y4,z4,x5,y5,z5
align with x(:) :: fx(:),fy(:),fz(:)
```

The data parallel structure of the computations, and the absence of "stencil effects" implies a total absence of communications.

The step (2) computes the distances `xx/yy/zz/rr2(n,n)` (bidimensional arrays) from `x,y,z`; it involves the construction of the "Verlet lists" `list/list1(n,n)` (together with the filling marker vectors `nlst/nlst1(n)`), which keep track of the simmetry and sparsity of the values stored in `xx/yy/zz/rr2` and are consequently used to reduce the amount of computation when accessing the latter arrays.

The optimal data layout for these data structures is their alignment with the arrays of coordinates, and thus their distribution (cyclic) over the first coordinate:

```
align list(:,*) with x(:) (i.e.: distribute (cyclic,*) :: list )
align list1 with list
align nlst(:) with list(:,*)
align nlst1(:) with list1(:,*)
align with list :: xx,yy,zz,rr2
```

The iterations of the double loop nests which perform the computation of distances and verlet lists would be distributed accordingly. Unfortunately, this data layout is in conflict with the optimal data layout for the last step of step (5), the Hellman-Feynman forces computation.

This step updates the data structures `fhfx,fhfy,fhfz(n)`, monodimensional arrays which represent the Hellman-Feynman forces, and utilizes the auxiliary bidimensional arrays `fhffx,fhffy,fhffz(n,n)`. The elements of these arrays are computed by sweeping over the bidimensional distance arrays and, for each distance (i, j), by reducing the $i-th$ to $(i+4)-th$ and $j-th$ to $(j+4)-th$ rows of the Hamiltonian matrix `h(n4,n4)`.

Thus, for each sweep of a row of the distance matrices, there is a sweep of the whole matrix `h`. If we followed the data layout prescribed above for the distance matrices (i.e. cyclic distribution of their rows over the processors) we should replicate the whole matrix `h` over the processors, in order to avoid a huge amount of communications. This is unfeasible, because this matrix is by far the largest structure of the whole computation, 16 times bigger than the distance matrices.

Thus the only way to cope with this constraint is to give up on distributing the distance matrices (thus replicating them over the processors), and distribute the hamiltonian `h` by the second dimension (i.e. distribute its columns) instead.

The sweep over the elements of distance matrices is thus replicated, but the reduction of the 8 rows of h, for each distance (i,j), can be performed in parallel: each processor sweeps the portion of rows of h assigned to it, and performs a partial reduction of those elements; the partial reduction results coming from each processor are then composed and the final reduction result is then broadcasted to them. The amount of distribution of the iterations of the overall loop nest (and thus the degree of parallelization) remains the same as in the case of the alternative data layout (distances distributed, and hamiltonian replicated). With the latter scheme, there is an additional communication overhead due to the communication of the partial results of the reduction, but this overhead turns out to be negligible with respect to the computation overhead, and thus the efficiency gained by the parallel execution.

Instead of redistributing the distance matrices from the optimal data layout of step (2) to the whole replication needed in this step was too high, we have preferred to keep these matrices always replicated (thus also during step (2)): even though the computation is in this way replicated among the processors during that phase, the communication overhead occurring during the redistribution of those matrices would be much more costly than the computation overhead of the replication of step (2).

In step (3) and (4) the computation and diagonalization of the Hamiltonian bidimensional matrix `h(n4,n4)`, and computation of its eigenvalues in the monodimensional array `eval(n4)` are performed. The PESSL routine SYEVX is called for this task.

The Parallel ESSL is a parallel mathematical library specifically tuned for the IBM SP computers and can be called both from message passing routines and HPF framework. In the latter case the efficiency of the matrix distribution and of the start up of the diagonalization are lower than in the message passing framework, but the implementation is simpler. This allows a sharp tuning of the parameters needed for the parallel routine.

The use of the PESSL routine introduces extra inter-node communications because the SYEVX routine needs the matrix to be distributed in both dimensions as cyclic. We thus need to redistribute the hamiltonian matrix h during this step, and to restore its distribution over the columns as needed in the Hellmann-Feynman routine.

4 Results: benchmarks of the parallel code

To validate the porting and the data distribution adopted we have performed several runs on two IBM SP2 : the first one is located in ENEA Research Centre in Frascati (Italy) and the other in the IBM Centre of Poughkeepsie (USA). The first has 390-series processors with 66 MHz and 128 MB RAM, while the second has 397-series with 160 MHz and 256 MB RAM.

The run have been performed on 2,4,6 and 8 processors. A clear speedup till 8 nodes is showed both using the SP2 in the ENEA Centre and that one in the IBM Centre (see fig.1(a) and fig.(1(b)).

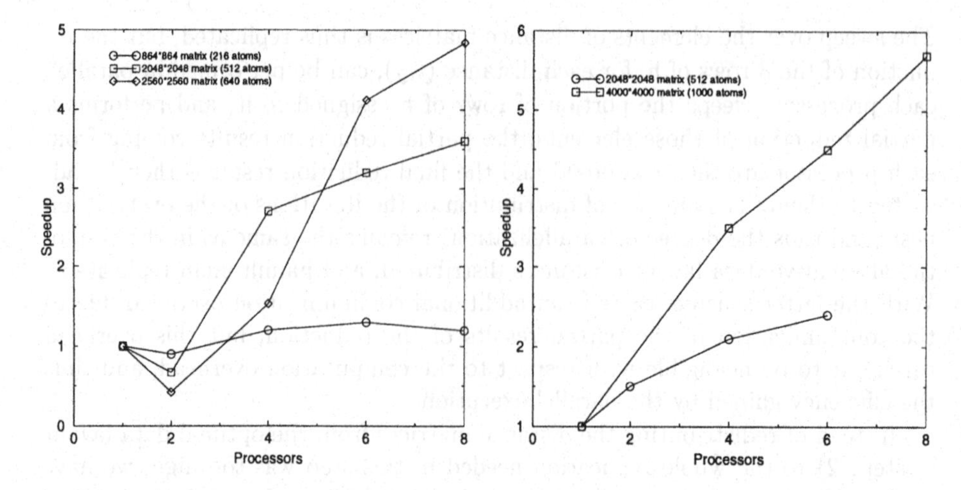

Fig. 1. *Speedup obtained using HPF+PESSL parallel code (a) (left) on the IBM SP2 located at ENEA Frascati Research Centre and (b) (right) on the IBM SP2 located in IBM Centre of Poughkeepsie (USA).*

The point at one processor is measured only for the 216 and for the 512-atoms case and extrapolated for the 640 and 1000-atoms cases. Some points corresponding to two processors lie below unity due to the large RAM occupancy of the data on the single nodes, which lead to swapping on disk.

When the number of atoms increases, all the routines (especially the routine for atomic distance computation and the Feynman routine for the attractive force computation) become more cpu time consuming relatively to the diagonalization routine. In this respect we can observe:

- The approach described in the last section consists in distributing all the atomic coordinates obtaining in the first and in the last phase (predictor and corrector algorithm, see section (2)) a speedup equal to the number of the processors.
- The Feynman routine, even if the atomic coordinates are distributed all over the processors (thus requiring a communication among all the processors), in the case of 640 atoms shows a clear speedup of 3 going from 4 to 6 processors.

5 Conclusions

In this paper we have described the porting of a code for the study of problems in material science on a parallel computer using the HPF language and the parallel optimized libraries PESSL. The approach described can be considered successfull for two reasons:

1. even if HPF is not completely suitable for irregular problems, like MD simulations, we have obtained good speedup and efficiency;

2. even if different data distributions (one for the matrix diagonalization routine and the other for the MD routines) are needed, the overall performance are good because the new parallelization strategy adopted can minimize the data communication.

Using the approach described in the preceding sections the target of simulations with a number of atoms of order 10^3 is reached: the parallel code can perform simulations in a reasonable amount of wall clock time and the RAM occupancy of the whole code can be managed.

References

1. M.P.Allen and D.J.Tildesley, "Computer simulation of liquids", Clarendon Press, Oxford, 1987.
2. R.Car and M.Parrinello, Phys.Rev.Lett. **55**, 2471 (1985).
3. L.Colombo in: Annual Review of Computational Physics IV, edited by D.Stauffer (World Scientific, Singapore, 1996) p.147.
4. B.Di Martino, S.Briguglio, G.Vlad and P.Sguazzero, "Parallel Plasma Simulation through Particle Decomposition Techniques", Lecture Notes in Computer Science n. 1401, Springer-Verlag, April 1998 (Proc. of *High Performance Computing and Networking '98* Conference, Amsterdam (Nl), Apr. 1998).
5. C.W.Gear, "The numerical integration of ordinary differential equations of various orders". Report ANL 7126, Argonne National Laboratory (1966); C.W. Gear, "Numerical initial value problems in ordinary differential equations". Prentice-Hall, Englewood Cliffs, NJ (1971).
6. High Performance Fortran Forum, "High Performance Fortran Language Specification", Version 1.1, 1994.
7. High Performance Fortran Forum, "High Performance Fortran Language Specification", *Scientific Programming,* **2**(1-2), 1-170 (1993).
8. I.Kwon, R.Biswas, C.Z.Wang, K.M.Ho, C.M.Soukoulis, Phys.Rev.B **49**, 7242 (1994).
9. M.Parrinello and A.Rahman, J. Chem. Phys. **72**, 2662 (1982); S. Nosé and M.L. Klein, Molecular Physics, **50** 1055-1076 (1983).
10. C.Z.Wang, K.M.Ho, and C.T.Chan, Computational Mat. Sci. **2**, 93 (1994) and references therein.

Parallel Computation of Multidimensional Scattering Wavefunctions for Helmholtz/Schroedinger Equations

Åke Edlund, Ilan Bar-On, and Uri Peskin

Department of Chemistry, Technion-Israel Institute of Technology,
Haifa, 32000 Israel
ake@parallelsystems.se, baron@cs.technion.ac.il,
uri@chem.technion.ac.il
http://www.technion.ac.il/technion/chemistry/staff/

Abstract. Multidimensional scattering wavefunctions are calculated using a new fully distributed parallel solver for the Helmholtz/Schroedinger equations. The solver is based on a parallel preconditioner which is derived from the generic structure of the Helmholtz/Schroedinger partial differential equations with scattering boundary conditions. The approach is useful for a broad range of scientific applications, e.g. in nanoelectronics, fiber optics and multidimensional quantum scattering calculations. With minor changes, the solver can be applied as an exponential propagator for time-dependent Helmholtz/Schroedinger initial value problems. Examples are given for 3D models of a wave propagation in a discontinuous waveguide and of electron transmission through a water barrier.

1 Introduction

We consider the solution of the Helmholtz and Schroedinger equation with relation to the design of new electro-optical and nanoelectronic devices[1–6] and to the study of the single-electron elastic scattering in disordered media, electron-electron dynamical correlations in mesoscopic system, and the trapping mechanisms of electrons in molecular systems. Such problems are especially important due to new experimental devices in the fields of electro-optics and nanoelectronics (in particular, the scanning tunneling or the atomic force microscopies) in which the quantum mechanical regime is reached and multidimensional wave-function description can no longer be avoided. In all these applications, the computational "bottle neck" is associated with solving an inhomogeneous wave equation of the type

$$\hat{A}\chi = \phi$$

where $\hat{A}$ is a second order, multidimensional (partial) differential operator, and ϕ contains the (known) asymptotic scattering wave-function multiplied by the appropriate boundary operator[7, 8]. These are then discretized on a spatial grid using high order finite difference schemes for compactness [9, 10], resulting in matrices of the generic form[4–6]

$$\mathcal{A} = \sum_{l=1}^{n} \mathcal{A}_l - \mathcal{V}, \ \mathcal{A}_l = \mathbf{I}_1 \otimes \mathbf{I}_2 \otimes \cdots \otimes \mathbf{I}_{l-1} \otimes \mathbf{A}_l \otimes \mathbf{I}_{l+1} \otimes \cdots \otimes \mathbf{I}_n, \tag{1}$$

where n is the dimensionality of the problem (the number of coupled degrees of freedom), $\{\mathbf{A}_l\}$ are complex symmetric (non Hermitian) matrices of order N_l representing one-dimensional (one coordinate) operators, and $\mathcal{V}$ is a diagonal matrix which represents interaction between different degrees of freedom. The matrices $\mathcal{A}_l$ are essentially block diagonal in their appropriate basis, and their combined structure in a selected basis is highly sparse which makes them suitable for the use of efficient iterative solvers. Note that as the computational effort and memory storage grow exponentially with the number of dimensions, the range of problems we could tackle on sequential machines is quite restricted. However, on massively parallel supercomputers such as the IBM SP2, with potentially thousands of processors and Tera bytes of memory, problems which could not have been realized so far, are becoming feasible. We propose an efficient parallel algorithm for solving multidimensional Helmholtz/Schroedinger equations for both time-independent and time-dependent problems. At the heart of the algorithm is a new parallel preconditioner[4–6]. Note that standard preconditioner techniques, such as Jacobi, SOR, SSOR and ILU's [11] are not usually amenable for massively parallel implementation. The new preconditioner on the other hand, is fully scalable, both in terms of memory storage and arithmetic operations, and offers linear speedup with increasing number of processors. Examples and applications, illustrating the applicability of the approach and the speedup (obtained on the IBM SP2 with 64 processors) are given below.

2 Methodology.

We can write the linear system (see (1)), as

$$\left(\sum_{l=1}^{N} (\mathcal{A}_l + \alpha_l \mathbf{I}_l) - (\mathcal{V} + \alpha \mathbf{I})\right) \chi = \phi, \ \left(\sum_{l=1}^{N} \mathcal{A}'_l - \mathcal{V}'\right) \chi = \phi. \tag{2}$$

The parallel preconditioner is based on an explicit formula for the inverse of the separable part of $\mathcal{A}$, i.e.,

$$\mathcal{G} = \left(\sum_{l=1}^{N} \mathcal{A}'_l\right)^{-1}.$$

Let $\mathbf{A}'_l = \mathbf{U}_l \mathbf{D}_l \mathbf{W}_l$ denote the spectral decomposition of the respective dense (high order finite difference) matrices as in (1). Then, $\mathcal{A}'_l = \mathcal{U}_l \mathcal{D}_l \mathcal{W}_l$ with

$$\begin{aligned}
\mathcal{U}_l &= \mathbf{I}_1 \otimes \mathbf{I}_2 \otimes \cdots \otimes \mathbf{I}_{l-1} \otimes \mathbf{U}_l \otimes \mathbf{I}_{l+1} \otimes \cdots \otimes \mathbf{I}_n, \\
\mathcal{D}_l &= \mathbf{I}_1 \otimes \mathbf{I}_2 \otimes \cdots \otimes \mathbf{I}_{l-1} \otimes \mathbf{D}_l \otimes \mathbf{I}_{l+1} \otimes \cdots \otimes \mathbf{I}_n, \\
\mathcal{W}_l &= \mathbf{I}_1 \otimes \mathbf{I}_2 \otimes \cdots \otimes \mathbf{I}_{l-1} \otimes \mathbf{W}_l \otimes \mathbf{I}_{l+1} \otimes \cdots \otimes \mathbf{I}_n,
\end{aligned}$$

and

$$\mathcal{G} = (\prod_{l=n}^{1} \mathcal{W}_l^{-1})\mathcal{D}^{-1}(\prod_{l=1}^{n} \mathcal{U}_l^{-1}), \ \mathcal{D}^{-1} = (\sum_{l=1}^{n} \mathcal{D}_l)^{-1}.$$

We then apply the *Green function* $\mathcal{G}$ to the linear system (2) to obtain

$$(\mathcal{I} - \mathcal{G}\mathcal{V})\chi = (\mathcal{I} - \mathcal{H})\chi = \varphi = \mathcal{G}\phi.$$

where the norm of $\mathcal{H}$ could be made relatively small by the proper choice of the weight factors α_l in (2). We can then solve the sparse system using any efficient Krylov or Lancsoz based iterative methods [11–13].

Here we consider the application of the above algorithm on massively parallel machines such as the IBM SP2, where independent processors can exchange information via a fast interconnection network. There are two stages to the algorithm:

- Spectral factorizations of the block $\mathcal{A}_l$, distribution of the coefficients among the processors, and the construction of the initial modified system as above.
- The gradual iterative solution of the linear system by the parallel application of the sparse routine.

The diagonalization of the independent complex symmetric (non Hermitian) matrices $\mathcal{A}_l$ can be carried out on different processors in parallel. Moreover, we have shown recently[14] that one can diagonalize complex symmetric matrices rather efficiently, and since $N_l \ll N = \Pi N_l$ this stage typically becomes computationally negligible. What remains is therefore to form the linear system and apply the sparse solver, which boils down to the efficient implementation of the matrix vector product, e.g.,

$$\varphi = \mathcal{G}\phi = (\prod_{l=n}^{1} \mathcal{W}_l^{-1})\mathcal{D}^{-1}(\prod_{l=1}^{n} \mathcal{U}_l^{-1})\phi.$$

For that purpose we propose the following multidimensional matrix vector product algorithm[6]. We envision the vector ϕ of order $N = \Pi_{l=1}^{n} N_l$, as a multidimensional cube of order n, with sides N_l. We similarly envision the system of parallel processors as a cube of order n, with sides P_l such that $\Pi_{l=1}^{n} P_l = P$ is the number of processors in the system. Hence, the vector ϕ could be mapped into the smaller processor cube in such a way that each processor gets a corresponding sub block of order $M = \Pi_{l=1}^{n} M_l, M_l = N_l/P_l$. The actual application of the matrix product is then carried out in parallel on each cube dimension: Each matrix $\mathcal{W}_l$ or $\mathcal{U}_l$ is shuffled to a block diagonal form[6] and the block matrix multiplication is done using the standard ring matrix vector algorithm [15]. The communication is each step is limited to P_l processors associated with the lth dimension. We note that in this way we can essentially eliminate the communication overhead time and obtain a practical linear time algorithm, scalable to any number of processors. For example, on a perfect hypercube in which $P_l = P/n$, the communication time is reduced by a factor of $P/\log(P)$.

3 Applications

As an application we first consider an electron wave in a rectangular electron wave guide ("quantum wire") in which the electron motion is quantized along the lateral (x, y) directions. Along the propagation (z) direction the free electrons are scattered over a barrier associated with an abrupt change of the Fermi energy. This problem is identical to the problem of light distribution near a discontinuity in a rectangular optical waveguide which was studied in ref.([4]). A two dimensional cut through the three dimensional wave-function is plotted in Fig.1. Figure 2 illustrates the convergence of the calculated wave function

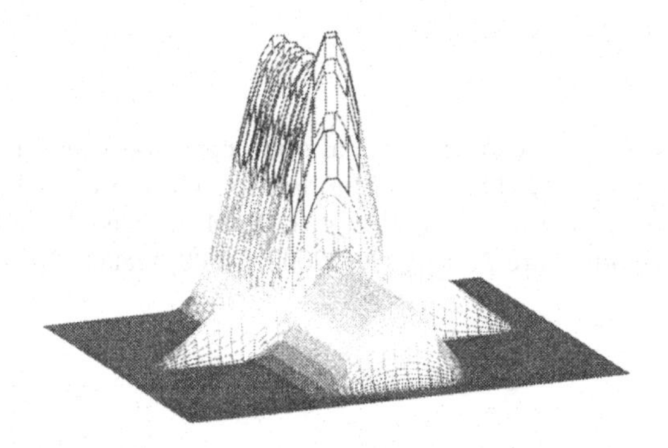

Fig. 1. A cut through the three dimensional stationary wave function distribution $|\Psi(x, y_0, z)|^2$ near a waveguide discontinuity $(y_0 = 0)$, illustrating transmission, reflection and leaking of the incoming wave due to scattering

near the discontinuity area as the number of processors increases. Convergence is obtained when the volume in which the wave function is sampled is sufficiently large so that the error due to the absorbing boundaries vanishes in the physically interesting area. However, as the accuracy of the solution improves with increasing number of processors (sampling volume) the computational task increases as well. Nonetheless, this is compensated by a linear speedup in the computational rate as illustrated in Figure 3. The Mflops per a single iteration of the solver (the QMR algorithm) increases linearly with up to 64 processors on the IBM-SP2 machine.

As a second application, we consider the calculation of electron transmission through thin molecular films. In particular, we consider the tunneling of electron currents between two metal plates (electrodes) separated by a narrow (≈ 10 Å) water potential barrier. Figure 4 illustrates a 2D cuts through the three dimensional effective potential energy surface. The details of the effective potential can be found in refs.[16, 17]. A flux of electrons with energy below the potential barrier is typically represented by a probability wavefunction as shown in figure 5. The wavefuction accumulates in front of the water potential barrier, and only residual (insignificant) tunneling current is transmitted between the electrodes in this case

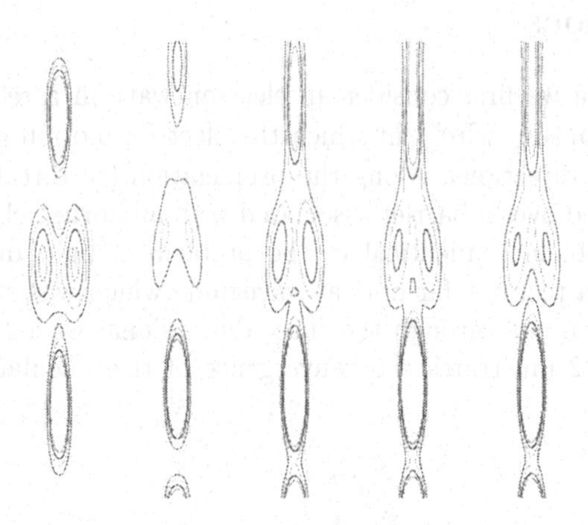

Fig. 2. Contour plots of the three dimensional stationary wave function distribution in the (x, z) plane $(|\Psi(x,0,z)|^2)$. The plots from left to right represent the convergence due to increasing number of processors, $P = 4, 8, 16, 24, 32$, where the corresponding grid sizes (sampling volumes) are $N = 16000, 32000, 64000, 128000, 256000, 512000$

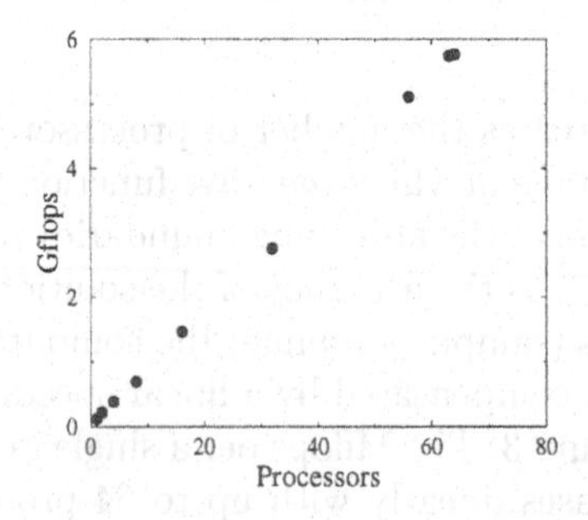

Fig. 3. Flops vs. processors, illustrating a linear speedup of the solver with increasing number of processors. Flops were averaged over 20 iterations of the parallel QMR algorithm

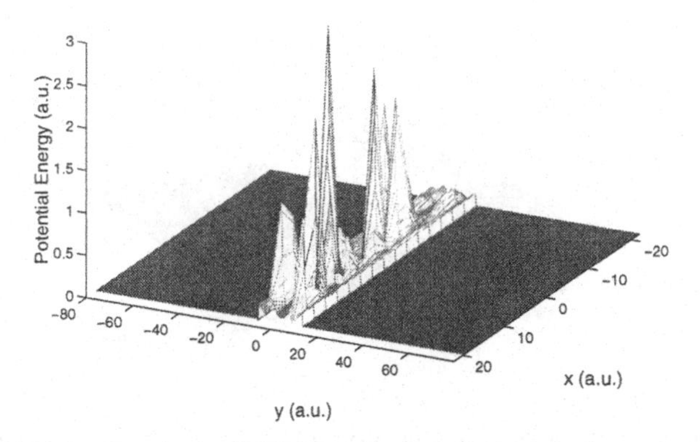

Fig. 4. A 2D cut through an effective three dimensional potential energy surface for an electron in the vicinity of a water barrier ("noisy" area) between two metal electrodes

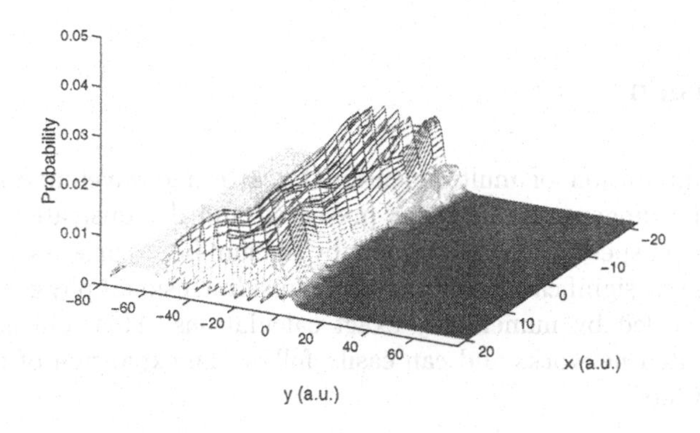

Fig. 5. A 2D cut through a stationary electron wavefunction representing an electron flux (propagating from left to right along the y direction, with impact energy 4.0 eV)

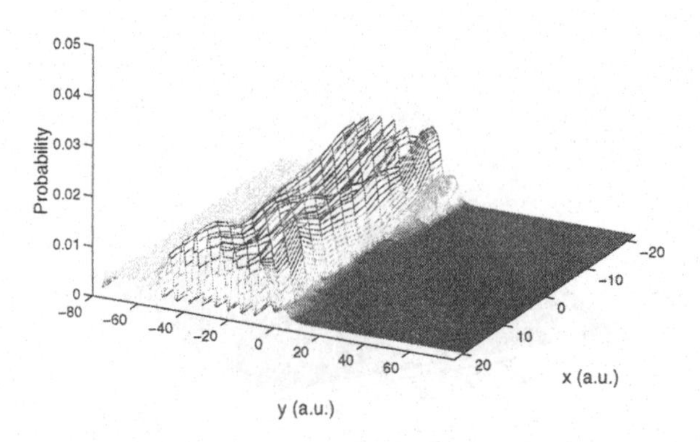

Fig. 6. The same as fig.5 with a slightly different impact energy (4.4 eV)

Figure 6. represents a different wavefunction corresponding to a slightly different electron energy, still below the potential barrier. Here a totally different behavior is illustrated where the electron probability density is strongly localized inside the barrier. This localization due to a quantum resonance of the water barrier results in a significant enhancement of the tunneling currents and illustrates the importance of the fine molecular structure on the electron transmission probability.

4 Conclusions

A parallel computation of multidimensional scattering wavefunctions for the Helmholtz/Schrodinger equations was presented and demonstrated. The algorithm is limited in speed and memory only by the number of processors available, and thus increases significantly the range of multidimensional physical problems that can be tackled by numerically exact calculations. This tool is appropriate for distributed networks and can easily follow the expansion of distributed parallel computers.

5 Acknowledgments

This research was supported by the Israel Science Ministry, by the Israel Science Foundation founded by the Israel Academy of Sciences and Humanities, and by the foundation for promotion of research at the Technion. Å. E. acknowledges support by the Göran Gustafsson and the Goldshmidt Foundations. U.P. is an Yigal Alon Fellow.

References

1. Peskin U. and Miller W. H.: Reactive scattering theory for field induced transition. J. Chem. Phys. **102**, (1995) 4084 .
2. Peskin U., Miller W. H. and Edlund Å.: Quantum time-evolution in time-dependent fields and time-independent reactive-scattering calculations via an efficient Fourier grid preconditioner. J. Chem. Phys. **103**, (1995) 10030 .
3. Vorobeichik I., Moiseyev N., Neuhauser D., Orenstein M. and Peskin U.: Calculation of light distribution in optical devices by a global solution of an inhomogeneous scalar wave equation. IEEE J. Quant. Elec. **33** (1997) 1236.
4. Edlund Å., Vorobeichik I. and Peskin U.: High order perturbation theory for Helmholtz/Schrödinger equations via a separable preconditioner. J. Comput. Phys. **138** (1997) 788-800.
5. Edlund Å. and Peskin U.: A parallel Green's operator for multidimensional quantum scattering calculations". Int. J. of Quant. Chem. (to appear).
6. Bar-On I., Edlund Å. and Peskin U.: A fully distributed parallel solver for multidimensional Helmholtz/Schrödinger equations. (submitted).
7. Neuhauser D.: State-to-state reactive scattering amplitudes from single-arrangement propagation with absorbing potentials. Chem. Phys. Lett.**200** (1990) 7836.
8. Seideman T. and Miller W. H.: Calculation of the cumulative reaction probability via a discrete representation with absorbing boundary conditions. J. Chem. Phys. **96** (1992) 4412.
9. Colbert D. T. and Miller W. H.: A novel discrete variable representation for quantum mechanical reactive scattering via the S-matrix Kohn method. J. Chem. Phys. **96** (1992) 1982.
10. Kosloff R.: in Numerical Grid Methods and Their Application to Schrödinger's Equation. NATO ASI Series C 412, edited by C. Cerjan (Kluwer, Academic, 1993).
11. Saad Y.: Iterative methods for sparse linear systems. PWS Publishing, (1996).
12. Ratowsky R. P., Fleck Jr. J. A., and Feit M. D.: Helmholtz beam propagation in rib waveguides and couplers by iterative Lanczos reduction. Opt. Soc. Am. A. **9** (1992) 265 .
13. Freund R. W., Golub G. H. and Nachtigal N.: Iterative Solution of linear systems. Acta Numerica, 1-44, (1992).
14. Bar-On I. and Ryaboy V.: Fast Diagonalization of Large and Dense Complex Symmetric Matrices, with Applications to Quantum Reaction Dynamics. SIAM J. of Scientific and Stat. Comp. **28** (1997), 5.
15. G. H. Golub and C. f. Van Loan *Matrix Computations* (The John Hopkins University Press, 1989).
16. Mosyak A., Graf P., Benjamin I. and Nitzan A., J. Phys. Chem. (to appear).
17. Benjamin I., Evans D. and Nitzan A.: Electron tunneling through water layers: Effect of layer structure and thickness. J. Chem. Phys. **106**(1997), 6647-6654 .

New Serial and Parallel Recursive *QR* Factorization Algorithms for SMP Systems

Erik Elmroth[1] and Fred Gustavson[2]

[1] Department of Computing Science and HPC2N, Umeå University, S–901 87 Umeå, Sweden.
`elmroth@cs.umu.se`

[2] IBM T.J. Watson Research Center, P.O. Box 218, Yorktown Heights, NY 10598, U.S.A.
`gustav@watson.ibm.com`

Abstract. We present a new recursive algorithm for the QR factorization of an m by n matrix A. The recursion leads to an automatic variable blocking that allow us to replace a level 2 part in a standard block algorithm by level 3 operations. However, there are some additional costs for performing the updates which prohibits the efficient use of the recursion for large n. This obstacle is overcome by using a hybrid recursive algorithm that outperforms the LAPACK algorithm DGEQRF by 78% to 21% as $m = n$ increases from 100 to 1000. A successful parallel implementation on a PowerPC 604 based IBM SMP node based on dynamic load balancing is presented. For 2, 3, 4 processors and $m = n = 2000$ it shows speedups of 1.96, 2.99, and 3.92 compared to our uniprocessor algorithm.

1 Introduction

LAPACK algorithm DGEQRF requires more floating point operations than LAPACK algorithm DGEQR2, see [1]. Yet, DGEQRF outperforms DGEQR2 on a RS/6000 workstation by nearly a factor of 3 on large matrices. Dongarra, Kaufman and Hammarling, in [7], and later Bishof and Van Loan, in [3], and still later, Schreiber and Van Loan, in [9], demonstrated why this is possible by aggregating the Householder transforms before applying them to a matrix C. The result of [3] and [9] was the k way aggregating WY Householder transform, and the k way aggregating storage efficient Householder transform. In the latter, the WY representation of $Q = I - YTY^T$. Here lower trapezoidal Y is m by k and upper triangular T is k by k.

Our recursive algorithm, RGEQR3, starts with a block size k of 1 and doubles k in each step. If we would allow this to continue for a large number of columns the performance would eventually degrade as the additional floating point operations (FLOPS) grow cubically in k. Thus, to avoid this occurring, RGEQR3 should not be used until $k = n/2$. Instead, we propose to use a hybrid recursive algorithm, RGEQRF, which is a slight modification of a standard level 3 block algorithm in that it calls RGEQR3 for factorizing block columns. Hence, RGEQRF is a modified version of Alg. 4.2 of Bishof and Van Loan, in [3] and RGEQR3 is a recursive level 3 counterpart of their Alg. 4.1.

In Section 2 we describe our new recursive serial algorithms RGEQR3 and RGEQRF and compare their performance with LAPACK's serial algorithms DGEQR2 and DGEQRF. In Section 3 we describe our new parallel recursive algorithm. Section 3.1 shows performance results of near perfect speedups for large matrices of size m = n = 2000. We also describe an anomaly (10% performance loss on one processor) with the `!SMP$ do parallel` construct.

2 Recursive QR Factorization

In our recursive algorithm, the QR factorization of an $m \times n$ matrix A

$$\begin{pmatrix} A_{11} & A_{12} \\ A_{21} & A_{22} \end{pmatrix} = Q \begin{pmatrix} R_{11} & R_{12} \\ 0 & R_{22} \end{pmatrix},$$

is initiated by a recursive factorization of the left-hand side $\lfloor n/2 \rfloor$ columns, i.e.,

$$Q_1 \begin{pmatrix} R_{11} \\ 0 \end{pmatrix} = \begin{pmatrix} A_{11} \\ A_{21} \end{pmatrix}.$$

The remaining part of the matrix is updated,

$$\begin{pmatrix} R_{12} \\ \tilde{A}_{22} \end{pmatrix} \longleftarrow Q_1^T \begin{pmatrix} A_{12} \\ A_{22} \end{pmatrix}, \tag{1}$$

and $\tilde{A}_{22}$ is recursively factorized into $Q_2 R_{22}$. The recursion stops when the matrix to be factorized consists of a single column. This column is factorized using an elementary Householder transformation. We use the storage efficient WY form, i.e., $Q = I - YTY^T$, to represent the orthogonal matrices, and the update by Q_1^T in (1) is performed as a series of matrix multiplications with general and triangular matrices (i.e., by calls to level 3 BLAS DGEMM and DTRMM).

In Figure 1, we give the details of the recursive algorithm, called RGEQR3.

RGEQR3 $A(1{:}m, 1{:}n) = (Y, R, T)$! Note, Y and R replaces A
if $(n = 1)$ **then**
  compute Householder transform $Q = I - \tau u u^T$ such that $QA = (x, 0)^T$
  return (u, x, τ) ! Note, $u = Y$, $x = R$, and $\tau = T$
else
  $n1 = n/2$ and $j1 = n1 + 1$
  call RGEQR3 $A(1{:}m, 1{:}n1) = (Y_1, R_1, T_1)$ where $Q_1 = I - Y_1 T_1 Y_1^T$
  compute $A(1{:}m, j1{:}n) = Q_1^T A(1{:}m, j1{:}n)$
  call RGEQR3 $A(j1{:}m, j1{:}n) = (Y_2, R_2, T_2)$ where $Q_2 = I - Y_2 T_2 Y_2^T$
  compute $T_3 = T(1{:}n1, j1{:}n) = -T_1 (Y_1^T Y_2) T_2$.
  set $Y = (Y_1, Y_2)$! Y is m by n unit lower trapezoidal
  return (Y, R, T), where $R = \begin{pmatrix} R_1 & A(1{:}n1, j1{:}n) \\ 0 & R_2 \end{pmatrix}$ and $T = \begin{pmatrix} T_1 & T_3 \\ 0 & T_2 \end{pmatrix}$
endif

Fig. 1. Recursive QR factorization routine RGEQR3.

We assume A is m by n where $m \geq n$. In the "else" clause there are two recursive calls, one on matrix $A(1{:}m, 1{:}n1)$, the other on matrix $A(j1{:}m, j1{:}n)$, and the computations $Q_1^T A(1{:}m, j1{:}n)$ and $-T_1(Y_1^T Y_2)T_2$. These two computations consist mostly of calls to either DGEMM or DTRMM. Our implementation of RGEQR3 is made in standard Fortran 77 which requires the explicit handling of the recursion.

In Figure 2 we give annotated descriptions of algorithms DGEQRF and DGEQR2 of LAPACK. See [1] for full details. The routine DGEQRF calls DGEQR2 which is a level 2 version of DGEQRF. In the annotation we have assumed $m \geq n$.

```
DGEQRF(m, n, A, τ, work)
do j = 1, n, nb      ! nb is the block size
   jb = min(n − j + 1, nb)
   call DGEQR2(m − j + 1, jb, A(j, j), τ(j) )
   if (j + jb .LE. n) then
      compute T(1:jb, 1:jb) in work via a call to DLARFT
      compute (I − YT^T Y^T)A(j:m, j + jb:n) using work and T via
      a call to DLARFB
   endif
enddo
```

```
DGEQR2(m, n, A, τ)
do j = 1, n
   compute Householder transform Q(j) = I − τuu^T such that
   Q(j)^T A(j:m, j) = (x, 0)^T via a call to DLARFG
   if (j .LT. n) then
      apply Q(j)^T to A(j:m, j + 1:n) from the left by calling DLARF
   endif
enddo
```

Fig. 2. DGEQRF and DGEQR2 of LAPACK.

2.1 Remarks on the recursive algorithm RGEQR3

The algorithm RGEQR3 requires more floating point operations than algorithm DGEQRF which requires more floating point operations than DGEQR2. Dongarra, Kaufman and Hammarling, in [7], showed how to increase performance by increasing the FLOP count when they aggregated two Householder transforms before they were applied to a matrix C. The computation they considered was

$$C = Q_1 Q_2 C, \tag{2}$$

where C is m by n and Q_1 and Q_2 are Householder matrices of the form $I - \tau_i u_i u_i^T$, $i = 1, 2$. Their idea was to better use high speed vector operations and thereby gain a decrease in execution time. Bishof and Van Loan, in [3], generalized (2) by using the *WY* transform. They represented the product of k Householder transforms $Q_i, i = 1, \ldots, k$, as

$$Q = Q_1 Q_2 \cdots Q_k = I - WY^T. \tag{3}$$

They used (3) to compute $Q^T C = C - YW^T C$. Later on, Schreiber and Van Loan, in [9], introduced a storage-efficient WY representation for Q:

$$Q = Q_1 Q_2 \cdots Q_k = I - YTY^T, \tag{4}$$

where T is an upper triangular k by k matrix. In both of these cases performance was enhanced by increasing the FLOP count. Here the idea was to replace the matrix-vector type computations by matrix-matrix type computations. The decrease in execution time occurred because the new code, despite requiring more floating point operations, made better use of the memory hierarchy.

In (4) Y is a trapezoidal matrix consisting of k consecutive Householder vectors, $u_i, i = 1, \ldots, k$. The first component of each u_i is one, where the one is implicit and not stored. These vectors are scaled with τ_i. For $k = 2$, the T of equation (4), is

$$T = \begin{pmatrix} \tau_1 & -\tau_1 u_1^T u_2 \tau_2 \\ 0 & \tau_2 \end{pmatrix}. \tag{5}$$

Suppose $k_1 + k_2 = k$ and T_1 and T_2 are the associated triangular matrices in (4). We have

$$Q = (Q_1 \cdots Q_{k_1})(Q_{k_1+1} \cdots Q_k) = (I - Y_1 T_1 Y_1^T)(I - Y_2 T_2 Y_2^T) = I - YTY^T, \tag{6}$$

where $Y = (Y_1, Y_2)$ is formed by concatenation. Thus a generalization of (5) is

$$T = \begin{pmatrix} T_1 & -T_1 Y_1^T Y_2 T_2 \\ 0 & T_2 \end{pmatrix}, \tag{7}$$

which is essentially a level 3 formulation of (5), (6).

Schreiber and Van Loan and LAPACK's DGEQRF compute (6) via a bordering technique consisting of a series of level 2 operations. For each $k_1 = 1, \ldots, k-1$, k_2 is chosen to be 1. However, as (6) and (7) suggests, $Q = I - YTY^T$ can be done recursively as a series of $k-1$ matrix-matrix computations. Also, the FLOP count in doing the T computation in (6) by this matrix-matrix computation is the same as the FLOP count of doing the bordering computation to compute T.

Algorithm RGEQR3 can be viewed as starting with with $k = 1$ and doubling k until $k = n/2$. If this doubling was allowed to continue performance would drastically degrade due to the cubically increasing FLOP count in the variable k. To avoid this occurring RGEQR3 should *not* be used for large n. Instead, the current algorithm, DGEQRF, should be revised and used with RGEQR3. In Figure 3 we give our hybrid recursive algorithm which we name RGEQRF. The routine DQTC applies $Q^T = I - YT^T Y^T$ to a matrix C as a series of DGEMM and DTRMM operations, i.e., $Q^T C = C - Y(T^T(Y^T C))$. Note that RGEQRF has no call to LAPACK routine DLARFT. The DLARFT routine computes the upper triangular matrix T via level 2 calls. Instead, routine RGEQR3 computes T via our matrix-matrix approach in addition to computing τ, Y and R.

RGEQRF($m, n, A, lda, \tau, work$)
do $j = 1, n, nb$! nb is the block size
$jb = \min(m - j + 1, nb)$
call RGEQR3 $A(j : m, j + jb - 1) = (Y, R, T)$! note T is stored in work
if ($j + jb$ .LE. n) **then**
compute $(I - YT^TY^T)A(j : m, j + jb : n)$ using *work* and T via a call to DQTC
endif
enddo

Fig. 3. Hybrid algorithm RGEQRF.

2.2 Uniprocessor performance results

Figures 4 and 5 show that RGEQRF outperforms DGEQRF of LAPACK on one processor by 20% for large matrices and up to 78% for small matrices. The gain in performance is partially explained by the increased performance of RGEQR3 compared to DGEQR2. Additional benefits occur because the faster RGEQR3 leads to an increased optimal block size in RGEQRF compared to DGEQRF, which in turn improves the performance of the level 3 BLAS updates in DQTC.

The performance results are obtained on a 4-way 112 MHz PowerPC 604 based IBM SMP High Node, using IBM XL Fortran 5.1.

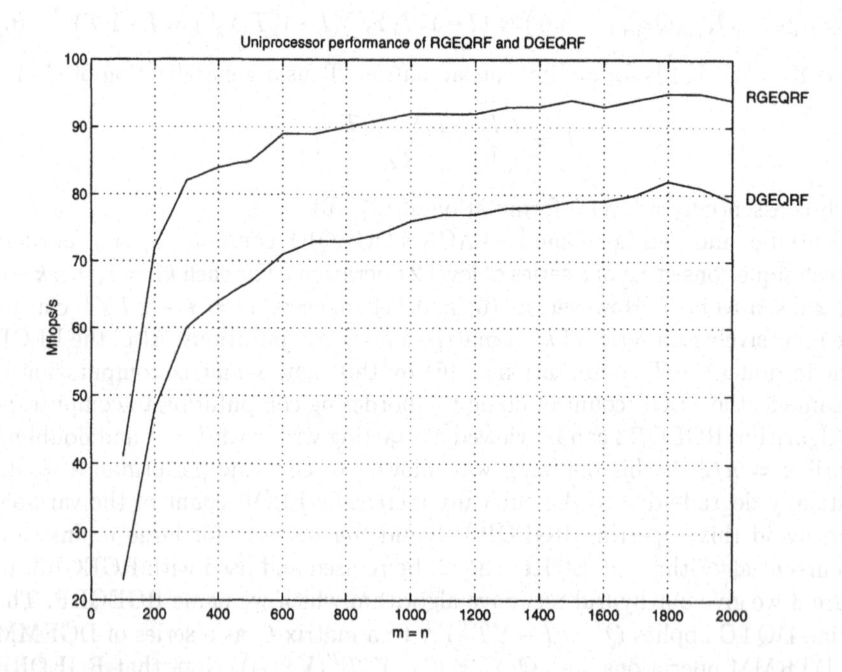

Fig. 4. Uniprocessor performance results in *Mflops* for the hybrid recursive algorithm RGEQRF and DGEQRF of LAPACK, using optimal block sizes 48 and 32, respectively.

$m=n$	100	200	300	400	500	600	700	800	900	1000	1500	2000
Ratio	1.78	1.50	1.37	1.31	1.27	1.25	1.22	1.23	1.23	1.21	1.19	1.19

Fig. 5. Ratio of uniprocessor performance results for RGEQRF and DGEQRF.

3 Parallel Hybrid Recursive QR Factorization

In the parallel version of RGEQRF, a "node program", NODEQR, is called once for each available processor, as shown in Alternative 1 in Figure 6. In the NODEQR routine, each processor repeatedly performs three operations. It finds a new block to update through a call to routine GETJOB, it updates that block through a call to routine DQTC, and it factorizes the block through a call to RGEQR3, if required. The processors returns from NODEQR when the *QR* factorization is completed. When using one processor, one might argue that Alternative 2 in Figure 6 should be equivalent to Alternative 1. However, as we will see in Section 3.1, the `parallel do` construction appears to introduce a significant amount of system overhead when run on one processor.

Alternative 1

```
!SMP$ do parallel
      do me = 1, NUMPROCS
         call NODEQR(me, A, ...)
      end do
```

Alternative 2

```
if (NUMPROCS .EQ. 1) then
   call NODEQR(me, A, ...)
end if
```

Fig. 6. Code segments for calling the routine NODEQR.

The critical part of the parallel *QR* factorization is done by the GETJOB routine. GETJOB implements a distributed version of the pool-of-tasks principle [4] for dynamically assigning work to the processors. The GETJOB algorithm is sketched in Figure 7, and is described in the next paragraph.

In the pool-of-tasks implementation only one critical section is entered per job (to find a new task, to tell what submatrix is reserved for this task, to update variables after the task is completed, etc). There are three variables associated with each processor that keep track of an iteration index and two indices equal to the first and last column the processor is working on. The size of the blocks to update depends on the size of the remaining matrix and the number of processors available. Near the end of the computation, a form of adaptive blocking is used [2]. As the remaining problem size decreases, both the number of processors and the sizes of the blocks being updated are decreased. A processor that will both update and factorize a block will only update the columns it will factor. There is no fixed synchronization in the algorithm; it is an asynchronous algorithm. This means that sometimes processors may be working on different iterations. Therefore more than one T matrix is needed. At least three T matrices are required to avoid contention. However, using 4 appears to further improve performance.

```
while (I have not yet found a new task, i.e., a new block to update)
    Enter Critical section
    If I did factor in my last task, update global variables
    If (remaining problem is too small for the current # processors) then
        Update global variables and terminate
    else
        Find the next matrix block to update
        Test if it is OK to start working on this block, i.e. test that:
            - no one is writing on any column in this block
            - the block I will read is computed
            - if I will factor: is it safe to overwrite one of the T matrices?
        If (it is OK to start working on this block) then
            Update global variables
        else
            Update variables to show that no block is reserved
        endif
    endif
    Leave Critical section
end while
```

Fig. 7. Algorithm GETJOB does the pool-of-tasks implementation.

3.1 Parallel Performance Results

As mentioned above, there seem to be a significant amount of system overhead introduced by the `parallel do` construction in Figure 6 (Alternative 1). In performance comparisons on one processor, Alternative 2 outperforms Alternative 1 by around 10% for large matrices. Since the `parallel do` construction has to be used for running on more than one processor, this is a bit discouraging, i.e., it suggests that $4 \times 90\%$ of the best uniprocessor performance is the best we can achieve on 4 processors. With the Alternative 2 approach, the parallel RGEQRF shows approximately the same performance as the serial variant, indicating that the overhead costs for additional blocking etc is negligible. The only significant extra cost in the parallel algorithm is associated with the 10% overhead of the `parallel do` construction.

Fortunately, it seems that the amount of overhead for the `parallel do` construction is reduced when the number of processors is increased. Performance results for 1 to 4 processors are shown in Table 1. *Mflops* are given for the Alternative 1 approach. The speedup is shown both relative to the uniprocessor results using the Alternative 1 approach, denoted S_1, and using the Alternative 2 approach, denoted S_2.

The speedup S_2 which is relative to the best uniprocessor implementation of the algorithm, i.e., using Alternative 2 in Figure 6, is almost optimal for large matrices. The fact that the speedup increases by more than one when adding another processor (from one to two and from two to three processors) indicates that the amount of overhead per processor is decreased when the number of processors is increased. We think the speedup S_1 (up to 4.29) is remarkable and the speedup S_2 (up to 3.92) is excellent.

Table 1. Parallel performance results and speedup for PRGEQRF.

	#proc = 1			#proc = 2			#proc = 3			#proc = 4		
$m = n$	*Mflops*	S_1	S_2	*Mflops*	S_1	S_2	*Mflops*	S_1	S_2	*Mflops*	S_1	S_2
100	40	*1.00*	*0.95*	55	*1.38*	*1.31*	55	*1.38*	*1.31*	54	*1.35*	*1.29*
200	65	*1.00*	*0.96*	99	*1.52*	*1.46*	126	*1.94*	*1.85*	139	*2.14*	*2.04*
300	74	*1.00*	*0.94*	128	*1.73*	*1.62*	170	*2.30*	*2.15*	201	*2.72*	*2.54*
400	77	*1.00*	*0.93*	145	*1.88*	*1.75*	194	*2.52*	*2.34*	235	*3.05*	*2.83*
500	78	*1.00*	*0.92*	149	*1.91*	*1.75*	209	*2.68*	*2.46*	255	*3.27*	*3.00*
600	81	*1.00*	*0.92*	161	*1.99*	*1.83*	227	*2.80*	*2.58*	288	*3.56*	*3.27*
700	81	*1.00*	*0.91*	166	*2.05*	*1.87*	238	*2.94*	*2.67*	307	*3.79*	*3.45*
800	82	*1.00*	*0.91*	169	*2.06*	*1.88*	248	*3.02*	*2.76*	318	*3.88*	*3.53*
900	83	*1.00*	*0.91*	173	*2.08*	*1.90*	253	*3.05*	*2.78*	330	*3.98*	*3.63*
1000	83	*1.00*	*0.90*	175	*2.11*	*1.90*	258	*3.11*	*2.80*	334	*4.02*	*3.63*
1100	83	*1.00*	*0.91*	178	*2.14*	*1.96*	262	*3.16*	*2.88*	341	*4.11*	*3.75*
1200	85	*1.00*	*0.92*	179	*2.11*	*1.95*	267	*3.14*	*2.90*	349	*4.11*	*3.79*
1300	84	*1.00*	*0.89*	181	*2.15*	*1.93*	268	*3.19*	*2.85*	352	*4.19*	*3.74*
1400	84	*1.00*	*0.88*	179	*2.13*	*1.88*	271	*3.23*	*2.85*	352	*4.19*	*3.71*
1500	84	*1.00*	*0.90*	183	*2.18*	*1.97*	273	*3.25*	*2.94*	357	*4.25*	*3.84*
1600	84	*1.00*	*0.90*	182	*2.17*	*1.96*	275	*3.27*	*2.96*	360	*4.29*	*3.87*
1700	85	*1.00*	*0.90*	183	*2.15*	*1.95*	275	*3.24*	*2.93*	362	*4.26*	*3.85*
1800	86	*1.00*	*0.91*	186	*2.16*	*1.96*	278	*3.23*	*2.93*	364	*4.23*	*3.83*
1900	85	*1.00*	*0.90*	183	*2.15*	*1.95*	279	*3.28*	*2.97*	365	*4.29*	*3.88*
2000	85	*1.00*	*0.91*	182	*2.14*	*1.96*	278	*3.27*	*2.99*	365	*4.29*	*3.92*

4 Conclusions

We have shown that a significant increase of performance can be obtained by replacing the level 2 algorithm in a LAPACK-style block algorithm for *QR* factorization by a recursive algorithm that essentially performs level 3 operations.

Recursion has previously been successfully applied to *LU* factorization [8, 10]. For the *LU* case, no extra FLOPS are introduced from recursion and the performance significantly exceeds that of LAPACK block algorithms. In this contribution we show that despite the extra flops, the hybrid *QR* factorization benefits even more from using recursion than the *LU* factorization does.

The parallel speedup compares well with previously published results for *QR* factorization routines for shared memory systems. The algorithm presented here show better speedup then what is presented for the IBM 3090 VF/600J in [6] and the Alliant fx2816 in [5] and it is similar to what is presented for IBM 3090 VF/600J in [5].

5 Acknowledgements

We thank Olov Gustavsson for sharing his experiences in parallel program development for the PowerPC 604 based IBM SMP High Nodes which, e.g., led to our decision to only use one critical section per iteration due to the overhead associated with the implementations of critical sections.

This research was conducted using the resources of High Performance Computing Center North (HPC2N).

References

1. E. Anderson, Z. Bai, C. Bischof, J. Demmel, J. Dongarra, J. Du Croz, A. Greenbaum, S. Hammarling, S. McKenney, S. Ostrouchov, and D. Sorensen. *LAPACK Users' Guide - Release 2.0.* SIAM, Philadelphia, 1994.
2. C. Bischof. Adaptive blocking in the QR factorization. *The Journal of Supercomputing*, 3:193–208, 1989.
3. C. Bischof and C. Van Loan. The *WY* representation for products of householder matrices. *SIAM J. Scientific and Statistical Computing*, 8(1):s2–s13, 1987.
4. A. Chalmers and J. Tidmus. *Practical Parallel Processing.* International Thomson Computer Press, UK, 1996.
5. K. Dackland, E. Elmroth, and B. Kågström. A ring-oriented approach for block matrix factorizations on shared and distributed memory architectures. In R. F. Sincovec et al, editor, *Proceedings of the Sixth SIAM Conference on Parallel Processing for Scientific Computing*, pages 330–338, Norfolk, 1993. SIAM Publications.
6. K. Dackland, E. Elmroth, B. Kågström, and C. Van Loan. Parallel block matrix factorizations on the shared memory multiprocessor IBM 3090 VF/600J. *International Journal of Supercomputer Applications*, 6(1):69–97, 1992.
7. J. Dongarra, L. Kaufman, and S. Hammarling. Squeezing the most out of eigenvalue solvers on high performance computers. *Lin. Alg. and its Applic.*, 77:113–136, 1986.
8. F. Gustavson. Recursion leads to automatic variable blocking for dense linear-algebra algorithms. *IBM Journal of Research and Development*, 41(6):737–755, 1997.
9. R. Schreiber and C. Van Loan. A storage efficient *WY* representation for products of householder transformations. *SIAM J. Scientific and Statistical Computing*, 10(1):53–57, 1989.
10. S. Toledo. Locality of reference in *LU* decomposition with partial pivoting. *SIAM J. Matrix. Anal. Appl.*, 18(4):1065–1081, 1997.

Visualization of CFD Computations

Jonas Engström

Parallelldatorcentrum (PDC),
Royal Institute of Technology, S-100 44 Stockholm, Sweden.
jonase@pdc.kth.se

Abstract. An application for visualization of results from 3D CFD computations has been developed, utilizing VR technology to improve the level of interaction with the target data set. The application displays geometry and solution data in a format used by a Navier-Stokes solver called VOLSOL developed by Volvo Aero Corp. By designing a visualization program specifically for a ccNUMA computer with powerful computation and graphics capabilities, a high level of interaction can be achieved in combination with fast traversal of large data sets.

1 Introduction

A software package called EnVis, or *Engine Visualizer* has been designed and implemented as part of a CFD project carried out in cooperation between Parallel and Scientific Computing Institute (PSCI), Volvo Aero Corporation (VAC) and PDC. The application is designed to do visualization of results from CFD computations and utilizes immersive VR technology and parallel computations to provide an interactive 3D visualization environment (an example is shown in Fig. 1) with capabilities to provide fast traversal of large data sets.

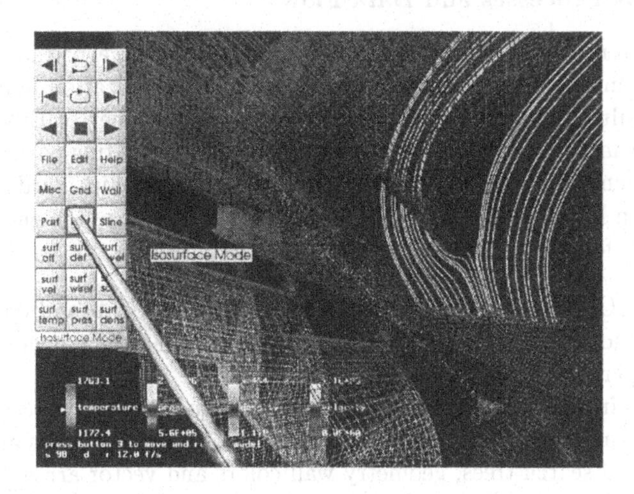

Fig. 1. Screen shot from EnVis.

2 Application Overview

The intention with EnVis is to provide basic visualization of large data sets while maintaining a high level of interaction and immersion. The application and its user interface is designed to run in a CAVE-like[1] environment [2]. EnVis supports data sets produced by a Navier-Stokes solver called *VOLSOL* [3], developed by VAC.

Intended users are people with large VOLSOL-generated data sets and access to high performance graphics computers. EnVis is designed to run on a high end ccNUMA[2]-architecture computer [1] computer (in this case, SGI), equipped with a projection-based VR system. The major application characteristics are:

VOLSOL Data Sets — EnVis uses VOLSOL grid, command- and solution files as input.

CAVE Operation and User Interface — all geometry viewing, visualization and user user interface operations are performed in a 3D projection-based virtual environment (real or simulated).

Solution Sequences — fast animation of solution file sequences along with associated visualization functions is the primary feature of EnVis.

Visualization — support for basic visualization functions such as *vector arrows*, *streamlines* and *traced particles*, colored with scalar values like temperature or pressure. In addition, *isosurfaces* can be rendered to show scalar value level boundaries.

More information about EnVis is available in *EnVis User's Guide* [5], *EnVis Implementation Reference* [4] and the full version of this paper [6].

2.1 EnVis Processes and Data Flow

For performance reasons, EnVis uses multiple processes, a private solution file format and multiple cache files for navigation and visualization functions.

The resulting data is then streamed through a number of EnVis processes which perform different tasks such as data importing, updating particle positions, isosurface generation and transfer of final data to graphics hardware. This is illustrated in Fig. 2 below. The data path from VOLSOL to the end user can be described as follows:

1. A VOLSOL computation produces solution data. Grid- and command files exists prior to execution and the process of creating them is out of scope for this paper.
2. EnVis is invoked once in file conversion mode to create preprocessed solution data suitable for fast importing. At the same time, cache files are generated for spatial search trees, geometry wall colors and vector arrow fields. These files are then used directly by all subsequent invocations of the program.

---

[1] CAVE Automated Virtual Environment

[2] Cache-coherent Non-Uniform Memory Access

3. When all data files are in place, EnVis creates the following processes:
   - An importer process which will take care of all data loading.
   - A number of render processes, which will perform all visualization computations not done by graphics hardware.
   - A number of display processes to handle user interaction and transfer of graphics primitives to hardware. The number of processes is equal to the number of walls in the CAVE.
   - The CAVE software, *CAVElib*, creates a number of additional processes for hardware interaction etc.

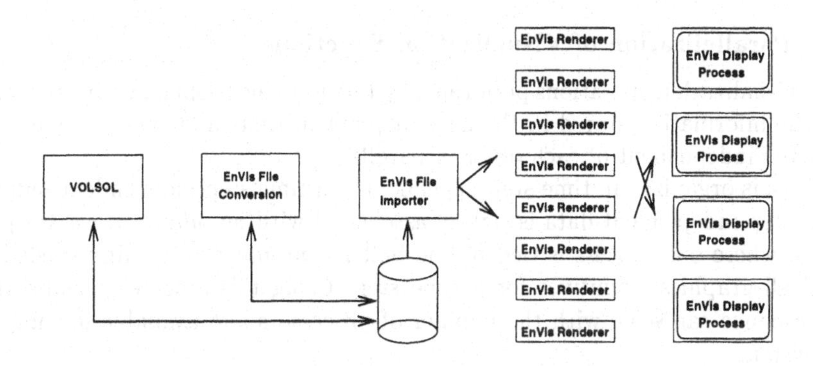

Fig. 2. EnVis processes and data flow.

2.2 Grid Coordinate Transformations

The standard problem of navigating from Cartesian (x, y, z) to grid coordinates describing the indices of a particular cell has been addressed using a common approach for 3D spatial search problems; an *oct-tree* is used to perform binary searching in three dimensions [8]. The leaves of the tree contains relatively short lists of intersecting or enclosed cells that have to be searched.

2.3 Data Organization for Graphics Performance

In order to keep the *redraw loop* [9] free from time-consuming operations such as computations of streamlines, particle positions, isosurfaces etc, as much time-consuming work as possible is done by other rendering processes, and the redraw loop is "served" buffers containing points, line segments and polygons.

An example of this is the implementation of *wall coloring* where the walls in the geometry are colored using adjacent temperature or pressure. This means that the coloring of the walls change every time step, but the positions of the colored polygons remain fixed. The last wall property is used in combination with utilization of the texture mapping hardware to perform wall coloring (using a one-to-one mapping between polygon colors and pixels in texture memory). By precomputing all polygon colors into a memory mapped cache file, the update procedure is reduced to a single API call to copy data into a region of texture memory.

3 Results

The major results of this project can be divided into 3D user interface development experiences and achieved visualization performance. This paper introduces the later by discussing data organization and parallelization.

3.1 Parallelization of Visualization Functions

Most visualization operations performed by EnVis are not dependent on anything but the information stored in the current point in time, and time steps can be rendered independent of each other in parallel.

Data is organized in time step-dependent and time step-independent buffers. Time step-independent data is tightly associated with *visualization worker* processes, where one process is tied to one buffer containing everything needed to build all graphics primitives for a time step. Using this scheme, visualization performance can scale with the number of processors and available memory in the system.

In order to maximize overall performance, rendering of time step-independent data is performed at the coarsest possible level using one process per time step. This organization gives efficient use of multiple CPUs and available memory bandwidth at the cost of larger memory requirements.

The other category of visualization functions use information from sequences of time steps to illustrate flow, typically by combining data from the previous time step with the current. This means that only a single time step can be rendered at a time due to data dependencies. Thus, parallelization must be implemented at a more fine-grained level. In order to achieve a high degree of parallelism provided that work can be distributed to a number of potentially idling processes, a job queue for partitioned render work has been implemented. The rendering process puts all units of work in the queue, and all idling processes assist the renderer by processing work in the queue.

An example of achieved total speedups when traversing a data set (1000 time steps, 8 GB total) with high numbers of visualization functions activated is shown in Table 1.

N_{proc}	2	3	4	5	6	7
Time step independent	1.5	2.0	2.5	2.9	3.4	3.8
Time step dependent	1.3	1.7	2.2	2.7	3.2	3.7

Table 1. Examples of speedup for different data dependencies.

4 Future Work

Future work consists mostly of a number usability and performance enhancements. The single most requested feature is support for a common data file format such as the widely used PLOT3D [7] format. The second most wanted enhancement is a 2D GUI version of EnVis which runs on more affordable computer systems, such as high end desktop workstations. Other plans include multi-user networked operation together with support for networked data input.

Support for multiple simultaneous models has been considered together with local parameter scopes (using small tool box gadgets) as methods to enhance application usability. The current visualization functionality can be extended with *Surface-Attached Streamlines Contour Surfaces* and *Clipping Planes*. A natural extension to the current particle tracing functions is *Trajectory Lines*, showing the path of emitted particles. Level-of-Detail (LOD) performance optimizations have the potential to increase rendering speed dramatically and have already been partially implemented.

5 Acknowledgments

The author would like to thank the following people for their valuable efforts during the implementation of this project: Anders Ålund, Nils Jönsson, Johan Ihrén, Kai-Mikael Jää-Aro, Francois Moyroud, Hans Mårtensson, Björn Torkelsson and Per Öster.

References

1. Laudon J., Lanoski D.: *The SGI Origin: A highly scalable server,* Silicon Graphics Inc, 2011 North Shoreline Boulevard, Moutain View, California 94043
2. Cruz-Neira, C., Sandin, D.J., and DeFanti, T.A.: *Surround-Screen Projection-Based Virtual Reality: The Design and Implementation of the CAVE,* In Proceedings of SIGGRAPH '93 Computer Graphics Conference, ACM SIGGRAPH, August 1993, pp. 135-142.
3. Rydén R., Groth, P.: *VOLSOL V2.6 User's guide,* Volvo Aero Corporation, Combustor Division, S-461 81 Trollhättan, Sweden
4. Engström, J.: *EnVis Implementation Reference,* Center for Parallel Computers, Royal Institute of Technology, S-100 44 Stockholm, Sweden
5. Engström, J.: *EnVis User's Guide*, Center for Parallel Computers, Royal Institute of Technology, S-100 44 Stockholm, Sweden
6. Engström, J.: *Visualization of CFD Computations,* Center for Parallel Computers, Royal Institute of Technology, S-100 44 Stockholm, Sweden
7. P.P. Walatka, P.G. Buning, L. Pierce, and P.A. Elson: *PLOT3D User's Manual,* NASA TM 101067, March 1990.
8. Warren M.S.: Salmon J.K.: *A Parallel Hashed Oct-Tree N-Body Algorithm* In Proceedings of Supercomputing '1993, Theoretical Astrophysics, Mail Stop B288, Los Alamos National Laboratory, Los Alamos, NM 87545
9. OpenGL Architecture Review Board: *OpenGL Reference Manual,* Chapter 2, Second Edition, Addison-Wesley Developers Press, ISBN 0-201-46140-4

Improving the Performance of Scientific Parallel Applications in a Cluster of Workstations*

A. Flores and J.M. García

Dept. de Ingeniería y Tecnología de Computadores, University of Murcia,
Campus de Espinardo s/n,
30080 Spain
{aflores, jmgarcia}@dif.um.es

Abstract. Recent improvements in LANs make network of workstations a good alternative to traditional parallel computers in some applications. However, in this platform the communication performance is over two orders of magnitude inferior to state-of-art multiprocessors. Currently networking technologies have put the pressure in the software overhead. Because of this, applications could not take advantage of this communication performance potential. In this paper, we present an implementation of Virtual Circuit Caching that reduces the software overhead by allocating communication resource once and re-using them for multiple messages to the same destinations. With this approach, the communication overhead is reduced by approximately a 35% for long messages; this reduction should enable the extensive use of networks of workstations for scientific parallel applications.

1 Introduction

Research in parallel computing has traditionally focused on multicomputers and shared memory multiprocessors. Currently, networks of workstations (NOWs) are being considered as a good alternative to parallel computers. That is due to there are high performance workstations with microprocessors that challenge custom-made architectures. This class of workstations is widely available at relatively low cost. Furthermore, these networks provide the wiring flexibility, scalability and incremental expansion capability required in this environment.

Communication performance at the application level depends on the collaboration of all components in the communication system, especially the network interface hardware and the low-level communication software that bridges the hardware and the application. In principle, scientific parallel applications should be coded in a portable message-passing library, as MPI [7] or PVM [6]. Usually, these libraries are based in the TCP/IP layer. Unfortunately, TCP/IP layer imposes large software overhead [11](several hundred to several thousand instructions per message). As a result, only embarrassingly parallel applications

* This work was supported in part by the Spanish CICYT under Grant TIC97-0897-C04-03

(that is, applications that almost never communicate) can make use of workstation clusters. To solve this problem, we can design better layers or we can improve the TCP/IP layer. In both cases, the final objective is the same, that is, improving the network interface rather than the network switches and links [3].

Currently, in the first approach there are new layers with reduced overheads (that is, on the scale of tens of instructions per message). Active Messages [5] is one of this. This layer offers simple, general-purpose communication primitives as a minimal layer over the raw hardware. However, it is not intended for direct use by application programmers and it is not portable. In fact, Active Messages provide low level services from which communication libraries can be built. The current prototype of Active Messages is the GAMMA project [2]. GAMMA exploits a more general Active Message-like model inspired to Thinking Machines' CMAML for the CM-5, where each process has a number of *communication ports* and may attach both a receiver handler and a data structure to a single communication port in order to handle all messages incoming through that port. This technique is suitable for MIMD as well as SPMD programming, as it does not require a common address space among cooperating processes.

The second approach is more interesting form point of view of portability. The main objective in this approach is to improve the network performance via to reduce the latency of the network. Traditionally, latency and throughput are the two parameters used to indicate the performance of an interconnection network. Although these parameters are very related, it is easier to increase the peak throughput -achievable for long messages-, but it is harder to reduce the latency -because the software overhead-. Recently, several works have been developed in this way as Pupa [12] or Beowulf [1]. These approaches have been implemented over off-the-shelf network hardware and they co-exist with legacy communication software such as TCP/IP.

Our research line is in the later approach. Nowadays, it does make sense from a technological as well as financial point of view to use off-the-shelf network hardware with existing software, usually based on TCP/IP. So, it is very important to study this protocol and try to reduce the large software overhead that it imposes. In this paper, we focus on analyzing the TCP/IP layer to detect the main software overhead points, and to solve this problem via the Virtual Circuit Caching technique. In this way, we can significantly reduce the communication latency in the network.

The rest of the paper is structured as follows. In the next section we analyze sources of software overhead in the communication system. In section 3, we describe the techniques used to reduce software communication overhead. The results of running test programs are analyzed in section 4. Finally, some conclusions and ways of future work are drawn.

2 Analysis Software Overhead Sources

As Local Area Networks improve the performance, their interfaces, resources, organization, and integration into their host computer become increasingly important. Recent papers have pointed out that announced performance for such as LANs could suffer a severe degradation due to overhead in communication software. Currently, with the most widespread use of ATM [8] in LANs and the announcement of Ethernet Gigabit standard [9], bottleneck studies and alternative solutions become essential.

To make an exhaustive study of software overhead, we must evaluate the communication channel performance at several layers in the protocol stack in order to identify bottlenecks. Using this technique we can measure the added overhead of each layer, the total software overhead and the network congestion.

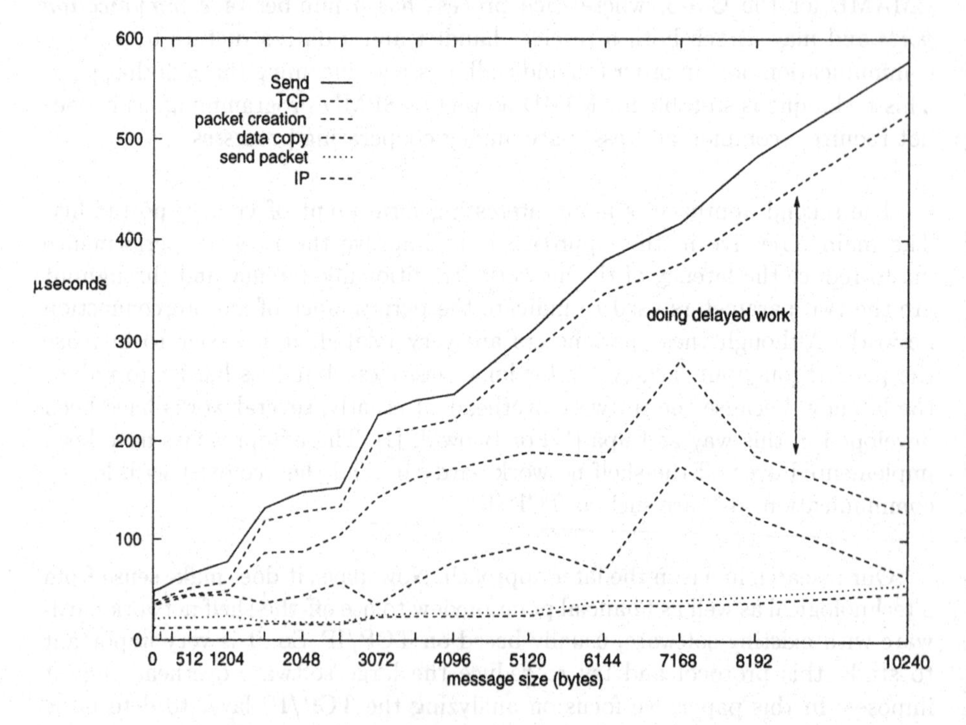

Fig. 1. Detailed study of software overhead in TCP/IP layer.

Figure 1 show software overhead breakdown in a TCP send call. Data was obtained take advantage of TCP/IP source code available in LINUX operating system. By introducing measurement instructions in appropriate points of TCP/IP implementation, we could obtain real overhead of critical sections of TCP/IP protocols. To obtain an accurate measurement we measure a portion of TCP/IP source code at once. The measurement overhead was estimate in

1μsecs. Measurement results were retrieve via an additional system call that we introduce for this purpose.

As it can be seen in the figure, the send latency is composed by various factors. This figure shows two important facts. Firstly, packet creation and data copy are the main causes of the send software overhead as expected. The second thing is that exists a significant difference between TCP layer overall overhead and the breakdown of this overhead. That is, the overhead obtained by adding the measured overhead of critical sections of TCP/IP communication layer. Although there are lines of code that are not taken into account, their impact in the overhead is very small. The real reason is the behavior of LINUX TCP/IP protocols with high workloads. In this case, most of the work is delayed in order to avoid saturation in the receiver side. This delayed work will be done as soon as possible, interrupting the normal program execution and introducing a software overhead that have been labeled as "*doing delayed work*" in figure 1.

Above, we show packet creation and data copy were the main causes of software overhead in a send system call. So, it should be possible to improve the network performance by reducing these two overhead sources. In the next section, we show an approach that minimizes these two overhead causes.

3 Latency Tolerance and Reduction Techniques

In previous section, we identified two main sources of software overhead in communication events. Namely, overhead due to allocation of resources and overhead due to data copy from/to user memory space.

The first factor can be greatly alleviated by implementing software techniques of resource reutilization. In [4], authors propose a new technique named Virtual Circuit Caching (VCC) that enable reduction in communication overheads by allocating communication resources once and re-using them form multiple messages to the same destination.

In the MPI/RT Standard Draft [10], is proposed a similar technique in buffer management, allowing to the operating system buffer re-use. This technique also allows zero copy data transfer at the expense of limited buffer length.

We propose a transparent software implementation that allows buffer reutilization at top level of protocol stack. Figure 2 show the basic idea of resource utilization. In this case, the resource to re-use is packet structure used by operating system.

The idea is the following. When a packet is sent, the packet structure (protocols headers and data space) is not freed but insert in a pool of free packets. In this way, they can be reused to hold new data, avoiding the overhead of handling the creation and deallocation of messages.

The second factor (software overhead due to data copy from/to user memory space) is more difficult to hide because the overhead can not be distributed among several messages. However, a precommunication technique can be applied in some cases obtaining good results. Next, we show an example of communication pattern when this technique can be applied.

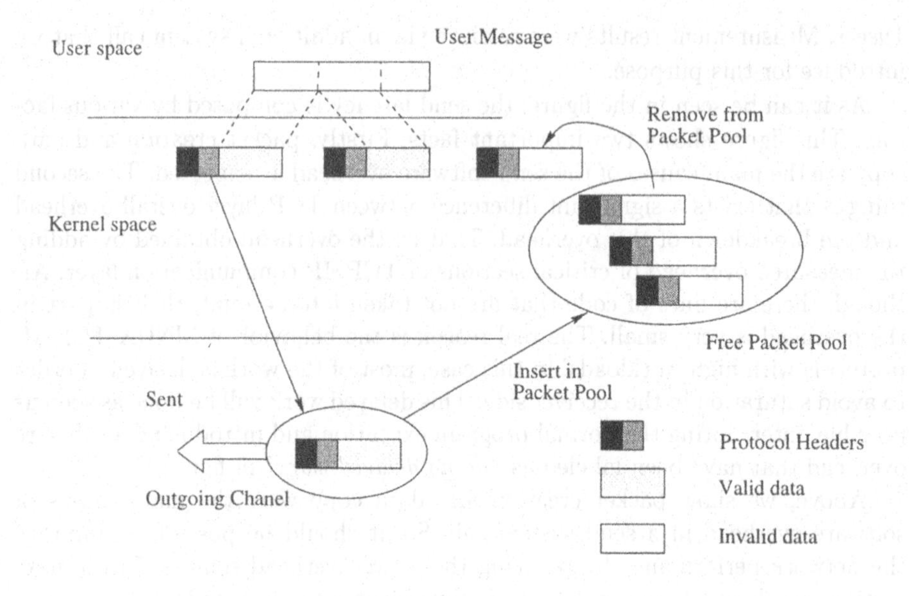

Fig. 2. Overview of VCC software implementation (Opaque to users).

Example of no dependence pattern

```
compute message A;
recv(token,length(token));  /* Waiting for permission.      */
send(dest,A,length(A));     /* Permission granted, send A. */
```

We have observed this communication pattern in programs with global communication events, such as barriers, where all processes wait in the barrier until last process arrive. We can use this idle time in order to process subsequent send events in the assumption of no data dependence. In order to deal with this communication pattern, we have include a new system call, named *indication*, that allows an earlier processing of the *send* system call. Then, the modified code with the precommunication technique is seen as follows:

Precommunication of message A

```
compute message A;
indication(dest,A,length(A)); /* There is a send with no      */
                              /* dependence.                  */
recv(token,length(token));    /* Waiting for permission;      */
                              /* begin to process A sending.  */
send(dest,A,length(A));       /* Permission granted, send A.  */
```

4 Result Evaluation

4.1 Hardware and Software platforms

We have performed our tests on a cluster of workstations with Intel Pentium 200Mhz processor, 32 Mbytes main memory, 256KBytes cache memory, 1GByte IDE hard disk and Fast Ethernet SMC EtherPower 10/100 adapter. The operating system used is Linux 2.0.32.

4.2 The test program

To evaluate performance parameters, we make use of Ping-Pong test. This test program is written in standard C. The best compiler option is always applied, unless otherwise noted. The test Ping-Pong is a simple echo between two adjacent nodes. A receiving node simply echoes back whatever it is sent, and the sending node measures round-trip time. When precommunication technique is used, we suppose no dependence between send and receive messages, although program order is maintain (a message is not sent until the process receives the matching message). Times are collected for some number of iterations (we have used 10000) over various messages sizes and after a transient period of 500 iterations. The minimal time, the maximal time and the mean time from all processes are collected. To interpret the results, we focus on the average time, because it is more representative of the performance the user can obtain from the machine.

4.3 Results

Figure 3 show software overhead for messages up to 5K. Results are obtain by subtract delay due to physical transmission from round-trip time. In the ideal case, hiding of almost all software overhead, results will be near to zero. Results are shown for three cases: original TCP/IP protocols, TCP/IP with VCC (Virtual Circuit Caching), and TCP/IP with precommunication and packet VCC.

From figure 3 several interesting conclusions follow. First of all, using a transparent technique such as packet reuse mechanism we reduce software overhead in a range that vary from 5-10% for short messages up to 15-35% for long ones.

Another remarkable fact is the software overhead reduction when using precommunication. For messages up to 1K, utilization of both techniques: precommunication and VCC packet show worse performance that packet reuse technique. This is due to new software overhead sources. Namely, *indication* system call and modifications introduced in *recv* system call to deal with precommunication. For longer messages, this overhead is overwhelmed with benefits of an earlier buffer copy.

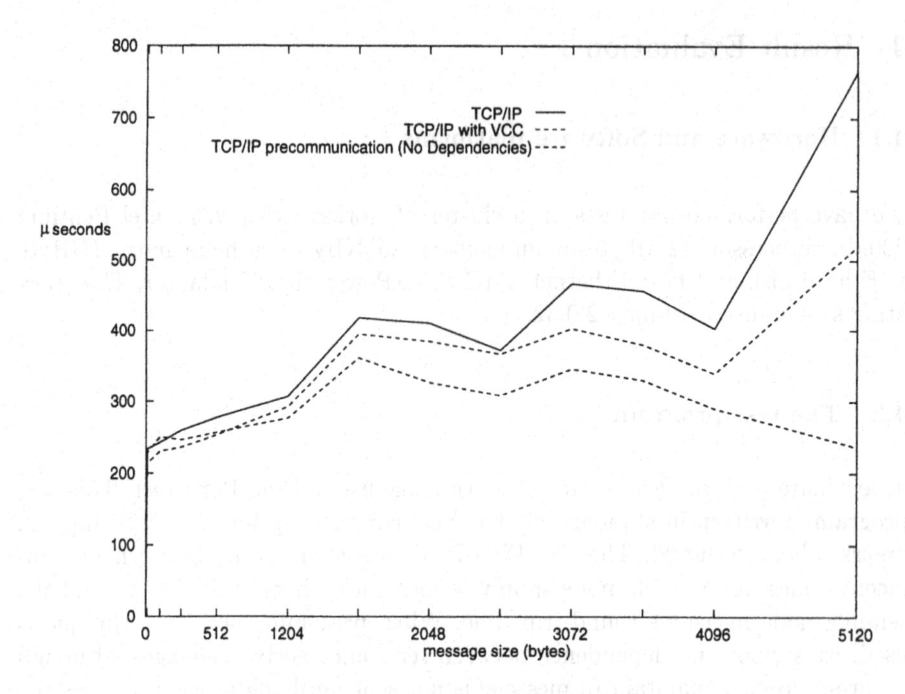

Fig. 3. Software overhead in TCP/IP protocol.

5 Conclusions and future work

The emergence of high-bandwidth, low-latency networks makes the use of workstation clusters attractive for parallel computing applications. Up to date, the main problem in this environment is the high latency compared to integrated custom multiprocessors.

In this paper, we have presented a software implementation of Virtual Circuit Caching (VCC) that enables reductions in communication overheads by allocating communication resource once and re-using them for multiple messages to the same destinations. The software overhead is reduced in a range that varies from 5-10% for short messages up to 15-35% for long ones. In same cases, further reduction in communication overheads can be achieved using stalls in process execution to process messages ahead in program order.

Our future work is focused on testing these techniques with usual scientific applications (FFT, LU, etc). Furthermore, we are currently examining the latest generation of ATM network interfaces and the Gigabit Ethernet standard to study the application of our techniques to these new interfaces.

References

1. D. Becker, T. Sterling, D. Savarese, J. Dorband, U. Ranawake, C. Packer: Beowulf: A Parallel Workstation for Scientific Computation. Procc. of Int. Conference on Parallel Processing, 1995.

2. G. Ciaccio: Optimal Communicationn Performance on Fast Ethernet with GAMMA. Procc. of the Workshop on Personal Computer Based Networks of Workstations, IPPS/SPDP, 1998.
3. D.E. Culler, L.T. Liu, R.P. Martin and C.O. Yoshikawa: Assessing Fast Network Interfaces. IEEE Micro, Vol. 16, No. 1, pp. 35-43, Feb. 1996.
4. B.V. Dao, S. Yalamanchili, and J. Duato: Architectural Support for Reducing Communication Overhead in Multiprocessor Interconnection Networks. Procc. of the Third Int. Symp. on High Performance Computer Architecture, February 1997.
5. T. Von Eicken *et al*: Active Messages: A Mechanism for Integrated Communication and Computation. Procc. of the 19th ISCA, pp. 256-266, May 1992.
6. A. Geist, A. Beguelin, J. Dongarra, W. Jiang, R. Manchek and V. Sunderam: PVM: Parallel Virtual Machine. The MIT Press, 1994.
7. W. Gropp, E. Lusk and A. Skjellum: Using MPI. The MIT Press, 1996.
8. R. Handel, M.N. Huber and S. Schroder: ATM Networks, Concepts, Protocols, Applications. Addison-Wesley 1994, 2/e.
9. Gigabit Ethernet Alliance. DRAFT Document for IEEE 802.3z Gigabit Ethernet Standard. *http://www.gigabit-ethernet.org.*
10. Real-time Message Passing Interface (MPI/RT) Forum: DRAFT Document for the Real-time Message Passing Interface (MPI/RT) Standard. *http://www.mpirt.org*, May 1998.
11. J. Piernas, A. Flores, J. M. García: Analyzing the Performance of MPI in a Cluster of Workstations base on Fast Ethernet. Procc. of 4th European PVM/MPI Users' Group Meeting, pp. 17-24, November 1997.
12. M. Verma, T. Chiueh: Pupa: A low Latency Communication System for Fast Ethernet. Procc. of the Workshop on Personal Computer Based Networks of Workstations, IPPS/SPDP, 1998.

On the Parallelisation of Non-linear Optimisation Algorithms for Ophthalmical Lens Design

Enric Fontdecaba Baig[1,2], José M. Cela Espín[1], and Juan C. Dürsteler Lopez[2]

[1] Universitat Politecnica de Catalunya.
{enricf, cela}@ac.upc.es
[2] Industrias de Óptica S.A.
Dus@indo.es

Abstract. This paper presents the parallelization of a non linear non constrained optimization code used in a industrial design, two different approaches are presented and the results of the comparison is shown.

1 Introduction

In this paper we will discuss an industrial design problem, we will show the difficulties encountered and why a parallel approach was needed. Furthermore the parallel algorithm will be described, and the performance obtained also will be presented.

Industrias de Optica S.A. is the biggest Spanish lens manufacturer, the flagship product of the company is the progressive lens. This kind of lens is used to compensate the presbiopya, resulting from the aging of the eye. This product is growing its market share and becoming the most important product for the ophthalmic lens manufacturers.

A progressive lens has three different vision zones. With the first zone the user can see distant objects. With the second zone the user can see at intermediate distances, this is due to a progressive change of optical power. The last zone is used for near vision. It is known that there is no analytical solution that gives the best possible progressive lens, so it is mandatory to use an optimization algorithm [2].

In addition to these three zones, used in phoveal vision, there is a fourth zone, the lateral zone. All the effort in the optimisation process is devoted in reducing the astigmatism in this zone, improving the overall lens performance. In figure 1 the different zones can be observed.

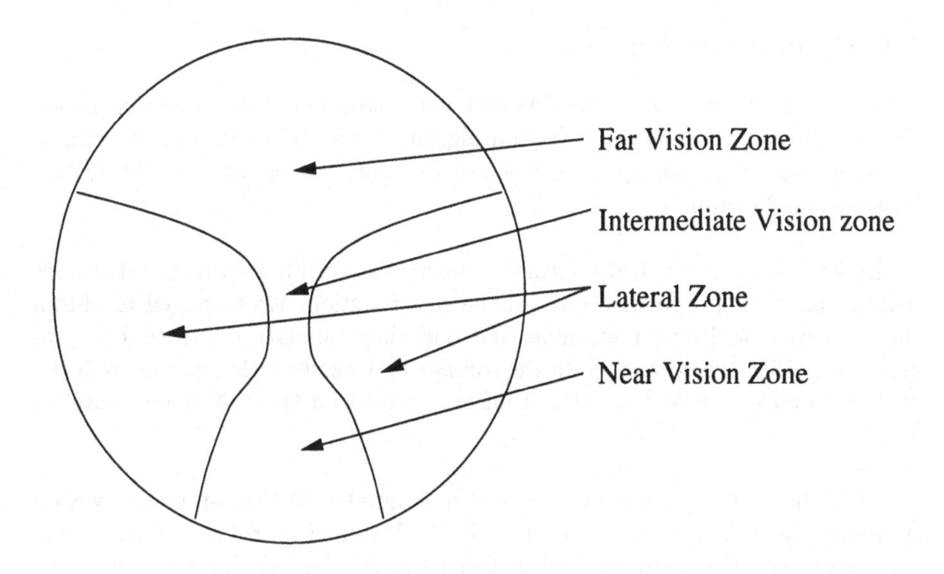

Fig. 1. Progressive Addition Lens vision zones

In the Progressive Addition Lens design process, it is necessary to optimize the lens surface for every set of design parameters (ophthalmic power distribution). This is an iterative process, so it is very important to use the fastest possible algorithm. This is the motivation that led us to a parallel approach.

2 Mathematical approach

2.1 Lens Modeling

The issue in this optimisation problem is to build an appropiate surface, thus a surface modeling tool is needed. We use B-Splines as the basis to model the optical surface. With this basis, two interesting properties are achieved:

Local control. When moving a control point of a B-Splines model, only the neighbouring zones are modified. A control point of the surface is a degree of freedom in our optimisation problem. The local control behaviour leads to a quasi-regular sparse structure in the hessian matrix of the objective function.

C^2 **class surface.** The B-Splines are twice derivable. This property is needed in order to calculate the surface curvature. The optical power and the astigmatism, the lens properties measured, are related to the surface curvature. The computation of this curvature is the source of the non-linearity.

The B-Splines modelling allows enough flexibility. Also, as our algorithms improve, it is possible to obtain a more accurate model increasing the number of control points [3].

2.2 Optimization Algorithm

After a complete review of the different non linear non constrained optimisation algorithms available, the Newton algorithm was selected. The algorithms reviewed were, Polytope Method, Steepest Descent, Conjugate Gradient, Quasi-Newton and Newton.

In the numerical test it was clear the enormous possible gain we could achieve with an analytical derivative form of the cost function. We managed to obtain the derivative with exact formulas and had those analytical expressions programmed. With this modification the computing time for a Hessian was reduced enough to make the Newton algorithm preferable to a Quasi-Newton approach [6] [4].

In further developments we managed to compute the Hessian matrix with a simplified finite diferences approach, taking advantage of the special cost function structure. This modifications reduced the Hessian computing cost to the equivalence of computing five gradients.

3 Parallelization approaches

The targeted platform was a workstation cluster, so we choose PVM as the message passing environment for the new application [8].

In order to start the parallelization, a profile of the sequential version was performed. As a result of this profile it was clear that the biggest part of the CPU time was spent on building the Hessian. Those routines where the first ones to be parallelized. With the first parallel program, performance measurements were done to study its behaviour. We used the analysis tools available on the CEPBA (European Center for Parallel Computing of Barcelona) [7], the Dimemas and Paraver tools, to perform those tests.

3.1 Objective Function Parallelization

The numerical test showed that a very important part of the calculation time was spent in computing the objective function. Furthermore, the most important part is the Hessian computation. So, the first parallel approach faced the reduction of this time.

The Hessian is computed by finite differences of the gradient. In order to improve the performance, an analytical gradient routine was implemented. It is notable that symbolical mathematical packages like Mathematica or Maple failed to compute this analytic derivative.

In order to obtain a finite difference Hessian approach, it is necessary to calculate $n+1$ (n is the problem dimension) function gradients. Those calculations are independent, so they are splitt among the different available processors. A master-slave approach is used. The other computations needed by the algorithm, the linear search and the linear equations system, are computed by the master. In table 1 the speed-up results of different problem sizes are shown. The tests were performed for 2,4,8,12 and 16 processors in order to study the algorithm scalability.

As we mentioned before, the Hessian is computed with a simplified finite difference scheme. In this scheme we take advantage of the special form of the objective function. This approach can be parallelised in a similar way.

Studying the code and profiles, it was clear that as the function evaluation costs diminished due to its parallelisation, the linear solver costs became more important in the total computing time. The traces obtained in our performance analysis tool corroborate this conclusion. In order to improve the scalability, the parallelization of the linear system solver was decided upon.

Table 1. Speed-Up with Function Parallelisation

Dimension	2 Proc	4 Proc	8 Proc	12 Proc	16 Proc
70	1.69	2.63	3.75	3.38	3.95
140	1.85	3.16	5.00	6.13	6.96
390	1.92	3.56	6.12	8.10	9.18
1390	1.95	3.78	6.90	9.08	12.39

3.2 Linear Solver Parallelization

In order to achieve a better scalability we parallelized the linear solver. We used preconditioned Krylov subspace iterative methods as linear solvers (GMRES(m)). The selected preconditioners are a set of different Incomplete Factorizations. The parallelization of the linear solvers is based on a Domain Decomposition data distribution. [1] The main bottleneck of the linear solver is the solution of the sparse triangular linear system arising from the preconditioner.

The communication requirements of this operation depend on the block structure of the triangular factors. In order to minimize this bottleneck two strategies are used:

1. Control the fill-in at the block level with a different criteria than at the element level. Each new non-zero block implies a new communication between processors.
2. Perform a coloring of the domains which minimizes the fill-in at the block level and ensures the maximum parallelism.

Because the granularity of the Hessian assembly and the linear solver is quite different, we use a different number of processes in each phase. This means that additional communications are required to redistribute the data before and after the linear system solution phase. We must find for each problem size the optimum number of processes of each part in order to obtain the minimum execution time. In this way we can improve the scalability of the whole application.

The results are shown in table 2 and table 3. The results with the smaller data sets are not shown because due to their size they did not achieve any reasonable speed-up.

Table 2. Speed-Up with 2 processors for the Linear Solver

Dimension	2 Proc	4 Proc	8 Proc	12 Proc	16 Proc
390	1.84	2.91	4.08	4.63	4.20
1390	1.97	3.58	6.06	7.88	9.27

Table 3. Speed-Up with 4 processors for the Linear Solver

Dimension	2 Proc	4 Proc	8 Proc	12 Proc	16 Proc
390	1.73	2.58	3.45	4.02	3.34
1390	*No convergence*				

In comparisson with the results shown in table 1 this new parallelisation introduces a slow-down in the total solving time. All the parallel performace is obtained by the first approach.

We found no way to avoid this behaviour. In one hand, the condition number of the Hessian matrix is very high (even without any reordering) and this number rapidly increases with the problem size. In order to obtain convergence with any iterative method with this kind of matrix it is necessary to use a good preconditioner. A better preconditioner increases the filling on the Incomplete Factorisation used. A more dense matrix implies more communcation cost and longer execution time.

On the other hand, using several processors implies a data distribution, it is important to reorder the matrix in order to minimise the communication between processors but the reordeing affects the numerical stability of the decomposition.

Summing up, without any reordering in the Hessian matrix the computing time is too big, due to the communication cost. With reordering in the matrix the iterative solver does not converge.

So, we achieve no increases in speed in parallelising the linear solver. Analysing the results and the code, we find two reasons for this behaviour:

- As we use an iterative method, the number of iterations needed in order to solve the linear system is a key parameter. The parallelisation, involving a matrix reordering increased the number of iterations. Futhermore, in the bigger case (when we expected some performance improvements), the reordering affected the algorithm convergence in such a way that made it diverge.
- With the solver parallelisation, the number of communications is greatly increased. In the Hessian parallelisation there are two communications, at the beginning and the end of the parallel phase. With the linear solver, there are comunications in each linear solver iteration.

Summarising, the linear system involved in the optimisation algorithm is too small and too badly conditioned to be solved with a parallel iterative method.

4 Conclusion and Future Work

The speed-ups obtained are satisfactory for the industrial process. It is not expected to use more than 12 machines at the same time. In fact INDO is installing a network of 6 DEC Alpha workstation with a Fast Ethernet switch. Taking the previous results into account, with the targeted platform, the first parallel approach is the most suitable for the company.

It is also interesting to remark that the problems with the parallel linear solver. In our previous experience with linear systems from numerical simulations we have never found such a bad conditioned problem. In order to overcome this behaviour we are thinking about new reordering methods.

The future work includes an upgrade of the basic sequential algorithm, and the changes needed by this improved approach. This new approach could consist in a non-linear least square formulation of the problem.

5 Acknockledgements

We would like to thanks all the R & D staff of INDO for their support in the development of this project. Specially to Roberto Villuela for their contribution to the tests and in developing key parts of the code.

References

1. Cela, José M.; Alfonso, J. M.; Labarta, J.: *PLS: A Parallel Linear Solvers library for domain decomposition methods*: EUROPVM'96, Lecture Notes in Computer Science 1156, Springer-Verlag, 1996.
2. Dürsteler, Juan Carlos.: *Sistemas de Diseño de Lentes Progresivas Asistido por Ordenador.*: PhD Thesis. Universitat Politecnica de Catalunya, 1991.
3. Farin, Gerald.: *Curves and Surfaces for Computer Aided Geometric Design. A Practical Guide.*: Second Edition. Academic Press, 1990.
4. Gill, Philip E.; Murray, Walter & Wright, Margaret H.: *Practical Optimization.*: Academic Press, 1981.
5. Dennis & Schnabel.: *Numerical Methods for Unconstrained Optimization and Non Linear Equations.*: Prentice-Hall, 1983.
6. Nemhauser, G.L.; Rinnooy Kan, A.H.G. & Todd, M.J.: *Optimization.*: Elsevier Science Publishers B.V. 1989.
7. Labarta, J., Girona, S., Pillet, V., Cortes, T., Cela, J.M.: *A Parallel Program Developement Environment*: CEPBA/UPC Report No. RR-95/02 (1995)
8. Gueist, A., Beguelin, A., Dongarra, J., Jiang, W., Manchel, R., Sunderam, V.: *PVM 3 User's Guide and Reference Manual*: Oak Ridge National Laboratory, TM-12187 (May 1994)

Modelica – A Language for Equation-Based Physical Modeling and High Performance Simulation

Peter Fritzson

PELAB, Dept. of Computer and Information Science
Linköping University, S-58183, Linköping, Sweden
petfr@ida.liu.se

Abstract. A new language called Modelica for hierarchical physical modeling is developed through an international effort. Modelica 1.0 [http://www.Dynasim.se/Modelica] was announced in September 1997. It is an object-oriented language for modeling of physical systems for the purpose of efficient simulation. The language unifies and generalizes previous object-oriented modeling languages and techniques.

Compared to the widespread simulation languages available today this language offers two important advances: 1) non-causal modeling based on differential and algebraic equations; 2) multidomain modeling capability, i.e. it is possible to combine electrical, mechanical, thermodynamic, hydraulic etc. model components within the same application model.

A class in Modelica may contain variables (i.e. instances of other classes), equations and local class definitions. The multi-domain capability is partly based on a notion of connectors, which are classes just like any other entity in Modelica.

Simulation models can be developed using a graphical editor for connection diagrams. Connections are established just by drawing lines between objects picked from a class library. The Modelica model is translated into a set of constants, variables and equations. Equations are sorted and converted to assignment statements when possible. Strongly connected sets of equations are solved by calling a symbolic and/or numeric solver. The C/C++ code generated from Modelica models is quite efficient.

High performance parallel simulation code can be obtained either at the coarse-grained level by identifying fairly independent submodels which are simulated in parallel, or at the fine-grained level by parallelizing on clustered expression nodes in the equation graph. Preliminary results using the coarse-grained approach have been obtained in an application on simulating an autonomous aircraft watching car traffic.

1 Introduction

The use of computer simulation in industry is rapidly increasing. This is typically used to optimize products and to reduce product development cost and time. Whereas in the past it was considered sufficient to simulate subsystems separately, the current trend is to simulate increasingly complex physical systems composed of subsystems from multiple domains such as mechanic, electric, hydraulic, thermodynamic, and control system components.

1.1 Background

Many commercial simulation software packages are currently available. The market is divided into distinct domains, such as packages based on block diagrams (block-oriented tools, such as SIMULINK, System Build, ACSL), electronic programs (signal-oriented tools, such as SPICE, Saber), multibody systems (ADAMS, DADS, SIMPACK), and others. With very few exceptions, all simulation packages are strong only in one domain and are not capable of modeling components from other domains in a reasonable way. However, this is a prerequisite to be able to simulate modern products that integrate components from several domains such as for example electric, mechanic, hydraulic and control.

1.2 Problems

To summarize the current situation, there are at least three serious problems:

- High performance simulation of complex multi-domain systems is needed. Current widespread methods cannot cope with serious multi-domain modeling and simulation in an economical way.
- Simulated systems are increasingly complex. Thus, system modeling has to be based primarily on combining reusable components instead of re-inventing the wheel. A better technology is needed in creating easy-to-use reusable components.
- It is hard to achieve truly reusable components in object-oriented programming and modeling.

1.3 Proposed Solution

The goal of the Modelica project[8] is to provide practically usable solutions to these problems, based on techniques for mathematical modeling of reusable components.

Several first generation object-oriented mathematical modeling languages and simulation systems (ObjectMath [4], Dymola [2], Omola [1], NMF, gPROMS, Allan, Smile, etc.) have been developed during the past few years. These languages were applied in areas such as robotics, vehicles, thermal power plants, nuclear power plants, airplane simulation, real-time simulation of gear boxes, etc.

Several applications have shown that object-oriented modeling techniques is not only comparable to, but outperform special purpose tools on applications that are far beyond the capacity of established block-oriented simulation tools.

However, the situation of a number of different incompatible object-oriented modeling and simulation languages was not satisfactory. Therefore in the fall of 1996 a group of researchers (see Sect. 3.5) from universities and industry started work towards standardization and making this object-oriented modeling technology widely available.

The new language is called Modelica and designed for modeling dynamic behavior of engineering systems, intended to become a *de facto* standard.

Modelica is superior to current technology mainly for the following reasons:

- *Object-oriented modeling.* This technique makes it possible to create physically relevant and easy-to-use model components, which are employed to support hierarchical structuring, reuse, and evolution of large and complex models covering multiple technology domains. More details on object-orientation in Modelica can be found in [5] and [8].
- *Non-causal modeling.* Modeling is based on equations instead of assignment statements as in traditional input/output block abstractions. Direct use of equations significantly increases re-usability of model components, since components adapt to the data flow context in which they are used. This generalization enables both simpler models and more efficient simulation. However, for interfacing with traditional software, algorithm sections with assignments as well as external functions/procedures are also available in Modelica.
- *Physical modeling of multiple domains.* Model components can correspond to physical objects in the real world, in contrast to established techniques that require conversion to "signal" blocks with fixed input/output causality. In Modelica the structure of the model becomes more natural in contrast to block-oriented modeling tools. For application engineers, such "physical" components are particularly easy to combine into simulation models using a graphical editor.

1.4 Modelica view of object-oriented mathematical modeling

Traditional procedural languages such as Fortran or C, and object-oriented languages like C++, Java, Smalltalk and Simula support programming with operations on state. The state of the program includes variable values and object data. The number of objects changes dynamically. The Smalltalk view of object orientation is of sending messages between (dynamically) created objects. The Modelica view is different since the Modelica language emphasizes *structured* mathematical modeling. Object-orientation is primarily viewed as a structuring concept that is used to handle the complexity of large system descriptions. A Modelica model is primarily a declarative mathematical description, which allows analysis and equational reasoning. Dynamic system properties are expressed in a declarative way through equations.

1.5 Traditional High-Performance Software

High-performance software is traditionally developed in languages such as Fortran, C or C++, often combined with parallel libraries e.g. MPI, ScalaPack, etc. Although this paper primarily presents new aspects of Modelica examplified through non-causal classes using equations, it is possible to write algorithmic code in Modelica, and to integrate traditional software components with Modelica code. In fact, Modelica's strong view of software components [11] opens up new possibilities for component based software design and system integration.

2 A Modelica overview

Modelica programs are built from *classes*. Like in other object-oriented languages, a class contains variables, i.e. class attributes representing data. The main difference compared with traditional object-oriented languages is that instead of functions (methods) we primarily use *equations* to specify behavior. Equations can be written explicitly, like `a=b`, or be inherited from other classes. Equations can also be specified by the `connect` statement. The statement `connect(v1, v2)` expresses coupling between variables *v1* and *v2*. These variables are called *connectors* and belong to the connected objects. This gives a flexible way of specifying topology of physical systems described in an object-oriented way using Modelica.

In the following sections we introduce some basic and distinctive syntactical and semantic features of Modelica, such as connectors, encapsulation of equations, inheritance, declaration of parameters and constants.

2.1 Modelica model of an electric circuit

As an introduction to Modelica we will present a model of a simple electrical circuit as shown in Fig. 1. The system can be broken into a set of connected electrical standard components. We have a voltage source, two resistors, an inductor, a capacitor and a ground point. Models of such components are available in Modelica class libraries.

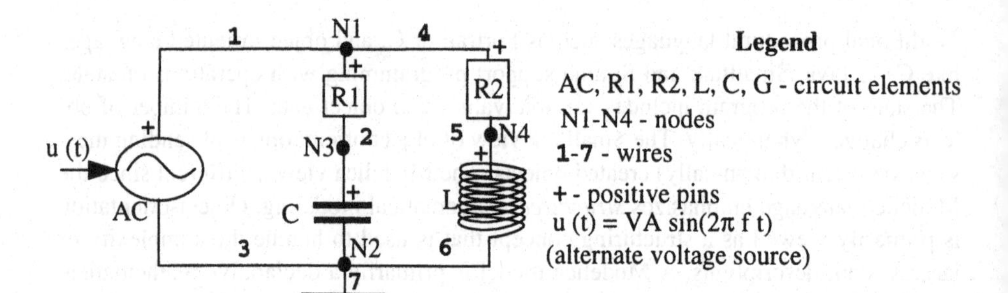

Fig. 1. A connection diagram of the simple electrical circuit example. The numbers of wires and nodes are used for reference in Table 1.

A declaration like the one below specifies `R1` to be of class `Resistor` and sets the default value of the resistance, `R`, to `10`.

```
Resistor R1(R=10);
```

A Modelica description of the complete circuit appears as follows:

```
class circuit
  Resistor  R1(R=10);
  Capacitor C(C=0.01);
  Resistor  R2(R=100);
  Inductor  L(L=0.1);
```

```
    VsourceAC AC;
    Ground    G;

  equation
    connect (AC.p, R1.p);      // Wire 1
    connect (R1.n, C.p);       // Wire 2
    connect (C.n,  AC.n);      // Wire 3
    connect (R1.p, R2.p);      // Wire 4
    connect (R2.n, L.p);       // Wire 5
    connect (L.n,  C.n);       // Wire 6
    connect (AC.n, G.p);       // Wire 7
  end circuit;
```

A composite model like the circuit model described above specifies the system topology, i.e. the components and the connections between the components. The connections specify interactions between the components. In some previous object-oriented modeling languages connectors are referred to cuts, ports or terminals. The keyword `connect` is a special operator that generates equations taking into account what kind of interaction is involved as explained in Sect. 2.3.

Variables declared within classes are public by default, if they are not preceded by the keyword `protected` which has the same semantics as in Java. Additional `public` or `protected` sections can appear within a class, preceded by the corresponding keyword.

2.2 Library classes

The next step in introducing Modelica is to explain how library model classes can be defined.

A connector must contain all quantities needed to describe an interaction. For electrical components we need the variables `voltage` and `current` to define interaction via a wire. The types to represent those can be declared as:

```
  class Voltage = Real;
  class Current = Real;
```

where `Real` is the name of a predefined variable type. A `Real` variable has a set of default attributes such as unit of measure, initial value, minimum and maximum value. These default attributes can be changed when declaring a new class, for example:

```
  class Voltage = Real(unit="V", min=-220.0, max=220.0);
```

In Modelica, the basic structuring element is a `class`. There are seven restricted class categories with specific keywords, such as `type` (a class that is an extension of built-in classes, such as `Real`, or of other defined types) and `connector` (a class that does not have equations and can be used in connections). In any model the `type` and `connector` keywords can be replaced by the `class` keyword giving a semantically equivalent model. Other specific class categories are `model`, `package`, `function` and `record` of which `model` and `record` can be replaced by `class`.

The idea of restricted classes is advantageous because the modeler does not have to learn several different concepts, except for one: the class concept. All properties of a class, such as syntax and semantics of definition, instantiation, inheritance, and generic properties are identical to all kinds of restricted classes. Furthermore, the construction of Modelica translators is simplified considerably because only the syntax and semantics of a class have to be implemented along with some additional checks on restricted classes. The basic types, such as `Real` or `Integer` are built-in type classes, i.e., they have all the properties of a class. The previous two definitions can be expressed as follows using the keyword `type` which is equivalent to `class`, but limiting the defined type to be extension of a built-in type, record or array.

```
type Voltage = Real;
type Current = Real;
```

2.3 Connector classes

A connector class is defined as in the example below:

```
connector Pin
  Voltage      v;
  flow Current i;
end Pin;
```

Connection statements are used to connect instances of connection classes. A connection statement `connect(Pin1,Pin2)`, with `Pin1` and `Pin2` of connector class `Pin`, connects the two pins so that they form one node. This implies two equations, namely:

```
Pin1.v = Pin2.v
Pin1.i + Pin2.i = 0
```

The first equation says that the voltages of the connected wire ends are the same. The second equation corresponds to Kirchhoff's current law saying that the currents sum to zero at a node (assuming positive value while flowing into the component). The sum-to-zero equations are generated when the prefix `flow` is used. Similar laws apply to flow rates in a piping network and to forces and torques in mechanical systems.

2.4 Virtual classes

A common property of many electrical components is that they have two pins. This means that it is useful to define an "interface" model class,

```
class TwoPin              "Superclass of elements
                           with two electric pins"
  Pin p, n;
  Voltage v;
  Current i;
equation
```

```
    v = p.v - n.v;
    0 = p.i + n.i;
    i = p.i;
end TwoPin;
```

that has two pins, p and n, a quantity, v, that defines the voltage drop across the component and a quantity, i, that defines the current into the pin p, through the component out from pin n (Fig. 2).

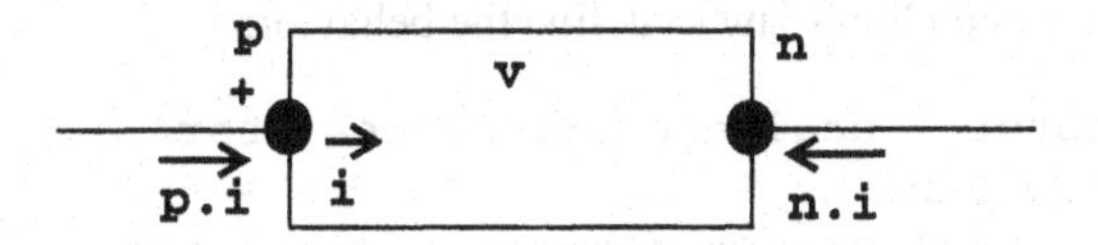

Fig. 2. Generic TwoPin model.

The equations define generic relations between quantities of a simple electrical component. In order to be useful a constitutive equation must be added.

2.5 Equations and non-causal modeling

Non-causal modeling means modeling based on equations instead of assignment statements. Equations do not specify which variables are inputs and which are outputs, whereas in assignment statements variables on the left-hand side are always outputs (results) and variables on the right-hand side are always inputs. Thus, the causality of equations-based models is unspecified and becomes fixed only when the corresponding equation systems are solved. This is called non-causal modeling.

The main advantage with non-causal modeling is that the solution direction of equations will adapt to the data flow context in which the solution is computed. The data flow context is defined by telling which variables are needed as *outputs* and which are external *inputs* to the simulated system.

The non-causality of Modelica library classes makes these more reusable than traditional classes containing assignment statements where the input-output causality is fixed.

For example, regard the equation below from the Resistor class:

```
R*i = v;
```

which can be used in two ways. The variable v can be computed as a function of i, or the variable i can be computed as a function of v as shown in the two assignment statements below:

```
i := v/R;
v := R*i;
```

In the same way the following equation from the class TwoPin:

```
v = p.v - n.v
```

can be used in three ways:

```
v   := p.v - n.v;
p.v := v + n.v;
n.v := p.v - v;
```

2.6 Inheritance, parameters and constants

To define a model for a resistor we exploit `TwoPin` and add a definition of a `parameter` for the resistance and Ohm's law to define the behavior:

```
class Resistor "Ideal electrical resistor"
  extends TwoPin;
  parameter Real R(unit="Ohm") "Resistance";
equation
    R*i = v;
end Resistor;
```

The keyword `parameter` specifies that the variable is constant during a simulation run, but can have its value initialized before a run. This means that `parameter` is a special kind of constant, which is implemented as a static variable that is initialized once and never changes its value during a specific execution. A `parameter` is a variable that makes it simple for a user to modify the behavior of a model.

A Modelica `constant` never changes and can be substituted inline.

The keyword `extends` specifies the parent class. Variables, equations and `connects` are inherited from the parent. Multiple inheritance is supported in Modelica.

Analogously to C++, variables, equations and connections of the parent class cannot be *removed* in the subclass.

In C++ a virtual function can be *replaced* by a function with the same name in the child class. In Modelica 1.0 equations cannot be named and therefore in general inherited equations cannot be replaced/specialized in a child class[1]. However, equations may be replaced in the special case where they have a single identifier on When classes are inherited, equations are accumulated. This makes the equation-based semantics of the child classes consistent with the semantics of the parent class.

2.7 Time and model dynamics

Dynamic systems are models where behavior evolves as a function of *time*. We use a predefined variable `time` which steps forward during system simulation.

A class for the voltage source in our circuit example can thus be defined as:

```
class VsourceAC "Sin-wave voltage source"
  extends TwoPin;
```

1. In the ObjectMath language, one of the precursors to Modelica, equations in general can be named and thus specialized through inheritance.

```
    parameter Voltage  VA               = 220 "Amplitude";
    parameter Real     f(unit="Hz")  = 50   "Frequency";
    constant  Real     PI               = 3.141592653589793;
  equation
    v = VA*sin(2*PI*f*time);
  end VsourceAC;
```

A class for an electrical capacitor can also reuse the `TwoPin` as follows:

```
  class Capacitor "Ideal electrical capacitor"
    extends TwoPin;
    parameter Real  C(unit="F")   "Capacitance";
  equation
      C*der(v) = i;
  end Capacitor;
```

The notation `der(v)` means the time derivative of `v`.

During system simulation the variables `i` and `v` evolve as functions of time. The solver of differential equations (see Sect. 3.2) computes the values of $i(t)$ and $v(t)$ (t is `time`) so that $C\,v'(t)=i(t)$ for all values of t.

Finally, we define the ground point as a reference value for the voltage levels:

```
  class Ground "Ground"
    Pin p;
  equation
    p.v = 0;
  end Ground;
```

More details on other Modelica constructs are presented in [8].

3 Implementation

3.1 Flattening of equations

Classes, instances and equations are `translated` into flat set of equations, constants and variables (see Table 1). As an example, we translate the `circuit` model from Sect. 2.1.

The equation `v=p.v-n.v` is defined by the class `TwoPin`. The `Resistor` class inherits the `TwoPin` class, including this equation. The `circuit` class contains a variable `R1` of type `Resistor`. Therefore, we include this equation instantiated for `R1` as `R1.v=R1.p.v-R1.n.v` into the set of equations.

The wire labelled `1` is represented in the model as `connect(AC.p, R1.p)`. The variables `AC.p` and `R1.p` have type `Pin`. The variable `v` is a *non-flow* variable representing voltage potential. Therefore, the equality equation `AC.p.v=R1.p.v` is generated. Equality equations are always generated when non-flow variables are connected.

Table 1: Equations generated from the simple circuit model

AC	`0=AC.p.i+Ac.n.i`
	`AC.v=Ac.p.v-AC.n.v`
	`AC.i=AC.p.i`
	`AC.v=AC.VA*`
	`sin(2*PI*AC.f*time);`
		L	`0=L.p.i+L.n.i`
			`L.v=L.p.v-L.n.v`
			`L.i=L.p.i`
			`L.v = L.L*L.der(i)`

R1	`0=R1.p.i+R1.n.i`
	`R1.v=R1.p.v-R1.n.v`
	`R1.i=R1.p.i`
	`R1.v = R1.R*R1.i`
		G	`G.p.v = 0`

R2	`0=R2.p.i+R2.n.i`
	`R2.v=R2.p.v-R2.n.v`
	`R2.i=R2.p.i`
	`R2.v = R2.R*R2.i`
		wires	`R1.p.v=AC.p.v // wire 1`
			`C.p.v=R1.n.v   // wire 2`
			`AC.n.v=C.n.v   // wire 3`
			`R2.p.v=R1.p.v // wire 4`
			`L.p.v=R2.n.v   // wire 5`
			`L.n.v=C.n.v    // wire 6`
			`G.p.v= AC.n.v // wire 7`

C	`0=C.p.i+C.n.i`
	`C.v=C.p.v-C.n.v`
	`C.i=C.p.i`
	`C.i = C.C*C.der(v)`
		flow at node	`0=AC.p.i+R1.p.i+R2.p.i // 1`
			`0=C.n.i+G.i+AC.n.i+L.n.i // 2`
			`0=R1.n.i+ C.p.i // 3`
			`0 =R2.n.i + L.p.i // 4`

Notice that another wire (labelled 4) is attached to the same pin, `R1.p`. This is represented by an additional connect statement: `connect(R1.p.R2.p)`. The variable `i` is declared as a `flow` variable. Thus, the equation `AC.p.i+R1.p.i+R2.p.i=0` is generated. Zero-sum equations are always generated when connecting flow variables, corresponding to Kirchhoff's current law.

The complete set of equations generated from the `circuit` class (see Table 1) consists of 32 differential-algebraic equations. These include 32 variables, as well as `time` and several parameters and constants.

3.2 Solution and simulation

After flattening, all the equations are sorted. Simplification algorithms remove most equations. If two syntactically equivalent equations appear, only one copy of the equations is kept.

Then independent equations are converted to assignment statements. If a strongly connected set of equations appears, this set is transformed by a symbolic solver which performs a number of algebraic transformations to simplify the dependencies between the variables. It can also solve a system of differential equations if it has a symbolic solution. Finally, C/C++ code is generated, and linked with a numeric solver.

The initial values can be taken from the model definition. If necessary, the user specifies the parameter values (Sect. 2.6). Numeric solvers for differential equations (such as LSODE, part of ODEPACK[7]) give the user possibility to ask about the value of specific variable at a specific moment in time. As the result a function of time, e.g.

`R2.v(t)` can be computed for a time interval $[t_0, t_1]$ and displayed as a graph or saved in a file. This data presentation is the final result of system simulation.

In most cases (but not always) the performance of generated simulation code (including the solver) is similar to hand-written C code. Often Modelica is more efficient than straightforwardly written C code, because additional opportunities for symbolic optimization are used.

3.3 Current status and future work

As a result from 8 meetings in the period September 1996 - September 1997 the first full definition of Modelica 1.0 was announced in September 1997 [8, 3]. The work by the Modelica Committee on the further development of Modelica and tools is continuing. Current issues include definition of Modelica standard libraries, efficient compilation of Modelica, parallel simulation, experimentation environments. The next revision, Modelica 1.1, is expected during the fall of 1998, including among other things more precise semantic definitions, standard libraries, etc.

3.4 Conclusion

A new object-oriented language Modelica designed for physical modeling takes some distinctive features of object-oriented and simulation languages. It offers the user a tool for expressing non-causal relations in modeled systems. Modelica is able to support physically relevant and intuitive model construction in multiple application domains. Non-causal modeling based on equations instead of procedural statements enables adequate model design and a high level of reusability.

Traditional high-performance computational software written in languages such as Fortran, C, or C++, possibly including parallel libraries, is viewed as external functions or computational blocks by Modelica and can be linked together with compiled Modelica code. Furthermore, Modelica's strong component-based view of software facilitates integration and reuse of software components in both traditional languages and in Modelica.

There is a fairly straightforward algorithm translating Modelica classes, instances and connections into a flat set of equations. Further symbolic optimizations reduces the size of the equation system and prepares for generation of efficient code. A choice of several numeric DAE solvers are available for linking with the generated code for computation of simulation results. Experience shows that Modelica is an adequate tool for design of simulation models in several different application domains, as well as for complex multi-domain models.

3.5 Acknowledgments

The Modelica definition has been developed by the Eurosim Modelica technical committee (Hilding Elmqvist, Francois Boudaud, Jan Broenink, Dag Brück, Thilo Ernst, Peter Fritzson, Alexandre Jeandel, Kaj Juslin, Matthias Klose, Sven-Erik Mattson, Martin Otter, Per Sahlin, Hubertus Tummescheit, Hans Vangheluwe) under the chairmanship of Hilding Elmqvist. The work by Linköping University has been supported by the Wallenberg foundation as part of the WITAS project.

References

[1] Mats Andersson: *Object-Oriented Modeling and Simulation of Hybrid Systems.* Ph.D. thesis ISRN LUTFD2/TFRT--1043--SE, Department of Automatic Control, Lund Institute of Technology, Lund, Sweden, December 1994.

[2] Hilding Elmqvist, Dag Brück, and Martin Otter: *Dymola — User's Manual.* Dynasim AB, Research Park Ideon,Lund, Sweden, 1996, http://www.Dynasim.se

[3] Hilding Elmqvist, Sven-Erik Mattsson: Modelica - The Next Generation Modeling Language - An International Design Effort. In *Proceedings of First World Congress of System Simulation,* Singapore, September 1-3 1997.

[4] Peter Fritzson, Lars Viklund, Dag Fritzson, Johan Herber. High-Level Mathematical Modelling and Programming, *IEEE Software,* 12(4):77-87, July 1995, http://www.ida.liu.se/labs/pelab/omath

[5] Peter Fritzson, Vadim Engelson. Modelica - A Unified Object-Oriented Language for System Modeling and Simulation, In *Proceedings of ECOOP-98,* Brussels, July 1998, LNCS 1445, Springer Verlag.

[6] Dag Fritzson, Patrik Nordling. Solving Ordinary Differential Equations on Parallel Computers Applied to Dynamic Rolling Bearing Simulation. In *Parallel Programming and Applications*, P. Fritzson, L. Finmo, eds., IOS Press, 1995

[7] A.C. Hindmarsh. ODEPACK, A Systematized Collection of ODE Solvers, *Scientific Computing,* R. S. Stepleman et al. (eds.), North-Holland, Amsterdam, 1983 (Vol. 1 of IMACS Transactions on Scientific Computation), pp. 55-64, also http://www.netlib.org/odepack/index.html

[8] Modelica Home Page. http://www.Dynasim.se/Modelica

[9] ObjectMath Home Page. http://www.ida.liu.se/labs/pelab/omath

[10] Martin Otter, C. Schlegel, and Hilding Elmqvist. Modeling and Real-time Simulation of an Automatic Gearbox using Modelica. In Proceedings of ESS'97—European Simulation Symposium, Passau, Oct. 19-23, 1997.

[11] Clemens Szyperski. *Component Software—Beyond Object-Oriented Programming*. Addison-Wesley, 1997.

Distributed Georeferring of Remotely Sensed Landsat-TM Imagery Using MPI

J.D. García-Consuegra, J.A. Gallud, G. Sebastián

Departamento de Informática,
Remote Sensing Section - IDR,
Universidad de Castilla-La Mancha,
Campus Universitario, 02071 Albacete, Spain
jgallud@info-ab.uclm.es

Abstract. Remotely-sensed images processing requires a lot of computational resources. Distributed processing is applied in remote sensing in order to reduce spatial or temporal cost by means of using the message passing interface (MPI). In this paper we have implemented a parallel approach to the rectification and restoration process of Landsat images on a cluster of NT workstations. We have defined a strategy to distribute and collect the Landsat multispectral image.

1 Introduction

Spatial data handling is an interesting application area for distributed computing techniques, which has become more and more attractive both for the spatial resolution increase of the new sensor generation, and for the emerging use of Geographical Information Systems (GIS) [1] in environment studies. The large data volumes involved and consequent processing bottleneck may indeed reduce their effective use in many real situations, and hence the need for exploring the possibility of splitting both the data and processing over several storing and computing units [5].

Remote sensed imagery processing involves several steps that begin when a image is sensed by the satellite, and continue with the image analysis and processing to obtain, finally, a variety of results. One of these steps consists in the process of georeferring the remotely-sensed image. As sensor resolution increases, the size of the sensed image increases to cover the same scanned zone and the time required to georeference it increases too. As an example, a typical Landsat image has 7020x5761 pixels which means about 200 million float point operations needed to georeference it.

In this paper, we describe a distributed algorithm to georeference Landsat-TM (satellite Landsat V with Thematic Mapper sensors) images on a cluster of low-cost workstations by using MPI functions. We will show how the computation time required by georeferring can be reduced without using high-cost parallel machines. The distributed application is implemented by splitting the original images in several small ones, which are processed in each node parallely.

Our workbench is integrated by a cluster of several NT workstations connected by means of a fast ethernet.

The following section explains how the georeference algorithm works when a Landsat image arrives from a satellite. In section 3 we show the parallel approach to the algorithm of section 2. In section 4 we analyze the performance evaluation of our parallel approach and, finally in section 5 the conclusions and future work are presented.

2 The sequential algorithm

When a image is remotely sensed, a number of different distorsions appear and these images contain such significant errors that they cannot be used as maps. The process of georeferring a satellite image consists in the application of a set of mathematical operations on the original image to obtain a geometrically corrected image. So, the purpose of geometric correction is to compensate for the distortions introduced by different factors (earth curvature, relief displacement, and so on) so that the corrected image will have the geometric integrity of a map [4].

This process is usually implemented as a two-step procedure [6, 3]. In the first one, those distorsions that are predictable are considered, and then, those distorsions that are essentially random or unpredictable are considered. In this paper we have focused in the second step related to random distorsions. This kind of distorsion is corrected by analyzing well-distributed ground control points ocurring in an image. The GCP (ground control points) are features of known ground location that can be accurately located on the digital imagery. These values are then submitted to a least-squares regression analysis to determine coefficients for two coordinate transformation equations that can be used to interrelate the geometrically correct coordinates and the distorted image coordinates. Once the coefficients for these equations are determined, the distorted image coordinates for any map position can be precisely estimated by using this expression:

$$X = f1(x, y)$$
$$Y = f2(x, y)$$

Where (X,Y) is the distorted image coordinates, (x,y) are the correct coordinates (map) and f1 and f2 are the transformation functions. The process is, actually, made reversely. An undistorted output matrix of empty map cells is defined, and then each cell is filled in with the gray level of the corresponding pixel (digital number or DN), or pixels depending on the employed method, in the distorted image.

That is, the process is performed using the following operations:

1. The coordinates of each element in the undistorted image are transformed to determine their corresponding location in the original distorted image.
2. The intensivity value or DN of the undistorted image is determined by using one of these usual methods: nearest neighbour, bilinear interpolation and cubic convolution.

In this paper we have implemented the nearest neighbour method in both the sequential process and parallel version to determine the digital number.

3 The distributed geometric correction of Landsat imagery

At this point, we are going to explain a first approach for the georeference distributed algorithm. The basic aim consists in reducing the response time of the sequential algorithm by using many tasks that work collectively to solve the problem. The message-passing model is used over a high-speed local area network that connects several general-purpose workstations, each running one or more of the above mentioned tasks.

In our system we have a collection of closely related processes executing the same code and performing computations on different portions of the original data [7]. The message-passing model admits several submodels, we have adopted the master-slave schema to implement the distributed system. The master is responsible for process spawning, initialization, collection and display of results. The slave nodes perform the actual computation related to the geometric correction explained above (figure 1).

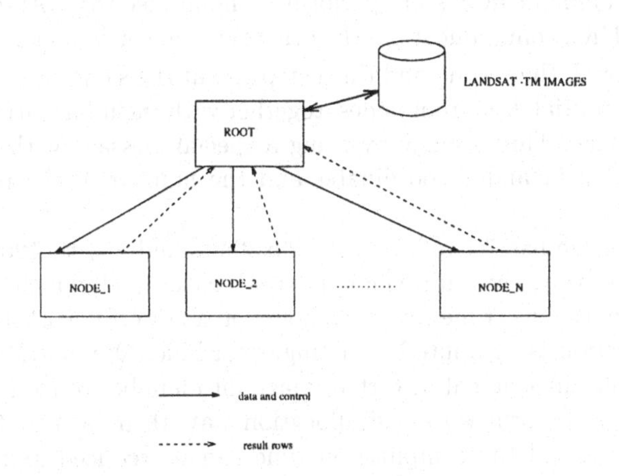

Fig. 1. Schematic view of the distributed georeferring

Our distributed approach can be structured into three steps:

1. The distorted image is broadcasted into the nodes.
2. The computations are performed to get the corrected image.
3. The resultant image is restored on the master computer.

In order to confer a greater generality to our development, and with a view to further applications, two types of messages have been defined: control and data. The first type of message is used to specify the activity we wish to carry out and the parameters necessary for the georeference, as well as for the control of the MPI. The second type of message is used for the passing of file data or for sending georeferred pixel sequences.

The first step is made by the master process, which opens a file that contains the information of the task to be solved. The root process is initiated with a parametric file, in which all the activities to be performed and their parameters are specified in a structural way. The master process sends such information to all nodes involved. Thus, each nodo knows what it must do (correction method) and how much information must receive (number of bands, resolution, the transformation functions, etc). At this point, the master process sends the Landsat image to all the nodes. The special features of the georeference algorithm force us to send all data to all nodes, as can be seen in figure 2-method B. As we have explained above, the resultant image is a combination of rotations and translations of the original distorted image. So that, the figure 2-method A is not valid because the rotation of the original image. On the other hand the MPI broadcast function allows us to distribute the image easily, but unfortunately at a high cost.

The second step is executed in each node, which acts on the region of the distorted image where the computations must do the corrections. As soon as a row of the resultant image is computed, it is sent to the master process, so computations and communications are performed simultaneously . And so, when a node is involved in a communication, the others can do correction computations.

As can be noted, the second and third step run at the same time. The master process receives partial data from nodes, together with the information to restore the resultant image. This is made by using a special message with information about the location (x and y coordinates and the number of the band) of the received data.

A Landsat image has 7 bands, 6 with 20x30mts and 1 with 120x120 mts of resolution respectively, with aproximately 40Mb each band, which explains the high value of the response time (over two hours on a powerful workstation) when geometric correction is computed in a single machine. Our parallel algorithm works by splitting the original distorted image into number-of-rows/number-of-nodes blocks in a uniform workload allocation way, then each node computes a small submatrix and the computation time can be reduced and the overall performance of the algorithm can be improved.

4 Experimental Results

In this section we explain the results obtained from a program for the geometric correction of a distorted satellite image executed following both the sequential and distributed model according to the considerations made in the previous sections. We have used different image sizes to compare both algorithms. All executions improve the whole time of computation of the parallel algorithm related to the size of each image.

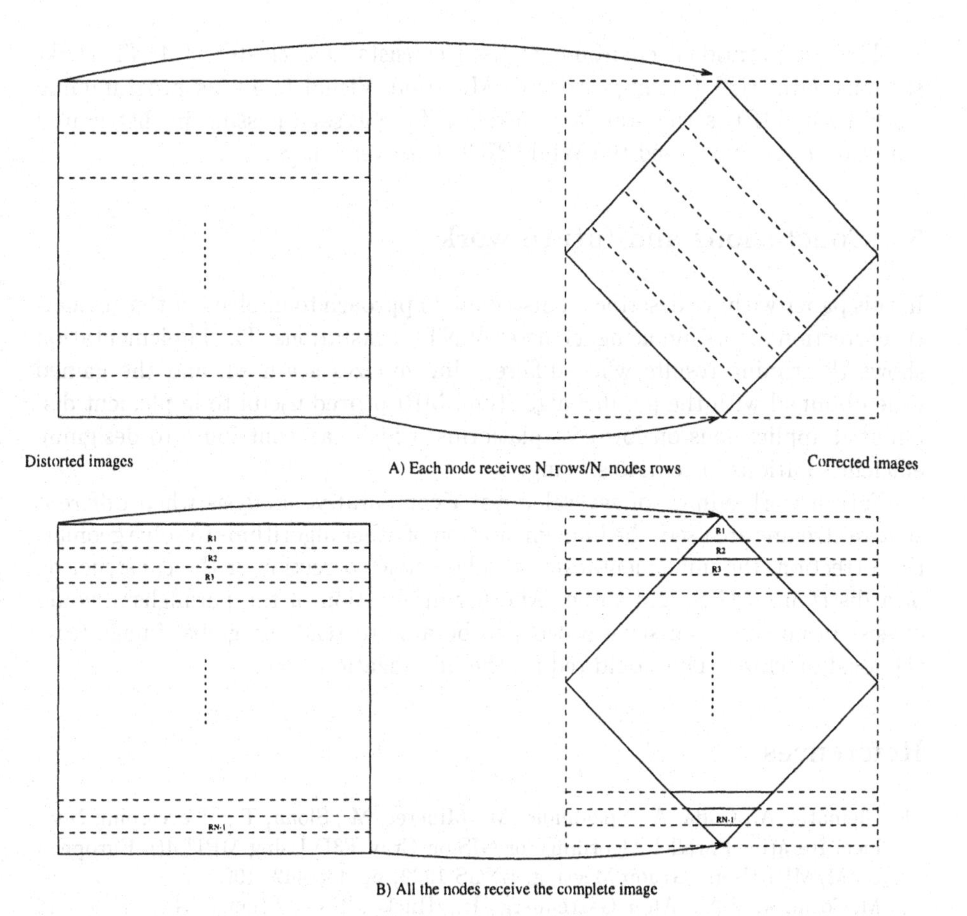

Fig. 2. Workload distribution methods

The computational environment used consists of a cluster of 4 NT workstations with MPI (Winmpich) with Microsoft Visual C 4.0 as programming workbench. MPI is the standard interface for message passing in distributed application, we have used the WinMPICH 0.9b version [8].

5 Conclusions and future work

In this paper we have described a distributed approach to implement the geometric correction algorithm using a cluster of NT workstations. The implementation shows the timing results when different image sizes are used, and the gained time obtained with the parallel algorithm. MPI proved useful to implement distributed applications on low-cost platforms, which can contribute to designing efficient solutions in remote sensing.

Future work admits of several ways: a comparative analysis when different sizes of streams are used, the implementation of other algorithms to solve geometric correction, the implementation of radiometric correction or the construction of a distributed GIS, and so on. Another application of the parallelism would consist in getting each slave process to be able to read the global image from the local machine. This could reduce the initialization time.

References

1. Clematis, A., Coda, A., Spagnuolo, M., Mineter, M., Sloan, T., "Developing Non-local Iterative Parallel Algorihms for GIS on Cray T3D Using MPI" 4th European PVM/MPI Users' Gruop Meeting, LNCS 1332, pp 435-442, 1997.
2. McCormick, J.A., Alter-Gartenberg, R., Huck , F.O., "Image Gathering and restoration: information and visual quality", Journal of Optical Society of America, Vol 6,No. 7, pp 987-1005, 1989
3. Lillesand, T.M., Kiefer, R.W., Remote Sensing and Image Interpretation 2nd Edition, J.Wiley & Sons
4. Markham, B.L., "The Landsat Sensors' Spatial Responses" IEEE Transactions on Geoscience and Remote Sensing, Vol. GE-23,No. 6, November 1985
5. Andersen, J.D., Digital Image Processing: A 1996 Review, Applied Parallel Computing - 3rd International Workshop Para 96. LNCS 1184, 1996.
6. Mather, P.M. Computer Processing of Remotely-Sensed Images, John Wiley & Sons
7. MPI Forum MPI:A message-passing interface standard, International Journal of Supercomputer Applications 8(3/4), 1994
8. Gropp, Lusk, Doss, Skjellum. A High-Performance, Portable Implementation of the MPI Message Passing Interface Standard. Parallel Computing, 1996.

Parallel Test Pattern Generation Using Circuit Partitioning in a Shared-Memory Multiprocessor

Consolación Gil[1], Julio Ortega[2], Jose Luis Bernier[2], and Maria Dolores Gil[1]

[1] Dept. de Arquitectura de Computadores y Electrónica. Universidad de Almería. La Cañada de San Urbano s/n, 04120 Almería, Spain.
cgil@peke.ualm.es
mari@iron.ualm.es

[2] Dept. de Arquitectura y Tecnología de Computadores. Universidad de Granada. Campus de Fuentenueva s/n, Granada, Spain
jortega@atc.ugr.es
jbernier@atc.ugr.es

Abstract. This paper presents the results obtained by a new parallel procedure that generates the patterns for testing digital circuits when it is implemented in a shared-memory multiprocessor. The procedure is based on a new sequential algorithm which mixes both the Boolean difference and digital spectral techniques, thus being different from other parallel methods proposed up to now. First, it uses a static circuit partitioning procedure and later a dynamic load balancing scheme to distribute the load among the processors.

1 Introduction

The test problem consists of searching for the set of assignments to the circuit inputs that allows us to distinguish between a faulty and a fault-free circuit [1]. It usually requires a great amount of time even in moderate sized circuits. Nevertheless, by using techniques such as LSSD [1], the problem can be reduced to the generation of tests for combinational circuits. It is important to have efficient procedures to solve the problem of testing combinational circuits, which despite being considered a subproblem of the sequential test problem, is an NP-complete problem. The most frequently used model of fault is the single stuck-at fault that fixes the faulty line to the logical value 0 (stuck-at 0 fault) or 1 (stuck-at 1 fault). A great deal of research has been done trying to develop more efficient algorithms to search for test patterns through the space of all input combinations [2, 3]. Thus, as the increase in the size and complexity of the circuits has been faster than the improvements achieved with the new serial procedures [1], it would very useful to speed up test generation by using parallel computers [4-8].

The parallel test pattern generators reported up to now are based on algorithms such as PODEM and its derivatives, that perform an implicit enumeration of the space of input patterns, guided by local information about the circuit. The strategies that can be used to parallelize these test algorithms can be classified as [4-8]: a) *Fault partitioning*, where the set of faults is divided among the processors that generate the patterns for each fault in their corresponding fault list; b) *Search-Space partitioning*, where the space of assignments to the inputs, the space of assignments to the lines, or both, is divided among the processors which work together in the search of a pattern for each fault; and c) *Circuit partitioning* where the circuit is distributed among the processors that apply the corresponding test algorithm to each subcircuit.

Other kinds of procedures different from implicit enumeration techniques are those based on the Boolean difference or a variation of it. Thus, given a circuit with n inputs, $y = (y_0, y_1, \ldots, y_{n-1})$ and ℓ lines, $i = 0, \ldots, \ell-1$, the function implemented by the line i is termed as $f_i(y) = f_i(y_0, \ldots, y_{n-1})$. The function synthesized by a given output of the circuit is noted as $\mathcal{H}(y) = \mathcal{H}(y_0, \ldots, y_{n-1})$. For any internal line i, it is also possible to describe $\mathcal{H}(y)$ as a function of both the inputs, y, and f_i, $\mathcal{H}(y) = \mathcal{H}_i(f_i, y) \equiv \mathcal{H}_i(f_i, y_0, \ldots, y_{n-1})$. If the line i presents a stuck-at α fault, then $f_i(y) = \alpha$ for all values of $y = (y_0, \ldots, y_{n-1})$, and the set of input patterns that will allow us to detect this fault is the one verifying

$$(f_i(y_0, \ldots, y_{n-1}) \oplus \alpha)(\partial\mathcal{H}/\partial f_i) = 1 \ . \quad (1)$$

where $\partial\mathcal{H}/\partial f_i = \mathcal{H}_i(\alpha, y_0, \ldots, y_{n-1}) \oplus \mathcal{H}_i(\overline{\alpha}, y_0, \ldots, y_{n-1})$ is called the Boolean difference of $\mathcal{H}$ with respect to f_i, and represents the set of inputs for which the value of $\mathcal{H}$ depends on the value of f_i. The set of solutions of equation (1) is the set of test patterns that allows the detection of the stuck-at α fault in line i due to a change in the output line that synthesizes $\mathcal{H}(y)$. Although the first implementations of these methods were quite slow, recently several Boolean methods have been reported [3] to overcome this drawback. Nevertheless, to the best of our knowledge, the parallelization of these algorithms has not been considered yet.

The parallel alternative we describe here is based on a procedure [9] which formulates the test equation (1) at node i in terms of two node functions f_i and g_i, and then operates in the domain of the Reed-Muller spectral coefficients of these two functions not only to quickly obtain the equation that the patterns for a given fault must verify, but also to formulate it in such a way that it allows us to determine one of its solutions (i.e., a test pattern) in an easy way.

2 Description of the Parallel Procedure

The parallel procedure allows each processor to independently compute the relative test equation of the nodes assigned to it. The procedure starts by applying a new partitioning algorithm [10] which mixes simulated annealing and tabu search to divide the circuit under test into separate subcircuits with at least one logic gate in each. The number of subcircuits must be equal to the number of processors to be used and the goal is to minimize the amount of communication among the subcircuits, as the work to do with each subcircuit is equally distributed to balance the load of the processors.

Thus it is possible for all the processors to concurrently apply the test generation algorithm in order to determine functions f and g at the nodes of the corresponding subcircuit. These functions are termed *relative functions*, and are noted with the superindex *r*, f^r and g^r. The processors also synchronize to determine functions *f* and *g* (also called *absolute functions*) for the whole circuit at the nodes of their partition.

Procedural description of the algorithm.

```
program circuit-partitioning
  begin
    Divide-circuit;
    for i=0 to numproc-1
      Generate information of i-th subcircuit;
      Computing-relative-functions;
      Dynamic-allocation-lines;
      Computing-absolute-functions;
      Determine-test-pattern;
end.
```

The procedure Divide-circuit makes a static division of the circuit under test using the hybrid heuristic MSATS [10]. The procedure Computing-relative-functions determines the functions f^r and g^r corresponding to the nodes in the subcircuit assigned to each processor. These functions are relative to the corresponding subcircuit. Computing-absolute-cuts determines the absolute functions of the nodes where the cuts have been made to divide the circuit. The Dynamic-allocation-lines procedure performs a dynamic balancing of the load to compute the absolute functions. Although in the static partitioning of the circuit a minimum number of cuts is obtained, the lines in each subcircuit are usually near the input of another one thus leading to a not very good workload balancing. As a shared memory multiprocessor is used, and all the data (i.e. basically the description of the circuit) can be accessed by all the processors, they can dynamically take the lines of the circuit as they conclude their present computation, thus providing optimal load balancing. Procedure Computing-absolute-functions computes the absolute functions *f* and *g* corresponding to the nodes assigned to the processor while the procedure Determine-test-pattern builds the test equations and solves them as indicated in [9].

3 Experimental Results

The parallel procedure has been implemented in C and run in a SGI Power Challenge computer with 4 R8000 CPUs. Some of the experimental results are better than those obtained in a multicomputer [11] as Table 1 shows. This derives from the use of a dynamic load balancing scheme, which is time-intensive in a multicomputer. This improvement accelerates the computation of the test equations and facilitates their solution in order to determine the test patterns. , if a test pattern generator can work concurrently within the partitions defined in the circuit, the test generation process can be accelerated. The use of this strategy might, however, present some problems for the most frequently used sequential test generators, and the circuit partitioning approach has not been used to parallelize the test generation. In [12] it is analyzed a parallel test pattern generator which uses topological partitioning combined with the PODEM algorithm. The experimental results obtained shown higher runtimes for the parallel test generator with respect to those required when only one processor is used. As the authors of the paper indicate, this increase in the run time is due to the overhead resulting from the implementation of the PODEM algorithm across the circuit partitioning, because the number of elemental operations required by the PODEM algorithm is not decreased by the parallelization. These problems do not appear in the parallel generator proposed here because it is based on the new test algorithm we have developed, which is different from PODEM.

Table 1. Speedup and coverage for some ISCAS circuits with 2 , 3 and 4 processors.

Circuit	faults	S_2	S_3	S_4	Cover
ALU	384	1.97	2.65	3.48	100%
c432	864	1.75	2.47	3.78	98%
c499	998	1.79	2.54	3.81	99%
c880	1760	1.78	2.56	3.91	100%
c1355	2710	1.98	2.74	3.47	98%
c1908	3816	1.96	2.85	3.39	98%
c3540	7080	1.81	2.69	3.79	98%
c6288	12570	1.76	2.67	3.52	97%

4 Conclusions

The parallelization alternative presented is very interesting. Dividing the circuit among the processors allows the memory of the computer to be used in a very efficient way, and enables test patterns to be generated for circuits that cannot be analyzed by serial procedures due to the great amount of time required, or by parallel procedures that would need a lot of memory. The results show good communication/computation rates, and thus high efficiency when the parallel procedure is applied to the circuits used as benchmarks.

By mixing a static circuit partitioning algorithm and a dynamic load balancing procedure, the speedup values obtained are better than those obtained by the parallel implementation of the same procedure in a distributed memory multicomputer. To implement such a dynamic load balancing procedure in a distributed memory multicomputer would result in a higher cost due to the computation requirements, while the speedup figures would be lower than those obtained by using only the static partitioning procedure.

Acknowledgements: This paper has been partially supported by project TIC97-1194 (CICYT, Spain).

References

1. Klenke, R.H., Williams, R.D., Aylor, J.H.: Parallel-Processing Techniques for Automatic Test Pattern Generation. IEEE Computer (January 1992) 71-84
2. Goel, P.: An Implicit Enumeration Algorithm to Generate Tests for Combinational Logic Circuits. IEEE Transactions on Computers 30 (3) (1981) 215-222
3. Stanion, R.T., Bhattacharya, D., Sechen, C.: An Efficient Method for Generating Exhaustive Test Sets. IEEE Transactions on CAD 14 (12) (1995) 1516-1525
4. Fujiwara, H., Inoue, T.: Optimal Granularity of Test Generation in a Distributed System. IEEE Transactions on Computers-Aided Design 9 (8) (1990) 885-892
5. Patil, S., Banerjee, P.: Fault Partitioning Issues in an Integrated Parallel Test Generation/Fault Simulation Enviroment. International Test Conference (1989) 718-726
6. Gil, C., Ortega, J.: A Parallel Test Pattern Generator based on Spectral Techniques. 5th Euromicro on PDP, IEEE (1997) 199-204
7. Patil, S., Banerjee, P.: A Parallel Branch and Bound Algorithm for Test Generation. IEEE Trans. on CAD 9 (3) (1990) 313-322
8. Smith, S., Underwood, B., Mercer, M.R.: An Analysis of several approaches to Circuit Partitioning for Parallel Logic Simulation. Proceedings IEEE Int'l Conf. Computer-Design. CS Press, Los Alamitos, Calif. (1987) 664-667
9. Gil, C., J. Ortega: Algebraic Test-Pattern generation based on the Reed-Muller Spectrum. To appear in: IEE Proc. Computer and Digital Techniques 145 (4) (July 1998)
10. Gil, C., Ortega, J., Diaz, A.F., Montoya, M.G.: Meta-heuristics for Circuit Partitioning in Parallel Test Generation. First Workshop on Biologically Inspired Solutions to Parallel Processing Problems. Springer-Verlag, (1998)
11. Gil, C., Ortega, J.: Parallel Test Pattern Generation Using Circuit Partitioning and Spectral Techniques, 6th Euromicro on PDP, IEEE (1998) 264-270
12. Klenke, R. H., Williams, R. D., Aylor, J. H.: Parallelization Methods for Circuit Partitioning Based Parallel Automatic Test Pattern Generation. IEEE VLSI Test Symposium, (1993) 71-78.

Parallel Adaptive Mesh Refinement for Large Eddy Simulation Using the Finite Element Method

Darach Golden, Neil Hurley and Sean McGrath

Hitachi Dublin Laboratory, Trinity College, Dublin 2, Ireland,
darach@hdl.ie

Abstract. This paper describes work in progress at Hitachi Dublin Laboratory to develop a parallel adaptive mesh refinement library. The library has been designed to be linked with a finite element simulation engine for solving three-dimensional unstructured turbulent fluid dynamics problems, using large eddy simulation. The library takes as input a distributed mesh and a list of mesh elements to be refined, carries out the refinement in parallel on the distributed data structure, redistributes the computational load and passes the updated mesh back to the main simulation engine. The library has been implemented and tested on the distributed memory parallel machine, the Hitachi SR2201. Results of performance and scalability of the code are given.

1 Introduction

The computational power available in modern distributed memory parallel machines is making the simulation of realistic three dimensional fluid dynamics problems more and more feasible. Nevertheless, the development of simulation techniques which can accurately model turbulent flows is still a matter of some research. One computationally intensive method, of which much is expected, is large eddy simulation (LES). There are still many issues which need investigation regarding the application of LES to practical problems. One such issue, as pointed out in [Dailey 1993], for example, is how to handle complex geometries. The question of interest in our research programme is whether the finite element method (FEM) applied over an unstructured three dimensional mesh can be used as a basis for accurate LES. Examples of the type of application for which the software is being developed include the analysis of aero-acoustic noise generation from a train as it passes through a tunnel, and the design of turbochargers and pumps. In order to obtain sufficient accuracy, a high grid resolution is required in critical regions of the domain. When the solution is changing more rapidly in some areas of the domain than in others, it is more computationally and memory efficient to apply a non-uniform mesh, whose resolution follows the solution requirements. Adaptive mesh refinement techniques [Flaherty 1989] in which the mesh is refined (or de-refined) during the course of the computation, according to local error estimates on the elements of the domain, have proven to be an effective way of obtaining high accuracy without resorting to a uniformly fine mesh across the entire domain.

In this paper we do not discuss the parallel simulation engine, which is being developed elsewhere, nor the method for calculating local error estimates, but rather we focus on the adaptive construction of a three dimensional, unstructured tetrahedral mesh on a parallel machine.

The outline of the paper is as follows: In section 2 we give a description of the adaptive algorithm we are using. The parallel algorithm is dealt with in section 3. Implementation details are given in section 4. Some results are given in section 5, and finally some avenues for future work are given in section 6. The method described in the paper has been implemented on the Hitachi SR2201 distributed memory parallel machine.

2 Adaptive Mesh Refinement

The following discussion applies to tetrahedral meshes only. By the phrase *adaptive mesh refinement* we mean what is described in (Fig. 1), where we concentrate on the refinement step, and in the parallel case, on the rebalancing step. It is assumed that a list of elements has already been chosen by some error estimator.

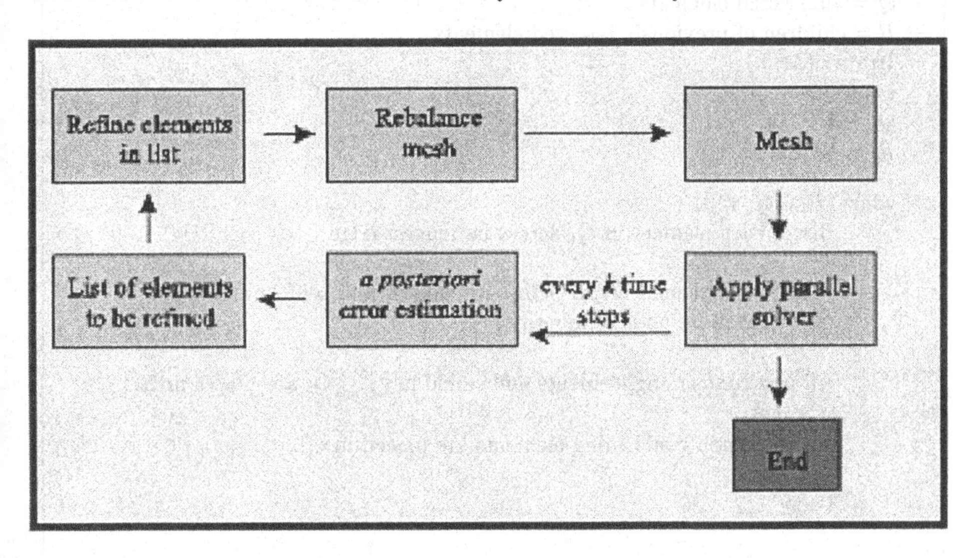

Fig. 1. Parallel Adaptive Meshing Overview

We use bisection as the method of refining elements. An element is refined by placing a node at the centre of one of its edges, and creating new edges between it and the two opposite vertices. This creates two tetrahedra from the original one. For meshes with more than one element, placing a node on an edge will render all other elements sharing that edge illegal (*non-conforming*). At some later stage in the process these non-conforming elements must themselves be bisected on the illegal edge. Thus more elements will finally be refined than were initially marked for refinement. A further consideration is the quality of new elements created. Bisecting elements on their shortest edges, for example, is likely to lead to thin, elongated new elements, and to accuracy problems in the solution. Whereas one may chose to bisect initially marked elements on a suitable edge, subsequent illegal elements cannot choose their bisected edge. To address these issues, we use the *longest edge bisection algorithm* of [Rivara 1984], [Rivara 1989], [Rivara 1992]. The strategy is to bisect non-conforming elements across their longest edge, before bisecting them along their non-conforming edges. Although this means that elements may be bisected more than once, it ensures that the tetrahedra remain reasonably well shaped (if the starting mesh is of good quality). The algorithm assumes an initial conforming tetrahedral mesh, and iteratively bisects the mesh until a new, conforming mesh is obtained (with non-conforming meshes appearing in the intermediate steps). We give the (sequential) algorithm in (Fig. 2)([Jones I 1994], [Jones II 1994]).

Fig. 2. Sequential Algorithm

- M = mesh
- L = list of elements to be refined
- Q = unbisected elements
- R = children of previously bisected elements
- **Input:** (M, L)

  $i = 0$

  $Q_i = L$

  $R_i =$ NULL

  while $(R_i \cup Q_i \neq \emptyset)$ {

  Bisect each element in Q_i across its longest edge

  Bisect each element in R_i across its longest non conforming edge (ties broken by node numbering)

  All non conforming elements embedded in $\bigcup_{j=0}^{i} Q_i$ are places in R_{i+1}

  All other non conforming elements are placed in Q_{i+1}

  $i{+}{+}$

  }

3 Parallel Adaptation

The parallel adaptive algorithm mirrors the progress of the sequential longest edge bisection algorithm. The algorithm is based on a 2D algorithm given in [Jones I 1994]. The adaptive routine takes as input a distributed, conforming mesh and a (global) list of elements to be bisected. The mesh is assumed well distributed (partitioned) on to the parallel machine. At the moment we are using the ParMetis [Karypis 1995] code to produce initial partitions. The adaptive code proceeds in an iterative fashion like the sequential case. However, the distributed nature of the problem means that:

1. Suitable notification must be communicated between processors for bisections taking place at boundaries between processors
2. *Contested bisections* must be dealt with

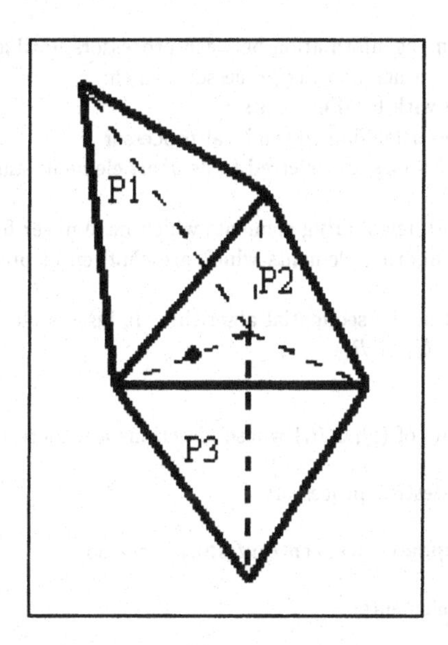

Fig. 3. Example of contested bisection

By *contested bisections* we mean the situation described in (Fig. 3), where two (or more) elements assigned to different processors are attempting to bisect on the same edge. If both processors bisect independently, then two new nodes will be added to the mesh, where only one is needed. In [Jones I 1994] contesting elements are handled (in 2D) by splitting elements into independent sets. For the 3D case we have used an alternative approach which is described below.

3.1 Parallel Algorithm

In (Fig. 4) details of the parallel algorithm are given.

The algorithm in (Fig. 4) involves two separate send / receives in each iteration of the adaptation - first those connected with contested bisections, and then all other sends and receives. In the case of contested bisections the conflict is resolved using random numbers to decide which processor labels the new node on the contested edge. This information is then broadcast to all contesting processors in the first send / receive stage. Potentially there is quite a large amount of communication during each bisection iteration, depending on the set of elements initially marked for bisection, and the partitioning of the mesh. However, in the

Fig. 4. **Parallel Algorithm (steps carried out on each processor)**

- **Input** (M, L)
  - M is *local* subset of mesh for each processor
    It is partitioned according to elements *and* nodes
  - L is a *global* list of elements to be refined
- **Setup** (M)
  - Send / receive element information between processors until all processors are aware of elements on neighboring processors which:
    * share an edge with local elements
    * contain a node belonging to the local processor
  - Such elements and nodes are referred to as *ghost* elements and nodes
- **Main** (M, L)
  - Q_i: list of local and neighboring elements which have never been bisected
  - R_i: local and neighboring elements which are children of previously bisected elements
  - *Bisect* below is as in the sequential algorithm. It bisects elements depending on whether it is in Q_i or R_i

while $(Q_i \cup R_i \neq \emptyset)$ {

Let C_i = elements of $(Q_i \cup R_i)$ whose bisections are contested between processors

Bisect C_i on contesting processors

Send / receive updates between *contesting* processors

Bisect all other elements

Send / receive all remaining updates

Q_{i+1} = non conforming elements not embedded in $(Q_i \cup R_i)$

R_{i+1} = all other non conforming elements

}

later iterations of the process there are less elements marked for bisection, and generally less communication. This combined with the relatively low number of iterations of the process (usually 10 - 15 iterations) means that the performance of the algorithm is satisfactory.

4 Implementation Details

The parallel and sequential libraries are implemented in C. The LES simulation engine has been developed by the Hitachi Mechanical Engineering Research laboratory using Fortran77. The parallel implementation uses the MPI message passing library.

4.1 Data Structures

The parallel and sequential codes use the same basic data structures with extra features in the parallel case. The fundamental structures are intuitive; being

- Mesh
- Element
- Node
- Edge
- Parallel extras:
  - local and global numbering used
  - partition information stored

The *Mesh* structure contains pointers to all mesh information. *Element* edge sharing information is accessed via neighbouring element lists kept by all nodes. The *Edge* structure contains three node pointers (two old nodes, and a new node added to the centre of the edge). This means that lists of non conforming elements created by the bisection of an edge are easily obtained from the relevant *Edge* structure. Bisected edges are stored in a hash table.

4.2 Hardware

The code has been implemented and tested on a Hitachi SR2201, a distributed memory massively parallel machine, which can be scaled from 8 to 2048 processors. Each processor runs a UNIX microkernel and delivers a peak performance of 300MFLOPS. Processors are connected by a high-performance three-dimensional crossbar, with a maximum bandwidth of 300Mb per second.

5 Results

Tests reported in this paper have been carried out on a 16 node machine (a Hitachi SR2201), with 256MB of RAM per node. The meshes we tested on are two industrial meshes (see Fig. 5). The meshes are

1. A mesh for examining the flow of fluid around a turbo charger (35007 nodes, 165027 elements)
2. A mesh of a pump (82853 nodes, 461035 elements)

Both of the meshes are application meshes used by the Hitachi Mechanical Engineering Laboratory CFD group for product development.

We evaluate the method of bisection with respect to:

- Comparison between sequential and parallel case
- The number of additional non-conforming tetrahedra that must be bisected as a result of the bisection of the initial set of elements.
- Timings

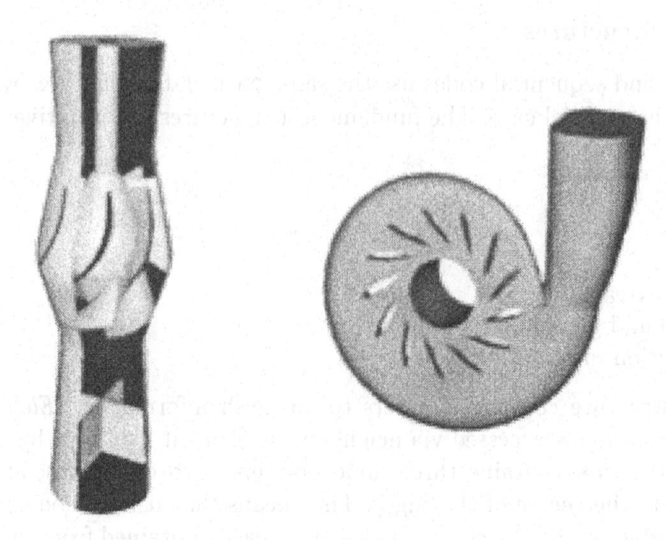

Fig. 5. Pump and turbo charger meshes

5.1 Sequential / Parallel Comparison

Since the parallel implementation tries to mirror the effect of the sequential, one would expect great similarity between the two. We show in (Fig. 6) the results of a sample adaptation of the turbo charger mesh. The *before* mesh shows the elements initially marked for refinement. The *after* mesh shows the result. This pictorial result looks the same for both the parallel and the sequential case. There are 25 more elements in the sequential case. The difference is under investigation, but the similarity is reassuring.

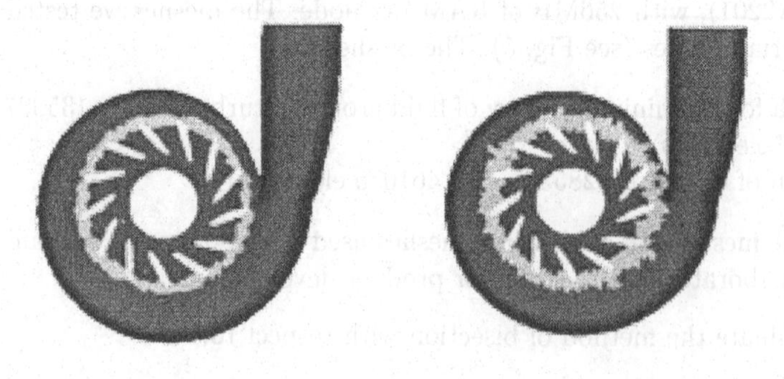

Fig. 6. Before and after refinement

5.2 Additional Tetrahedra

As can be seen in (Fig. 6) there is a significant difference between the number of elements initially marked for refinement, and the final number of elements refined. The amount of new elements added for each initially marked element varies from mesh to mesh, and is dependent on the number of elements which share edges with the marked elements. We give (widely different) results for both the turbo charger, and the pump meshes (Fig. 7) and (Fig. 8). The slopes of the pump and turbo charger graphs are ~ 28 and ~ 4.4 respectively.

5.3 Timings

Timings are given here for up to 12 processors in (Fig. 9). Since the number of processors is very small, they represent encouraging early results, and are not interpreted as any conclusive evidence of scaleability. The timings are taken for the turbo charger mesh. Between 2000 and 10000 elements are initially marked for bisection, and the times are shown.

6 Conclusions and Future Work

The work described in this paper represents work in progress. There are a number of things which have to be completed, and a number of directions for future work. For example

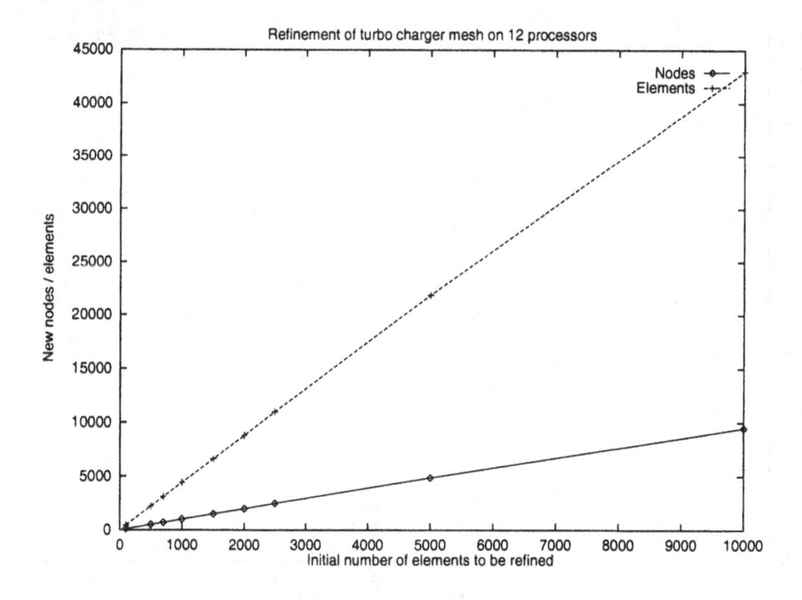

Fig. 7. Elements added to turbo charger due to refinement

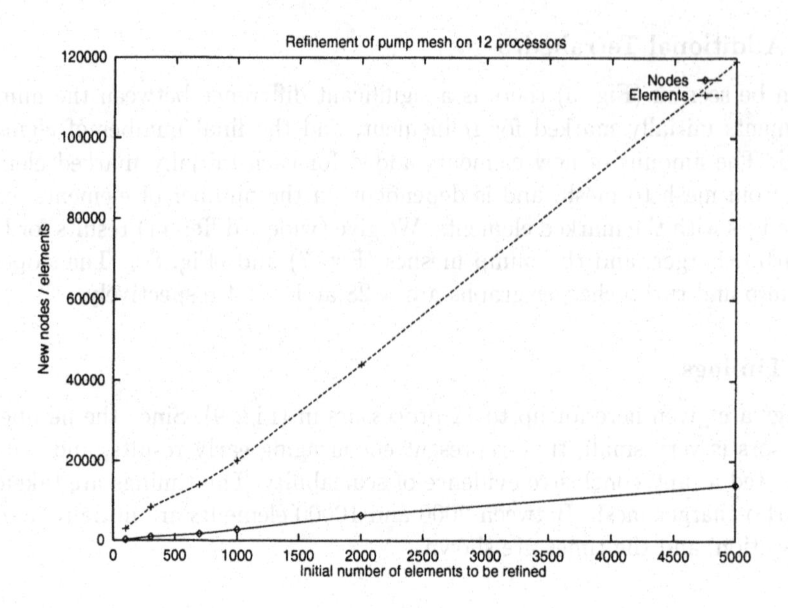

Fig. 8. Elements added to pump due to refinement

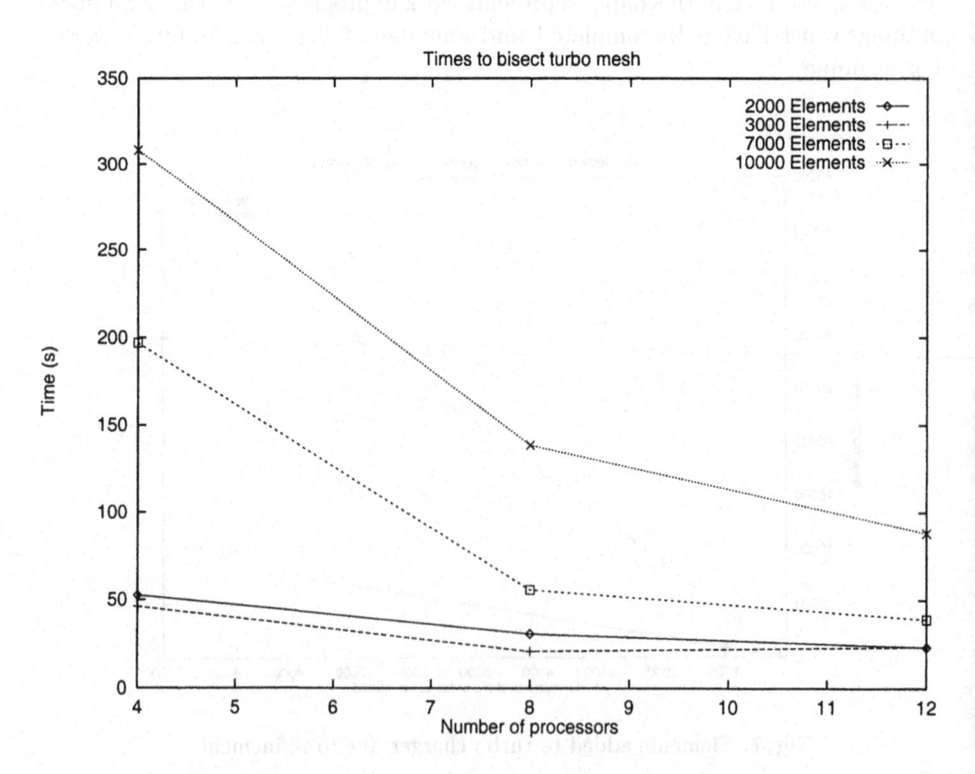

Fig. 9. Timings for turbo charger mesh on 4-12 processors

1. The code will be fully linked with a LES solver. This will provide a good testbed for further development.
2. Dynamic load balancing: After one (or more) refinement steps in which new elements are being added to processors at different rates, load on the parallel machine is likely to become unbalanced. This can seriously affect the performance of the simulation. In order to address this problem the refinement code will be linked to a parallel load balancing routine which takes care of redistributing the mesh elements (eg the METIS library [Karypis 1995]).
3. Further investigation into communication properties of algorithm / implemention. Although the timings show a good trend, we feel it is likely it will become communication bound quite quickly. Further research is needed to better understand and improve the sitution.
4. As it stands now, the algorithm for adaptation is potentially unbalanced. It is evident that one could choose elements which are all on the same processor as an initial list to be refined. A fast, intermediate repartitioning step would be desirable here.
5. Implementation of de-refinement of the mesh. At present the mesh can only be refined.
6. Implementation of types of mesh transformation other than bisection.

References

[Jones I 1994] Jones, M.T., Plassman, P.E.: Parallel Algorithms for Adaptive Mesh Refinement Preprint MCS-P421-0394 Mathematics and Computer Science Division, Argonne National Laboratory, Argonne, III

[Jones II 1994] Jones, M.T., Plassman, P.E.: Adaptive Refinement of Unstructured Finite Element Meshes Preprint Mathematics and Computer Science Division, Argonne National Laboratory, Argonne, III

[Dailey 1993] Dailey, L., Simons, T., Pletcher, P.: Large Eddy Simulation of Isotropic Decaying Turbulence with Structured and Unstructured Grid Finite Volume Methods. Dept. of Mech. Eng. Iowa State Univerity.

[Flaherty 1989] Flaherty, J.E., Paslow, P.J., Shepherd, M.S., Vasilakis, J.D.: Adaptive Methods for Partial Differential Equations. Society for Industrial and Applied Mathematics, Philadelphia.

[Rivara 1984] Rivara, M.C.: Mesh Refinement Processes based on the Generalised Bisection of Simplices. SIAM Journal of Numerical Analysis **21** 604–613

[Rivara 1989] Rivara, M.C.: Selective Refinement / Derefinement Algorithms for Sequences of Nested Triangulations. Int. J. Num. Meth. Eng **28** 2889–2906

[Rivara 1992] Rivara, M.C., Levin, C.: A 3D refinement algorithm suitable for adaptive and multi-grid techniques. Comm. Appl. Numer. Methods **8 (1992)** 281–290

[Karypis 1995] Karypis, Kumar, V. Parallel Multilevel Graph Partitioning, University of Minnesota, Dept. of Comp. Sc. Technical Report 95-036

WSSMP: A High-Performance Serial and Parallel Symmetric Sparse Linear Solver

Anshul Gupta[1], Mahesh Joshi[2], and Vipin Kumar[2]

[1] IBM T.J. Watson Research Center
P.O. Box 218, Yorktown Heights, NY 10598, U.S.A.
anshul@watson.ibm.com
[2] Department of Computer Science
University of Minnesota, Minneapolis, MN 55455, U.S.A.

Abstract. The Watson Symmetric Sparse Matrix Package, *WSSMP*, is a high-performance, robust, and easy to use software package for solving large sparse symmetric systems of linear equations. It can can be used as a serial package, or in a shared-memory multiprocessor environment, or as a scalable parallel solver in a message-passing environment, where each node can either be a uniprocessor or a shared-memory multiprocessor. *WSSMP* uses scalable parallel multifrontal algorithms for sparse symmetric factorization and triangular solves. Sparse symmetric factorization in *WSSMP* has been clocked at up to 210 MFLOPS on an RS6000/590, 500 MFLOPS on an RS6000/397 and in excess of 20 GFLOPS on a 64-node SP with RS6000/397 nodes. This paper gives an overview of the algorithms, implementation aspects, performance results, and the user interface of *WSSMP*.

1 Introduction

Solving large sparse systems of linear equations is at the core of many problems in engineering and scientific computing. A direct method for solving a sparse linear system of the form $Ax = b$ involves explicit factorization of the sparse coefficient matrix A into the product of lower and upper triangular matrices L and U. This is a highly time and memory consuming step; nevertheless, direct methods are important because of their generality and robustness and are used extensively in many application areas. Direct methods also provide an effective means for solving multiple systems with the same coefficient matrix and different right-hand side vectors because the factorizations needs to be performed only once. A wide class of sparse linear systems arising in practical applications have symmetric coefficient matrices, whose factorization is numerically stable with any symmetric permutation. The Watson Symmetric Sparse Matrix Package, *WSSMP*, is a high-performance, robust, and easy to use software package for solving large sparse linear systems of this type. *WSSMP* can be used as a serial package, or in a shared-memory multiprocessor environment with threads, or as a scalable parallel solver in a message-passing environment, where each node can either be a uniprocessor or a shared-memory multiprocessor. It uses a modified multifrontal algorithm and scalable parallel algorithms for sparse symmetric factorization and triangular solves. The ordering/permutation of matrices is also parallelized. Sparse symmetric factorization in *WSSMP* has been clocked at up to 210 MFLOPS on an RS6000/590, 440 MFLOPS on an RS6000/397 and in excess of 20 GFLOPS on a 64-node SP with RS6000/397 nodes. This paper gives an overview of the algorithms, implementation aspects, performance results, and the user interface of WSSMP.

In the paper, the term *node* refers to a uniprocessor or a multiprocessor computing unit with shared memory. A node may consist of one or more ***processors*** or CPUs. Nodes communicate with other nodes only via message-passing over the SP high-speed switch. In *WSSMP*, parallelism within a multiprocessor node is exploited by threads. In order to keep the definition of a node unique, we would refer to a meeting point of edges of a graph or a tree as a *vertex*. Accordingly, the term *supervertex* is used to describe a set consecutive columns of the sparse matrix with identical sparsity pattern, which is more commonly referred to as a *supernode* in literature.

2 Algorithms

The process of obtaining a direct solution of a sparse symmetric system of linear equations of the form $Ax = b$ consists of the following four phases: **Ordering**, which determines a symmetric permutation of the coefficient matrix A such that the factorization incurs low fill-in; **Symbolic Factorization**, which determines the structure of the triangular matrices that would result from factoring the coefficient matrix resulting from the ordering step; **Numerical Factorization**, which is the actual factorization step that performs arithmetic operations on the coefficient matrix A to produce a lower triangular matrix L such that $A = LL^T$ or $A = LDL^T$; and **Solution of Triangular Systems**, which produces the solution vector x by performing forward and backward eliminations on the triangular matrices resulting from numerical factorization. In this section, we give an overview of the algorithms used in these four phases.

2.1 Ordering

Finding an optimal ordering is an NP-complete problem and heuristics must be used to obtain an acceptable non-optimal solution. Two main classes of successful heuristics have evolved over the years: (1) Local greedy heuristics, and (2) Global graph-partitioning based heuristics. In [5], we presented an optimally scalable parallel algorithm for factoring a large class of sparse symmetric matrices. This algorithm works efficiently only with graph-partitioning based ordering. Although, traditionally, local ordering heuristics have been preferred, recent research [8, 1, 6, 3] has shown the graph-partitioning based ordering heuristics can match and often exceed the fill-reduction of local heuristics.

The ordering heuristics that *WSSMP* uses have been described in detail in [5]. Basically, the sparse matrix is regarded as the adjacency matrix of an undirected graph and a divide-and-conquer strategy (also known as nested dissection) is applied recursively to label the vertices of the graph by partitioning it into smaller subgraphs. At each stage, the algorithm attempts to find a small vertex-separator that partitions the graph into two disconnected subgraphs satisfying some balance criterion. The vertices of the separator are numbered after the vertices in the subgraphs are numbered by following the same strategy recursively.

There are two main approaches to parallelizing this algorithm. In the first approach, the process of finding the separator is performed in parallel [9]. The advantages of this approach are that a reasonably good speedup on ordering can be obtained and the graphs (both original and the subsequent subgraphs) are stored on multiple nodes, thereby avoiding excessive memory use on any single node in a distributed-memory environment. A disadvantage of this approach is that the refinement of partitions becomes less effective as the number of nodes in the parallel computer increases. As a result, the quality of ordering for a given sparse matrix gradually degrades as the number of nodes increases. The second approach is more conservative and exploits only the natural parallelism of nested-dissection ordering. A single node finds the first separator, then two different nodes independently work on the two subgraphs, and so on. An advantage of this approach is that the quality of ordering is independent of the number of nodes in the parallel computer. The disadvantage is that this approach is not scalable with respect to memory use and speedup.

After doing some analysis and studying practical problems, we decided to implement the second approach in parallel *WSSMP*. In a majority of applications, ordering is performed only once for several factorizations of matrices with identical structure, but different numerical values. Therefore, the speed of ordering is often unimportant relative to the quality of ordering, which determines the speed of factorization. The inability of this approach to scale in terms of speedup is not a serious handicap given the fact that it can yield good quality orderings even for a large number of nodes. The memory requirement of ordering is also not a major practical concern because usually the amount of memory required for ordering does not exceed the amount of memory required for subsequent factorization. For instance, we were able to order a linear system of a million equations derived from a $100 \times 100 \times 100$ finite-difference discretization on a single RS6000/397 node with 1 Gigabytes of RAM, but it took a 64-node SP with 1 Gigabytes of RAM on each node to actually solve the system. In addition, in many finite-element applications, each vertex of the discretized domain has multiple degrees of freedom and the graph corresponding to the original system can be compressed into a much smaller graph with identical properties. The ordering algorithm works on the smaller compressed graph and uses much less time and memory than ordering the original graph.

2.2 Symbolic Factorization

The phase of processing between ordering and numerical factorization in *WSSMP* does much more than the traditional symbolic factorization task of computing the nonzero pattern of the triangular factor matrix and allocating data-structures for it. Since this phase takes a very small amount of time and, like ordering, is usually performed only once for several numerical factorization steps, it is performed serially in *WSSMP*. In this phase, we first compute the elimination tree, then perform a quick symbolic step to predict the amount of potential numerical computation associated with each part of the elimination tree, and then adjust the elimination tree (as described in [5]) in order to balance the distribution of numerical factorization work among nodes. The elimination tree is also manipulated to reduce amount of stack memory required for factorization [10]. If the parallel machine has heterogeneous nodes in terms of processing power or the amount of memory, the symbolic phase takes that into account while assigning tasks to nodes. Finally, the vertices of the modified elimination tree are renumbered in a post-ordered sequence, the actual symbolic factorization is performed, respective information on the structure of the portion of the factor matrix to reside on each node is conveyed to all nodes, which, upon receipt of this information, allocate the data-structures for storing their portion of the factor.

2.3 Numerical Factorization

WSSMP can perform either Cholesky (LL^T) or LDL^T factorization. We use a highly scalable parallel algorithm for this step, the detailed description and analysis of which can be found in [5].

The parallel numerical factorization in *WSSMP* is based on the multifrontal algorithm [10]. Given a sparse matrix and the associated elimination tree, the multifrontal algorithm can be recursively formulated as follows. Consider an $N \times N$ matrix A. The algorithm performs a postorder traversal of the elimination tree associated with A. There is a frontal matrix F^k and an update matrix U^k associated with any vertex k. The row and column indices of F^k correspond to the indices of row and column k of L, the lower triangular Cholesky factor, in increasing order. In the beginning, F^k is initialized to an $(s+1) \times (s+1)$ matrix, where $s+1$ is the number of non-zeros in the lower triangular part of column k of A. The first row and column of this initial F^k is simply the upper triangular part of row k and the lower triangular part of column k of A. The remainder of F^k is initialized to all zeros. After the algorithm has traversed all the subtrees rooted at a vertex k, it ends up with a $(t+1) \times (t+1)$ frontal matrix F^k, where t is the number of non-zeros in the strictly lower triangular part of column k in L. The row and column indices of the final assembled F^k correspond to $t+1$ (possibly) noncontiguous indices of row and column k of L in increasing order. If k is a leaf in the elimination tree of A, then the final F^k is the same as the initial F^k. Otherwise, the final F^k for eliminating vertex k is obtained by merging the initial F^k with the update matrices obtained from all the subtrees rooted at k via an extend-add operation. The extend-add is an associative and commutative operator on two update matrices such the index set of the result is the union of the index sets of the original update matrices. After F^k has been assembled, a single step of the standard dense Cholesky factorization is performed with vertex k as the pivot. At the end of the elimination step, the column with index k is removed from F^k and forms the column k of L. The remaining $t \times t$ matrix is

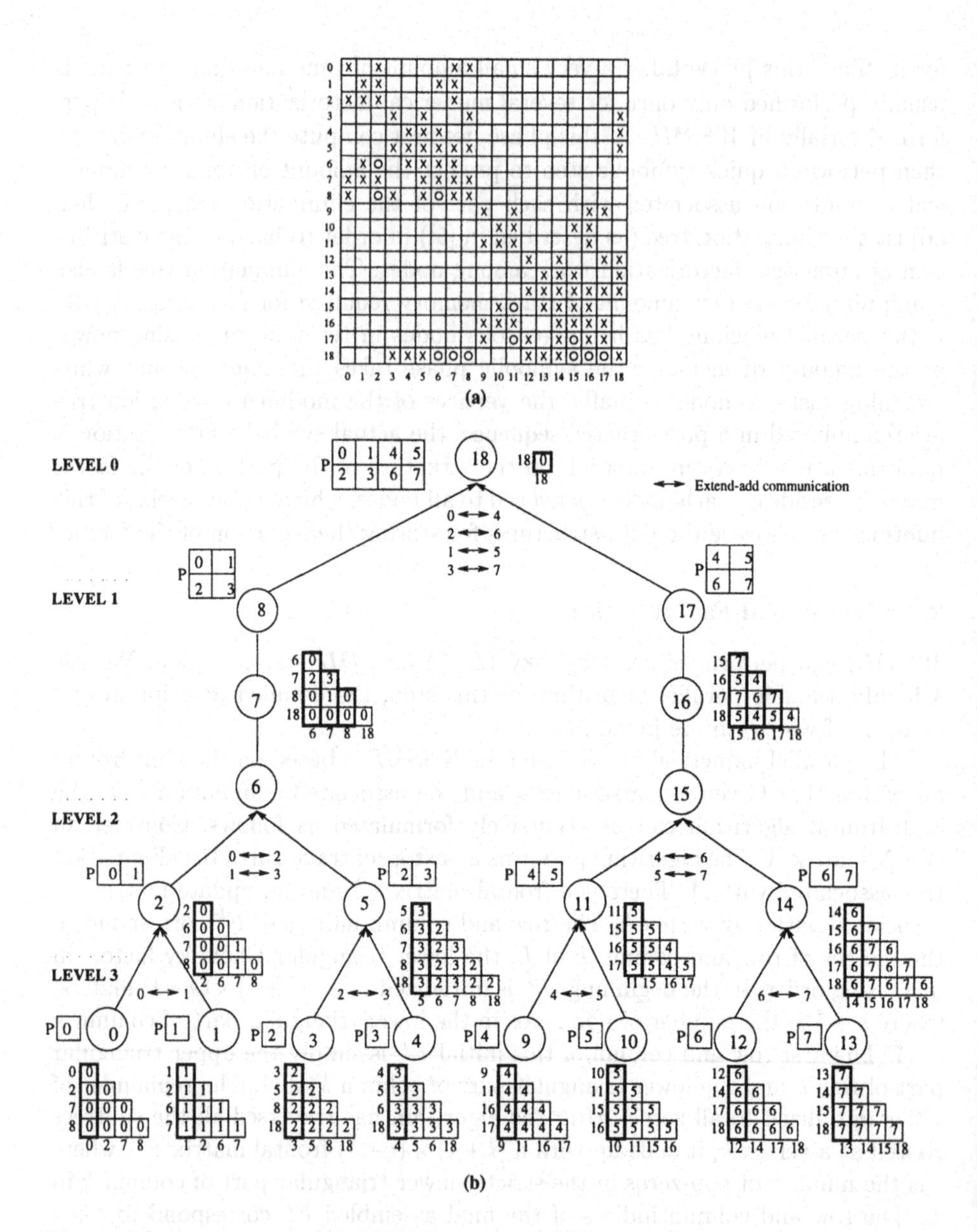

Fig. 1. (a). An example symmetric sparse matrix. The non-zeros of A are shown with symbol "×" in the upper triangular part and non-zeros of L are shown in the lower triangular part with fill-ins denoted by the symbol "o". **(b).** The process of parallel multifrontal factorization using 8 nodes. At each supervertex, the factored frontal matrix, consisting of columns of L (thick columns) and update matrix (remaining columns), is shown.

called the update matrix U^k and is passed on to the parent of k in the elimination tree.

We assume that the supernodal tree is binary in the top $\log p$ levels. The portions of this binary supernodal tree are assigned to the nodes using a subtree-to-subcube strategy illustrated in Figure 1(b), where eight nodes are used to factor the example matrix of Figure 1(a). The subgroup of nodes working on various subtrees are shown in the form of a logical mesh labeled with P. The frontal matrix of each supervertex is distributed among this logical mesh using a bit-mask based block-cyclic scheme [5]. Figure 1(b) shows such a distribution for unit block size. This distribution ensures that the extend-add operations required by the multifrontal algorithm can be performed in parallel with each node exchanging roughly half of its data *only* with its partner from the other subcube. Figure 1(b) shows the parallel extend-add process by showing the pairs of nodes that communicate with each other. Each node sends out the shaded portions of the update matrix to its partner. The parallel factor operation at each supervertex is a pipelined implementation of the dense block Cholesky factorization algorithm.

2.4 Triangular Solves

The solution of triangular systems involves a forward elimination $y = L^{-1}b$ followed by a backward substitution $x = (L^T)^{-1}y$ to determine the solution $x = A^{-1}b$. Our parallel algorithms for this phase are guided by the supernodal elimination tree. They use the same subtree-to-subcube mapping and the same two-dimensional distribution of the factor matrix L as used in the numerical factorization.

Figure 2(a) illustrates the parallel formulation of the forward elimination process. The right hand side vector b, is distributed to the nodes that own the corresponding diagonal blocks of the L matrix as shown in the shaded blocks in Figure 2(a). The computation proceeds in a bottom-up fashion. Initially, for each leaf vertex k, the solution y_k is computed and is used to form the update vector $\{l_{ik}y_k\}$ (denoted by "U" in Figure 2(a)). The elements of this update vector need to be subtracted from the corresponding elements of b, in particular $l_{ik}y_k$ will need to be subtracted from b_i. However, our algorithm uses the structure of the supernodal tree to accumulate these updates upwards in the tree and subtract them only when the appropriate vertex is being processed. For example consider the computation involved while processing the supervertex {6,7,8}. First the algorithm merges the update vectors from the children supervertex to obtain the combined update vector for indices {6,7,8,18}. Note that the updates to the same b entries are added up. Then it performs forward elimination to compute y_6, y_7 and y_8. This computation is done using a two dimensional pipelined dense forward elimination algorithm. At the end of the computation, the update vector on node 0 contains the updates for for b'_{18} due to y_6, y_7 and y_8 as well as the updates received from supervertex {5}. In general, at the end of the computation at each supervertex, the accumulated update vector resides on the column of nodes that store the last column of the L matrix of that supervertex. This

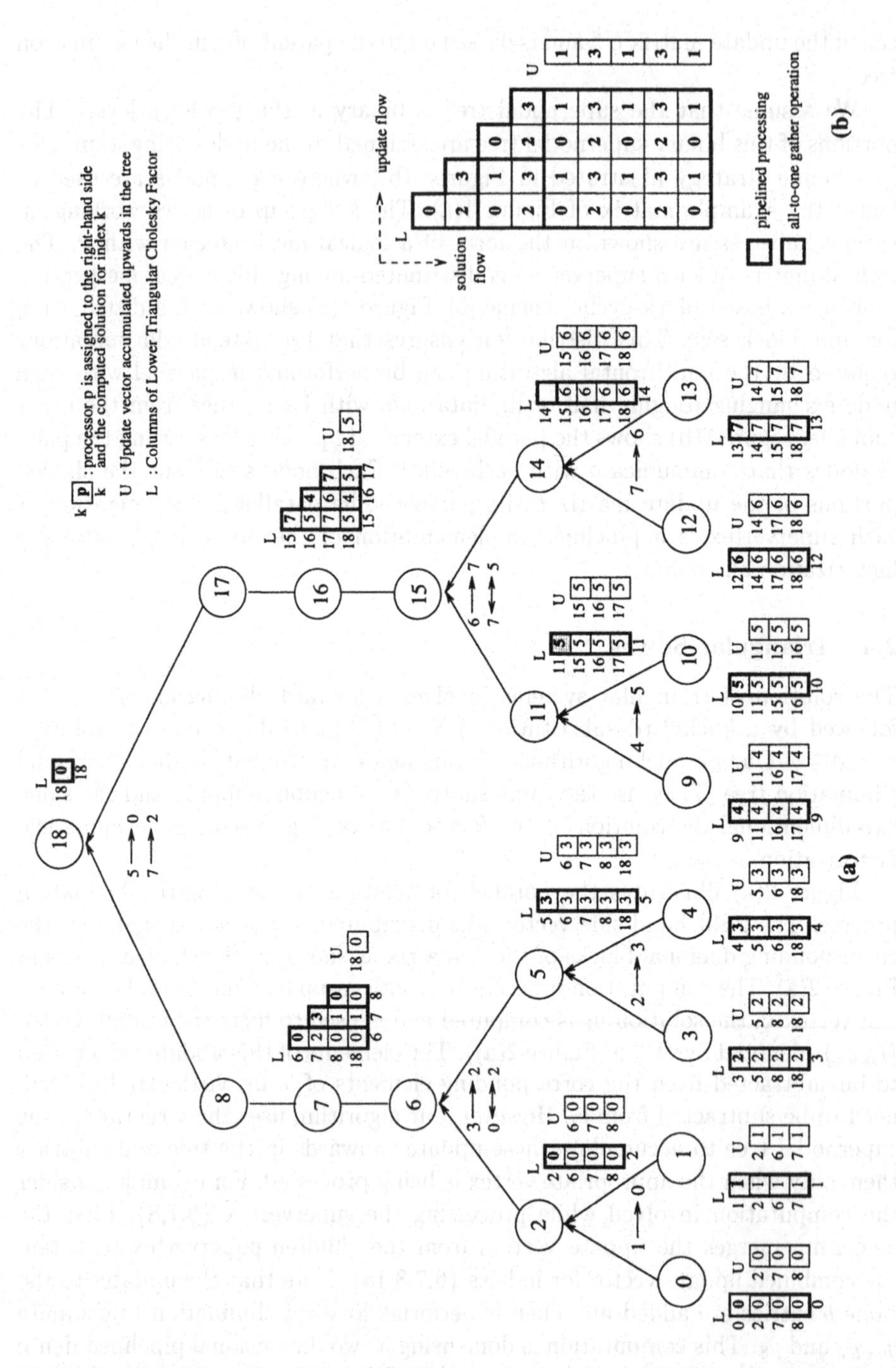

Fig. 2. Parallel Triangular Solve. **(a).** Entire process of parallel forward elimination for the example matrix. **(b).** Processing within a hypothetical supernodal matrix for forward elimination.

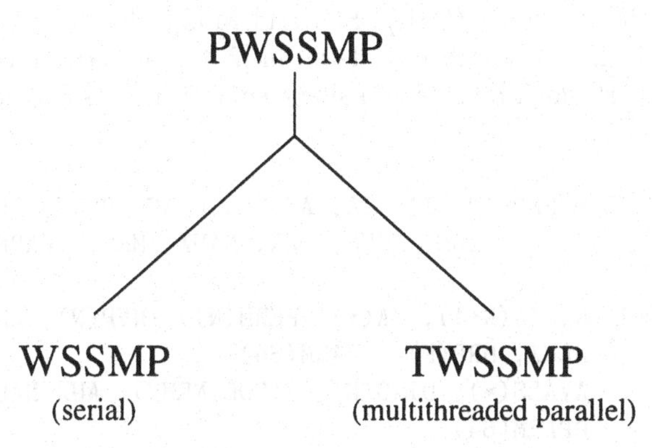

Fig. 3. The two-tier organization of the *WSSMP* software. The message-passing layer can work on top of either a serial or a multithreaded layer.

update vector needs to be sent to the nodes that store the first column of the L matrix of the parent supervertex. Because of the bit-mask based block-cyclic distribution, this can be done by using at most two communication steps [7].

The details of the two-dimensional pipelined dense forward elimination algorithm are illustrated in Figure 2(b) for a hypothetical supervertex. The solutions are computed by the nodes owning diagonal elements of L matrix and flow down along a column. The accumulated updates flow along the row starting from the first column and ending at the last column of the supervertex. The processing is pipelined in the shaded regions and in other regions the updates are accumulated using a reduction operation along the direction of the flow of update vector.

Our algorithm for parallel backward substitution is similar.

3 The Software

The *WSSMP* software is organized into three libraries (see Figure 3). The user's program needs to be linked with the *WSSMP* library on a serial work-station and with the *TWSSMP* library on an SMP work-station. The latter is a multithreaded library and uses *Pthreads* to exploit parallelism on a multiprocessor work-station. On an SP with uniprocessor nodes, the user's program needs to be linked with both *WSSMP* and *PWSSMP* libraries. The latter contains all the message-passing code. On an SP with multiprocessor nodes, *WSSMP* must be replaced with *TWSSMP*. Thus, depending on the configuration of the SP, the message-passing library *PWSSMP* is used in conjunction with either *WSSMP* or *TWSSMP*.

In this section, we give a brief description of the main routines of the package. The details of the user interface can be found in [4]. *WSSMP* accepts as input a triangular portion of the sparse coefficient matrix. Both C-style (indices starting from 0) and Fortran-style (indices starting from 1) numbering is supported. *WSSMP* supports two popular formats for the coefficient matrix: Compressed Sparse Row/Column (CSR/CSC) and Modified Sparse Rows/Column (MSR/MSC), which are described in detail in [2,4]. All major functions of the package can be performed by calling a single subroutine. Its calling sequence is as follows:

```
      SUBROUTINE WSSMP (N, IA, JA, AVALS, DIAG, PERM, INVP, B,
     +                 LDB, NRHS, AUX, NAUX, MRP, IPARM, DPARM)

      INTEGER*4  N, IA(N+1), JA(*), PERM(N), INVP(N), LDB, NRHS,
     +           NAUX, MRP(N), IPARM(64)
      REAL*8     AVALS(*), DIAG(N), B(LDB,NRHS), AUX(NAUX),
     +           DPARM(64)
```

N is the number of equations in the system. In CSR/CSC format, *IA, JA,* and *AVALS* contain the coefficient matrix; in MSR/MSC format, the coefficient matrix is contained in *IA, JA, AVALS,* and *DIAG*. *IA* contains pointers to row/column indices in *JA*, which contains the actual indices, and *AVALS* contains the values corresponding to the indices in *JA*. In MSR/MSC format, the diagonal entries are stored separately in *DIAG*. *PERM* and *INVP* contain the permutation and inverse permutation vectors, respectively, to reorder the matrix for reducing fill-in. This permutation can either be supplied by the user, or generated by *WSSMP*. *B* is the *LDB* × *NRHS* right-hand side matrix (vector, if *NRHS* = 1), which is overwritten by the solution matrix/vector by *WSSMP*. *AUX* is optional working storage that the user may supply to *WSSMP* and *NAUX* is the size of *AUX* in 8-byte words. *MRP* is an integer vector in which the user can request output information about the pivots. *IPARM* and *DPARM* are integer and double precision arrays that contain the various input-output numerical parameters that direct program flow (input parameters) and convey useful information to the user (output parameters). The most important of these parameters control which task(s) *WSSMP* will be performing in a given call. *WSSMP* can perform any set of consecutive tasks from the following list:

Task # 1:	Ordering
Task # 2:	Symbolic Factorization
Task # 3:	Cholesky or LDL^T Factorization
Task # 4:	Forward and Backward Elimination
Task # 5:	Iterative Refinement

The user specifies a starting task and an ending task and *WSSMP* performs these tasks and all the tasks between them. The user can obtain the entire solution in one call or perform multiple factorizations for different matrices with identical structure by performing ordering and symbolic factorization only once, or solve for multiple RHS vectors with a given factorization, or perform multiple steps of iterative refinement for a given solution. The user can supply *PERM* and *INVP* from any other source and skip the first task. All the information that is generated by one task and is needed in subsequent tasks is maintained and stored internally by *WSSMP*.

The distribute memory parallel subroutine *PWSSMP* is almost identical:

```
      SUBROUTINE PWSSMP (N_i, IA_i, JA_i, AVALS_i, DIAG_i, PERM,
     +                  INVP, B_i, LDB_i, NRHS, AUX_i, NAUX_i,
     +                  MRP_i, IPARM, DPARM)

      INTEGER*4  N_i, IA_i(N_i+1), JA_i(*), PERM(*), INVP(*),
     +           LDB_i, NRHS, NAUX_i, MRP_i(N_i), IPARM(64)
      REAL*8     AVALS_i(*), DIAG_i(N_i), B_i(LDB_i,NRHS),
     +           AUX_i(NAUX_i), DPARM(64)
```

In the distributed version, an argument can be either *local* or *global*. A global array or variable must have the same size and contents on all nodes. The size and contents of a local variable or array vary among the nodes. In the context of *PWSSMP*, global does not mean globally shared, but refers to data that is replicated on all nodes. In the above calling sequence, all arguments with a subscript are local.

N_i is the number of columns/rows of the matrix A and the number of rows of the right-hand side B residing on node P_i. The total size N of system of equations is $\Sigma_{i=0}^{p-1} N_i$, where p is the number of nodes being used. The columns of the coefficient matrix can be distributed among the nodes in any fashion (see [4] for details). $N_0 = N$ and $N_i = 0$ for $i > 0$ is also an acceptable distribution. Thus, it is rather easy to migrate from using *WSSMP* on a single node to using it in the distributed-memory environment of an SP.

4 Performance Results

The performance *WSSMP* on a given number of processors for a problem depends on several factors, such as fill-in, load-imbalance, size of supernodes, etc. Many of these factors depend on ordering and the structure of the sparse matrix being factored. Therefore, the performance on the same number of processors can vary widely for different problems of the same size (in terms of the total amount of computation required). In Tables 1 and 2, we present some performance results of *WSSMP* on two practical classes of sparse matrices on up to 64 nodes of an SP. We feel that a reader may find these results more interpretive than those for a large number of unrelated problems. The first column of the tables gives some relevant information about the matrices, including the number of nonzeros in the lower triangular factor (in millions) and the number of floating point operations required for factorizations (in billions). These values are only averages; the actual values vary depending on the number of nodes used because the ordering is different on different number of nodes.

Each node of the SP is a 160 MHz Power-2 processor (RS6000/370) with 1 Gigabytes of RAM. At the time of preparing this document, the multi-threaded implementation of *WSSMP* for SMP machines was not yet complete.

Problem Description	Number of Nodes 1	2	4	8	16	32	64	
Cube-35	19.0	9.67	4.69	3.43	2.15	1.38	0.88	F_time
F_opc = 9 B	471	897	1572	2667	4205	6521	9076	F_mflops
nnz_L = 10 M	.373	.197	.111	.081	.057	.049	.044	S_time
	110	202	368	510	726	828	946	S_mflops
Cube-39	37.2	19.8	10.4	6.09	3.59	2.55	1.56	F_time
F_opc = 18 B	484	911	1626	2947	4469	6948	11372	F_mflops
nnz_L = 16 M	.617	.342	.174	.136	.081	.072	.055	S_time
	107	208	397	569	890	950	1217	S_mflops
Cube-44	75.9	40.0	22.4	11.9	7.24	5.11	2.58	F_time
F_opc = 37 B	498	951	1696	3156	5217	7329	13975	F_mflops
nnz_L = 28 M	.909	.516	.281	.165	.103	.086	.072	S_time
	123	216	397	671	1085	1302	1496	S_mflops
Cube-49	146.	74.0	41.1	21.8	13.2	9.46	5.36	F_time
F_opc = 74 B	508	973	1796	3290	5588	7820	13871	F_mflops
nnz_L = 43 M	1.37	.726	.496	.223	.169	.134	.107	S_time
	129	239	354	767	1104	1289	1648	S_mflops
Cube-55		150.	83.0	42.9	25.3	16.0	9.40	F_time
F_opc = 149 B		1016	1829	3555	5819	9312	15836	F_mflops
nnz_L = 71 M		1.20	.623	.346	.242	.205	.185	S_time
		241	464	845	1190	1383	1517	S_mflops
Cube-62			165.	88.6	53.3	29.6	16.0	F_time
F_opc = 320 B			1939	3601	5790	10615	20065	F_mflops
nnz_L = 120 M			1.01	.560	.365	.238	.192	S_time
			477	859	1333	1992	2525	S_mflops
Cube-70				168.	91.2	52.6	28.9	F_time
F_opc = 604 B				3522	6532	11458	21480	F_mflops
nnz_L = 185 M				.818	.500	.330	.194	S_time
				870	1462	2133	3186	S_mflops
Cube-79				372.	214.	114.	60.2	F_time
F_opc = 1402 B				3772	6471	12265	22436	F_mflops
nnz_L = 337 M				1.27	.854	.561	.358	S_time
				821	1580	2410	3762	S_mflops
Cube-89					409.	231.	123.	F_time
F_opc = 2892 B					6902	12751	23725	F_mflops
nnz_L = 559 M					1.31	.836	.508	S_time
					1699	2718	4373	S_mflops
Cube-100							255.	F_time
F_opc = 6133 B							24028	F_mflops
nnz_L = 931 M							.827	S_time
							4502	S_mflops

Table 1. *WSSMP* performance for factorization and solution on IBM SP/370 for sparse matrices arising from finite-difference discretization of 3-D domains. A matrix Cube-n is obtained from an $n \times n \times n$ 3-D mesh. F_time is factorization time in seconds, F_mflops is factorization megaflops, S_time is solution time in seconds, S_mflops is solution megaflops, F_opc is number of factorization floating point operations in billions, and nnz_L is the number of nonzeroes (fill) in the triangular factor in millions.

Table 1 shows the performance of factorization and triangular solves for sparse matrices arising from three-dimensional finite-difference grids discretized using a 5-point stencil. The grid dimension n is increased in such steps as to increase the factorization work by a factor of two. For this, we have used the fact that with an optimal ordering (finding which is an NP-complete problem), the the number of floating point operations required to factor a sparse matrix derived from an $n \times n \times n$ grid is $O(n^6)$.

Problem Description	1	2	4	8	16	32	64	
Sheet-20	26.8	14.6	8.47	5.64	3.33	2.30	1.45	F_time
F_opc = 12 B	444	814	1402	2109	3564	5160	8185	F_mflops
nnz_L = 23 M	0.55	0.32	0.22	0.15	0.10	0.07	0.06	S_time
	167	288	418	612	919	1312	1531	S_mflops
Sheet-40		41.3	24.1	14.9	9.11	5.67	3.61	F_time
F_opc = 36 B		874	1495	2424	3959	6361	9991	F_mflops
nnz_L = 52 M		0.84	0.48	0.29	0.19	0.12	0.09	S_time
		248	433	717	1095	1733	2311	S_mflops
Sheet-60			42.6	25.4	16.5	9.40	5.99	F_time
F_opc = 70 B			1643	2756	4242	7447	11667	F_mflops
nnz_L = 85 M			0.69	0.37	0.22	0.15	0.11	S_time
			493	918	1545	2266	3091	S_mflops
Sheet-80			70.5	38.1	22.6	14.0	8.52	F_time
F_opc = 118 B			1674	3097	5221	8429	13850	F_mflops
nnz_L = 106 M			0.96	0.52	0.29	0.19	0.14	S_time
			442	815	1462	2232	3029	S_mflops
Sheet-100			94.8	50.7	29.3	18.0	10.9	F_time
F_opc = 152 B			1604	3031	5188	8449	13948	F_mflops
nnz_L = 153 M			1.24	0.65	0.38	0.26	0.20	S_time
			493	941	1610	2353	3060	S_mflops

Table 2. *WSSMP* performance for factorization and solution on IBM SP/370 for sparse matrices generated by a 2-D finite-element model. The matrix Ford-n is obtained from an n-element model and has roughly $4.5 \times n$ equations. F_time is factorization time in seconds, F_mflops is factorization megaflops, S_time is solution time in seconds, S_mflops is solution megaflops, F_opc is number of factorization floating point operations in billions, and nnz_L is the number of nonzeroes (fill) in the triangular factor in millions.

Table 2 shows the performance of factorization and triangular solves for sparse matrices arising from finite-element meshes in a sheet metal forming application. In all the five problems, the same surface is discretized with different degrees of fineness (and thus, with different number of elements) to control the problem size.

5 Concluding Remarks

In this paper, we have described a scalable parallel solver (*WSSMP*) for sparse symmetric systems of linear equations, which works on serial, message-passing parallel, and shared-memory parallel machines, or any combination of these. By means of experimental performance results, we have shown that the solver scales well on up to dozens of nodes of an SP. On large enough problems, the factorization performance exceeds 20 Gigaflops on 64 nodes, thereby obtaining about half of the peak theoretical performance of each node. Even on a moderate size problem (e.g., Sheet-80, Sheet-100) that takes only 11 seconds in factorization, a 64-node ensemble operates at about one-third of the peak theoretical performance of each node and at about 50% efficiency with respect to the serial performance of the solver on the same type of problem. As a result, *WSSMP* is already in use in several commercial applications, in many of which, it is solving problems that are so large that they could not be solved until recently with a direct solver.

References

1. C. Ashcraft and J. W.-H. Liu. Robust ordering of sparse matrices using multisection. Technical Report CS 96-01, Department of Computer Science, York University, Ontario, Canada, 1996.
2. E. Chu et al. Users guide for SPARSPAK–A: Waterloo sparse linear equations package. Technical Report CS-84-36, University of Waterloo, Waterloo, IA, 1984.
3. A. Gupta. Fast and effective algorithms for graph partitioning and sparse matrix ordering. *IBM Journal of Research and Development*, 41(1/2):171–183, 1997.
4. A. Gupta, M. Joshi, and V. Kumar. WSSMP: Watson symmetric sparse matrix package: The user interface. RC 20923 (92669), IBM T. J. Watson Research Center, Yorktown Heights, NY, July 17, 1997.
5. A. Gupta, G. Karypis, and V. Kumar. Highly scalable parallel algorithms for sparse matrix factorization. *IEEE Transactions on Parallel and Distributed Systems*, 8(5):502–520, May 1997.
6. B. Hendrickson and E. Rothberg. Improving the runtime and quality of nested dissection ordering. SAND96-0868J, Sandia National Laboratories, Albuquerque, NM, 1996.
7. M. Joshi, A. Gupta, G. Karypis, and V. Kumar. Two dimensional scalable parallel algorithms for solution of triangular systems. TR 97-024, Department of Computer Science, University of Minnesota, Minneapolis, MN, July 1997.
8. G. Karypis and V. Kumar. A fast and high quality multilevel scheme for partitioning irregular graphs. TR 95-035, Department of Computer Science, University of Minnesota, 1995.
9. G. Karypis and V. Kumar. Parmetis: Parallel graph partitioning and sparse matrix ordering library. TR 97-060, Department of Computer Science, University of Minnesota, 1997.
10. J. W.-H. Liu. The multifrontal method for sparse matrix solution: Theory and practice. *SIAM Review*, 34:82–109, 1992.

Recursive Blocked Data Formats and BLAS's for Dense Linear Algebra Algorithms

Fred Gustavson[1], André Henriksson[2], Isak Jonsson[2], Bo Kågström[2], and Per Ling[2]

[1] IBM T.J. Watson Research Center, P.O. Box 218, Yorktown Heights, NY 10598, U.S.A.
gustav@watson.ibm.com
[2] Department of Computing Science and HPC2N, Umeå University, S–901 87 Umeå, Sweden
{andropov,isak,bokg,pol}@cs.umu.se

Abstract. Recursive blocked data formats and recursive blocked BLAS's are introduced and applied to dense linear algebra algorithms that are typified by LAPACK. The new data formats allow for maintaining data locality at every level of the memory hierarchy and hence providing high performance on today's memory tiered processors. This new data format is hybrid. It contains blocking parameters which are chosen so that the associated submatrices of a block-partitioned A fit into level 1 cache. The recursive part of the data format chooses a linear order of the blocks that maintains a two-dimensional data locality of A in a one-dimensional tiered memory structure. We argue that, out of the NB factorial choices of ordering the NB blocks, our recursive ordering leads to one of the best. This is because our algorithms are also recursive and will do their computations on submatrices that follow the new recursive data structure definition. This is in analogy with the well known principle that the data structure should be matched to the algorithm. Performance results in support for our recursive approach are also presented.

1 Introduction

Today block-based algorithms are the standard in state-of-the-art software for dense linear algebra computations [2]. These algorithms perform most of their computations in calls to high-performance level 3 BLAS [3, 10]. The level 3 BLAS obtain most of their performance by data rearrangements and the use of high-performance atomic kernel routines. The current BLAS are designed to exploit the architecture design while maintaining the functionality of the BLAS and thereby guarantee high performance and portability of dense linear algebra codes. The level 3 factorization algorithms typified by LAPACK make repeated calls to the level 3 BLAS with matrix operands equal to submatrices of fixed size. This results in multiple data copying on operands that are related. How can this be challenged? An answer is to change the data structure of the BLAS input matrices to reflect this relationship, thereby removing any data copying from the BLAS. The obvious choice is to store matrices as blocks, i.e., submatrices of block-partioned matrices, instead of the standard column-major (Fortran) or row-major (C) orderings.

One way is to store these blocks in block row or block column formats, with the individual blocks stored in the conventional way. This will give an overall performance improvement since the new BLAS codes become simpler and perform no redundant data copying. However, since the data structure and the algorithms will only partially match, these revised codes will not perform at peak performance.

Recursion is a key concept for matching an algorithm and its data structure. A recursive algorithm leads to automatic blocking which is variable and "squarish" [6]. This layered and variable blocking allow for good data locality, which makes it possible to approach peak performance on today's memory tiered processors. Using the current standard data layouts and applying recursive blocking have led to faster algorithms for the Cholesky, LU and QR factorizations [6, 11, 4]. However, as we will show here even better performance can be obtained when recursive dense linear algebra algorithms are expressed using a recursive data structure and recursive kernels.

Before we go into any further details we outline the rest of the paper. Section 2 introduces the new recursive blocked data formats. We describe two atomic formats for rectangular and triangular matrices, respectively, that are tailored for recursive dense linear algebra algorithms. In Section 3 we introduce the recursive blocked BLAS's that support the recursive data formats. We use the packed Cholesky factorization of a positive definite matrix A as a case study. Section 4 describes our design principles for a recursive level 3 algorithm for the general matrix multiply and add (GEMM) operation resulting in a recursion tree with kernel routine calls at the leaves. Finally, in Section 5 we present some performance results from our ongoing developments that strongly support our new recursive approach.

2 Recursive Blocked Data Formats

We introduce a new set of data formats for storing block-partitioned dense matrices. The set is a hybrid of two "addressing" techniques. At the block level each submatrix is stored in the standard column-major (or row-major) order and the size of each block is constrained so that a few of them will simultaneously fit in level 1 cache. Blocks stored in this fashion are operands for the kernel routines. Next we describe the layout of the blocks in memory. To allow for efficient utilization of a memory hierarchy, including efficient cache reuse, we propose a novel method for storing blocks of matrices recursively. We describe two recursive matrix formats: the rectangle and the isosceles right triangle. With these data formats in hand we show that the performance of dense linear algebra computations can be improved.

2.1 Rectangular Recursive Data Format

The metrics of a matrix A of size $M \times N$ stored in ordinary column-major order (Fortran format) include M, N and the leading dimension lda. The latter is used for a proper specification of a subarray of A. For block-partitioning of A we also use two parameters specifying the block sizes, mb and nb, where $1 \leq mb \leq M$ and $1 \leq nb \leq N$. A block-partitioned A consists of $NB = m \cdot n$ blocks A_{ij} of size $mb \times nb$, where $m = \lceil M/mb \rceil$ and $n = \lceil N/nb \rceil$. We use the convention that the last block row and/or block column are padded with zero elements when M and/or N are not multiples of mb and nb. However, no computations on these zero elements are performed. Each submatrix A_{ij} is stored in column-major or row-major order. Notice that padding leads to many economies in code production, e.g., no fix-up code are required in the atomic kernels.

Now, these submatrices can be ordered in $NB!$ different ways. We claim that a recursive ordering is an effective ordering when performing linear algebra operations on the matrix. In fact, we claim that it is better than block row or block column ordering of the blocks. The reason for this is that a recursive blocked format allow for maintaining the two-dimensional data locality at every level of the one-dimensional tiered memory structure, while the block row and block column orderings only maintain data locality at a submatrix level. Irrespective of how the block sizes are set, the block row or block column format can only automatically match one level of the memory hierarchy, e.g., level 1 cache.

The recursive block ordering is determined by always dividing the rectangular dimension of the submatrix that is the largest. The choice when there is a tie leads to two formats: *recursive block row* (RBR): always divide the row dimension; and *recursive block column* (RBC): always divide the column dimension. When dividing an odd number of rows the middle row is assigned to the block at the bottom. For an odd number of columns, the middle one is assigned to the block to the right. The reason is that the block to the right or at the bottom may contain submatrices that are not entirely filled, so this strategy keeps the difference in number of used elements between the two blocks after the splitting to a minimum.

For illustration we consider A of size 496×380 and $mb = nb = 100$, giving $m = 5$ and $n = 4$. Using a block column format, the blocks are mapped as in (1) of Figure 1. The numbers 0–19 denote contiguous blocks in memory. Notice that not all elements in submatrices 4, 9, 14–19 match elements in A, i.e., these submatrices are padded with zero elements.

Using the RBR format for assigning contiguous blocks in memory, we first observe that $5 = m > n = 4$, so the splitting is below the second row ($\lfloor m/2 \rfloor = 2$), assigning the block numbers 0–7 to the upper part and 8–19 to the lower. Since the number of columns (4) is greater than the number of rows (2) in the upper part, the next splitting is vertical, assigning block numbers 0–3 to the left hand side part. This submatrix is now square so the row dimension is split. The complete block assignment associated with the RBR format is displayed in (2) of Figure 1. The procedure is similar when applying the RBC assignment of contiguous blocks, except when splitting square submatrices, see (3). We have

$$\begin{bmatrix} A_{11} & A_{12} & \cdots & A_{14} \\ A_{21} & \ddots & & A_{24} \\ \vdots & & & \vdots \\ A_{51} & \cdots & & A_{54} \end{bmatrix} \sim \begin{pmatrix} 0 & 5 & 10 & 15 \\ 1 & 6 & 11 & 16 \\ 2 & 7 & 12 & 17 \\ 3 & 8 & 13 & 18 \\ 4 & 9 & 14 & 19 \end{pmatrix}_{(BC)} \tag{1}$$

$$\sim \left(\begin{array}{cc|cc} 0 & 1 & 4 & 5 \\ 2 & 3 & 6 & 7 \\ \hline 8 & 9 & 14 & 15 \\ 10 & 11 & 16 & 17 \\ 12 & 13 & 18 & 19 \end{array}\right)_{(RBR)} \tag{2}$$

$$\sim \left(\begin{array}{cc|cc} 0 & 2 & 4 & 6 \\ 1 & 3 & 5 & 7 \\ \hline 8 & 9 & 14 & 15 \\ 10 & 12 & 16 & 18 \\ 11 & 13 & 17 & 19 \end{array}\right)_{(RBC)} \tag{3}$$

Fig. 1. The mapping of a 5×4 block matrix in block column order (BC), recursive block row order (RBR), and recursive block column order (RBC).

also marked the result of the first recursive splittings for the RBR and RBC formats, see (2) and (3).

The choice of mb and nb is crucial and closely linked to the memory hierarchy of the target architecture. In the extreme case of $mb = M, nb = N$, the entire matrix is stored in one block, which corresponds to the conventional column-major and row-major orderings with their deficiences like bad data locality in either the row or the column dimension. If $mb = nb = 1$, we apply recursive splitting down to the single elements. This results in good data locality, but the submatrix operations become very inefficient since the kernels will only operate on single elements. For example, register blocking is prohibited. The best choice of mb and nb depends on the size of the level 1 cache. One should strive for fitting one or a few submatrices in level 1 cache. The consequence of having bad data locality in one dimension of the submatrix does not hinder performance, since the level 1 cache is random access. Thus, we have linear addressing in the kernels, which simplifies loop unrolling, preloading and register blocking. The time to calculate the address of an element is $O(\log M + \log N)$. However, this information may be stored in tables, which gives greater flexibility in the placement of the blocks and constant time addressing.

Another benefit of the tables is that the space allocated for each block, S, may be greater than the space required, i.e., the block addresses may be padded so that cache coherency problems may be avoided. For example, the allocation strategy may be to let all blocks begin at a line or page boundary.

2.2 Triangular Recursive Data Format

Since all matrix factorizations can be expressed in terms of rectangular and isosceles triangular matrices it is enough to consider the isosceles case. For an isosceles right triangle of order N, the splitting procedure resembles the rectangular case. Let $nb \times nb$ be the size of the submatrices, giving a block-partitioned triangular A consisting of $n(n+1)/2$ blocks where $n = \lceil N/nb \rceil$. Now divide the triangle into one sub-rectangle and two isosceles sub-triangles. For a lower triangle, the upper left triangle is assigned block numbers 0 to $\lfloor n/2 \rfloor \lfloor n/2+1 \rfloor/2 - 1$, the lower right triangle is assigned block numbers $\lfloor n/2 \rfloor \lfloor n/2+1 \rfloor/2 + \lfloor n/2 \rfloor \lceil n/2 \rceil$ to $n(n+1)/2 - 1$, and the blocks inside the rectangle will use block numbers $\lfloor n/2 \rfloor \lfloor n/2+1 \rfloor/2$ to $\lfloor n/2 \rfloor \lfloor n/2+1 \rfloor/2 + \lfloor n/2 \rfloor \lceil n/2 \rceil - 1$. The interior ordering of the blocks in the triangles are determined by applying the algorithm recursively, and the block ordering is either RBR or RBC as in the rectangular case. Figure 2 illustrates the triangular recursive blocked orderings for triangular matrices of order 450, and block size $nb = 100$. The diagonal blocks 0, 2, 9, 12, and 14 are stored in the conventional full format. We use full format since it is easier to write high-performance kernel routines using this format. The blocks in the last block row or block column are possibly padded. The zero blocks in the upper or lower parts, respectively, are not stored, so the space overhead is linear to the matrix order.

$$\begin{bmatrix} A_{11} & & & \\ A_{21} & A_{22} & & \\ \vdots & & \ddots & \\ A_{51} & \cdots & & A_{55} \end{bmatrix} \sim \left(\begin{array}{cc|ccc} 0 & & & & \\ 1 & 2 & & & \\ \hline 3 & 4 & 9 & & \\ 5 & 6 & 10 & 12 & \\ 7 & 8 & 11 & 13 & 14 \end{array}\right)_{(RBR)} \sim \left(\begin{array}{cc|ccc} 0 & & & & \\ 1 & 2 & & & \\ \hline 3 & 4 & 9 & & \\ 5 & 7 & 10 & 12 & \\ 6 & 8 & 11 & 13 & 14 \end{array}\right)_{(RBC)}$$

$$\begin{bmatrix} A_{11} & A_{12} & \cdots & A_{15} \\ & A_{22} & & \vdots \\ & & \ddots & \\ & & & A_{55} \end{bmatrix} \sim \left(\begin{array}{cc|ccc} 0 & 1 & 3 & 5 & 6 \\ & 2 & 4 & 7 & 8 \\ \hline & & 9 & 10 & 11 \\ & & & 12 & 13 \\ & & & & 14 \end{array}\right)_{(RBR)} \sim \left(\begin{array}{cc|ccc} 0 & 1 & 3 & 5 & 7 \\ & 2 & 4 & 6 & 8 \\ \hline & & 9 & 10 & 11 \\ & & & 12 & 13 \\ & & & & 14 \end{array}\right)_{(RBC)}$$

Fig. 2. The RBR and RBC orderings associated with triangular recursive blocked data formats. The splittings from the first recursion step are shown.

From Figure 2 we see that the RBR ordering of a symmetric matrix in lower triangular storage format is equivalent to the RBC ordering of a symmetric matrix in upper triangular storage format, and vice versa.

3 Recursive Blocked BLAS's

A main motivation for introducing the recursive blocked data format is to be able to match the data structure with the recursive algorithm and thereby avoiding data copying. However, this requires a new set of level 3 BLAS that support the new recursive storage format. In the following we use the Cholesky factorization for illustration. Here we assume that the lower triangular part of A is stored in recursive block row (RBR) format. In Figure 3 we show the labeling of the blocks of size 100×100 associated with the RBR format applied to A of size 800×800. Block $i, i = 0, \ldots, 35$ of size 100×100 starts in position $100^2 i$ of the one-dimensional array that stores A. The individual blocks can optionally be stored in column-major or row-major order. Notice that the diagonal blocks are stored in full format which results in a small overhead in storage compared to a standard packed format besides possible paddings of the last block row.

$$A \sim \left[\begin{array}{cccc|cccc} 0 & & & & & & & \\ 1 & 2 & & & & & & \\ 3 & 4 & 7 & & & & & \\ 5 & 6 & 8 & 9 & & & & \\ \hline 10 & 11 & 14 & 15 & 26 & & & \\ 12 & 13 & 16 & 17 & 27 & 28 & & \\ 18 & 19 & 22 & 23 & 29 & 30 & 33 & \\ 20 & 21 & 24 & 25 & 31 & 32 & 34 & 35 \end{array}\right] \sim \left[\begin{array}{c|c} A_{11} & \\ \hline A_{21} & A_{22} \end{array}\right] \tag{4}$$

Fig. 3. Recursive blocked row format of lower triangular A ($N = 800, nb = 100$).

The submatrices A_{ij} in Figure 3 are associated with the first level of recursion in the recursive blocked-Cholesky factorization algorithm (see Figure 4). We get

function $[A(1:N, 1:N)]$ = **RB-Cholesky** $(A(1:N, 1:N), n)$
if $n = 1$ **then** *{Leaf node; n = # diagonal blocks in A}*
  Cholesky factor $A(1:N, 1:N)$; *{Use Cholesky kernel}*
else
  $N1 = N/2$; $N2 = N - N1$; $J1 = N1 + 1$;
  $n1 = n/2$; $n2 = n - n1$; *{# blocks in next recursive calls}*
  RB-Cholesky $(A(1:N1, 1:N1), n1)$;
  Solve $A(1:N1, 1:N1)X = A(1:N1, J1:N)$; *{Use* **RB-TRSM***}*
  Update $A(J1:N, J1:N) = A(J1:N, J1:N) - X^T X$; *{Use* **RB-SYRK***}*
  RB-Cholesky $(A(J1:N, J1:N), n2)$
end

Fig. 4. Recursive blocked-Cholesky algorithm.

two branches in the recursion tree that correspond to recursive calls to the Cholesky algorithm with the arguments A_{11} and A_{22}, respectively. In between these calls there are calls to the level 3 BLAS routines for triangular solve of multiple right hands sides (TRSM) and symmetric rank-k update (SYRK). These operations are straightforwardly derived from an appropriate block-partitioning of A and L in $A = LL^T$:

$$\begin{bmatrix} A_{11} & A_{21}^T \\ A_{21} & A_{22} \end{bmatrix} = \begin{bmatrix} L_{11} & 0 \\ L_{21} & L_{22} \end{bmatrix} \begin{bmatrix} L_{11}^T & L_{21}^T \\ 0 & L_{22}^T \end{bmatrix} = \begin{bmatrix} L_{11}L_{11}^T & L_{11}L_{21}^T \\ L_{21}L_{11}^T & L_{21}L_{21}^T + L_{22}L_{22}^T \end{bmatrix}. \quad (5)$$

Notice that L overwrites A in algorithm RB-Cholesky. Since the two level 3 operations (TRSM and SYRK) are performed on the A_{ij} blocks in RBR format, we also need recursive algorithms for these operations that match with the blocked recursive data format. We use the convention to add the prefix RB- to the name of a recursive level 3 operation or algorithm. For example, RB-TRSM and RB-SYRK are the two recursive blocked BLAS operations explicitly called in RB-Cholesky.

In this example we have four levels of recursion $(0, 1, 2,$ and $3)$. Level 3 corresponds to the eight leaves of the binary tree, each calling the Cholesky kernel to factorize one (updated) diagonal block of A in RBR format (labeled $0, 2, 7, 9, 26, 28, 33,$ and 35). All other nodes perform RB-TRSM and RB-SYRK operations. The four nodes on level 2 work on

$$\left[\begin{array}{c|c} 0 & \\ \hline 1 & 2 \end{array}\right], \quad \left[\begin{array}{c|c} 7 & \\ \hline 8 & 9 \end{array}\right], \quad \left[\begin{array}{c|c} 26 & \\ \hline 27 & 28 \end{array}\right], \quad \left[\begin{array}{c|c} 33 & \\ \hline 34 & 35 \end{array}\right]. \quad (6)$$

Similarly, the two nodes on level 1 work on

$$\left[\begin{array}{cc|cc} 0 & & & \\ 1 & 2 & & \\ \hline 3 & 4 & 7 & \\ 5 & 6 & 8 & 9 \end{array}\right], \quad \left[\begin{array}{cc|cc} 26 & & & \\ 27 & 28 & & \\ \hline 29 & 30 & 33 & \\ 31 & 32 & 34 & 35 \end{array}\right]. \quad (7)$$

Finally, on level 0 there is the root of the tree which is working on the A_{ij} blocks in Figure 3.

The recursive blocked algorithms for the RB-TRSM and RB-SYRK use the GEMM-based approach [9, 5, 10] and therefore call RB-GEMM, which in turn perform all computations at the leaves of its recursion tree by calling register-based GEMM kernel routines with operands that fit in level 1 cache. Our recursive algorithm for RB-GEMM is further discussed in the next section.

4 Recursive Blocked DGEMM

The recursive blocked DGEMM, RB-GEMM, uses a similar divide and conquer algorithm as the RB-Cholesky algorithm (see Figure 6). However, since there are more problem dimensions than for the Cholesky case (one, n, for Cholesky as compared to three, m, n, and k for GEMM), the splitting of the nodes can take place in one of the three dimensions (see Figure 5). By always splitting the largest dimension, the problem is kept "squarish", i.e., the ratio between the number of operations made on subblocks and the number of subblocks is maintained as high as possible. Nevertheless, the "conquer" part in GEMM is trivial, since the addition of the results is made implicitly by the leaf kernels. The leaf kernels used are high-performance, prefetching unrolled kernels which utilize the linear addressing with fixed leading dimensions inherited from the block formats. The fixed leading dimensions guarantee that there is no or only small performance differences between different transpose arguments of the GEMM operation (AB, A^TB, AB^T, A^TB^T). For a thorough explanation of these kernels, see [7] and [8]. In our view the input matrices which an application (algorithm) will use define how their submatrices will be loaded. Knowing the full algorithm and the machine architecture allow us to choose nb so that misalignment never or rarely occur.

$$\begin{bmatrix} C_{11} & C_{12} \\ C_{21} & C_{22} \end{bmatrix} + \begin{bmatrix} A_{11} & A_{12} \\ A_{21} & A_{22} \end{bmatrix} \begin{bmatrix} B_{11} & B_{12} \\ B_{21} & B_{22} \end{bmatrix} =$$

$$= \begin{bmatrix} \begin{bmatrix} C_{11} & C_{12} \end{bmatrix} + \begin{bmatrix} A_{11} & A_{12} \end{bmatrix} \begin{bmatrix} B_{11} & B_{12} \\ B_{21} & B_{22} \end{bmatrix} \\ \begin{bmatrix} C_{21} & C_{22} \end{bmatrix} + \begin{bmatrix} A_{21} & A_{22} \end{bmatrix} \begin{bmatrix} B_{11} & B_{12} \\ B_{21} & B_{22} \end{bmatrix} \end{bmatrix} =$$

$$= \left[\begin{bmatrix} C_{11} \\ C_{21} \end{bmatrix} + \begin{bmatrix} A_{11} & A_{12} \\ A_{21} & A_{22} \end{bmatrix} \begin{bmatrix} B_{11} \\ B_{21} \end{bmatrix}, \begin{bmatrix} C_{12} \\ C_{22} \end{bmatrix} + \begin{bmatrix} A_{11} & A_{12} \\ A_{21} & A_{22} \end{bmatrix} \begin{bmatrix} B_{12} \\ B_{22} \end{bmatrix} \right] =$$

$$= \begin{bmatrix} C_{11} & C_{12} \\ C_{21} & C_{22} \end{bmatrix} + \begin{bmatrix} A_{11} \\ A_{21} \end{bmatrix} \begin{bmatrix} B_{11} & B_{12} \end{bmatrix} + \begin{bmatrix} A_{12} \\ A_{22} \end{bmatrix} \begin{bmatrix} B_{21} & B_{22} \end{bmatrix}$$

Fig. 5. Splitting the matrix multiplication on the m, n, and k dimensions, respectively.

Notice that if $m = n = k > 1$, or two of them are equal, there is a choice on which dimension to split. Some experimental results have shown that this choice is not so crucial. This is due to the fact that a less than optimal ordering of the splittings will only impair performance in one or two levels of recursion. Further down the recursion tree, the impact of previous choices vanishes. In our prototype implementations, we have focused on square subblocks and mostly square matrices. When multiplying submatrices that cross block boundaries, the case of misalignment with the data structure and the algorithm will arise. This is a serious threat to good performance. One way to rectify this problem is to make nb an even power of two, and adjust the algorithm thereafter. This might not handle all cases of misalignment, but should treat the common linear algebra operations.

Implementations have been done in Fortran 77 and in C. The implementation in C is somewhat more readable, since the C language supports recursion, but by doing clear notations of the explicit stack, the Fortran 77 code is not hard to read.

Level 3 algorithmic prefetching [7, 1] is more difficult to apply in the recursive algorithm, since it is a non-trivial task to calculate the next set of kernel blocks to be multiplied given the current set of blocks. We have looked at three ways to solve this problem. One way is based on the divide-at-even-power-of-two scheme. If we have fixed division points, it is easier to calculate the next set of blocks by using bit arithmetics. Another way of doing prefetching is simply to also include prefetching information (next set of blocks) in the argument list to the RB-GEMM subroutine. A third way of doing prefetching is to in advance build a batch list of the multiplications to be performed. Using this list, finding the next set is trivial. The last method is used in our C implementation [8].

function $[C(1:M,1:N)] =$ **RB-GEMM** $(C(1:M,1:N), A(1:M,1:K),$
$B(1:K,1:N), m, n, k)$
if $m = 1$ **and** $n = 1$ **and** $k = 1$ **then** *{Leaf node}*
  DGEMM$(C(1:M,1:N), A(1:M,1:K), B(1:K,1:N))$; *{Call kernel routine }*
else if $m = \max\{m, n, k\}$ **then**
  $m1 = m/2$; $m2 = m - m1$; *{# blocks in next recursive calls}*
  RB-GEMM $(C(1:m1 \cdot mb, 1:N), A(1:m1 \cdot mb, 1:K),$
  $B(1:K,1:N), m1, n, k)$ *{mb is the size of submatrix blocks }*
  RB-GEMM $(C(m1 \cdot mb+1:M, 1:N), A(m1 \cdot mb+1:M, 1:K),$
  $B(1:K,1:N), m2, n, k)$
else if $n = \max\{n, k\}$ **then**
  $n1 = n/2$; $n2 = n - n1$; *{# blocks in next recursive calls}*
  RB-GEMM $(C(1:M, 1:n1 \cdot nb), A(1:M,1:K),$
  $B(1:K, 1:n1 \cdot nb), m, n1, k)$ *{nb is the size of submatrix blocks }*
  RB-GEMM $(C(1:M, n1 \cdot nb+1:N), A(1:M,1:K),$
  $B(1:K, n1 \cdot nb+1:N), m, n2, k)$
else
  $k1 = k/2$; $k2 = k - k1$; *{# blocks in next recursive calls}*
  RB-GEMM $(C(1:M,1:N), A(1:M, 1:k1 \cdot kb),$
  $B(1:k1 \cdot kb, 1:N), m, n, k1)$ *{kb is the size of submatrix blocks }*
  RB-GEMM $(C(1:M,1:N), A(1:M, k1 \cdot kb+1:K),$
  $B(k1 \cdot kb+1:K, 1:N), m, n, k2)$
end

Fig. 6. The recursive blocked-GEMM algorithm. This is one of the six (3!) ways of making splitting decisions when all or two of m, n, k are equal to the largest of these dimensions.

Another good characteristic of the recursive algorithm is that it accommodates Strassen's algorithm in a natural way.

5 Performance Results

We present some results from our ongoing software developments. All performance results are obtained on an IBM SMP Node (PowerPC 604, 112 MHz), which is a symmetric shared memory multiprocessing node with four processors. There are two levels of cache memories for each processor. The level 1 cache (L1) is 16 kB on-chip, and the level 2 cache (L2) is a 512 kB off-chip and non-shared memory. The SMP node has a shared 256MB main memory.

In Figure 7 the recursive blocked Cholesky algorithm assuming recursive blocked storage format has the label RBPPF (no copy). The similar algorithm assuming A originally stored in packed lower triangular format is labeled RBPPF (2 copy). In this case, the recursive algorithm is preceded and succeeded by a copy operation between the two data storage formats. Finally, ESSL DPPF denotes the level 3 Cholesky factorization algorithm for packed data from the IBM ESSL library. From these preliminary results we see a great benefit in using the recursive approach (up to 75% and 95%, respectively for large problems using one processor on the IBM SMP Node). There is an additional performance gain when the original data is stored in recursive blocked data format (up to 50% for small problems and around 10% for large problems).

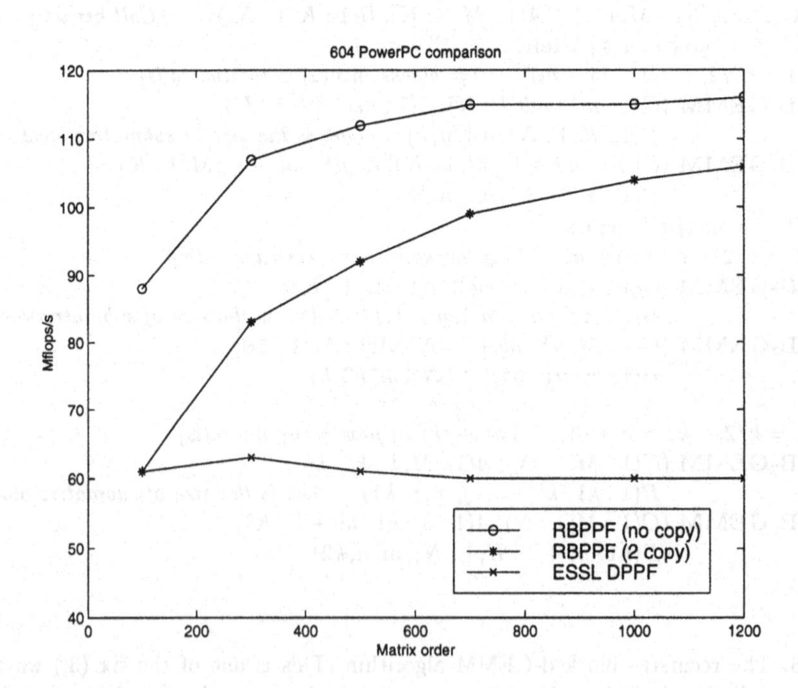

Fig. 7. Performance results for recursive blocked and ESSL packed Cholesky factorization algorithms.

In Figure 8 the recursive GEMM algorithm is compared to a blocked algorithm where the blocks are stored in column block order, i.e., no copying between conventional column-major order and column block order is included. The column blocked algorithm is optimized for efficient level 1 cache utilization using one level of blocking, 4×4 register blocking and prefetching techniques [8, 7]. The uniprocessor results on the SMP High Node show that the recursive algorithm outperforms the column blocked algorithm for matrices larger than 200×200. For smaller matrices the overhead for building the recursive tree is showing up. The main reason for the big difference in the performance results for larger problems is that the recursive blocked data format and algorithm provide operands to the kernel routines that are blocked for a memory hierarchy (including level 2 cache and main memory) automatically. In order to get similar performance results for the column blocked algorithm we would have to introduce at least another level of blocking. Such techniques are discussed in [7], where we also present some parallel performance results using a multithreaded recursive GEMM algorithm.

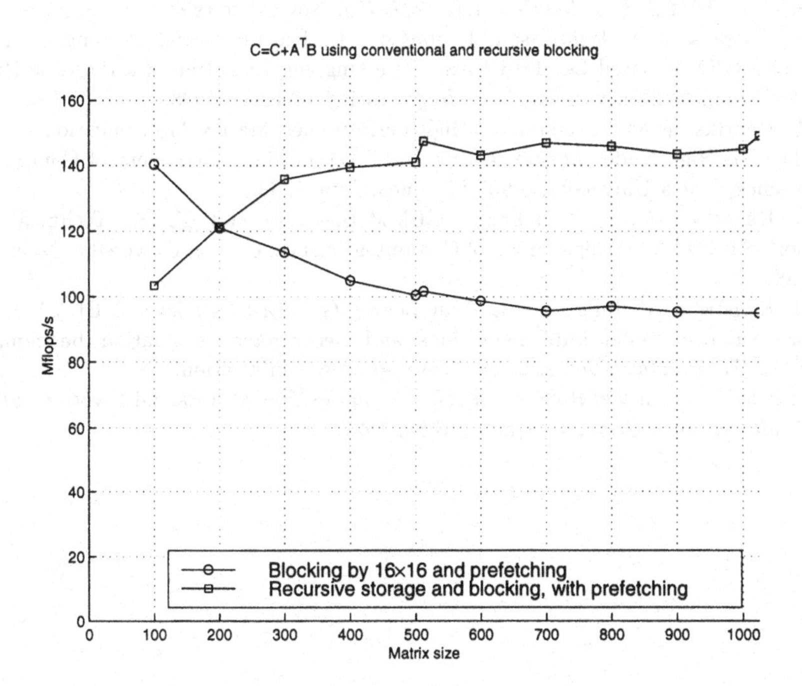

Fig. 8. Performance results for a hand-tuned blocked algorithm using block column format and the recursive GEMM algorithm.

In conclusion, the initial performance results verify the great potential of our recursive approach for the efficient handling of today's memory tiered processors. Our future work will include further developments in support for the recursive blocked data formats and BLAS's.

This research was conducted using the resources of the High Performance Computing Center North (HPC2N).

References

1. R. C. Agarwal, F. G. Gustavson, and M. Zubair. Improving performance of linear algebra algorithms for dense matrices, using algorithmic prefetch. *IBM J. Res. Develop*, 38(3):265–275, May 1994.
2. E. Anderson, Z. Bai, J. Demmel, J. Dongarra, J. Du Croz, A. Greenbaum, S. Hammarling, A. McKenney, S. Ostrouchov and D. Sorensen. *LAPACK Users' Guide*, Second Edition. SIAM Publications, Philadelphia, 1995.
3. J. Dongarra, J. DuCroz, I. Duff, and S. Hammarling. A Set of Level 3 Basic Linear Algebra Subprograms. *ACM Trans. Math. Softw.*, 16(1):1–17, March 1990.
4. E. Elmroth and F. Gustavson. New Serial and Parallel Recursive QR Factorization Algorithms for SMP Systems, *This Proceedings*, Springer Verlag, 1998.
5. IBM. *Engineering and Scientific Subroutine Library, Guide and Reference*, January 1994. SC23-0526-01.
6. F. Gustavson. Recursion leads to automatic variable blocking for dense linear algebra. *IBM J. Res. Develop*, 41(6):737–755, November 1997.
7. F. Gustavson, A. Henriksson, I. Jonsson, B. Kågström and P. Ling. Superscalar GEMM-based Level 3 BLAS - The Ongoing Evolution of a Portable High-Performance Library. *This Proceedings*, Springer Verlag, 1998.
8. A. Henriksson and I. Jonsson. High-Performance Matrix Multiplication on the IBM SP High Node. *Master Thesis*, UMNAD 98.235, Department of Computing Science, Umeå University, S-901 87 Umeå, June 1998.
9. B. Kågström and C. Van Loan. GEMM-Based Level-3 BLAS. Technical Report CTC91TR47, Department of Computer Science, Cornell University, December 1989.
10. B. Kågström, P. Ling, and C. Van Loan. GEMM-based level 3 BLAS: High-performance model implementations and performance evaluation benchmark. *ACM Trans. Math. Software*, 1997. Accepted for publication.
11. S. Toledo. Locality of Reference in LU Decomposition with Partial Pivoting. *SIAM J. Matrix Anal. Appl.*, 18(4):1065–1081, 1997.

Superscalar GEMM-based Level 3 BLAS – The On-going Evolution of a Portable and High-Performance Library

Fred Gustavson[1], André Henriksson[2], Isak Jonsson[2], Bo Kågström[2], and Per Ling[2]

[1] IBM T.J. Watson Research Center, P.O. Box 218, Yorktown Heights, NY 10598, U.S.A.
gustav@watson.ibm.com

[2] Department of Computing Science and HPC2N, Umeå University, S–901 87 Umeå, Sweden.
{andropov,isak,bokg,pol}@cs.umu.se

Abstract. Recently, a first version of our GEMM-based level 3 BLAS for superscalar type processors was announced. A new feature is the inclusion of DGEMM itself. This DGEMM routine contains inline what we call a level 3 kernel routine, which is based on register blocking. Additionally, it features level 1 cache blocking and data copying of submatrix operands for the level 3 kernel. Our other BLAS's which possess triangular operands, e.g., DTRSM, DSYRK use a similar level 3 kernel routine to handle the triangular blocks that appear on the diagonal of the larger input triangular operand. Like our previous GEMM-based work all other BLAS's perform the dominating part of the computations in calls to DGEMM. We are seeing the adoption of our BLAS's by several organizations, including the ATLAS and PHiPAC projects on automatic generation of fast DGEMM kernels for superscalar processors, and some computer vendors. The evolution of the superscalar GEMM-based level 3 BLAS is presented. Also, we describe new developments which include techniques that make the library applicable to symmetric multiprocessing (SMP) systems.

1 Introduction

The level 3 Basic Linear Algebra Subprograms (BLAS) [4] are a de facto standard for various matrix multiply and triangular system solving computations and are successfully used as building blocks for the development of high-performance dense linear algebra library software. Due to the complex hardware organization of state-of-the-art computer architectures the development of optimal level 3 BLAS code is costly and time consuming. However, as [8, 9] shows it is possible to develop a portable and high-performance level 3 BLAS library mainly relying on a highly optimized GEMM, the routine for the general matrix multiply and add operation, $C \leftarrow \beta C + \alpha AB$, where C, A and B are matrices of size $m \times n, m \times k$ and $k \times n$, respectively, and α, β are scalars. With suitable partitioning, all the other level 3 BLAS can be defined in terms of GEMM and a small amount of level 1 and level 2 computations [8]. Whenever new high-performance architectures or extensions and modifications of existing ones are introduced, only a few underlying routines need to be optimized for the target architecture, assuming no radical changes in the architecture and memory system. Most important is the routine that implement the level 3 GEMM operation.

Before we go into any further details we outline the rest of the paper. Section 2 gives a short overview of the general GEMM-based level 3 BLAS library. In Section 3 we present the further development of the GEMM-based library targeted to superscalar processors and describe the technique of register blocking used in our kernel routines. Section 4 presents some of our design principles and techniques used for reducing delays in a multi-level memory system. Performance results from our new developments, including parallelization using threads, are also presented.

2 GEMM-based level 3 BLAS

Kågström, Ling and Van Loan developed the model implementations in Fortran 77 of the GEMM-based level 3 BLAS [9, 10]. They also developed a performance evaluation benchmark, which is a tool for evaluating and comparing different implementations of the level 3 BLAS with the GEMM-based model implementations. This software can handle all four data types and is designed to be easy to install and use on different platforms. Each of the GEMM-based routines has a few system-dependent parameters that specify internal block sizes, cache characteristics, and branch points for alternative code sections, which are given as input to a program that facilitates the tuning of these parameters. For convenience, sample values for some common architectures are also provided.

The GEMM-based model implementations are designed to reach high performance through efficient cache reuse. The level 1 and level 2 BLAS used, are not in general associated with cache reuse. However, through appropriate blocking and use of local arrays, cache reuse is achieved during multiple calls to the level 1 and level 2 BLAS routines, and thereby their use approach level 3 performance. This idea works especially well on vector machines, provided that the underlying BLAS are highly optimized. The GEMM-based approach is also adopted in [5].

3 Superscalar GEMM-based level 3 BLAS

To approach peak performance on state-of-the-art superscalar microprocessors it is necessary to attain extensive register reuse. In general, multiple calls to the level 1 and level 2 BLAS routines prohibit an efficient register reuse.

Recently, Kågström and Ling announced the first version of the superscalar GEMM-based level 3 BLAS. They have also developed a superscalar DGEMM that currently is used with the library. The superscalar library has essentially the same overall structure, with similar blocking, as the regular GEMM-based level 3 BLAS. The main difference in the design is that all calls to underlying level 1 and level 2 BLAS have been removed. As before, the dominating part of all floating point operations take place in calls to DGEMM. The remaining computations that take care of triangular diagonal blocks are handled by "in-line" code optimized for efficient register reuse.

3.1 Unrolling for register reuse

One common technique for register reuse is loop unrolling [2]. We illustrate how loop unrolling is used for a GEMM operation, $C \leftarrow \beta C + \alpha AB$, operating on submatrices. For clarity we assume $\alpha = \beta = 1$.

Assume that we have a processor with at least 24 floating point registers for numbers and that the instruction set includes an instruction for floating point multiply and add (MAA), $z \leftarrow z + xy$, on register operands. Let LOAD and STORE denote the instructions for loading and storing data between memory and registers. The MAA instructions can be arranged so that a single MAA is completed every machine cycle after initial startup.

We use the method illustrated in Figure 1 for a blocked matrix multiplication. The two outermost loops (i- and j-) are unrolled four levels. Within the inner-

```
for j = 1 to n step 4
    for i = 1 to m step 4
        r1 = c_{i,j}
        r2 = c_{i+1,j}
        ...
        r16 = c_{i+3,j+3}
        for l = 1 to k
            r1 = r2 + a_{i,l} b_{l,j}
            r2 = r2 + a_{i+1,l} b_{l,j}
            ...
            r16 = r16 + a_{i+3,l} b_{l,j+3}
        end
        c_{i,j} = r1
        c_{i+1,j} = r2
        ...
        c_{i+3,j+3} = r16
    end
end
```

Fig. 1. Blocked matrix multiply and add operation using unrolling in two dimensions.

most l-loop, a 4×4 block of C, temporarily stored in the variables $r1, \ldots, r16$, is performing 16 MAAs which is the result of a matrix multiplication of blocks of A ($4 \times k$) and B ($k \times 4$) (see Figure 2). The use of the local scalar variables $r1, \ldots, r16$, for elements of C allow compilers to keep C in registers (and not load/store to cache/memory) inside the l-loop. Thus, in each iteration of the

$$\begin{bmatrix} r1 & r5 & r9 & r13 \\ r2 & r6 & r10 & r14 \\ r3 & r7 & r11 & r15 \\ r4 & r8 & r12 & r16 \end{bmatrix} \leftarrow \begin{bmatrix} r1 & r5 & r9 & r13 \\ r2 & r6 & r10 & r14 \\ r3 & r7 & r11 & r15 \\ r4 & r8 & r12 & r16 \end{bmatrix} + \begin{bmatrix} \cdots & \begin{bmatrix} a_{1\cdot} \\ a_{2\cdot} \\ a_{3\cdot} \\ a_{4\cdot} \end{bmatrix} & \cdots \end{bmatrix} \begin{bmatrix} \vdots \\ \begin{bmatrix} b_{\cdot 1} & b_{\cdot 2} & b_{\cdot 3} & b_{\cdot 4} \end{bmatrix} \\ \vdots \end{bmatrix}$$

Fig. 2. Sequence of 16 MAAs performed within the innermost l-loop.

l-loop, 16 MAA instructions and 8 LOAD instructions are performed. The four elements of A and B in Figure 2 are loaded into registers. Most of the time spent on LOADs is overlapped by the time consumed by 16 MAAs as they execute concurrently in separate functional units. A total of 24 floating point registers are used in the innermost loop.

This technique, to keep a small square block of C in registers and replace entries of A and B between consecutive iterations of the innermost loop, maximizes the ratio between the number of MAAs and the number of load and store instructions, used to transfer data to and from registers, i.e., # MAAs/(#LOADs+ #STOREs) is maximized. Notice, computing dot products at outer loop levels completely avoids the need for store instructions in the innermost loop. All of the MAAs in the innermost loop update different registers which means they execute in parallel and henced are safely pipelined. They can also be ordered arbitrarily to assist compilers in invoking multiple element load instructions, if available.

Generally, for efficient register reuse it is necessary to unroll in more than one dimension (use more than one loop) and hence matrix-matrix operations must be performed by these kernel routines. We have omitted the fix-up code required, when the dimensions m and n are not multiples of four.

3.2 Improved performance for the superscalar library

In the current release of the superscalar GEMM-based level 3 BLAS, 4×4 unrolling is used for the C matrix in DGEMM and 4×2 unrolling is used in the remaining routines. As for the GEMM-based model implementations all references are stride one which is implemented using work arrays and data copying prearranged so that the DGEMM kernel will run close to peak performance. The extra data copying allows the superscalar library to handle so called "critical" leading dimensions as well [9, 10]. The Fortran source code is publically available from netlib, see 'www.netlib.org/blas/gemm_based/ssgemmbased.tgz'.

Performance results from the GEMM-based level 3 BLAS performance benchmark on an IBM PowerPC 604 processor (112 MHz, IBM SP, SMP node) show substantial improvements for the current release of the superscalar library:

DSYMM	DSYRK	DSYR2K	DTRMM	DTRSM
+3%	+28%	+2%	+23%	+25%

These percentage numbers are for square matrices of size 500×500. We obtain up to 80% improvement for small matrices (32×32). The improvements are mainly for the routines that called level 2 routines in the model implementations [9, 10]. The GEMM-based algorithms for DSYMM and DSYR2K do not call any level 2 routines. The calculations are transformed to level 3 GEMM operations by copying the symmetric subblocks stored in triangular format to general full format subblocks in work arrays [11].

The ATLAS [12] and PHiPAC projects [3] use the superscalar GEMM-based level 3 BLAS together with their own automatically tuned DGEMM to provide a complete set of level 3 BLAS in double precision. The ATLAS project reports impressive performance results for several different machines where the combination of the superscalar GEMM-based level 3 BLAS and ATLAS DGEMM is often faster than the vendor supplied level 3 BLAS, see 'www.netlib.org/atlas'.

4 Techniques for More Complex Memory Hierarchies

In our on-going development of the superscalar GEMM-based level 3 BLAS we are investigating different techniques to reduce delays in cache and memory accesses. We focus on ways to handle memory hierarchies with multiple levels of cache memories efficiently. Furthermore, we are using threads for developing parallel versions of our routines. In the following we survey some of these techniques.

In this case study, all experiments are carried out on an SMP node of the IBM SP parallel system. Presumably, this SMP node has several characteristics which will be typical in future high performance computing systems. For example, several processors (currently four IBM PowerPC 604 with an MAA instruction), a memory hierarchy with more than one level of cache (currently L1 and L2 cache), and a single shared main memory. We sometimes use assembly language in these experiments since some of the techniques we investigate cannot be invoked via current Fortran or C compilers. One goal with this case study is to find strong evidence for introducing support for these techniques in future compilers.

4.1 Prefetching

Generally, when data does not reside in the top levels of a memory hierarchy (registers or L1 cache), significant delays occur when the data is referenced. A well known technique to overcome these delays, commonly used in compilers, is prefetching. This technique enables data to be brought to registers or cache, ahead of its use, so that when the data is needed, it is immediately available. Unfortunately, todays compilers do not perform prefetching as well as one would desire, especially for complex memory hierarchies. A software technique to include prefetching directly into Fortran programs called *algorithmic prefetching* is suggested by Agarwal, Gustavson and Zubair [1]. They used this technique successfully for the level 2 BLAS of the IBM ESSL library. However, the usefulness of this technique depends much on the optimization procedures performed by compilers. In our assembly programs we use a “touch” instruction, 'dcbt' to perform prefetching. It is currently not possible to invoke this instruction explicitly from Fortran or C. This instruction brings data up into L1 cache, essentially without disturbing the ongoing computations and data references.

4.2 Algorithmic preloading

We also use a technique that we call *algorithmic preloading* for prefetching data. We distinguish this technique from algorithmic prefetching, since the latter is characterized by statements or instructions in a program that can be removed without affecting the correctness of the program, while statements or instructions for algorithmic preloading cannot.

In Figure 3 we have modified the algorithm of Figure 1 to illustrate algorithmic preloading. Apart from loading entries of C into registers, via $r1, \ldots, r16$, we also load four elements of A and four elements of B into registers, via $r17, \ldots, r24$, before the l-loop starts. These elements are operands to the MAAs in the first iteration of the l-loop. Immediately after the last reading of each of $r17, \ldots, r24$ within the l-loop, they are preloaded with the entries from A and B that take part in the next iteration. This way, the operands of the MAAs are loaded into registers as early as possible, and thereby more likely to be available when the MAAs are ready to execute. Possibly all of the MAAs can execute without any delays, enabling for the l-loop to approach peak performance. The MAAs of the k-th iteration execute after the l-loop is completed.

4.3 Hierarchical blocking

The purpose of the blocking structure of the model implementations and the current version of the superscalar GEMM-based level 3 BLAS is to reuse the most frequently referenced matrix blocks in L1 cache as much as possible. However, for machines with several levels of cache memories a need for more advanced blocking strategies arises.

Explicit multi-level blocking: One way to handle the memory hierarchy explicitly is to reorganize loops such that each loop set matches a specific level of the memory hierarchy. However, this is a tedious work and requires a lot of knowledge about the architecture characteristics. For example, a blocking parameter is required for each level of the memory hierarchy. Some results of a two-level blocked matrix multiply for the PowerPC 604 are presented in Section 4.4. The two blocking parameters are tuned for L1 and L2 cache.

Automatic blocking via recursion: Recursion is a key concept for matching an algorithm and its data structure. In [6] we introduced recursive blocked data formats and recursive blocked algorithms for level 3 BLAS. The recursive algorithms lead to automatic blocking which is variable and "squarish". The blocking is variable and hierarchical and allow for maintaining the two-dimensional data locality at every level of the one-dimensional tiered memory structure. The only tuning parameter is the size of L1 cache. The size of each matrix block is constrained so that a few of them will simultaneously fit in L1 cache. For more details see [6].

for $j = 1$ to n step 4
    for $i = 1$ to m step 4
        Preload $C_{i:i+3,j:j+3}$ to $r_1, \ldots, r_{16}$
        Preload $A_{i:i+3,1}$ to $r_{17}, \ldots, r_{20}$
        Preload $B_{1,j:j+3}$ to $r_{21}, \ldots, r_{24}$
        for $l = 1$ to $k - 1$
            $r1 = r1 + r17r21$
            $\ldots$
            Preload: $r17 = a_{i,l+1}$
            $\ldots$
            Preload: $r21 = b_{l+1,j}$
            $\ldots$
            $r16 = r16 + r20r24$
            Preload: $r20 = a_{i+3,l+1}$
            Preload: $r24 = b_{l+1,j+3}$
        end
        $r1 = r1 + r17r21$
        $c_{i,j} = r1$
        $\ldots$
        $r16 = r16 + r20r24$
        $c_{i+3,j+3} = r16$
    end
end

Fig. 3. Blocked matrix multiply and add operation using unrolling in two dimensions and preloading of matrix entries one iteration ahead of their use.

Our assembly implementation consists of a recursively called routine that creates a list of calls to a computational kernel and then executes these calls. The list constitutes a precalculated *recursive task tree* with actual calculations at the leaf nodes. The computational kernel is designed to hold six matrix blocks in L1 cache simultaneously. Three of them are involved in the current computation while the other three blocks are being prefetched using the 'dcbt' instruction. When the current block computation is completed, a new computation starts involving the blocks just prefetched, and three new blocks are prefetched for the subsequent computations, and so on.

4.4 Some uniprocessor performance results

In Figure 4 we show the performance of an explicitly tuned two-level blocked algorithm and the recursive GEMM algorithm [6] executing on an IBM PowerPC 604. Both routines use the same atomic kernel that prefetches, register preloads, does 4×4 unrolling, and works on 16×16 submatrices. The performance results, which are very good, do not account for any data copying, i.e., the blocks are initially stored in column block order and recursive block order, respectively. The recursive algorithm performs around 7% better than the multi-level blocked

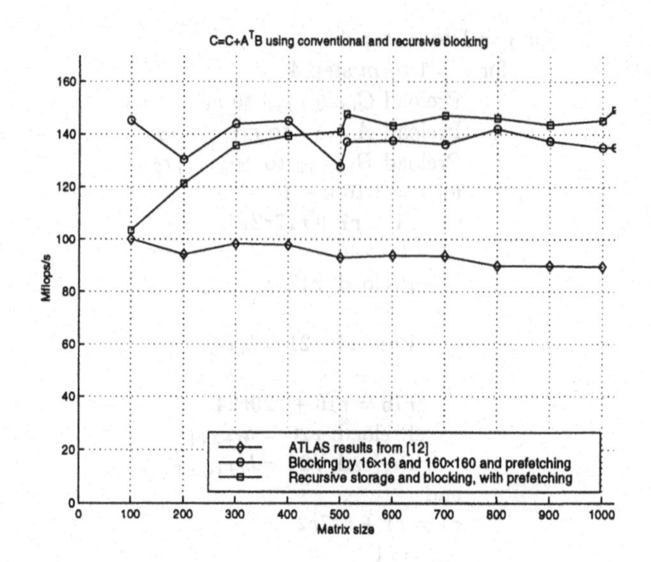

Fig. 4. Performance of two hierarchical blocking strategies.

algorithm for large enough problems (500×500), even though the tuning is much less explicit (only L1 cache versus L1 and L2 cache tuning for the multi-blocked algorithm).

4.5 Parallelization

The recursive task tree facilitates parallelization on shared memory machines. We simply divide the tree into subtrees and let different processes or threads (lightweight processes) execute on different subtrees. We use a single thread for each processor and divide the tree into one subtree more than the number of threads. The left-over subtree is divided among the threads when they become ready for additional work. This way of scheduling enables good load balancing on non-dedicated machines.

For multiple threads, we show performance results with the recursive task tree approach in Figure 5 [7]. The algorithm scales well. One reason is that the recursive task tree automatically keeps data references local to each thread.

This research was conducted using the resources of the High Performance Computing Center North (HPC2N).

References

1. R. C. Agarwal, F. G. Gustavson, and M. Zubair. Improving performance of linear algebra algorithms for dense matrices, using algorithmic prefetch. *IBM J. Res. Develop*, 38(3):265–275, May 1994.
2. R. C. Agarwal, F. G. Gustavson, and M. Zubair. Exploiting functional parallelism of POWER2 to design high-performance numerical algorithms. *IBM J. Res. Develop*, 38(5):563–576, September 1994.

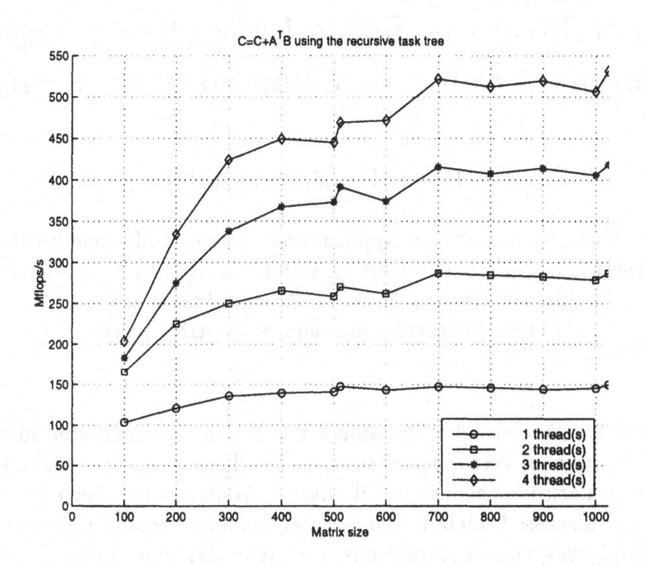

Fig. 5. Scheduling the matrix multiply problem on several threads.

3. J. Bilmes, K. Asanovic, C.-W. Chin, and J. Demmel. Optimizing matrix multiply using PHiPAC: A portable, high performance, ANSI C coding methodology. In *Proceedings of the 11th International Conference on Supercomputing (ICS-97)*, pages 340–347, New York, July 7–11 1997. ACM Press.
4. J. Dongarra, J. DuCroz, I. Duff, and S. Hammarling. A Set of Level 3 Basic Linear Algebra Subprograms. *ACM Trans. Math. Softw.*, 16(1):1–17, 18–28, March 1990.
5. M. J. Dayde, I. S. Duff, and A. Petitet. A parallel block implementation of level-3 BLAS for MIMD vector processors. *ACM Trans. Math. Softw.*, 20(2):178–193, June 1994.
6. F. Gustavson, A. Henriksson, I. Jonsson, B. Kågström and P. Ling. Recursive Blocked Data Formats and BLAS's for Dense Linear Algebra Algorithms. *This Proceedings*, Springer Verlag, 1998.
7. A. Henriksson and I. Jonsson. High-Performance Matrix Multiplication on the IBM SP High Node. *Master Thesis*, UMNAD 98.235, Department of Computing Science, Umeå University, S-901 87 Umeå, June 1998.
8. B. Kågström and C. Van Loan. GEMM-Based Level-3 BLAS. Technical Report CTC91TR47, Department of Computer Science, Cornell University, Dec. 1989.
9. B. Kågström, P. Ling, and C. Van Loan. GEMM-based level 3 BLAS: High-performance model implementations and performance evaluation benchmark. *ACM Trans. Math. Software*, 1997. To appear.
10. B. Kågström, P. Ling, and C. Van Loan. GEMM-based level 3 BLAS: Portability and optimization issues. *ACM Trans. Math. Software*, 1997. To appear.
11. P. Ling. A set of high-performance level 3 BLAS structured and tuned for the IBM 3090 VF and implemented in Fortran 77. *The Journal of Supercomputing*, 7(3):323–355, September 1993.
12. R. C. Whaley and J. J. Dongarra. Automatically tuned linear algebra software. Tech. Report TN 37996-1301, Computer Science Dept., Univ. of Tennessee, 1997.

Parallel Solution of Some Large-Scale Eigenvalue Problems Arising in Chemistry and Physics

David L. Harrar II and Michael R. Osborne

Centre for Mathematics and its Applications, School of Mathematical Sciences, Australian National University, Canberra ACT 0200, Australia
{David.Harrar, Michael.Osborne}@anu.edu.au
http://wwwmaths.anu.edu.au/~{dlh, mike}

Abstract. We consider the numerical solution on distributed-memory parallel arrays of vector processors of some large-scale eigenvalue problems which arise in chemistry and physics. Applications from a variety of areas are discussed, including molecular dynamics, quantum chemistry, optical physics, chemical reactions, and hydrodynamic stability.

1 Introduction

In this paper we consider some applications arising in chemistry and physics in which the solution of the eigenvalue problem (EVP)

$$Au = \lambda u \tag{1}$$

is the dominant computational component. In particular, we are interested in developing scientific subroutine library (SSL) software for solving these problems on distributed-memory parallel arrays of powerful vector processing elements (PEs), as exemplified by the Fujitsu VPP (Vector Parallel Processor) series of computers. On architectures such as these it is important to strive not only for parallelization but also for a high degree of vectorization. In general, the solution techniques are only briefly outlined here. A more detailed mathematical presentation can be found in [2] and the references therein.

2 Computational Quantum Chemistry

Electron interaction in protein molecules is governed by Schrödinger's equation, $\mathcal{H}\Psi = E\Psi$, where $\mathcal{H}$ is the Hamiltonian for the system, comprising nuclear and electronic kinetic and potential energies, E is the total energy, and Ψ is the electron wavefunction. Applying the variational method a set of coupled integro-differential equations is obtained, and using a known set of basis functions – a single "Slater-type" basis function for each valence atomic orbital (AO) with molecular orbitals expressed as linear combinations of AOs – the problem ultimately takes the form of an algebraic EVP for the "Fock matrix" F:

$$F\Psi = \epsilon\Psi. \tag{2}$$

Eigenvectors Ψ_i represent wave functions describing the electron orbitals and the corresponding eigenvalues ϵ_i give the energies of those orbitals.

The self-consistent field (SCF) method comprises an iterative approach to the solution of (2) in which an initial set of basis functions (AOs) is chosen, the corresponding Fock matrix F constructed, an eigendecomposition of F computed, the resulting orbitals and corresponding energies used to construct a new F, and so on until convergence. Solution of (2) is the most computationally expensive step. If β_i denotes the number of basis functions (AOs) associated with the i-th atom then the Fock matrix for a protein molecule consisting of N atoms has size $n = \sum_{i=1}^{N} \beta_i$. Despite the fact that protein molecules generally contain large numbers of hydrogen atoms, for which there is only one AO (i.e., $\beta = 1$), the Fock matrix dimension n is usually large since proteins typically consist of hundreds of atoms and may consist of tens of thousands.

The Fock matrices are symmetric and full, and to solve (2) we use the relatively standard technique of applying Householder transformations $U_1, \ldots, U_{n-2}$ to F in order to reduce the problem to a tridiagonal EVP for $T = UFU^T$, where $U = U_{n-2} \cdots U_1$. The reduction is parallelized using panel-wrapped storage [1]. In order to solve tridiagonal EVPs efficiently on VPP architectures we compute eigenvalues using multisection – a vectorized variant of bisection in which eigenvalue intervals are subdivided into greater than two subintervals – and subsequently determine corresponding eigenvectors with inverse iteration. The solution of tridiagonal linear systems required for inverse iteration is effectively vectorized by "wrap-around partitioning" [3]. In the final stage of the algorithm the eigenvectors of F are recovered from those for T using the U_i.

We consider five proteins: (1) pheromone protein from Euplotes Raikovi (573 atoms), $n = 1482$; (2) bovine pancreatic trypsin inhibitor (892 atoms), $n = 2254$; (3) bovine pancreatic ribonuclease A (1856 atoms), $n = 4709$; (4) dihydrofolate from Eshrichia Coli (2566 atoms), $n = 6460$; and (5) human leukocyte elastase (3291 atoms), $n = 8199$. Fock matrices are generated using MNDO94 [9]. In Table 1 we compare performance of our eigensolver with that obtained using the analogous routines from the LAPACK and ScaLAPACK libraries.

	SSL2VP(P)				(Sca)LAPACK			
$n \downarrow$ # PEs $\rightarrow$	1	2	4	8	1	2	4	8
1482	15.08	10.82	5.993	3.511	26.82	26.72	20.93	18.46
2254	45.31	32.42	16.21	9.035	87.80	72.98	50.65	47.31
4709	373.1	217.5	113.1	61.60	694.2	467.8	314.6	X
6460	934.5	526.6	268.6	143.5	1656.	X	X	X
8199	1832.	1096.	512.0	X	3400.	X	X	X

Table 1. Performance comparison for various proteins.

The LAPACK and ScaLAPACK routines also use Householder reduction. If all eigenvalues are required a QR procedure is used for T, but if $r \leq n-1$ are requested bisection and inverse iteration are used. Since the latter performs considerably worse – computing $n-1$ eigenpairs takes approximately three times as long as computing n [2], the LAPACK times are in some sense "best-case". Also, on vector architectures it is imperative to adjust a parameter within the bisection routine to enable vectorized bisection; otherwise it is a scalar computation and consequently extremely slow. Based on timing of individual algorithmic components, the (Sca)LAPACK implementations of the reduction and eigenvector recovery stages are more efficient than those in SSL2VP(P), but the more efficiently vectorized SSL2VP(P) tridiagonal eigensolver more than compensates for this. We initially had considerable difficulties in performing the ScaLAPACK runs due to problems with MPI; these difficulties persisted for larger problems and hence no results are reported for these. The SSL2VPP routines show excellent parallel efficiency; this is particularly true for the larger problem sizes, but the efficiency even for the smaller problem sizes is perhaps surprisingly acceptable, especially when compared with the analogous ScaLAPACK routines.

3 Molecular Dynamics

Schrödinger's equation is also used to describe molecular motion, and in this context also it cannot be solved analytically. Numerical simulations typically require $O(10^5)$ (or more) basis functions to model interesting reaction processes, and therefore construction and complete eigendecomposition of a full Hamiltonian matrix are not feasible. If a Lanczos procedure is used for these problems it is often the case that the size of the resulting tridiagonal EVP exceeds that of the original problem by a considerable margin. In cases like this eigendecomposition of a (symmetric) tridiagonal matrix becomes the most computationally expensive portion of the simulation.

In this sort of computation so-called "ghost eigenvalues" occur; these are either multiple approximations to the same eigenvalue or entirely spurious. Computationally, those corresponding to the same eigenvalue appear as multiple or tightly clustered eigenvalues, and it is generally the case that eigenvectors corresponding to these eigenvalues require some form of reorthogonalization. In a parallel environment this can be highly communication-intensive if eigenvectors corresponding to the same invariant subspace reside on different processors; hence, multiple or clustered eigenvalues should reside on individual processors. Ensuring this is nontrivial and can lead to a significant level of load-imbalance.

For the Fock matrices of the last section all eigenvalues were distinct so there were no difficulties – nor load-imbalance – associated with clustered eigenvalues. In order to illustrate the efficient redistribution of eigenvalues as implemented in the tridiagonal eigensolver we consider a tridiagonal matrix of size $n = 620,000$ obtained via the Lanczos method applied to a problem arising in a molecular dynamics application [8]. There are 11,411 eigenvalues of interest and these exhibit an awkward cluster distribution when computed to a tolerance of 10^{-8} – many of the lowest eigenvalues occur as multiple eigenvalues with widely varying multiplicities (2 to nearly 500), while the largest eigenvalues are all distinct. In Table 2 we show the resulting distribution of both the total number of eigenvalues (i.e., including multiplicities) and also the number of distinct eigenvalues when the tridiagonal eigensolver is run on different numbers of PEs.

#PEs	including multiplicities	distinct
1	11411	1763
2	5698, 5713	102, 1661
4	2842, 2856, 2862, 2851	18, 84, 446, 1215
8	1513,1329,1445,1411,1434,1428,1426,1425	8, 10, 27, 57, 138, 308, 747, 468

Table 2. Distribution of eigenvalues prior to computation of eigenvectors.

Note that the total number of eigenvalues on each processor is relatively consistent, whereas the number of distinct eigenvalues varies greatly. This results in considerable load-imbalance *during the eigenvalue computation* – note however that PEs with only a few distinct eigenvalues will be computing invariant subspaces for large clusters which, due to the required reorthogonalization, will take considerably longer than the eigenvector computation on processors with small (or no) clusters. The net effect is a relatively load-balanced computation even in the presence of large clusters.

4 Optical Physics

One way to compute guided mode solutions for optical waveguide problems is to solve the so-called vector wave equation for the electric field [5]:

$$\nabla^2 \mathbf{E} + \nabla \left(\mathbf{E} \cdot \frac{\nabla n^2}{n^2} \right) + n^2 k^2 \mathbf{E} = 0.$$

Here n is the refractive index of the guiding medium, $k = \omega\sqrt{\epsilon_0 \mu_0}$ with ϵ_0 and μ_0 the electric permittivity and magnetic permeability of vacuum and ω the angular frequency of the monochromatic field, and $\mathbf{E}$ is the electric field vector. The guided mode solutions of interest have z dependence $\exp(-i\beta z)$ and, since all field components can be expressed as functions of E_x and E_y, this equation reduces to a coupled system of two PDEs. Using a Galerkin approach these equations are transformed into a coupled system of algebraic equations, and these can be written as an EVP of the form (1) with eigenvalues $\lambda = (\beta/k)^2$.

Consider an optical fiber with indices of refraction n_i and n_o for the core and cladding regions, respectively. Eigenvalues corresponding to guided mode solutions are such that $\lambda \in [n_o^2, n_i^2)$. The matrix A is full, real, and non-symmetric; despite the lack of symmetry all eigenvalues are real. The size of A is determined by the numbers of terms, n_x and n_y, in the expansions of the electric field components: $n = 2n_x n_y$. In [5] the largest problem solved was $n = 450$ (15 expansion terms in each direction), and it is pointed out that this problem size does not involve enough terms to satisfactorily represent discontinuities at fiber core interfaces. With the considerable computational power and memory available on the VPP300 we are able to consider much larger problems.

We use a Newton-based method (see, e.g., [2]) to solve this EVP. The basic idea is to replace solution of the EVP with that of zero-finding for a single nonlinear function by embedding the problem (1) in the more general family

$$(A - \lambda I)v = \beta(\lambda)x, \quad s^T v = \kappa. \tag{3}$$

For a solution of the first equation to satisfy the second as $(A - \lambda I)$ becomes singular implies $\beta(\lambda) \to 0$, i.e., zeros of $\beta(\lambda)$ correspond to eigenvalues of A. It is possible to choose s and x dynamically, resulting in convergence rates of up to 3.56 in certain special cases (see Section 6). Differentiating equations (3) with respect to λ, the Newton update formula is

$$\lambda \longleftarrow \lambda + \frac{s^T v}{s^T (A - \lambda I)^{-1} v}.$$

Computationally, the most expensive aspect of the algorithm is inversion of the (generally complex) matrix $(A - \lambda I)$; currently this is the only component of the algorithm which is performed in parallel and uses a linear solver which has been highly optimized for Fujitsu VPPs [6]. Additional details, including a more general formulation in terms of generalized EVPs and discussion of other aspects of these methods, in particular deflation procedures, can be found in [7].

Although the Newton-based procedures are ultimately capable of very good convergence properties, they are subject to difficulties in the absence of good initial data. In order to circumvent this drawback we first apply an Arnoldi method to generate initial eigendata. Our implementation is still very much under development, but our current block Arnoldi routine uses restarting, shift-invert transformation, and an implicit deflation scheme. Currently, the Arnoldi method (in real arithmetic only) is used to generate estimates of the eigenvalues and Schur vectors of interest, then these are passed to the Newton-based procedures (complex arithmetic) to obtain the final eigendata. Preliminary results have shown the use of Arnoldi's method in this context to be worthwhile. Ideally, Arnoldi's method is used to obtain estimates which are refined to machine precision with only one additional Newton step since this should effectively double the accuracy, assuming we are within the region of quadratic convergence.

Matrices are generated using the software package NPL (Numerical Photonics Library [4]); this is written in C++ and has not been ported to the VPP300 so matrices are generated on a Sun Ultra SPARC 1. Our routines consistently locate eigenvalues which are missed by NPL's eigensolver [2]. The parallel version of the Newton-based procedures is still very much in the developmental stage, and the block Arnoldi solver has not yet been parallelized, nor optimally vectorized. Therefore we do not report solution times; instead, since the most significant computational component of the Newton-based procedures (and Arnoldi if shift-invert is used) is matrix inversion, we show in Table 3 the time required for complex matrix inversion on the VPP300. We consider two test problems: The first is an optical fiber with indices of refraction $n_i = 1.422$ in the core and $n_o = 1.308$ in the cladding region; the expansion uses 30 terms in each direction so that $n = 1800$. In the second problem $n_i = 1.415$ and $n_o = 1.265$, and 40 expansion terms are used in each spatial direction, i.e., $n = 3200$. Also shown in the table are the number of guided modes obtained. An 'X' indicates that restrictions on block-size parameter choices for the linear solver precluded solving that problem on that number of PEs.

n	1 PE	2 PEs	4 PEs	6 PEs	8 PEs	n_o	n_i	$\#\lambda$
1800	11.71	5.462	3.361	2.608	X	1.308	1.422	12
3200	62.92	28.50	15.56	X	9.270	1.265	1.415	14

Table 3. Times for complex matrix inversion and number of guided modes.

Based on the speed of complex matrix inversion, a more polished implementation of the other aspects of these methods should result in an efficient nonsymmetric eigensolver.

5 Chemical Reactions

A frequently used test case for eigensolvers is solution of the "Brusselator model" of the Belousov-Zhabotinski reaction in a tubular reactor. The model comprises a system of two coupled PDEs, and investigation of the stability of the steady-state leads to a coupled ODE eigenvalue problem,

$$\frac{\nu_x}{L^2}\frac{d^2U}{dt^2} + (\beta - 1)U + \alpha^2 V = \lambda U, \qquad \frac{\nu_y}{L^2}\frac{d^2V}{dt^2} - \beta U - \alpha^2 V = \lambda V,$$

in which ν_x, ν_y, α, and β are physical parameters and U and V are chemical concentrations of the two reactants.

The problem of most interest is to determine L such that the right-most eigenvalue pair is purely imaginary; this corresponds to a Hopf bifurcation and signals the onset of periodic behavior. The Newton-based procedures of the previous section are used to compute the bifurcation parameter L. Next, the partially-parallelized block Arnoldi-Newton method is used to compute the twelve eigenvalues nearest zero. Results are summarized in Figure 1, where a portion of the spectrum is shown, and the twelve eigenvalues of interest darkened.

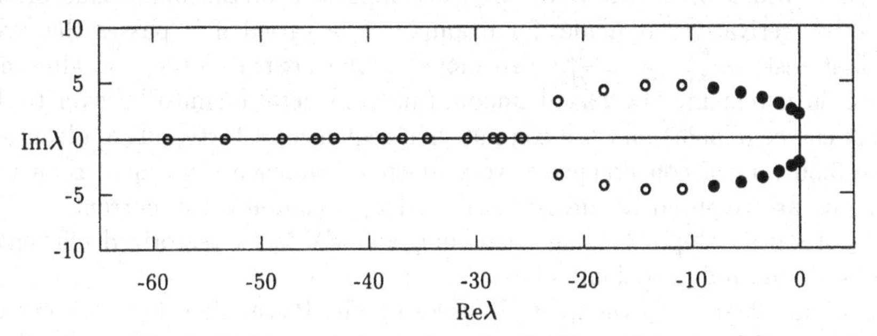

Fig. 1. Right-most eigenvalues for the Brusselator model.

6 Hydrodynamic Stability

Finally, we consider transition to turbulence and compute neutral stability curves for the Orr-Sommerfeld equation,

$$\frac{i}{\alpha R}\left(\frac{d^2}{dz^2}-\alpha^2\right)^2\phi+(U(z)-\lambda)\left(\frac{d^2}{dz^2}-\alpha^2\right)\phi-\frac{d^2U}{dz^2}\phi = 0$$

where $U(z)$ is the velocity profile in question, α is the wave number, and R is the Reynold's number. This is solved subject to the boundary conditions $\phi = d\phi/dz = 0$ at solid boundaries, and $\phi \sim \exp(-\alpha z)$, $z \to \infty$ for boundary layer flows. The neutral curve is the locus of points in the (α, R) plane which satisfy $\text{Im}\{\lambda(\alpha, R)\} = 0$. We consider two velocity profiles: Plane Poiseuille flow, corresponding to flow between two infinite plates, and Blasius flow, corresponding to flow over an infinite plate.

The differential equation is transformed into a first-order system

$$\left(\frac{d}{dz}+\alpha\right)v_1 = v_2, \quad \left(\frac{d}{dz}-\alpha\right)v_2 = v_3, \quad \left(\frac{d}{dz}+\alpha\right)v_3 = v_4,$$

$$\frac{i}{\alpha R}\left(\frac{d}{dz}-\alpha\right)v_4 = -(U(z)-\lambda)v_3 + \frac{d^2U}{dz^2}v_1.$$

Integrating this system using the trapezoidal rule, we obtain a block bidiagonal system of equations which, in combination with the boundary conditions, yields a generalized EVP for (λ, v, s):

$$(A(\alpha,R)-\lambda B)v = s^T(A(\alpha,R)-\lambda B) = 0.$$

This is solved using the Newton-based procedures of Section 4 which, for this formulation, possess an impressive convergence rate of 3.56 [7]. Newton's method is used to find a zero of $\text{Im}\{\lambda\}$ either as a function of α or of R using

$$\frac{\partial\lambda}{\partial\alpha} = \frac{s^T\frac{\partial A}{\partial\alpha}v}{s^TBv}, \quad \frac{\partial\lambda}{\partial R} = \frac{s^T\frac{\partial A}{\partial R}v}{s^TBv}.$$

To decide which of α and R to vary, a comparison of the magnitude of the respective derivatives is made; for example, α is varied if it passes the scale invariant test, $\left|\alpha\frac{\partial\lambda}{\partial\alpha}\right| \geq \left|R\frac{\partial\lambda}{\partial R}\right|$. To move on the neutral curve, the alternate variable is incremented by a fixed amount (due consideration must be given to the correct choice of sign) and the computation sequence restarted. The advantage of the 3.56 rate of convergence is very much in evidence – no more than two iterations are required at any stage in order to compute the current λ. The inverse iteration step, like that for tridiagonal EVPs, is vectorized efficiently using wrap-around partitioning [3].

In Figure 2 we show the neutral curves for the Poiseuille velocity profile on the left and Blasius on the right. Computations are carried out on three PEs, each one computing different portions of the curves (top, bottom, left). All results used 5000 finite difference grid points, corresponding to complex banded matrices of size $n = 20,000$. The solution of the block bidiagonal linear systems achieves about 92% vectorization for $n = 5000$; we have also performed experiments using up to 250,000 points, i.e. complex matrices of order $n = 1,000,000$, for which the linear solver takes under four seconds (in shared mode) and achieves over 99% vectorization. Further experiments are being initiated using higher-order differencing schemes.

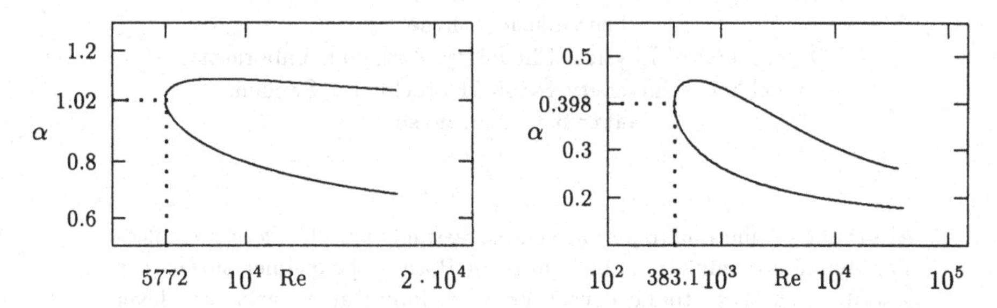

Fig. 2. Neutral stability curves for Poiseuille flow (Left) and Blasius flow (Right).

Acknowledgements

The authors thank Margaret Kahn, Anthony Rasmussen, Sean Smith, Andrey Bliznyuk, Francois Ladouceur, and David Singleton. This work was supported as part of the Fujitsu-ANU Parallel Mathematical Subroutine Library Project.

References

1. J. Dongarra and R. Van de Geijn, *Reduction to condensed form for the eigenvalue problem on distributed memory architectures*, Parallel Computing, 18 (1992), pp. 973–982.
2. D. Harrar II and M. Osborne, *Solving large-scale eigenvalue problems on vector parallel processors*, in Proc. of Int. Conf. on Vector and Parallel Processing: VECPAR'98, Berlin, Springer-Verlag. Submitted.
3. M. Hegland and M. Osborne, *Wrap-around partitioning for block bidiagonal systems*, IMA J. Num. Anal. to appear.
4. F. Ladouceur, 1997. Numerical Photonics Library, version 1.0.
5. D. Marcuse, *Solution of the vector wave equation for general dielectric waveguides by the Galerkin method*, IEEE J. Quantum Elec., 28(2) (1992), pp. 459–465.
6. M. Nakanishi, H. Ina, and K. Miura, *A high performance linear equation solver on the VPP500 parallel supercomputer*, in Proc. Supercomput. '94, 1994.
7. M. Osborne and D. Harrar II, *Inverse iteration and deflation in general eigenvalue problems*, Tech. Rep. Mathematics Research Report No. MRR 012-97, Australian National University. submitted.
8. A. Rasmussen and S. Smith, 1998. Private communication.
9. W. Thiel, 1994. Program MNDO94, version 4.1.

An Embarrassingly Parallel *ab initio* MD Method for Liquids

Fredrik Hedman[1] and Aatto Laaksonen[2]

[1] Parallelldatorcentrum (PDC),
Royal Institute of Technology, S-100 44 Stockholm, Sweden.
hedman@pdc.kth.se

[2] Department of Physical Chemistry, Arrhenius Laboratory,
Stockholm University, S-106 91 Stockholm, Sweden.
aatto@tom.fos.su.se

Abstract. A method to perform embarrassingly parallel *ab initio* molecular dynamics simulations of liquids on Born-Oppenheimer surfaces is described. It uses atomic energy gradient forces at an arbitrary level of quantum chemical methodology. The computational scheme is implemented with an MD program interfaced to a quantum chemistry package. Parallelization is done using the replicated data method.
Scaling results for up to 96 processors on an SP2 are presented. Results from simulations of liquid water at the *ab initio* SCF-MO Hartree-Fock level using single and double zeta basis function sets are compared with experimental radial distribution functions.

1 Introduction

Molecular dynamics (MD) computer simulations have over the last three decades become an important tool to study many-particle systems and nearly every possible aspect of condensed matter. The enabling factors in this development has been a combination of the growth in computer performance and capacity and more efficient simulation methods. With the arrival of massively parallel computer systems the raw computational capacity is abundant and this increased computing capacity can be used to improve the accuracy and reliability of molecular simulations.

Fundamentally there are three directions of development to follow: the time scales of the system, the size of the system and the interaction potentials of the system. In this communication we focus on a method that uses more realistic interaction potentials close to fundamental principles of molecular physics.

One such approach are so called polarizable interaction potentials and distributed multipoles [1]. With these models we are still in the domain of classical effective pair potentials, so even though the cost of calculating each interaction is greater, the simulation cost for a system containing N interaction sites is still of order $O(N \log N)$ to $O(N)$. By introducing electrons through quantum mechanical interactions between some or all particles in the simulation the accuracy of the simulation improves, but the computational cost will rise dramatically. The increased computational demand will be dominated by the level of sophistication of the quantum model (Semi-empirical, Hartree-Fock, DFT, MP2, etc).

We present a general method to perform *ab initio* MD simulations of liquid systems with periodic boundary conditions by first discussing the inclusion of first principles force fields in MD (Section 2). Then we go on to describe an implementation strategy which is based on extensive code and program reuse (Section 3). Scaling results for an embarrassingly parallel implementation using replicated data, running on up to 96 processors on an IBM SP2, is discussed (Section 4). Finally, some results from simulations of liquid water at the *ab initio* SCF-MO Hartree-Fock level using single and double zeta basis function sets are compared with experimental radial distribution functions (Section 5). We conclude with a summary (Section 6).

2 Including First Principles Force Fields in MD

To advance the MD computer simulation methods to studies of chemical processes, presently used force fields must be replaced. This is simply because the present models cannot treat phenomena such as dissociation, electron transfer or excited states. Fundamental quantum mechanics must be introduced into the simulation methods to replace the simple empirical interaction models. Unfortunately, to solve the time-dependent Schrödinger equation for condensed systems of some reasonable size is still far beyond current computational capacities.

One possible strategy to a quantum mechanical MD method, in which forces are calculated as gradients of the variational energy expressions, can be constructed by *replacing* the classical intra- and inter- molecular force calculations in a classical MD program by the corresponding quantum mechanical force calculations. This approach has been successfully tested in [2].

We have used a modified version of the computer simulation program "McMoldyn" [3] as the classical starting point. "McMoldyn' is an implementation of the classical MD method based on the neighbor list (NL) technique and uses periodic boundary conditions with minimum image criteria. The classical pairwise force calculations was replaced with a quantum mechanical (QM) calculation through and interface to a quantum chemistry package. To continue with periodic boundary conditions, each central molecule in every NL is treated equivalently. For each time step and for each molecule in the simulation we get one QM "cluster" calculation with its current NL included and minimum image criteria applied. To ensure energy conservation, all QM calculations are carried out to the same accuracy. Thus for each time step we get the following algorithm:

1. The MD program generates input files for the quantum chemistry program. The coordinates of the molecules, currently kept in the neighbor lists, are used. The molecules in each neighbor list become "clusters", which are input into a QM calculation. With N molecules in the simulation cell, N quantum mechanical cluster calculations are carried out in each time step.
2. From each cluster calculation we extract the forces acting on the central atom in the cluster. After having looped through all N clusters and calculated the

forces on each molecule, we perform a numerical time integration and update the position and velocity of each atom.

In addition to the sophistication level of the QM method employed, the cost of a time step will depend on the size of the system (N) and the number of functions (L) in the basis set used. Particularly, since the cluster of molecules on which we calculate can not be expected to contain any symmetries to reduce the number of operations, the computational complexity becomes $O(N * L^4)$ when using SCF Hartree-Fock.

3 Implementation of EP*ai*MD

In each time step, the quantum chemical calculations of molecular clusters are embarrassingly parallel. We have a number of large-grain tasks that only need to communicate very small amounts of data, and during each time step we only need one synchronization point. This problem naturally fits the Replicated Data (RD) parallelization.

To the conventional MD program, code was added to evenly distribute the cluster calculations on different processors and also to interface it to a quantum chemistry package. The parallel scheme is implemented using MPI. Cluster calculations are started from the MD code is done via the Unix system call *system.*

The problem complexity is $O(N * L^4)$. It follows that in a simulation containing N molecules we can not employ more than N processors when the QM code is serial. To efficiently use more processors than there are particles we must parallelize cluster calculations (the L dimension). We have tried two parallel QM programs: Gaussian94 [4] and GAMESS [5]. Since a simulation is made up of thousands of steps and each time step consists of N cluster calculations it must not take a lot of time to start each QM run. The parallel version of Gaussian94 available to us was ruled out because it took too long to start up.

In constructing a parallel program that in turn can start several parallel instances of GAMESS one has several options. We decided to assign the same number of processors to each instance of GAMESS. This can be achieved by arranging the available processors in an p × m matrix topology. Each column is one parallel "pool" assigned to each instance of GAMESS and the processors in the first row act as masters for each pool.

4 Scaling and Timing Results

We have made a number of runs varying the number of processors allocated by each GAMESS pool. The system used as a benchmark consists of 32 water molecules. Forces are calculated using SCF HF with a double zeta basis set and a non direct method. Timings were done using *timef* which gives wall clock elapsed time. Times reported in Table 1 are the fastest runs of each type measured on processor 0.

The speed up achieved is respectable for almost all cases. The most outstanding exception comes from using a pool size of two. The effect is puzzling; one possibility is that the overhead in using two nodes is not properly amortized by the increase in available memory compared to using a pool size of three.

It appears that a pool size of four strikes a good balance between efficiency and simulation time. Each "diagonal" in the table has the same number of cluster calculations to perform per processor In following a column we see the effect of load imbalance. As there are less cluster calculations per processor, variations in times to calculate these will result in load imbalance.

Table 1. Timings from runs using different number T2 nodes and "pool" sizes. Times are given in seconds per time step.

	GAMESS pool size				
# T2 nodes	2	3	4	5	6
16					
24		733.0			
32	1733.5		559.2		
40				497.6	
48		439.3			387.0
56					
64	873.5		286.6		
72					
80				247.9	
96		248.4			227.0

5 MD Simulation at the Hartree-Fock Level

A system of liquid water is prepared using classical MD. A cubic box is filled with 32 water molecules to a density of 1.0 g/cm^3, giving a box length of 9.86 Å. The system is equilibrated classically at 300 K. Temperature is maintained constant using the Nóse-Hoover method [6, 7].

The final configuration from the initial classical simulation was taken as an initial configuration for a run with a Hartree-Fock force field. The water system was first equilibrated for 2000 steps in order to get it adapted into a new environment of interactions, then a production run of 3000 steps was performed. The basis sets used were single and double zeta.

Figure 1 show $g(r_{OH})$ radial distribution functions from the two simulations and also experiment [8]. The striking feature here is the good result obtained, particularly with the double zeta basis set. We also obtain similarly good results for $g(r_{OO})$ and $g(r_{HH})$. This would indicate that the liquid near structure of water can be described well using as few as 32 water molecules.

6 Conclusion

This communication shows that it possible to construct an embarrassingly parallel *ab initio* MD simulation program using commonly available programs as building blocks. Our approach show good parallel scaling up to 96 nodes. Computational results show that the liquid near structure of water can be described well using as few as 32 water molecules. Knowing the capacity of high quality quantum chemical single point calculations, it should be possible to carry out very accurate MD simulations in the not too distant future.

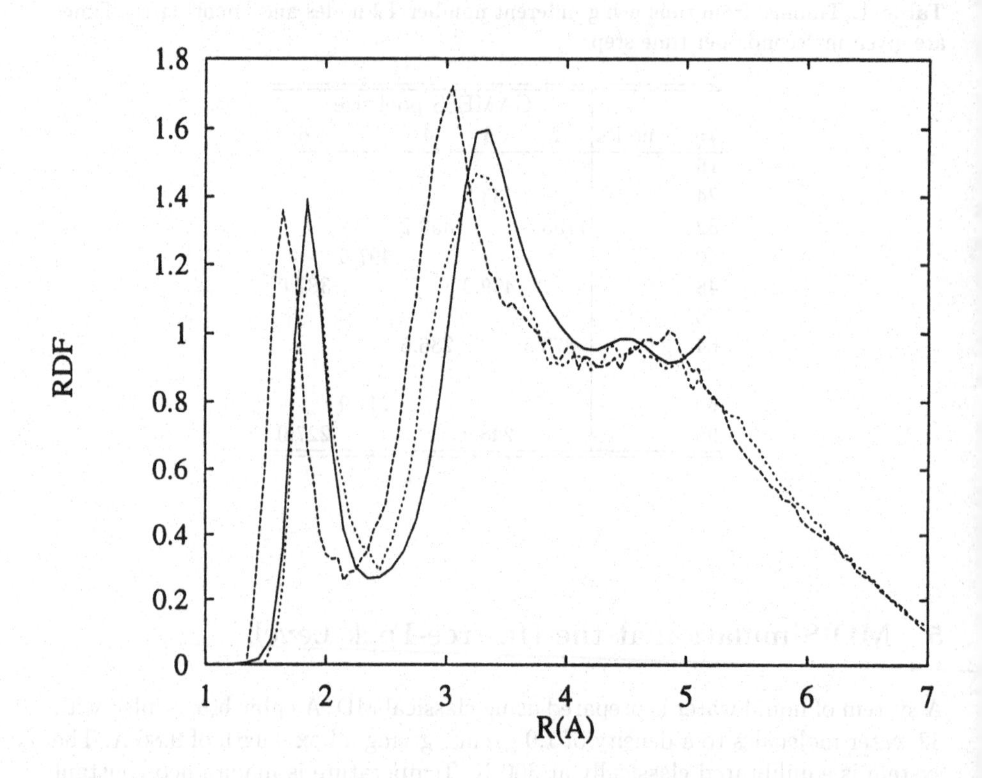

Fig. 1. Radial distribution function $g(r_{OH})$. Experimental [8] (*solid*), HF single zeta (*long dash*), HF double zeta (*short dash*)

References

1. A. J. Stone and M. Alderton. Distributed multipole analysis—methods and applications. *Molecular Physics*, 56:1047–1064, 1985.
2. Fredrik Hedman and Aatto Laaksonen. A parallel quantum mechanical md simulation of liquids. *Molecular Simulation*, 20:265–284, 1998.

3. A. Laaksonen. Computer simulation package for liquids and solids with polar interactions. I. McMOLDYN/H20: Aqueous systems. *Computer Physics Communications*, 42:271, 1986.
4. M.J. Frisch, G.W. Trucks, H.B. Schlegel, P.M.W. Gill, B.G. Johnson, M.A. Robb, J.R. Cheeseman, T.A. Keith, G.A. Peterson, J.A. Montgomery, K. Raghavachari, M.A. Al-Laham, V.G. Zakrewski, J.V. Ortiz, J.B. Foresman, J. Cioslowski, B.B. Stefanov, A. Nanaykkara, M. Challacombe, C.Y. Peng, P.Y. Ayala, W. Chen, M.W. Wong, J.L. Andres, E.S. Replogle, R. Gomperts, R.L. Martin, D.J. Fox, J.S. Binkley, D.J. Defrees, J. Baker, J.P. Stewart, M. Head-Gordon, C. Gonzales, and J.A. Pople. Gaussian 94, (revision b2), 1995. Gaussian, Inc., Pittsburgh, PA.
5. M.W. Schmidt, K.K. Baldridge, J.A. Boatz, S.T. Elbert, M.S. Gordon, J.H. Jensen, S. Koseki, N. Matsunaga, K.A. Nguyen, S. Su, T.L. Windus, M. Dupuis, and J.A. Montgomery, Jr. The general atomic and molecular electronic structure system. *Journal of Computational Chemsitry*, 14:1347–1363, 1993.
6. S. Nóse. A molecular dynamics method for simulations in the canonical ensemble. *Molecular Physics*, 52:255–268, 1984.
7. W.G. Hoover. Canonical dynamics: equilibrium phase-space distributions. *Phys. Rev. A*, 31:1695–1697, 1985.
8. A.K. Soper and M.G. Phillips. A determination of the structure of water at 25°C. *Journal of Chemical Physics*, 107:47–60, 1986.

A New Parallel Preconditioner for the Euler Equations

Lina Hemmingsson and Andreas Kähäri

Department of Scientific Computing, Uppsala University,
Box 120, SE-751 04 Uppsala, Sweden,
{Lina.Hemmingsson, Andreas.Kahari}@tdb.uu.se
http://www.tdb.uu.se/

Abstract. A new parallel preconditioner for the Euler equations has been developed. The preconditioner solve, which is based on fast modified sine transforms and the solution of narrow-banded systems of equations, is shown to be highly parallelizable.

1 Introduction

The work described in this paper has been carried out within the framework of the Parallel Scientific Computing Institute (PSCI). The project is a collaboration between VOLVO Aero Corporation (VAC) and the Department of Scientific Computing (TDB) at Uppsala University.

The aim of the project is to develop an efficient solution method for low Mach number flows. For an explicit discretization in time, the restriction on the time step often becomes too severe, and an implicit discretization must be used.

In this paper, we describe a new highly parallelizable preconditioner, that is efficient for solving the systems of equations arising from the above mentioned applications.

2 The Euler equations

The equations governing the unsteady, compressible, inviscid flow in two space dimensions are the Euler equations

$$q_t + f_x + g_y = 0, \tag{1}$$

where

$$\begin{aligned} q &= \left[\rho,\, \rho u,\, \rho v,\, e\right]^T, \\ f &= \left[\rho u,\, \rho u^2 + p,\, \rho uv,\, (e+p)u\right]^T, \\ g &= \left[\rho v,\, \rho uv,\, \rho v^2 + p,\, (e+p)v\right]^T. \end{aligned}$$

Here u and v are the velocities in the x- and y-direction, ρ is the density, p is the pressure. and e is the total energy given by $e = p(\gamma-1)^{-1} + \frac{1}{2}\rho\left(u^2+v^2\right)$, where γ is the ratio of specific heats.

3 Discretization

In time we discretize equation (1) with time-step Δt as

$$q^{n+1} + \theta \Delta t \left(f_x^{n+1} + g_y^{n+1} \right) = q^n - (1-\theta) \Delta t \left(f_x^n + g_y^n \right), \tag{2}$$

where $0 \leq \theta \leq 1$.

Consider a structured grid $(r_k(x,y), s_j(x,y))$ in space. The grid is numbered such that cell centers have integer indices, cell wall centers combined fractional-integer indices, and cell vertex points have fractional indices.

In space we use a third-order accurate, upwind-biased, cell centered finite volume method [1].

4 Newton's method and GMRES

For each time level $n+1$, we solve a non-linear system of equations

$$\boldsymbol{F}^{n+1} \equiv \boldsymbol{F}(\boldsymbol{q}^{n+1}) = 0, \tag{3}$$

where $\boldsymbol{q}^{n+1} = \left[q_{1,1}^{n+1}, q_{2,1}^{n+1}, \ldots, q_{m_k,1}^{n+1}, q_{1,2}^{n+1}, \ldots, q_{m_k,m_j}^{n+1} \right]^T$.

For simplicity we will drop the super-index $n+1$ and discuss the solution of

$$\boldsymbol{F}(\boldsymbol{q}) = 0. \tag{4}$$

We will solve (4) using Newton's method, which gives

$$\boldsymbol{J}(\boldsymbol{q}^\mu) \delta \boldsymbol{q}^{\mu+1} = -\boldsymbol{F}(\boldsymbol{q}^\mu), \tag{5}$$

where $\boldsymbol{J}$ is the Jacobian of $\boldsymbol{F}$. In each iteration we update the solution by the latest correction $\delta \boldsymbol{q}^{\mu+1}$ yielding $\boldsymbol{q}^{\mu+1} = \boldsymbol{q}^\mu + \delta \boldsymbol{q}^{\mu+1}$.

In this paper, we will solve the linear system of equations (5) using the Krylov subspace method GMRES [8]. To reduce the number of iterations we will precondition the system, and solve

$$\left(\boldsymbol{M}(\boldsymbol{q}^\mu) \right)^{-1} \boldsymbol{J}(\boldsymbol{q}^\mu) \delta \boldsymbol{q}^{\mu+1} = -\left(\boldsymbol{M}(\boldsymbol{q}^\mu) \right)^{-1} \boldsymbol{F}(\boldsymbol{q}^\mu).$$

The preconditioner system

$$\boldsymbol{M}\boldsymbol{x} = \boldsymbol{y} \tag{6}$$

should be easy to solve in parallel.

In the following section we define a preconditioner that is based on Toeplitz blocks, having an eigen-decomposition in modified sine matrices, [4]. Thus, the preconditioner solve is accomplished through fast modified sine transforms, [2], and the solution of narrow-banded systems of equations.

GMRES requires the product of a vector by the coefficient-matrix for each iteration. We will consider a so-called matrix-free version of GMRES based on the finite difference approximation $\boldsymbol{J}(\boldsymbol{q}^{\mu})\boldsymbol{x} \approx \sigma^{-1}(\boldsymbol{F}(\boldsymbol{q}^{\mu}+\sigma\boldsymbol{x})-\boldsymbol{F}(\boldsymbol{q}^{\mu}))$, where σ is a small scalar. This relation means that we do not have to form the Jacobian for the matrix–vector computation. In the construction of the preconditioner however, we will need to compute the Jacobian as we will see in Section 6.3. To avoid this expensive computation for each nonlinear Newton iteration, we intend use the same preconditioner for several Newton steps, see [7].

The preconditioner that we will construct is based on a symmetric, second-order flux-scheme [4, 6, 5, 3]. This is in order to make use of the fast solver defined in [2] for the preconditioner solve.

5 The basis for the preconditioner

In this section, we employ the symmetric, second-order flux-scheme and derive the entries of the Jacobian.

By defining A and B by $A = \frac{\partial f}{\partial q}$ and $B = \frac{\partial g}{\partial q}$, and

$$\begin{aligned}
\beta_{k,j} &= I_4 + \frac{\theta \Delta t}{2S_{k,j}} A_{k,j} \left(\Delta y_{k+\frac{1}{2},j} - \Delta y_{k-\frac{1}{2},j} - \Delta y_{k,j+\frac{1}{2}} + \Delta y_{k,j-\frac{1}{2}} \right) + \\
&\quad \frac{\theta \Delta t}{2S_{k,j}} B_{k,j} \left(\Delta x_{k+\frac{1}{2},j} - \Delta x_{k-\frac{1}{2},j} + \Delta x_{k,j+\frac{1}{2}} - \Delta x_{k,j-\frac{1}{2}} \right), \\
\alpha^{\bar{k}+}_{k,j} &= \frac{\theta \Delta t}{2S_{k,j}} \left(A_{k+1,j} \Delta y_{k+\frac{1}{2},j} + B_{k+1,j} \Delta x_{k+\frac{1}{2},j} \right), \\
\alpha^{\bar{k}-}_{k,j} &= \frac{\theta \Delta t}{2S_{k,j}} \left(A_{k-1,j} \Delta y_{k-\frac{1}{2},j} + B_{k-1,j} \Delta x_{k-\frac{1}{2},j} \right), \\
\alpha^{\bar{j}+}_{k,j} &= \frac{\theta \Delta t}{2S_{k,j}} \left(-A_{k,j+1} \Delta y_{k,j+\frac{1}{2}} + B_{k,j+1} \Delta x_{k,j+\frac{1}{2}} \right), \\
\alpha^{\bar{j}-}_{k,j} &= \frac{\theta \Delta t}{2S_{k,j}} \left(-A_{k,j-1} \Delta y_{k,j-\frac{1}{2}} + B_{k,j-1} \Delta x_{k,j-\frac{1}{2}} \right),
\end{aligned}$$

we can restate (5) as

$$\beta_{k,j}\delta q_{k,j} + \alpha^{\bar{k}+}_{k,j}\delta q_{k+1,j} - \alpha^{\bar{k}-}_{k,j}\delta q_{k-1,j} + \alpha^{\bar{j}+}_{k,j}\delta q_{k,j+1} - \alpha^{\bar{j}-}_{k,j}\delta q_{k,j-1} = -F_{k,j}.$$

6 Partitioning and preconditioning

6.1 The partitioning of the grid

We consider a partitioning of the grid into blocks, b_ℓ, $\ell = \{\ell^{\bar{k}}, \ell^{\bar{j}}\}$, see Figure 1(a). Let $\ell^{\bar{j}} \in [1, nb^{\bar{j}}]$ and $\ell^{\bar{k}} \in [\ell^{\bar{k}}_{\text{start}}(\ell^{\bar{j}}), \ell^{\bar{k}}_{\text{stop}}(\ell^{\bar{j}})]$, and $nb^{\bar{k}}(\ell^{\bar{j}}) = \ell^{\bar{k}}_{\text{stop}}(\ell^{\bar{j}}) - \ell^{\bar{k}}_{\text{start}}(\ell^{\bar{j}}) + 1$. Also introduce $\ell^{\bar{k}}_{\min} = \min_{\ell^{\bar{j}}} \ell^{\bar{k}}_{\text{start}}(\ell^{\bar{j}})$, $\ell^{\bar{k}}_{\max} = \max_{\ell^{\bar{j}}} \ell^{\bar{k}}_{\text{stop}}(\ell^{\bar{j}})$, and $nb^{\bar{k}} = \ell^{\bar{k}}_{\max} - \ell^{\bar{k}}_{\min} + 1$.

Denote by $\mathcal{L} = \{\ell; b_\ell \text{ exists in the grid considered}\}$ and ν_ℓ the total number of blocks in the grid. Each block is logically rectangular with $m^{\bar{k}}_\ell + 1 \times m^{\bar{j}}_\ell + 1$ grid-cells, see Figure 1(b). We require that the partitioning is made such that the neighboring blocks of block b_ℓ in direction $\bar{k}+$ and $\bar{k}-$ both have $m^{\bar{j}}_\ell + 1$

(a) The grid is partitioned into rectangular blocks.

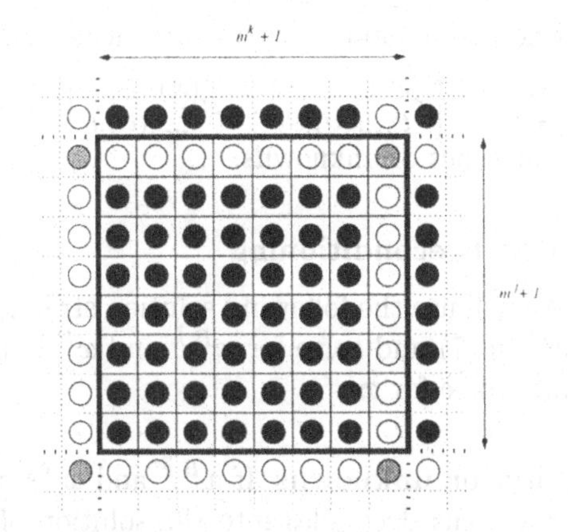

(b) The blocks consist of three kinds of cells; interior cells (black), boundary cells (white) and corner cells (gray).

Fig. 1. The partitioning of the grid and one of the blocks.

grid-cells along axis $\bar{\jmath}$ or that either one of them (or both) does not exist in the grid, and that the neighboring blocks in direction $\bar{\jmath}+$ and $\bar{\jmath}-$ either do not exist or have $m_\ell^{\bar{k}}+1$ grid-cells along axis $\bar{k}$. Also denote grid-cell lines $k = m_\ell^{\bar{k}}+1$ and $j = m_\ell^{\bar{\jmath}}+1$ by boundary lines, and the intersection between them the corner cell of the block.

Finally, we introduce the quantities $\ell^{\bar{k}\pm} = \{\ell^{\bar{k}} \pm 1, \ell^{\bar{\jmath}}\}$ and $\ell^{\bar{\jmath}\pm} = \{\ell^{\bar{k}}, \ell^{\bar{\jmath}} \pm 1\}$ that we will make use of in the following section.

6.2 The partitioning of the preconditioner system

We will consider a partitioned preconditioner system where we distinguish between interior unknowns $\boldsymbol{x}^I$, boundary unknowns $\boldsymbol{x}^{B,\bar{k}}$ and $\boldsymbol{x}^{B,\bar{\jmath}}$, and corner unknowns $\boldsymbol{x}^C$.

With these definitions the partitioned preconditioner system of equations (6) reads

$$\begin{bmatrix} \boldsymbol{M}^I & \boldsymbol{M}^{IB,\bar{k}} & \boldsymbol{M}^{IB,\bar{\jmath}} & 0 \\ \boldsymbol{M}^{BI,\bar{k}} & \boldsymbol{M}^{B,\bar{k}} & 0 & \boldsymbol{M}^{BC,\bar{k}} \\ \boldsymbol{M}^{BI,\bar{\jmath}} & 0 & \boldsymbol{M}^{B,\bar{\jmath}} & \boldsymbol{M}^{BC,\bar{\jmath}} \\ 0 & \boldsymbol{M}^{CB,\bar{k}} & \boldsymbol{M}^{CB,\bar{\jmath}} & \boldsymbol{M}^C \end{bmatrix} \begin{bmatrix} \boldsymbol{x}^I \\ \boldsymbol{x}^{B,\bar{k}} \\ \boldsymbol{x}^{B,\bar{\jmath}} \\ \boldsymbol{x}^C \end{bmatrix} = \begin{bmatrix} \boldsymbol{y}^I \\ \boldsymbol{y}^{B,\bar{k}} \\ \boldsymbol{y}^{B,\bar{\jmath}} \\ \boldsymbol{y}^C \end{bmatrix} \tag{7}$$

where the unknowns are arranged in the order $\boldsymbol{x}^{(X)} = \left[\boldsymbol{x}^{(X)}_{\bullet,1} \dots \boldsymbol{x}^{(X)}_{\bullet,nb^{\bar{\jmath}}}\right]^T$, $\boldsymbol{x}^{(X)}_{\bullet,\ell^{\bar{\jmath}}} = \left[\boldsymbol{x}^{(X)}_{\ell^{\bar{k}}_{start}(\ell^{\bar{\jmath}}),\ell^{\bar{\jmath}}} \cdots \boldsymbol{x}^{(X)}_{\ell^{\bar{k}}_{stop}(\ell^{\bar{\jmath}}),\ell^{\bar{\jmath}}}\right]^T$, $\ell^{\bar{\jmath}} = 1,\dots,nb^{\bar{\jmath}}$, and (X) is one of the four types of unknowns.

6.3 Preconditioning

We will use the following solution strategy: (Step 1) solve for $\boldsymbol{x}^C$, (Step 2) insert $\boldsymbol{x}^C$ in (7) and solve for $\boldsymbol{x}^{B,\bar{k}}$ and $\boldsymbol{x}^{B,\bar{\jmath}}$ and (Step 3) insert $\boldsymbol{x}^{B,\bar{k}}$ and $\boldsymbol{x}^{B,\bar{\jmath}}$ in (7) and solve for $\boldsymbol{x}^I$.

Interior unknowns If $\boldsymbol{x}^{B,\bar{k}}$ and $\boldsymbol{x}^{B,\bar{\jmath}}$ are known, the solution for the interior unknowns decouples into the solution of ν_ℓ independent systems of equations. For block b_ℓ we obtain the following system of equations

$$\boldsymbol{M}^I_\ell \boldsymbol{x}^I_\ell = \boldsymbol{y}^I_\ell - \boldsymbol{M}^{IB,\bar{k}}_\ell \boldsymbol{x}^{B,\bar{k}}_\ell - \boldsymbol{M}^{IB,\bar{\jmath}}_\ell \boldsymbol{x}^{B,\bar{\jmath}}_\ell \tag{8}$$

where

$$\boldsymbol{M}^I_\ell = \begin{bmatrix} \boldsymbol{D}_{\ell,1} & \boldsymbol{G}^{\bar{\jmath}+}_{\ell,1} & & & \\ -\boldsymbol{G}^{\bar{\jmath}-}_{\ell,2} & \boldsymbol{D}_{\ell,2} & \boldsymbol{G}^{\bar{\jmath}+}_{\ell,2} & & \\ & \ddots & \ddots & \ddots & \\ & & -\boldsymbol{G}^{\bar{\jmath}-}_{\ell,m^{\bar{\jmath}}_\ell-1} & \boldsymbol{D}_{\ell,m^{\bar{\jmath}}_\ell-1} & \boldsymbol{G}^{\bar{\jmath}+}_{\ell,m^{\bar{\jmath}}_\ell-1} \\ & & & -\boldsymbol{G}^{\bar{\jmath}-}_{\ell,m^{\bar{\jmath}}_\ell} & \boldsymbol{D}_{\ell,m^{\bar{\jmath}}_\ell} \end{bmatrix}, \tag{9}$$

$$\boldsymbol{D}_{\ell,j} = \begin{bmatrix} \beta_{\ell,1,j} & \alpha^{\bar{k}+}_{\ell,1,j} & & & \\ -\alpha^{\bar{k}-}_{\ell,2,j} & \beta_{\ell,2,j} & \alpha^{\bar{k}+}_{\ell,2,j} & & \\ & \ddots & \ddots & \ddots & \\ & & -\alpha^{\bar{k}-}_{\ell,m^{\bar{k}}_\ell-1,j} & \beta_{\ell,m^{\bar{k}}_\ell-1,j} & \alpha^{\bar{k}+}_{\ell,m^{\bar{k}}_\ell-1,j} \\ & & & -\alpha^{\bar{k}-}_{\ell,m^{\bar{k}}_\ell,j} & \beta_{\ell,m^{\bar{k}}_\ell,j} \end{bmatrix}, \quad j = 1,\dots,m^{\bar{\jmath}}_\ell, \tag{10}$$

$$\boldsymbol{G}^{\bar{\jmath}\pm}_{\ell,j} = \mathrm{diag}\left(\alpha^{\bar{\jmath}\pm}_{\ell,1,j},\dots,\alpha^{\bar{\jmath}\pm}_{\ell,m^{\bar{k}}_\ell,j}\right), \quad j = 1,\dots,m^{\bar{\jmath}}_\ell,$$

$$\text{and} \quad \boldsymbol{G}^{\bar{k}\pm}_{\ell,k} = \mathrm{diag}\left(\alpha^{\bar{k}\pm}_{\ell,k,1},\dots,\alpha^{\bar{k}\pm}_{\ell,k,m^{\bar{\jmath}}_\ell}\right), \quad k = 1,\dots,m^{\bar{k}}_\ell.$$

Now we will make the assumption that we have a smooth transformation (r_k, s_j) and that the solution q_ℓ is varying very slowly along a grid-line r_k within block ℓ. We introduce

$$\begin{aligned} \tilde{\beta}_{\ell,\star,j} &= \frac{1}{m^{\bar{k}}_\ell}\sum_{k=1}^{m^{\bar{k}}_\ell} \beta_{\ell,k,j}, \quad \tilde{\alpha}^{\bar{\jmath}\pm}_{\ell,\star,j} = \frac{1}{m^{\bar{k}}_\ell}\sum_{k=1}^{m^{\bar{k}}_\ell} \alpha^{\bar{\jmath}\pm}_{\ell,k,j}, \\ \tilde{\alpha}^{\bar{k}}_{\ell,\star,j} &= \frac{1}{2(m^{\bar{k}}_\ell - 1)} \sum_{k=1}^{m^{\bar{k}}_\ell-1} \left(\alpha^{\bar{k}+}_{\ell,k,j} + \alpha^{\bar{k}-}_{\ell,k+1,j}\right), \end{aligned} \tag{11}$$

and consider the following approximations of (9) and (10)

$$\tilde{\boldsymbol{D}}_{\ell,j} = \boldsymbol{I}_{m_\ell^{\bar{k}}} \otimes \tilde{\beta}_{\ell,j,\star} + \boldsymbol{R}_{m_\ell^{\bar{k}}} \otimes \tilde{\alpha}_{\ell,j,\star}^{\bar{k}}, \tag{12}$$

$$\tilde{\boldsymbol{G}}_{\ell,j}^{\bar{\jmath}\pm} = \boldsymbol{I}_{m_\ell^{\bar{k}}} \otimes \tilde{\alpha}_{\ell,j}^{\bar{\jmath}\pm}, \tag{13}$$

$$\tilde{\boldsymbol{M}}_\ell^I = \begin{bmatrix} \tilde{\boldsymbol{D}}_{\ell,1} & \tilde{\boldsymbol{G}}_{\ell,1}^{\bar{\jmath}+} & & & \\ -\tilde{\boldsymbol{G}}_{\ell,2}^{\bar{\jmath}-} & \tilde{\boldsymbol{D}}_{\ell,2} & \tilde{\boldsymbol{G}}_{\ell,2}^{\bar{\jmath}+} & & \\ & \ddots & \ddots & \ddots & \\ & & -\tilde{\boldsymbol{G}}_{\ell,m_\ell^{\bar{\jmath}}-1}^{\bar{\jmath}-} & \tilde{\boldsymbol{D}}_{\ell,m_\ell^{\bar{\jmath}}-1} & \tilde{\boldsymbol{G}}_{\ell,m_\ell^{\bar{\jmath}}-1}^{\bar{\jmath}+} \\ & & & -\tilde{\boldsymbol{G}}_{\ell,m_\ell^{\bar{\jmath}}}^{\bar{\jmath}-} & \tilde{\boldsymbol{D}}_{\ell,m_\ell^{\bar{\jmath}}} \end{bmatrix} \tag{14}$$

and replace the solution of (8) by the solution of

$$\tilde{\boldsymbol{M}}^I \boldsymbol{x}_\ell^I = \boldsymbol{y}_\ell^I - \boldsymbol{M}_\ell^{IB,\bar{k}} \boldsymbol{x}_\ell^{B,\bar{k}} - \boldsymbol{M}_\ell^{IB,\bar{\jmath}} \boldsymbol{x}_\ell^{B,\bar{\jmath}}. \tag{15}$$

Here $\boldsymbol{I}_{m_\ell^{\bar{k}}}$ is the identity matrix of order $m_\ell^{\bar{k}}$, and $\boldsymbol{R}_{m_\ell^{\bar{k}}}$ is a matrix of order $m_\ell^{\bar{k}}$ defined such that its first subdiagonal equals -1, its first superdiagonal equals 1, and its remaining entries are zero. This system is solved by means of modified sine transforms.

Block boundary unknowns Next we study the solution for the boundary unknowns. We start by eliminating the corner unknowns in (7). The resulting system of equations decouples into two independent systems of equations by, first assuming that $\boldsymbol{M}^{IB,\bar{\jmath}}$ is small and next assuming that $\boldsymbol{M}^{IB,\bar{k}}$ is small, yielding the corresponding system for $\boldsymbol{x}^I$ and $\boldsymbol{x}^{B,\bar{\jmath}}$. These two systems are both reduced by block Gaussian elimination to smaller system of equations for the block boundary unknowns

$$\boldsymbol{C}^{B,\bullet} \boldsymbol{x}^{B,\bullet} = \boldsymbol{y}^{B,\bullet} - \boldsymbol{M}^{BC,\bullet} \boldsymbol{x}^C - \boldsymbol{M}^{BI,\bullet} \left(\boldsymbol{M}^I\right)^{-1} \boldsymbol{y}^I, \tag{16}$$

where $\boldsymbol{C}^{B,\bullet} \equiv \boldsymbol{M}^{B,\bullet} - \boldsymbol{M}^{BI,\bullet} \left(\boldsymbol{M}^I\right)^{-1} \boldsymbol{M}^{IB,\bullet}$.

The systems defined by (16), the so-called block boundary Schur complement systems, decouple into $nb^{\bar{\jmath}}$ and $nb^{\bar{k}}$ independent systems respectively.

Now we make the same type of averaging along diagonals as in (11), (12), (13), and (14) for $\boldsymbol{M}_\ell^{B,\bar{k}}$, $\boldsymbol{M}_\ell^{BI,\bar{k}}$, $\boldsymbol{M}_\ell^{IB,\bar{k}}$, $\boldsymbol{M}_\ell^{B,\bar{\jmath}}$, $\boldsymbol{M}_\ell^{BI,\bar{\jmath}}$, and $\boldsymbol{M}_\ell^{IB,\bar{\jmath}}$. For this reason, we introduce the quantities

$$\hat{\beta}_{\ell,k,\star} = \frac{1}{m_\ell^{\bar{\jmath}}} \sum_{j=1}^{m_\ell^{\bar{\jmath}}} \beta_{\ell,k,j}, \quad \hat{\alpha}_{\ell,k,\star}^{\bar{k}\pm} = \frac{1}{m_\ell^{\bar{\jmath}}} \sum_{j=1}^{m_\ell^{\bar{\jmath}}} \alpha_{\ell,k,j}^{\bar{k}\pm}$$

$$\hat{\alpha}_{\ell,k,\star}^{\bar{\jmath}} = \frac{1}{2(m_\ell^{\bar{\jmath}}-1)} \sum_{j=1}^{m_\ell^{\bar{\jmath}}-1} \left(\alpha_{\ell,k,j}^{\bar{\jmath}+} + \alpha_{\ell,k,j+1}^{\bar{\jmath}-}\right).$$

From these definitions we obtain the following approximations

$$
\begin{aligned}
\hat{M}_\ell^{B,\bar{k}} &= I_{m_\ell^{\bar{\jmath}}} \otimes \hat{\beta}_{\ell,m_\ell^{\bar{k}}+1,\star} + R_{m_\ell^{\bar{\jmath}}} \otimes \hat{\alpha}^{\bar{\jmath}}_{\ell,m_\ell^{\bar{k}}+1,\star}, \\
\hat{M}_\ell^{BI,\bar{k}} &= \left[-I_{m_\ell^{\bar{\jmath}}} \otimes \left(e^{m_\ell^{\bar{k}}}_{m_\ell^{\bar{k}}}\right)^T \otimes \hat{\alpha}^{\bar{k}-}_{\ell,m_\ell^{\bar{k}}+1,\star},\ I_{m_\ell^{\bar{\jmath}}} \otimes \left(e_1^{m_\ell^{\bar{k}}}\right)^T \otimes \hat{\alpha}^{\bar{k}+}_{\ell,m_\ell^{\bar{k}}+1,\star}\right], \\
\hat{M}_\ell^{IB,\bar{k}} &= \left[-I_{m_\ell^{\bar{\jmath}}} \otimes e_1^{m_\ell^{\bar{k}}} \otimes \hat{\alpha}^{\bar{k}-}_{\ell,1,\star},\ I_{m_\ell^{\bar{\jmath}}} \otimes e^{m_\ell^{\bar{k}}}_{m_\ell^{\bar{k}}} \otimes \hat{\alpha}^{\bar{k}+}_{\ell,m_\ell^{\bar{k}},\star}\right], \\
\tilde{M}_\ell^{B,\bar{\jmath}} &= I_{m_\ell^{\bar{k}}} \otimes \tilde{\beta}_{\ell,\star,m_\ell^{\bar{\jmath}}+1} + R_{m_\ell^{\bar{k}}} \otimes \tilde{\alpha}^{\bar{k}}_{\ell,\star,m_\ell^{\bar{\jmath}}+1}, \\
\tilde{M}_\ell^{BI,\bar{\jmath}} &= \left[-\left(e^{m_\ell^{\bar{\jmath}}}_{m_\ell^{\bar{\jmath}}}\right)^T \otimes I_{m_\ell^{\bar{k}}} \otimes \tilde{\alpha}^{\bar{\jmath}-}_{\ell,\star,m_\ell^{\bar{\jmath}}+1},\ \left(e_1^{m_\ell^{\bar{\jmath}}}\right)^T \otimes I_{m_\ell^{\bar{k}}} \otimes \tilde{\alpha}^{\bar{\jmath}+}_{\ell,\star,m_\ell^{\bar{\jmath}}+1}\right], \\
\tilde{M}_\ell^{IB,\bar{\jmath}} &= \left[-e_1^{m_\ell^{\bar{\jmath}}} \otimes I_{m_\ell^{\bar{k}}} \otimes \tilde{\alpha}^{\bar{\jmath}-}_{\ell,\star,1},\ e^{m_\ell^{\bar{\jmath}}}_{m_\ell^{\bar{\jmath}}} \otimes I_{m_\ell^{\bar{k}}} \otimes \tilde{\alpha}^{\bar{\jmath}+}_{\ell,\star,m_\ell^{\bar{\jmath}}}\right].
\end{aligned}
$$

Here $e_j^{m_\ell^\bullet}$ is the jth unit vector of length $m_\ell^\bullet$.

Using the final approximations

$$\hat{\tilde{M}}_\ell^I = I_{m_\ell^{\bar{\jmath}}} \otimes \hat{\tilde{D}}_\ell + R_{m_\ell^{\bar{\jmath}}} \otimes \hat{\tilde{G}}_\ell,$$

$$\text{where} \quad \hat{\tilde{D}}_\ell = \frac{1}{m_\ell^{\bar{\jmath}}} \sum_{j=1}^{m_\ell^{\bar{\jmath}}} \tilde{D}_{\ell,j}, \quad \hat{\tilde{G}}_\ell = \frac{2}{(m_\ell^{\bar{\jmath}}-1)} \sum_{j=1}^{m_\ell^{\bar{\jmath}}-1} \left(\tilde{G}^{\bar{\jmath}+}_{\ell,j} + \tilde{G}^{\bar{\jmath}-}_{\ell,j+1}\right),$$

$$\text{we form} \quad \hat{C}^{B,\bullet} \equiv \hat{M}^{B,\bullet} - \hat{M}^{BI,\bullet}\left(\hat{\tilde{M}}^I\right)^{-1} \hat{M}^{IB,\bullet},$$

and solve

$$\hat{C}^{B,\bullet} x^{B,\bullet} = y^{B,\bullet} - M^{BC,\bullet} x^C - M^{BI,\bullet}\left(\hat{M}^I\right)^{-1} y^I. \tag{17}$$

using modified sine transforms and the solution of narrow-banded systems.

Corner unknowns Finally, we will study the solution for the corner unknowns. The system for these unknowns is difficult to solve, and we have to make several approximations to obtain a system of equations that is easy to solve.

We start by making the approximation $M^{BI,\bullet} \approx 0$, which gives a system of equations that is decoupled from the interior unknowns.

Using block Gaussian elimination we obtain

$$C^C x^C = y^C - M^{CB,\bar{k}}\left(M^{B,\bar{k}}\right)^{-1} y^{B,\bar{k}} - M^{CB,\bar{\jmath}}\left(M^{B,\bar{\jmath}}\right)^{-1} y^{B,\bar{\jmath}},$$

where $C^C \equiv M^C - M^{CB,\bar{k}}\left(M^{B,\bar{k}}\right)^{-1} M^{BC,\bar{k}} - M^{CB,\bar{\jmath}}\left(M^{B,\bar{\jmath}}\right)^{-1} M^{BC,\bar{\jmath}}$.

By making the approximation $M^{B,\bullet} \approx \hat{M}^{B,\bullet}$, we can use the eigen-decompositions of these approximations in modified sine matrices to compute all entries in $\hat{\tilde{C}}^C$ defined by

$$\hat{\tilde{C}}^C \equiv M^C - M^{CB,\bar{k}}\left(\hat{M}^{B,\bar{k}}\right)^{-1} M^{BC,\bar{k}} - M^{CB,\bar{\jmath}}\left(\tilde{M}^{B,\bar{\jmath}}\right)^{-1} M^{BC,\bar{\jmath}}.$$

This matrix is sparse with bandwidth $\mathcal{O}(\sqrt{\nu_\ell})$, and we solve

$$\hat{\tilde{C}}^C x^C = y^C + M^{CB,\bar{k}} \left(\hat{M}^{B,\bar{k}}\right)^{-1} y^{B,\bar{k}} + M^{CB,\bar{\jmath}} \left(\tilde{M}^{B,\bar{\jmath}}\right)^{-1} y^{B,\bar{\jmath}}$$

using a direct method.

7 Results

The code for the preconditioner has been implemented in Fortran 77 and PVM. It runs smoothly on a cluster of three Alpha Server 8200 (with a total of 12 Alpha EV5 processors), see Figure 2.

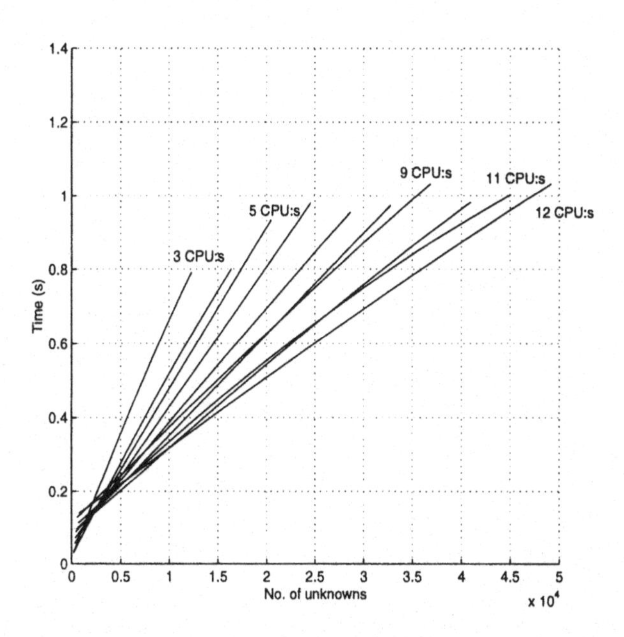

Fig. 2. Timing results for the preconditioner solve. The time for running the code on 3 to 12 CPU's with varying problem sizes are shown.

The results show that for a constant number of unknowns, doubling the number of processors reduces the elapsed time for the preconditioner solve by approximately one third.

References

1. L.-E. Eriksson. A third-order accurate upwind-biased finite-volume scheme for unsteady compressible viscous flow. Technical Report 9970-154, VAC Aerothermodynamics, 1995.

2. L. Hemmingsson. A fast modified sine transform for solving block-tridiagonal systems with Toeplitz blocks. *Numer. Algorithms*, 7:375–389, 1994.
3. L. Hemmingsson. A domain decomposition method for almost incompressible flow. *Comput. & Fluids*, 25:771–789, 1996.
4. L. Hemmingsson. Toeplitz preconditioners with block structure for first-order PDEs. *Numer. Linear Algebra Appl.*, 3:21–44, 1996.
5. L. Hemmingsson. Parallelization of a semi-Toeplitz preconditioner by domain decomposition. *BIT*, 1998. Submitted.
6. L. Hemmingsson and K. Otto. Analysis of semi-Toeplitz preconditioners for first-order PDEs. *SIAM J. Sci. Comput.*, 17:47–64, 1996.
7. D. A. Knoll, P. R. McHugh, and D. E. Keyes. Newton–Krylov methods for low-Mach-number compressible combustion. *AIAA J.*, 34:961–967, 1996.
8. Y. Saad and M. H. Schultz. GMRES: A generalized minimal residual algorithm for solving nonsymmetric linear systems. *SIAM J. Sci. Statist. Comput.*, 7:856–869, 1986.

Partitioning Sparse Rectangular Matrices for Parallel Computations of Ax and $A^T v$*

Bruce Hendrickson[1] and Tamara G. Kolda[2]

[1] Parallel Computing Sciences Department, Sandia National Labs, Albuquerque, NM 87185–1110.
bah@cs.sandia.gov.

[2] Computer Science and Mathematics Division, Oak Ridge National Laboratory, Oak Ridge, TN 37831–6367.
kolda@msr.epm.ornl.gov.

Abstract. This paper addresses the problem of partitioning the nonzeros of sparse nonsymmetric and nonsquare matrices in order to efficiently compute parallel matrix-vector and matrix-transpose-vector multiplies. Our goal is to balance the work per processor while keeping communications costs low. Although the symmetric partitioning problem has been well-studied, the nonsymmetric and rectangular cases have received scant attention. We show that this problem can be described as a partitioning problem on a bipartite graph. We then describe how to use (modified) multilevel methods to partition these graphs and how to implement the matrix multiplies in parallel to take advantage of the partitioning. Finally, we compare various multilevel and other partitioning strategies on matrices from different applications. The multilevel methods are shown to be best.

1 Introduction

In many parallel algorithms, we require numerous matrix-vector and matrix-transpose-vector multiplies with sparse matrices. Partitioning is used to distribute the nonzeros of the sparse matrix so that the work per processor is balanced and the communication costs are low. Of particular interest to us is the case when the matrix is square nonsymmetric or rectangular; we refer to both cases as *rectangular*. Specifically, given a rectangular matrix A, we find permutations P and Q so that the nonzero values of PAQ are clustered in the diagonal blocks as illustrated in Figure 1. As we will show in Sect. 2, a nearly block diagonal structure helps to reduce the memory requirements and communication cost in the matrix-vector products. Furthermore, we require the block rows and block columns to each have about the same number of nonzeros; this corresponds to balancing the floating point operations per processor.

* This work was supported by the Applied Mathematical Sciences Research Program, Office of Energy Research, U.S. Department of Energy, under contracts DE-AC05-96OR22464 and DE-AL04-94AL85000 with Lockheed Martin Energy Research Corporation.

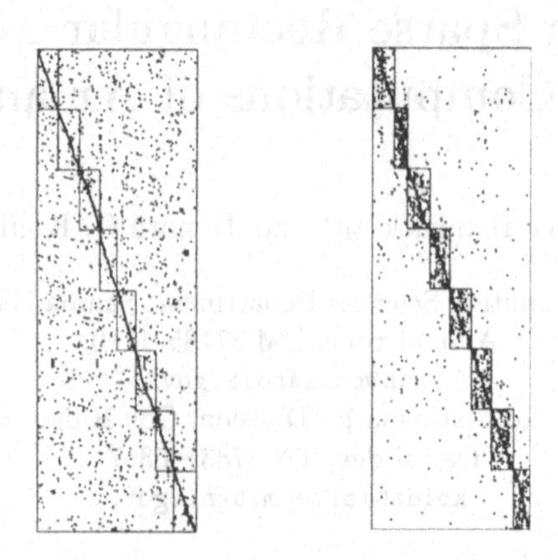

Fig. 1. Before and after partitioning.

The need to perform repeated matrix-vector multiplies with the same rectangular matrix (and its transpose) arises in numerous linear algebra algorithms. Important examples include applying QMR to nonsymmetric linear systems [3], using LSQR to solve least squares problems [14], solving the normal equations that arise in interior point methods using CG [15], or computing the truncated SVD of hypertext matrices in information retrieval [2].

Although matrix partitioning has been well-studied in the symmetric case, little work has been done in the rectangular case [13]. The symmetric problem is commonly phrased in terms of partitioning graphs. We will show that the rectangular problem can be conveniently phrased in terms of partitioning *bipartite* graphs. We will extend the work of Berry, Hendrickson, and Raghavan [2] and Kolda [13] for partitioning rectangular matrices by incorporating multilevel schemes which are already popular in the symmetric case [1,8,10,11]. Multilevel methods, described in Sect. 3, have three phases: coarsening, base-level partitioning, and un-coarsening with refinement. In Sect. 4, we will compare the (modified) multilevel methods with various refinements to non-multilevel methods on a set of test matrices.

Further information can be found in Hendrickson and Kolda [6], an extension of this research.

2 Parallel Multiplies

We propose the following parallel implementations for the matrix-vector and matrix-transpose-vector multiplications. Suppose that we have p processors. We partition A into a block $p \times p$ matrix,

$$A = \begin{bmatrix} A_{11} & A_{12} & \cdots & A_{1p} \\ A_{21} & A_{22} & \cdots & A_{2p} \\ \vdots & \vdots & \ddots & \vdots \\ A_{p1} & A_{p2} & \cdots & A_{pp} \end{bmatrix},$$

so that most of the nonzeros are in the diagonal blocks. Here block (i,j) is of size $m_i \times n_j$ where $\sum_i m_i = m$ and $\sum_j n_j = n$.

Matrix-Vector Multiply (Block Row). We do the following on each processor to compute $y = Ax$:

1. Let i denote the processor id. This processor owns the ith block row of A, that is, $\begin{bmatrix} A_{i1} \; A_{i2} \cdots A_{ip} \end{bmatrix}$, and x_i, the ith block of x of length n_i.
2. Send a message to each processor $j \neq i$ for which $A_{ji} \neq 0$. This message contains only those elements of x_i corresponding to nonzero *columns* in A_{ji}.
3. While waiting to receive messages, the processor computes the contribution from the diagonal matrix block, $y_i^{(i)} = A_{ii}x_i$. The block A_{ii}, while still sparse, may be dense enough to improve data locality.
4. Then, for each $j \neq i$ such that A_{ij} is nonzero, a message is received containing a sparse vector $\bar{x}_j$ that only has the elements of x_j corresponding to nonzero columns in A_{ij}, and $y_i^{(j)} = A_{ij}\bar{x}_i$, is computed. (We assume that processor i already knows which elements to expect from processor j.)
5. Finally, the ith block of the product y is computed via the sum $y_i = \sum_j y_i^{(j)}$. Block y_i is of size m_i.

Matrix-Transpose-Vector Multiply (Block Row). To compute $z = A^T v$, each processor does the following:

1. Let i denote the processor id. This processor owns v_i, the ith block of v of size m_i, and the ith block row of A.
2. Compute $z_j^{(i)} = A_{ij}^T v_i$, for each $j \neq i$ for which $A_{ij} \neq 0$. Observe that the number of nonzeros in $z_j^{(i)}$ is equal to the number of nonzero rows in A_{ij}^T, i.e., the number of nonzero columns in A_{ij}. Send the nonzero[1] elements of $z_j^{(i)}$ to processor j.
3. While waiting to receive messages from the other processors, compute the diagonal block contribution $z_i^{(i)} = A_{ii}^T v_i$.
4. From each processor j such that $A_{ji} \neq 0$, receive $\bar{z}_i^{(j)}$ which contains only the nonzero elements of $z_i^{(j)}$. (Again, we assume that processor i already knows which elements to expect from processor j.)
5. Compute the ith component of the product, $z_i = z_i^{(i)} + \sum_{j \neq i} \bar{z}_i^{(j)}$. Block z_i is of size n_i.

[1] Here we mean any elements that are guaranteed to be zero by the structure of A_{ij}. Elements that are zero by cancellation are still communicated.

Block column algorithms are analogous to those given for the block row layout. Observe that sparse off-diagonal blocks result in less message volume. See Hendrickson and Kolda [6] for more details on these algorithms.

3 Multilevel Partitioning of Rectangular Matrices

A rectangular $m \times n$ matrix $A = [a_{ij}]$ corresponds to an undirected *bipartite* graph $G = (R, C, E)$ with $R = \{r_1, \ldots, r_m\}$, $C = \{c_1, \ldots, c_n\}$ and $(r_i, c_j) \in E$ iff $a_{ij} \neq 0$ (see Fig. 2). The weight of each row vertex is the number of nonzeros in

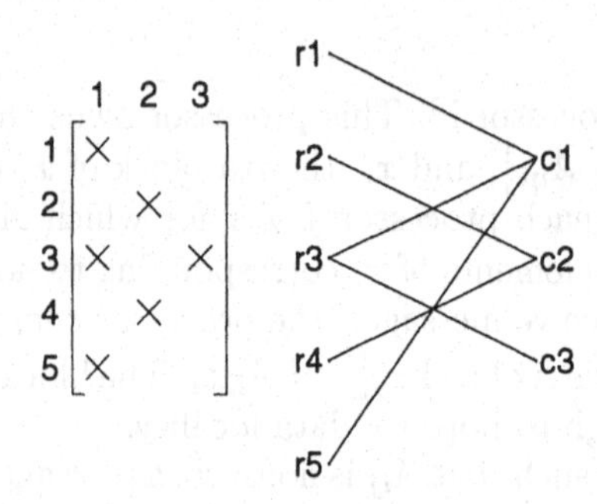

Fig. 2. Bipartite graph representation of a matrix.

the corresponding row; the column vertices are unweighted. The weight of each edge is initially set to one. We now wish to divide $R \cup C$ into p sets in such a way that the total row weight per set is balanced, and the total number of edges crossing between sets is kept small. The first criterion ensures load balance in the multiplies, while the third limits the communication volume. This partitioning problem is known to be NP-hard [4]. We will present methods to divide into $p = 2$ partitions. In order to divide into $p = 2^k$ sets, we recursively partition the block diagonals.

We will focus on multilevel methods to approximately solve this problem. This type of method has three phases. In phase 1, the graph is successively coarsened by merging vertex pairs. Row vertices merge only with row vertices, likewise for column vertices. The rows are merged as follows. A random eligible row is chosen, say i. It is merged with a randomly chosen, unmerged row node, say i', that is a path of length two away. That is, two row nodes can be merged if they are currently unmerged and some column has nonzero values in both row locations. (If no such row node exists, the i is ineligible for merging.) We choose another eligible row at random and continue the process until all row vertices have had the chance to be paired. We pair the column vertices via an analogous procedure. The vertex weight of a node in the coarse graph is the sum of the weights of its constituent pair. There is an edge between two nodes in the coarse graph if any pair of their constituent vertices had an edge. The weight of the edge is the sum of all the edge weights of the edges between the constituent vertices. The coarse graph maintains the bipartite structure of the original graph and has about half as many vertices as the original graph. To further coarsen, we repeat the process until we have reached the desired number of vertices.

In phase 2, the coarsest graph is partitioned randomly.

Lastly in phase 3, the graph is successively un-coarsened and the partition is refined each step. As we un-coarsen, the two constituent nodes of a merged vertex are initially in the same partition as the merged vertex. In the course of the refinement, one or both may switch partitions. We have experimented with three different refinement strategies. The first refinement option is a modified version of Kernighan-Lin [12] for bipartite graphs. This method moves one node at a time looking for a better partition. The second option is to use the alternating partitioning method presented in [13]. We start with a fixed column permutation that may be the result of some previous ordering. Given that the column partitioning is fixed, we must compute the best row partition. We then fix that row partition and compute the best column partition. We continue alternatingly fixing one partition and computing the other until the overall partition can no longer be improved. The final refinement option is to use alternating partitioning followed by Kernighan-Lin; this can be thought of as a rough refinement followed by a fine refinement. Note that in addition to being used as components of a multilevel approach, these refinement algorithms can improve a partition produced by any algorithm. More details on these methods can be found in Hendrickson and Kolda [6].

In Sect. 4, we will also give results for the Spectral method for bipartite graphs, described originally in Berry, *et al.* [2] and also in Hendrickson and Kolda [6].

4 Experimental Results

The software we are using is a modification of the Chaco package (written in C) developed by Hendrickson and Leland [7] for multilevel partitioning of symmetric matrices. The timings for the partitioning were done on a 300 MHz Pentium II.

We will give results for four matrices from different disciplines, each split over 16 processors. The alternating partitioning (AP) method starts with a random ordering; it has no multilevel component. The spectral method (Spectral) uses a multilevel Rayleigh Quotient Iteration/Symmlq eigensolver [1, 7]. The three multilevel methods differ in the refinement algorithm they apply after each uncoarsening step. ML-AP applies alternating partitioning, ML-KL uses Kernighan-Lin, and ML-AP+KL combines the two by applying first alternating partitioning and then Kernighan-Lin. We always coarsen to one hundred coarse vertices in the multilevel methods. In every case, the rows (or columns) of the matrix are divided among processors in such a way that there is less than 10% difference in the number of matrix nonzeros owned by different processors.

For all the tables, the format is as follows. *Edge Cuts* is the number of edges in the bipartite graph that are *cut* by the given partition. *Part Time* is the time (in seconds) to compute the partition. *Tot Msgs* and *Tot Vol* are, respectively, the total number of messages and total message volume for computing either Ax or $A^T v$. *Max Msg* and *Max Vol* are, respectively, the maximum number and maximum volume of messages handled by a *single* processor in the computation of Ax or $A^T v$, incoming or outgoing.

Table 1 shows the result of applying various partitioning methods to the $17,758 \times 17,758$ nonsymmetric `memplus` matrix[2] with 99,147 nonzeros for solving linear systems. Compared to a partitioning based upon the natural ordering, the number of edge cuts is substantially reduced by each method except Spectral, and the multilevel methods reduce the value by a factor of three. The overall communication volume is reduced by more than a factor of three by the multilevel methods, but not by the AP or Spectral methods. Only the Spectral method is expensive in terms of partition computation time, and that is because it had trouble converging. In fact, the Spectral method has problems in the next two examples as well. Note that the total number of messages increase because the message passing is more distributed; i.e., observe that the maximum message volume handled by a single processor is greatly reduced by the various reorderings.

Table 1. Communication pattern for row-based partitioning of the `memplus` matrix on 16 processors.

Method	Edge Cuts	Part Time	Total Msgs	Total Vol	Max Msgs	Max Vol
Natural	69381	0.26	74	30501	15	12826
AP	49535	1.66	216	28864	15	3152
ML-KL	18138	4.28	240	8991	15	877
ML-AP	20273	4.20	239	11040	15	1073
ML-AP+KL	18762	5.54	239	10128	15	1022
Spectral	59481	84.66	137	31709	15	7312

The $28,254 \times 17,284$ `pig-large` matrix [5,9] with 75,018 nonzeros arises from a least squares problems. The multilevel methods (see Table 2) are best, reducing the edge cuts, total message volume, and the processor message volume by factors of more than three in every case.

The $6,071 \times 12,230$ `df1001` matrix[3] with 35,632 nonzeros arises from linear programming. Here we partition the matrix *column-wise* since the matrix has some dense rows; this should yield a better partitioning. In Table 3, we see that again the number of edge cuts, the total message volume, and the maximum single processor volume are substantially reduced.

Results for the $1,853 \times 625$ `man1` matrix with 3,706 nonzeros are presented in Table 4. This is a hypertext matrix as described in Berry, *et al.* [2], and we

[2] Available from MatrixMarket (`http://math.nist.gov/MatrixMarket/`).
[3] Available from NETLIB (`http://www.netlib.org/lp`).

Table 2. Communication pattern for row-based partitioning of the `pig-large` matrix on 16 processors.

Method	Edge Cuts	Part Time	Total Msgs	Total Vol	Max Msgs	Max Vol
Natural	55332	0.26	78	23203	12	4740
AP	24016	2.44	229	14069	15	1335
ML-KL	11775	4.26	216	3796	15	494
ML-AP	14966	3.53	214	6694	15	1100
ML-AP+KL	12202	6.52	216	3632	15	416
Spectral	11726	193.14	187	6475	15	810

Table 3. Communication pattern for column-based partitioning of the `df1001` matrix on 16 processors.

Method	Edge Cuts	Part Time	Total Msgs	Total Vol	Max Msgs	Max Vol
Natural	33194	0.10	140	24636	15	8468
AP	14361	1.02	239	13804	15	1342
ML-KL	8663	2.54	238	7951	15	749
ML-AP	10388	1.80	239	9569	15	855
ML-AP+KL	8653	3.30	234	7919	15	742
Spectral	17067	44.15	228	13383	15	2056

want to compute a low-rank SVD for it. In this case, the number of messages is actually reduced, as was the total volume, and maximum processor volume. Here we finally have a case where the Spectral method does better than all the others, although it still takes the longest time to compute. Figure 1 shows the original matrix and the result of the ML-AP+KL partitioning for 8 processors.

5 Conclusions

We presented (modified) multilevel methods for partitioning sparse nonsymmetric and nonsquare matrices. We described how this can be used to implement efficient parallel matrix-vector and matrix-transpose-vector multiplies with reduced communication. We tested our methods on four matrices from four different mathematical applications. Our results show that partitioning clearly reduces the communication volume that multilevel partitioning with alternating partitioning plus Kernighan-Lin refinement is generally the best of the partitioners.

Acknowledgements

Thanks to Iain Duff and Michael Saunders for help in finding large matrices to work with. Also thanks to Michele Benzi for many helpful conversations.

Table 4. Communication pattern for column-based partitioning of the **df1001** matrix on 16 processors.

Method	Edge Cuts	Part Time	Total Msgs	Total Vol	Max Msgs	Max Vol
Natural	1497	0.01	236	1168	15	93
AP	1002	0.05	180	685	15	68
ML-KL	1045	0.13	137	556	15	103
ML-AP	811	0.08	151	521	14	62
ML-AP+KL	618	0.15	141	417	13	49
Spectral	556	1.75	119	306	12	39

References

1. Stephen T. Barnard and Horst D. Simon. A fast multilevel implementation of recursive spectral bisection for partitioning unstructured problems. *Concurrency: Practice and Experience*, 6:101–117, 1994.
2. Michael W. Berry, Bruce Hendrickson, and Padma Raghavan. Sparse matrix reordering schemes for browsing hypertext. In James Renegar, Michael Shub, and Steve Smale, editors, *The Mathematics of Numerical Analysis*, volume 32 of *Lectures in Applied Mathematics*, pages 99–122. American Mathematical Society, 1996.
3. Roland W. Freund and Noël M. Nachtigal. QMR: A quasi-minimal residual method for non-Hermitian linear systems. *Numer. Math.*, 60:315–339, 1991.
4. Michael R. Garey and David S. Johnson. *Computers and Intractability: A Guide to the Theory of NP-Completeness.* W. H. Freeman and Company, New York, 1979.
5. Markus Hegland. Description and use of animal breeding data for large least squares problems. Technical Report TR-PA-93-50, CERFACS, Toulouse, France, 1993.
6. Bruce Hendrickson and Tamara G. Kolda. Partitioning nonsquare and nonsymmetric matrices for parallel processing. Technical Memorandum TM-13657, Oak Ridge National Laboratory, Oak Ridge, TN 37831, 1998. Submitted to *SIAM J. Scientific Computing.*
7. Bruce Hendrickson and Robert Leland. The Chaco user's guide, version 2.0. Technical Report SAND95-2344, Sandia Natl. Lab., Albuquerque, NM, 87185, 1995.
8. Bruce Hendrickson and Robert Leland. A multilevel algorithm for partitioning graphs. In *Proc. Supercomputing '95.* ACM, 1995.
9. A. Hofer. *Schätzung von Zuchtwerten feldgeprüfter Schweine mit einem Mehrmerkmals-Tiermodell.* PhD thesis, ETH-Zurich, 1990. Cited in [5].
10. George Karypis and Vipin Kumar. A fast and high quality multilevel scheme for partitioning irregular graphs. Technical Report 95-035, Dept. Computer Science, Univ. Minnesota, Minneapolis, MN 55455, 1995.
11. George Karypis and Vipin Kumar. Parallel multilevel graph partitioning. Technical Report 95-036, Dept. Computer Science, Univ. Minnesota, Minneapolis, MN 55455, 1995.
12. B. W. Kernighan and S. Lin. An efficient heuristic procedure for partitioning graphs. *Bell System Technical J.*, 1970.
13. Tamara G. Kolda. Partitioning sparse rectangular matrices for parallel processing. In *Proc. 5th Intl. Symposium on Solving Irregularly Structured Problems in Parallel (Irregular '98)*, to appear.

14. Christopher C. Paige and Michael A. Saunders. LSQR: An algorithm for sparse linear equations and sparse least squares. *ACM Trans. Mathematical Software*, 8:43–71, 1982.
15. Weichung Wang and Dianne P. O'Leary. Adaptive use of iterative methods in interior point methods for linear programming. Technical Report CS-TR-3560, Dept. Computer Science, Univ. Maryland, College Park, MD 20742, 1995.

NetLink: A Modern Data Distribution Approach Applied to Transparent Access of High Performance Software Libraries

Ivar Holmqvist and Erik Lindström

Department of Computing Science
Umeå University
S-90187 Umeå, Sweden
{dpiht,karhu}@cs.umu.se

Abstract. In many situations, e.g. in academic and scientific computing, where neither time nor financial support allow developers to address more than the core problem, it is most important to have mechanisms that automatically serve for many important software engineering concerns. This paper describes a general data distribution architecture, very suitable for transparent access to software components, applied to the problem of sharing software libraries for high performance computers. The system has proved to be very efficient and simulations and tests have together shown that the scalability is very high due to extensive cache usage and several proactive components.

1 Introduction, Motivation, and Objectives

To solve scientific computing problems you need appropriate and efficient software together with access to CPU time. Many projects and research organizations, e.g. NetSolve [2], PDC [8], and HPC2N [7], provide easy access to high performance computers. For some types of problems, Internet-wide metacomputers [4, 5] are also a rising factor in accessing CPU-time. Collections of software can be found across the Internet. Sites like `www.download.com` and `www.shareware.com` provide users with a lot of software organized in fairly easy-to-search structures. In the area of scientific computing, the free software collection stored in NetLib [3] is maybe both the most complete and the most frequently accessed.

Few projects have addressed the problem of not only collecting and distributing software components, but also providing the components in a way that minimizes user-end software management. If we again turn our focus to scientific computing, we find many people spending a lot of time, compiling, testing, and tuning software libraries like LAPACK [1]. Our objective for this project is to find a data distribution architecture that, in a distributed manner, can help to centralize the library maintenance and tuning, avoiding that many users have to spend time and effort on things others already have completed. Instead of having to download, compile, configure, test, tune, and install each software library to use, it should be sufficient to install a NetLink access agent component, and thereby get transparent access to a whole variety of software libraries, all highly tuned for the target architecture.

The security aspects of automatic software distribution and updates are many. Authentifying the software distributor is from the user point of view, the most important security related concern. If everyone all over the Internet has the possibility to distribute software through the proposed system, one would probably want some guarantees before using a downloaded software component. On the other hand, from the distributor point of view, it is also important that only the intended end-users get access to published NetLink data. Authentifying the user through the access agent and providing a secure transportation method are two important objectives for our system to meet the distributor security needs.

NetLink does not propose any specific authentification mechanisms or encryption algorithms. Instead, the use of a general object access model that by a data issuer chosen authentification mechanism divides the agent collection into secure virtual domains [6].

2 Related Areas and Works

The NetLink architecture and communication protocols can in many terms be viewed as a new world-wide-web, where the traditional URL addresses have been replaced by searchable URN (universal resource name) addresses. The similarities with the well-known http-based world-wide web, enlighten the access agent role in the system. By changing the access agent capabilities and user interface, the system can be used for many different applications and types of data.

Distribution of software updates is an important, but often also very difficult and costly, task for software vendors. Internet is today a very common medium for the distribution, but the lack of automation is obvious. Often the download and update process must be explicitly trigged by the user. In NT5, we will see a concept of *active directories*, and Netscape has published a *SmartUpdate* system. Both these systems allow software to be automatically and securely installed on a user's machine. But they are both quite limited and the distributed database fundamentals and possible search strategies in NetLink are not addressed at all. SmartUpdate, though, has capabilities to handle payments and registrations, to help distributing commercial software components. NetLink does not explicitly address the payment and registration problems, but instead the system provides a security and authentification architecture, that can be used as a base for limiting or explicitly defining the distribution range of issued objects.

3 Architecture

In Figure 1 we have identified three types of agents. *Agent*, this is the most generic type of agent. An agent that both supports the creation of objects, has capability to resolve queries and cache objects to speed up the overall performance of NetLink. *Thin Agent*, sites that have users who require the same type

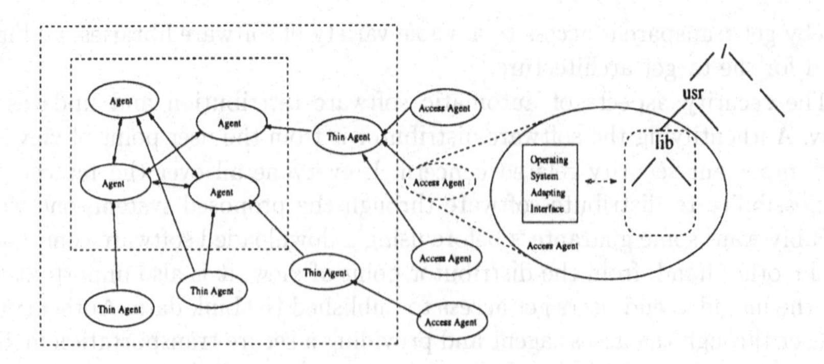

Fig. 1. Global System Overview

of objects can/will set up a thin agent that only has cache capability. Dramatically decreasing search times for the local users without interferring with the out side world. *Acces Agent*, agent only capable issuing queries. Should also provide a mechanism as shown in Figure 1 to transparently incorporate the results as a part of the local filesystem.

The architecture presented in Figure 2 is one solution to the problem of implementing the NetLink design [6]. We have several requirements on such an architecture. First, it must be general enough to handle all the requirements posed in the NetLink design [6]. Secondly, the architecture must put very few restrictions on the choice of central algorithms, such as searching and caching. This serves as a guarantee that the implementations can be made highly efficient.

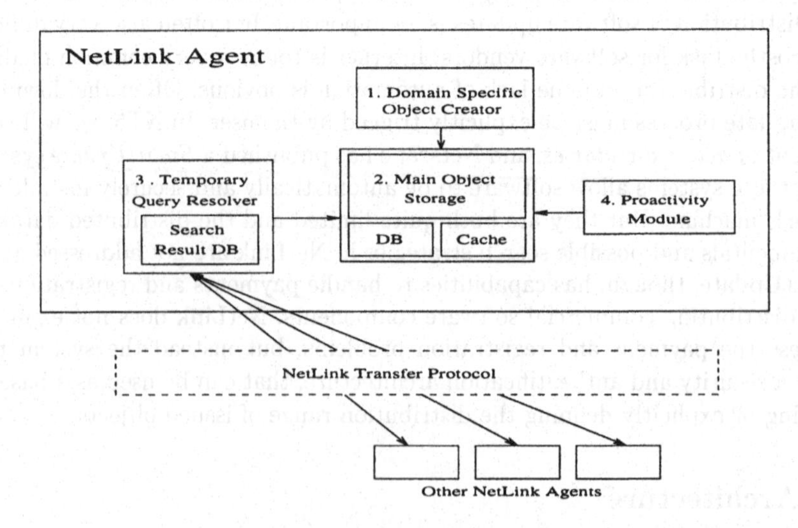

Fig. 2. Architecture Overview

As we can see in Figure 2, there are four main components in the main agents of the NetLink system, each supporting a solution to one of the different task that the NetLink agent is capable of, creating objects, storing objects, performing searches and keeping it self informed of changes in the outside world. In the following sections, we will take a brief look at each of the four central components and discuss their structures and the impact on the overall system performance different structural choices might have [6].

3.1 Site Specific Object Creator

This construct has a simple but basic task which is providning objects to the system. These objects vary from domain to domain. In one domain, the one implemented by the prototype, the objects might represent files and directories retrieved from a local file system tagged with optimize/architecture information.

3.2 Main Object Storage

The data storage part of the agent is, for mainly efficiency reasons, organized in a two-level database construction. The two levels are treated as two separate databases, `DB` and `Cache`. This storage organization allows for a object store and retrieve mechanism that can handle objects to and from other agents without claiming exclusive rights to the main storage database `DB`. This organization strongly increases the parallel capability of the system. Instead, a claim for exclusive rights to the much smaller `Cache` is needed, and the responsibility for consistency issues between the two databases is handled over to a separate proactive component of the agent, see Section 3.4.

The structure of the different databases allows for construction of very efficient algorithms for searching and merging databases. The main idea behind the design is to represent every object once only. For example, if we have a large number of objects sharing an equivalent parent, then the parent is stored in one memory location only, and the objects only contains a reference to the single parent-object. This applies all the way down to individual strings; each DB contains a structure that allows us to represent each string with a reference, thereby obtaining a naive and very effective $O(1)$ algorithm for comparing strings and objects.

3.3 Temporary Query Resolvers

The name of this agent component puts focus on that there occasionally might be several concurrently existing instances of this component , all with a limited life-time and single task purpose. A temporary query resolver is created when a query reaches the agent and terminated when the agent has done all in it's power to fulfill this request. Each resolver has a local database for storing the result from both the local search and the results from passing the query to the neighbor agents.

3.4 Permanent Proactivation Module

As opposed to the temporary query resolvers this component is a permanent part of the agent. It performs all necessary tasks to bring the system storage components consistent and updates the agents information about the outside world. This means that the proactivation component is responsible for merging Cache and DB, monitoring the communication between this agent and other agents to obtain objects suitable for caching, and obtaining names of other agents. The proactive agent component is also responsible for maintaining status information about other agents such as the speed of the connection, and what information we can expect to find in that agent.

3.5 Project Status

Parts of this NetLink architecture has been successfully implemented in form of a fully functional prototype and extensively described by a simulation model. The prototype effectively shows that the chosen algorithms are possible to implement and efficient to execute in a full scale working situation. The simulation model is mainly a analysis tool for studying what happens when the number of agents and objects increases. Although the prototype is fully functional and able to communicate (via the NetLink Transfer Protocol, NLTP) with other agents, it is not realistic to start a vast number of prototypes and analyze what happens. Instead we use the simulation environment to virtually start simulated agents. The simulation incorporates two simplifications of the real situation. First, $T_t(d, a_1, a_2)$ a function that approximates the time for agent a_1 to connect to agent a_2 and transfer d objects of a general size. This function is a very simple model of the agent-to-agent communication cost, but models the reality well in situations with moderate host and communication channel loads. Secondly, the simulation environment incorporates a simplified data model, where we represent all the objects in a domain by a bit-vector interpretation of sets. A local query then conforms to and:ing two bit-vectors and merging two data sets corresponds to or:ing the bit-vector interpretations.

4 Results

The simulation environment has proved to be an extraordinary tool in analyzing the scalability, stability and effectiveness of large systems. We have carried out several tests showing that the described architecture has a very effective structure. It is also simple to introduce new concepts in the simulation model, such as caching and replication.

From Figure 3, we can conclude that the relative number of agents that are visited is constant as we increase the number of agents in the network. Since these results are based on naive simulation models without any heuristics for improving the efficiency we can assume that this is an upper bound limit. We

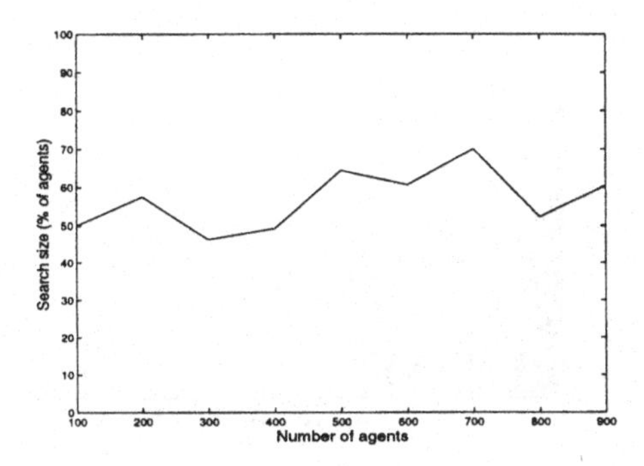

Fig. 3. % of the nodes visited

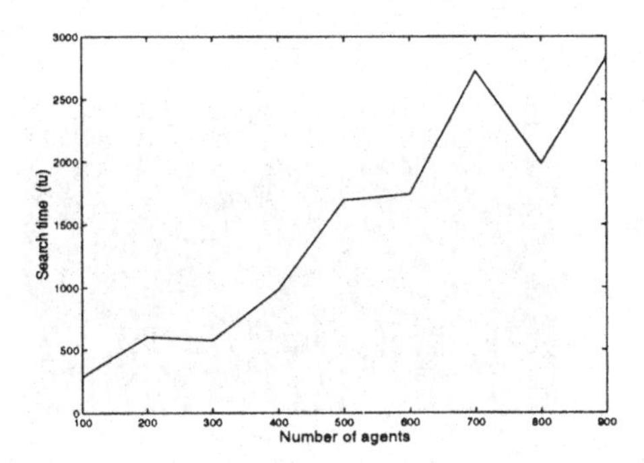

Fig. 4. Search time vs #nodes

expect several improvements when we apply more sofisticated algorithms. Another almost obvious result shown in Figure 4, is that normal search techniques result in search times linearly dependent on the number of agents.

In Figure 5 and 6, we can can observe the almost obvious fact that a search algorithm with caching is more efficient than one without. This is not the the main observation though; the interesting fact is that it is extremely easy to modify the simulation algorithm from cache to nocache, hence the simulation model can be used to compare algorithms in a much less difficult manner than modifying the prototype.

References

1. E. Anderson, Z. Bai, C. Bishof, J. Demmel, J. Dongarra, J. Du Croz, A. Greenbaum, A. Hammarling, A. McKenny, S. Ostrouchov, and D. Sorensen. *LAPACK User's Guide, Release 2.0.* SIAM Press, 1994.
2. H. Casanova and J. Dongarra. NetSolve: A Network-Enabled Server for Solving Computational Science Problems. *The International Journal of Supercomputer Applications and High Performance Computing,* 11(3):212–223, 1997.
3. J. Dongarra and E. Grosse ed. Netlib repository at utk and orn, accessed December 1997. http://www.netlib.org/.

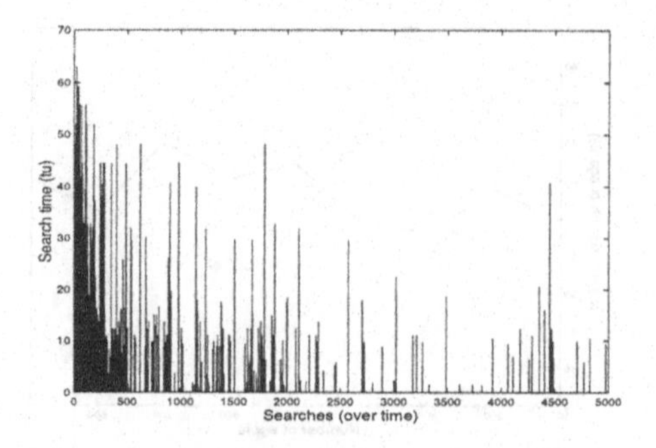

Fig. 5. Search w caching

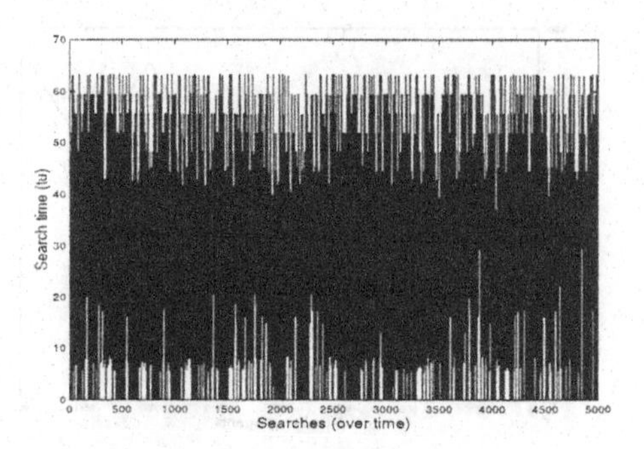

Fig. 6. Search w/o caching

4. G. C. Fox and W. Furmanski. Petaops and Exaops: Supercomputing on the Web. *IEEE Internet Computing*, 1(2), 1997.
5. I. Foster and C. Kesselman. Globus: A Metacomputing Infrastructure Toolkit. *International Journal of Supercomputer Applications (to appear)*, 1997.
6. I. Holmqvist. NetLink: Software libraries at your fingertips. Technical Report UMNAD 224.98, Department of Computer Science, Umeå University, 1998.
7. HPC2N. High Performance Computing center north, umea, sweden, accessed December 1997. http://www.hpc2n.umu.se.
8. PDC. Center for parallel computers, kth, sweden, accessed December 1997. http://www.pdc.kth.se.

Modernization of Legacy Application Software

Jeffrey Howe[1], Scott B. Baden[1], Tamara Grimmett[2], and Keiko Nomura[2]

[1] Department of Computer Science and Engineering,
University of California, San Diego,
9500 Gilman Drive, La Jolla CA 92093-0114, USA
{baden, jhowe}@cs.ucsd.edu
http://www-cse.ucsd.edu/users/baden
[2] Department of Applied Mechanics and Engineering Sciences
University of California, San Diego,
9500 Gilman Drive, La Jolla CA 92093-0411, USA
{tgrimmet, knomura}@opal.ucsd.edu

Abstract. Legacy application software is typically written in a dialect of Fortran, and must be reprogrammed to run on today's microprocessor-based multicomputer architectures. We describe our experiences in modernizing a legacy direct numerical simulation (DNS) code with the KeLP software infrastructure. The resultant code runs on the IBM SP2 with higher numerical resolutions than possible with the legacy code running on a vector mainframe.

1 Introduction

Legacy application software is typically written in a dialect of Fortran, and optimized for execution on a vector class architecture like the Cray C90 or T94. Such codes often run on just one processor and must be reprogrammed to run on today's microprocessor-based multicomputer architectures like the IBM SP2 or Cray T3E. A principal motivation for modernizing legacy code is increased memory capacity. For example, the Cray T94 at the San Diego Supercomputer Center has only about 2 Gigabytes of memory, but just 8 nodes of SDSC's 128-node IBM SP2 provide this same memory capacity.[1]

We will discuss our experiences in modernizing a legacy application code, called `DISTUF` [13, 7], using the KeLP software infrastructure developed by Fink and Baden [5]. The resultant code, called `KDISTUF`, can run at a higher numerical resolution than the legacy code, by virtue of increased memory and computational capacity. We discuss our methodology for parallelizing this application and present performance measurements on the IBM SP2.

2 Preliminaries

We first describe the starting point for our study: the `DISTUF` code. We then give a brief overview of the KeLP infrastructure, which we used to convert `DISTUF` to `KDISTUF`.

[1] See `http://www.npaci.edu/Resources/Systems/compute.html`.

2.1 DISTUF

DISTUF performs direct numerical simulations (DNS) of incompressible homogeneous sheared and unsheared turbulent flows. The significance of DNS is the elimination of turbulence closure models thus allowing recovery of the fundamental physics directly from simulation results. Complete resolution of all relevant scales is therefore required and the *total* number of computational grid points needed is proportional to the 9/4 power of the Reynolds number.[2]

DISTUF solves the three-dimensional, time-dependent Navier-Stokes, continuity, and energy (passive scalar) equations. The equations are discretized in an Eulerian framework using a second-order finite-difference method on a staggered grid. The Adams-Bashforth scheme is used to integrate the equations in time. Pressure is treated implicitly; DISTUF employs a fast Poisson solver [14] which combines fast Fourier transforms and Gaussian elimination to solve for the pressure field. Boundary conditions are periodic.

The data structure for DISTUF consists of a single logical 3-dimensional grid. Each point on the grid employs 9 double precision field variables: vector-valued velocity, scalar temperature and pressure, and 4 temporaries.

DISTUF is divided into two computational phases: (1) a finite difference computation to solve advection and diffusion equations for the velocity and temperature terms, and (2) a Fourier method to solve the Poisson equation. The computed velocity values from the first phase are then updated by the pressure solution to give the final velocities at the new time step.

DISTUF is written in Fortran 77 and has evolved over a period of 14 years. The original code was roughly 10,000 lines long including comments written in both English and German. We extracted a subset of this code, removing a post-processing phase that computed various statistics of the flow. The code we worked with was about 5,300 lines long excluding comments.

Nomura and co-workers have previously employed DISTUF in studies of the structure and dynamics of small-scale turbulence in sheared an unsheared flows [10–12]. A typical "large-scale run" employs a grid resolution of 128^3 and runs for 5.0 to 10.0 dimensionless time units–at 128 time steps per time unit–consuming 1.5 to 3.0 CPU hours on a Cray YMP. However, resolution is limited. Of critical importance to studies of small-scale turbulence is to analyze higher Reynolds number *Re* flows. As mentioned, the number of computation points required for DNS is proportional to $Re^{\frac{9}{4}}$; any significant increase in Reynolds number would cause computation and memory costs to increase beyond the capacity of current vector-class supercomputers. For example, the Cray T94 only has sufficient memory to support 256^3 resolution, whereas an IBM SP2 with 256 Megabytes of memory per node could run such a problem on just a small fraction of the machine–16 processors. Only the SP2 could run at 512^3 resolution.

[2] The Reynolds number is a flow parameter which effectively describes the ratio of the large scales to small scales of turbulence.

2.2 The KeLP Infrastructure

KeLP is publicly-available C++ class library that enables the rapid development of irregular and block structured applications on MIMD multicomputers. It provides a middle-level layer sitting on top of MPI to simplify global to local mappings and communication optimizations. KeLP applications have been shown to deliver performance comparable with the equivalent MPI codes, including stencil based methods and the FFT, which are employed in DISTUF [4].

KeLP provides a simple machine independent model of execution. It runs on multicomputers such as the IBM SP2, Cray T3E, SGI-Cray Origin 2000; on workstation clusters; and on single and multiple-processor workstations. KeLP defines only 7 new data types plus a small number of primitives to manipulate them. The casual programmer requires only basic knowledge for most of them. Space limits us to only a brief discussion of KeLP. Additional details may be found at the KeLP web page at **http://www-cse.ucsd.edu/groups/hpcl/scg/kelp**.

KeLP supports a task parallel model which isolates parallel coordination activities from numerical computation. This simplifies the design of application software and promotes software re-use. In particular, the KeLP programmer may employ existing serial numerical kernels with known numerical properties, and leverage mature compiler technology. These kernels may be written in Fortran 90 or Fortran 77 or any other language the programmer chooses.

KeLP provides two types of coordination abstractions: *structural abstraction* and *communication orchestration*. Structural abstraction permits the programmer to manipulate geometric sets of points as first class language objects and includes a calculus of geometric operations. Data layouts across processors may be managed at run time, which facilitates performance tuning. Structural abstraction simplifies the expression of global-to-local mappings and is also used to coordinate and optimize communication, a process known as communication orchestration. Communication orchestration permits the programmer to encapsulate details concerning data motion; data dependencies are expressed in high-level geometric terms freeing the user from having to manage the details of managing message passing activity.

KeLP has a minimalist design, and does not provide automated data decomposition facilities. The KeLP philosophy is to handle such capabilities change through layering: KeLP applications usually employ one or more Domain Specific Libraries, application programmer interfaces providing common facilities for a particular problem class. The DSL encapsulates or hides the expert's knowledge, freeing the user from having to manage low level implementation details. The KeLP programmer may therefore develop complicated applications in a fraction of the time required to code the application with explicit message passing, i.e. MPI, and as previously noted, at a comparable performance.

We used a DSL called DOCK to facilitate the implementation of KDISTUF. DOCK is written in KeLP and is packaged with the public distribution. DOCK supports HPF style BLOCK decompositions [8] that were used to parallelize DISTUF. (The details may be found in the KeLP User's Guide [3].) We may view the DOCK abstractions as specialized versions of the basic KeLP classes, which are constructed using the C++ inheritance mechanism. The organization of a typical KeLP application is as follows.

1. Top level KeLP/DOCK code that manages the data structures and the parallelism.
2. Lower level code written in C++, C, or especially Fortran to handle numerical computation.

3 Modernization of DISTUF

As mentioned previously, we simplified the DISTUF code prior to parallelizing it, removing a post-processor and modifying the initial conditions. The resultant code was 5,300 lines long. Howe and Grimmett began the code conversion in mid June 1997. Most of the 5,300 lines of code were converted by the end of September 1997 after approximately four man-months of effort. An additional three man-months of effort was required to tune performance. The conversion was completed in June 1998; this was a part time effort. In the process of converting the code, the 5,300 line Fortran 77 program shrank by nearly 40% to 3336 lines. A 492-line KeLP wrapper, written in C++, was added to handle parallelization. Other than an FFT computation, all parallelization was expressed at the KeLP level, which enhanced the modularity of the software.

The original DISTUF source code was divided into 52 separate files. The code conversion process left 42 of these files unchanged. The remaining ten files contained a total of 864 lines of code, or 26% of the total source code. However, not all this code actually changed, and we believe that a more accurate estimate of the fraction of code that changed is closer to 20%. The conversion process for the ten modules was facilitated by good coding style in the legacy code: these modules access data only within a plane, and therefore are not subject to dependence analysis, which was carried out by hand.

We represented the main computational data structure as a 4D array, adding a 4th index to specify the field variable number. As mentioned, most modules access the array sequentially in 2D slices. Therefore, the most logical way to distribute the data was to employ a 1D [*,*,BLOCK,*] partition. The KeLP code managed the data structures, including data distribution and ghost cell communication required by the advection-diffusion phase. The KeLP wrapper made calls to Fortran, converting KeLP data descriptors into a form that Fortran can understand. As discussed below, the KeLP library includes an external interface to Fortran 77 to facilitate this inter-operation.

Parallel control flow is expressed using a KeLP `nodeIterator` as shown in Fig. 1. The `XArray4< Grid4<double>>` represents a 4 dimensional distributed array of doubles. This array, distributed outside the `pprep01( )` routine, contains information about the distribution of the data across processors. Each element of the distributed array is a local 4-dimensional array living on one processor, and includes information about the local bounds of the array. The `nodeIterator` loop causes each processor execute the Fortran routine `pprep01( )` in SPMD fashion. Various macros and one datatype handle the details of the

KeLP-to-Fortran interface as described in the caption, including the global to local mapping of the data array and name-mangling. The fillGhost( ) routine (not shown) handles ghost cell communication. The convection-diffusion phase employed a stencil width spanning 3 planes in both directions. Thus, the ghost cell layers were three cells deep.

```
#define pprep01 FORTRAN_NAME(pprep01_, PPREP01, pprep01)
void pprep01(double *, FORTRAN_ARGS4, double *, FORTRAN_ARGS3);
void fillGhost(XArray4<Grid4<double>>& X);

void f_prep(XArray4<Grid4<double>>& A){ // XArray is a distributed array
  fillGhost(A);                          // Fill ghost cells (not shown)
  Array3<double> awork(NNI+2,NNK+2,15); // scratch array like an f90
        // allocatable array
  FortranRegion3 FW(awork.region());    // shape of the array for fortran

  for (nodeIterator ni(A); ni; ++ni) {  // Execute pprep01( ) in parallel
    int p = ni();                        // on each processor p

    FortranRegion4 FR(A(p).region());   // The region (bounding box) for
                                         // processor p's local subarray
    pprep01(FORTRAN_DATA(A(p)), FORTRAN_REGION4(FR),
            FORTRAN_DATA(awork),FORTRAN_REGION3(FW));
  }
}
```

Fig. 1. KeLP code for stencil-based computation. The KeLP-to-Fortran interface is managed by macros FORTRAN_REGION, FORTRAN_DATA, and FORTRAN_NAME, along with the FortranRegion datatype. FORTRAN_NAME handles name mangling, the remaining constructs handle the global-to-local mapping of the data.

The pressure solver has a non-localized communication structure, as it must compute two-dimensional Fourier transforms in planes orthogonal to those employed in the first phase. However, unlike the convection-diffusion code, we handled some aspects of parallelization within the Fortran module. We originally installed KDISTUF on the Cray T3E, and used the SCILIB's FFT, which was able to interoperate with distributed KeLP arrays. However, the IBM PESSL library did not have a compatible FFT, and so we wrote Fortran code to rearrange data in preparation for calls to the manufacturer's supplied parallel 2D FFT routine, pscfft2D.

A major difficulty in modernizing this legacy application was in contending with sequence association. This now deprecated practice has been used historically to emulate dynamic memory and to improve vector lengths on vector architectures. This infamous problem is well known to the HPF and Fortran 90 community [9, 2, 1], and it seriously impedes the process of modernizing of large legacy Fortran 77 codes in general. We therefore had to restructure the code significantly, de-linearizing 1-dimensional arrays and loops back to their multi-dimensional counterparts.

4 Results

We ran KDISTUF on an IBM SP2 with 160MHz POWER Super Chip (P2SC) processors. Each node had 256 Megabytes of memory. We report timings for various problems sizes and numbers of processors. Table 1 reports fixed-sized speedups with N=128. On 8 processors the running time is about 1200 sec. per unit of simulated time. Performance is comparable with the legacy DISTUF code running on the Cray C90, about 1100 sec.

Table 1. Timings in CPU seconds for KDISTUF running on the IBM SP2 with N=128. We ran for 256 timesteps, which corresponds to 1.0 units of simulated time.

Processors	1	2		4		8	
Time	CPU	CPU	%	CPU	%	CPU	%
Computation	7856	3661	87	1832	83	922	77
Communication	N/A	446	11	266	12	181	15
FillGhost	N/A	56	1.3	56	2.5	56	4.7
Transpose	N/A	390	9.2	210	9.5	125	10
Miscellaneous	N/A	112	2.6	103	4.7	108	9.0
Total	7856	4219	100	2201	100	1204	100
Efficiency	1.00	0.93	N/A	0.89	N/A	0.82	N/A

The timings in Table 1 are broken down into 3 parts: computation, communication, and miscellaneous. Communication is further broken down into the time to fill in ghost cells and the time to perform the transpose for the FFT. Miscellaneous work includes initialization and output. Computational work scales almost perfectly, initialization and output times are insignificant, and ghost cell communication is a modest constant. The bottleneck for this computation is the transpose, but if we scale the problem appropriately with the number of processors we may effectively run the larger problems that motivated this effort.

We also ran larger problem sizes: N=256 and N=384. Due to an as-yet unresolved memory allocation problem, we were unable to collect fixed size speedups and we report just one set of timing data for each problem size.

For N=256, KDISTUF runs on 16 processors at a rate of 4.3 hours of CPU time per unit of simulated time (a total of 512 timesteps.) The code spends 88% of its time in local computation. Communication accounts for just 13% of the total running time, with 70% of that time spend in transpose. We also ran a problem with N=384 on 64 processors. However data for this case are inconclusive due to the unresolved memory allocation problem. As a stopgap measure we were forced to allocate more processors than was indicated (32). Not surprisingly communication costs were significant, with the transpose accounting for 33% of the total running time, and local computation accounting for just 53%. We are currently investigating this problem so that we can reduce communication costs and hence run at higher resolution.

5 Discussion and Conclusions

Converting legacy software is an economically important activity, but is also a delicate process. Production users are reluctant to jeopardize an investment in working software. But, they are keenly aware of the need to periodically upgrade application software in order to leverage the latest technological advances in their computing environment. Though our experience has been a positive one, our results were obtained for a relatively small application code consisting of thousands of lines of Fortran 77. By comparison, large industrial or research codes are much bulkier, hundreds of thousands to millions of lines long. We have ignored management issues since our programming team comprised just two individuals. Nevertheless, some important lessons have been learned here, which generalize to larger scale applications.

Our approach to modernizing a legacy code resulted in good parallel speedups. We used the KeLP run time library which facilitated the process. The modernization went more quickly than anticipated, though we were slowed down by difficulties with sequence association, memory consumption–as-yet unresolved at this time–and the lack of an appropriate FFT routine. We cannot overemphasize the need for standardized FFTs that deliver portable performance across diverse computing platforms.

In "KeLPifying" or parallelizing the DISTUF Fortran 77 code, we employed an important underlying principle of "minimal disturbance." We attempted to work around difficulties in the Fortran 77 code at the KeLP (C++) level rather than recoding the Fortran. This strategy payed off: we had to change only about 20% to 25% of the original Fortran code and added just 500 lines of C++ code wrapper. However, we had to settle for less than optimal performance in the interest of conserving programming effort. For example, we did not experiment with 2-D partitionings, which may be more efficient than the 1-D partitionings on larger numbers of processors. The changes to the Fortran 77 code would have been extensive, though the KeLP wrapper supports alternative partitionings. Code re-use is extremely important and has led to success in other conversion efforts, for example using PCN [6].

A major consideration in modernizing DISTUF is to increase the amount of available memory in order to improve resolution. The resultant improved code, called KDISTUF, will be employed by Grimmett and Nomura on high performance parallel computers. Using this parallel code, they will be able to compute at higher resolutions than possible with the legacy code running on the Cray T94. Because KeLP applications are portable, KDISTUF may also run on various parallel computers such as the Cray T3E and on clusters of workstations.

Acknowledgments

This work was supported in part by U.S. Office of Naval Research Contract N00014-94-1-0657, the National Partnership for Advanced Computational Infrastructure (NPACI) under U.S. National Science Foundation contract ACI9619020, and a University of California, San Diego faculty startup award. The IBM SP2 used in this study is located at the San Diego Supercomputer Center (SDSC); computer time was supported both by SDSC and by a UCSD School of Engineering block grant. The authors wish to acknowledge the assistance of Stephen J. Fink, who designed and implemented KeLP.

References

1. T. Brandes and K. Krause. Porting to hpf: Experiences with dbetsy3d within pharos. In *Proc. 2nd Annual HPF User Group meeting*, June 1998.
2. M. Delves and H. Luzet. Semc3d code port to hpf. In *Proc. 2nd Annual HPF User Group meeting*, June 1998.
3. S. J. Fink and S. B. Baden. The kelp user's guide, v1.0. Technical report, Dept. of Computer Science and Engineering, Univ. Calif., San Diego, March 1996.
4. S. J. Fink, S. B. Baden, and S. R. Kohn. Run-time support for irregular block-structured applications. *Journal of Parallel and Distributed Computing*, 1998. In press.
5. S. J. Fink, S. R. Kohn, and S. B. Baden. Flexible communication mechanisms for dynamic structured applications. In *IRREGULAR '96*, Santa Barbara, California, August 1996.
6. I. Foster, R. Olson, and S. Tuecke. Productive parallel programming: The pcn approach. *J. of Sci. Prog.*, 1:51–66, 1992.
7. T. Gerz, U. Schumann, and S. Elghobashi. Direct simulation of stably stratified homogeneous turbulent shear flows. *J. Fluid Mech.*, 200:563–594, 1989.
8. High Performance Fortran Forum, Rice University, Houston, Texas. *High Performance Fortran Language Specification*, November 1994.
9. C. Koelbel. Making hpf work: Past success and future challenges. In *Workshop on HPF for Real Applications*, July 1996.
10. K. K. Nomura and S. E. Elghobashi. Mixing characteristics of an inhomogeneous scalar in isotropic and homogeneous sheared turbulence. *Phys. Fluids*, 4:606–625, 1992.
11. K. K. Nomura and G. K. Post. The structure and dynamics of vorticity and rate-of-strain in incompressible homogeneous turbulence. *(Submitted to J. Fluid Mech.)*, 1997.
12. K.K. Nomura, G.K. Post, and P. Diamessis. The interaction of vorticity and rate-of-strain in turbulent homogeneous shear flow. *(In Preparation)*, 1998.
13. U. Schumann. Dynamische datenblock-verwaltung in fortran. Technical report, Institut fur Reaktorentwicklung, Gessellschaft fur Kernforschung M.B.H., Karlsruhe, Germany, August 1974.
14. U. Schumann. Algorithms for direct numerical simulations of shear-periodic turbulence. In Soubbaramayer and J.P. Boujot, editors, *Lecture Notes in Physics*, volume 218, pages 492–496. Springer, 1985.

Parallel Methods for Fluid-Structure Interaction

Carl B. Jenssen[1], Trond Kvamsdal[1], Knut M. Okstad[1], and Jørn Amundsen[2]

[1] SINTEF Applied Mathemetics
N-7034 Trondheim, Norway
[2] Norwegian University of Science and Technology
N-7034 Trondheim, Norway

Abstract. A parallel CFD code capable of simulating flow within moving boundaries has been coupled to a beam element structural dynamics code. The coupled codes are used to simulate fluid- structure interaction for a class of applications involving long and slender structures, e.g. suspension bridges and offshore risers. Due to the difference in size and dimensionality of the 3D CFD problem on one side, and the essentially 1D structure problem on the other side, the bulk of the computations are carried out in the CFD code. The parallel efficiency of the coupled codes thus rest on the parallel performance of the CFD code, and on minimizing the amount of communication between the two codes. The CFD code uses implicit time stepping, and is parallelized by a multi-block technique based on a block-Jacobi iteration together with coarse grid correction. To reduce the amount of communication between the CFD code and the structure code, the mesh movement algorithm is split into two parts, where the most computationally intensive part is carried out in parallel within the CFD code. The resulting coupled system has a high parallel efficiency even if the structure code runs on a workstation and the CFD code runs on a parallel supercomputer provided that the size of the CFD problem is sufficiently large.

1 Introduction

The coupled problem of Fluid-Structure Interaction (FSI) describes many interesting problems in science and engineering. Essentially, FSI is caused by the pressure forces from the flow around a structure resulting in a deflection or deformation of the structure which in turn alters the flow in a dynamic fashion. We are considering a class of FSI problems characterized by long and slender structures where the local deformation is negligible compared to the local displacement. This is the case for e.g. offshore risers and suspension bridges, and means that the structure can be accurately modeled with a beam element formulation. The fluid flow is assumed to be truly tree-dimensional (3D), although when computing the flow over long and slender bodies, the variation of the flow along the structure must inevitably be assumed to be small due to the mesh resolution.

It has been the basis for this work that FSI should be simulated using a generic coupling of two existing codes for Computational Fluid Dynamics (CFD) and structure mechanics respectively. Although writing a completely new code for the simultaneous solution of the fluid flow and the structure mechanics could probably result in a more efficient algorithm, using two separate codes has several advantages:

- User familiarity: Simulation codes can be complicated and require a long time to learn, and most users have a preference for a particular code.
- Code maintenance: Simulation codes that are used for a wide range of problems, not only FSI, are more likely to be maintained and updated regularly.
- Specialized code: Many industries use codes that have been specially developed for their particular purposes.

Of course it must be assumed that the codes used fulfill certain requirements such as moving mess capability for the CFD code.

2 Components of the FSI Software

Our FSI software has been written on the basis of two existing codes developed at SINTEF, and thus consists of the following three parts:

- The CFD code CBJ (Concurrent Block Jacobi), an in-house code at SINTEF.
- The Structure beam element code USFOS, which is a commercially available code.
- The FSI coupler which was especially written for the task of coupling the two codes above, and controlling the computations.

The system is designed so that each of the three different codes can run on completely separate computers, and communicate using Parallel Virtual Machine (PVM) message passing. Typically, the CFD code would run on a large number of processors on a parallel machine, while the structure code would run on a workstation or a single node of the parallel machine. The coupler, which contains a user interface, is designed to run on the user's own workstation. A schematic view of the communications between the different codes is shown in Figure 1, and as we can see, all communication goes through the coupler. The coupler thus has the task of restricting the fluid forces acting on the surface of the structure as discretized by the CFD mesh, to the corresponding nodes connecting the beam elements in the structure model. Also, the coupler receives the displacements of the structure nodes and computes new mesh coordinates which are passed on to the CFD code. As described in the next section, to ensure computational efficiency and parallel scalability, only a small part of mesh movement computations are carried out by the coupler, while most of the computations are carried out within the CFD code.

3 Parallel Solution Procedure

Due to the difference in size and dimensionality of the 3D CFD problem on one side, and the essentially 1D structure problem on the other side, the bulk of

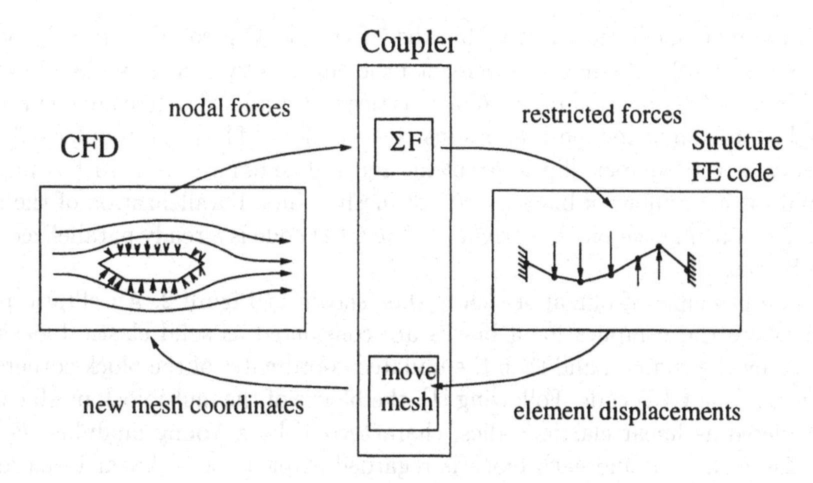

Fig. 1. Program Overview.

the computations are carried out in the CFD code. The parallel efficiency of the coupled codes thus rest on the parallel performance of the CFD code, provided that the additional mesh movement and communication through the coupler can be carried out in an efficient manner.

3.1 Mesh Movement Strategy

The problem of moving the mesh to conform with one or more moving bodies as well as a number of non-moving boundaries is global in nature. It is a problem of book-keeping to keep track of several moving bodies and their relative location, but also a mesh generation problem to create a smoothly moving mesh with as little distortion as possible. The global nature of the problem calls for some sort of elliptic procedure as commonly used in mesh generation. Both the elliptic problem and the book-keeping challenge makes it tempting to implement the whole mesh movement procedure as single processor code.

It is however clear that a system architecture as shown in Figure 1, would not be feasible if at each iteration, all mesh coordinates were to be updated by the coupler and than passed to the CFD code. If the 3D CFD problem has size N^3, such an approach would require the order of N^3 operations on a single processor as well as having to send $3N^3$ values over the network. Since in the CFD code, the computational cost and communication costs are of the order N^3 and N^2 respectively, the coupler would then create a scalar bottleneck and an increase of the communication cost of one order of magnitude.

Embedding the mesh movement algorithm completely within the parallel CFD code would eliminate both the communication overhead and scalar bottleneck, but would also introduce complicated book-keeping and require the implementation of an elliptic solver within the CFD code. Our solution to the problem has been to split the mesh movement problem into two parts, a global elliptic problem for the location of the block vertices, and a local independent problem for the location of the mesh nodes inside each block. Thus, the coupler only has to solve a very coarse elliptic problem, and the communication cost is proportional to the number of blocks, and not mesh points. Parallelization of the local problem within each block is trivial, as the CFD code is already parallelized over the blocks.

The principle of our approach is thus shown in Figure 2. An elliptic problem where the computational blocks are considered as solid elastic blocks are solved by the coupler, and then the updated coordinates of the block corners are passed to the CFD code. Following [1], the blocks of the multiblock mesh can be considered as linear elastic bodies, characterized by a Young modulus, E, and a Poisson ratio, ν and each block is regarded as bi- (or tri-) linear isoparametric element, e_i. The mesh movement problem then becomes an elliptic problem with only Dirichlet boundary conditions, and it can be solved using any linear solver. This method has the additional advantage that the Young modulus can vary form block to block, making it possible to keep blocks in the vicinity of the structure almost unchanged in shape, while most of the distortion is taken up by more distant blocks.

Once the global mesh movement problem is solved for the corners of the blocks, the updated values are sent to the CFD code, where Trans Finite Interpolation (TFI) is used to update the interior coordinates of the blocks. Note that we are using TFI on the displacement, or change between each iteration, of the mesh coordinates. Thus, the overall mesh quality remains as in the initial grid, and if the corner nodes of the block remain unchanged, the mesh within the block also stays unchanged. This ensures that the mesh retain the properties such as clustering and orthogonality as imposed by the original mesh generation and do not inherit properties of the TFI mesh generation.

3.2 Coarse Grid Correction Algorithm for the CFD Code

The CFD code solves the integral form of the Navier-Stokes equayions for a moving and deforming control volume as seen in Figure 3:

$$\frac{d}{dt}\int_{\Omega} U dV + \int_{\partial\Omega} (\boldsymbol{F} - U\dot{\boldsymbol{r}}) \cdot \boldsymbol{n}\, ds = 0 . \tag{1}$$

Here U is the conserved variable, $\boldsymbol{F}$ the flux vector and $\boldsymbol{n}$ the outwards pointing normal to the surface $\partial\Omega$ enclosing an arbitrary volume Ω, and $\dot{\boldsymbol{r}}$ the velocity of the point $\boldsymbol{r}$ on $\partial\Omega$. A second order upwind scheme is used for the convective part of the flux vector, while second order central differencing is used for the viscous terms. An Arbitray Eulerian Lagrangian (ALE) method [2] is used to allow for a moving mesh.

Implicit second order time integration is achieved by an A-stable two-step method usually referred to as second order backwards differentiation [3] which is

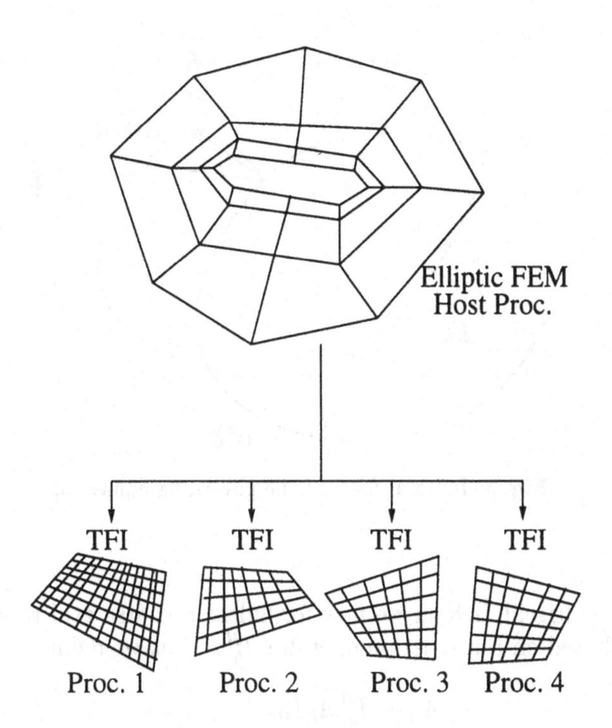

Fig. 2. Mesh movementy strategy.

iterated to convergence at each time step using a Newton procedure. Our code is based on a structured mesh and multiblock approach. Thus, for each iteration of the Newton procedure we have to solve a linear system of equations that is septa-diagonal in each block. Ignoring the block interface conditions, this system is solved concurrently in each block using a line Jacobi procedure [4]. A coarse grid correction scheme (CGCS) [5], [6] is used to compensate for ignoring the block interface conditions by adding global influence to the solution.

The CGCS is briefly reviewed here. Using the subscript h to denote the fine mesh and H for the coarse mesh, the method is described in the following. In matrix form, we can write the linear system of equations that we wish to solve at each Newton iteration as

$$A_h \Delta U_h = R_h \tag{2}$$

However, by ignoring the coupling between blocks, we actually solve a different system that we symbolically can write:

$$\widetilde{A}_h \widetilde{\Delta U}_h = R_h \tag{3}$$

Defining the error obtained by not solving the correct system as $\varepsilon_h = \Delta U_h - \widetilde{\Delta U}_h$ we have

$$A_h \varepsilon_h = R_h - A_h \widetilde{\Delta U}_h \tag{4}$$

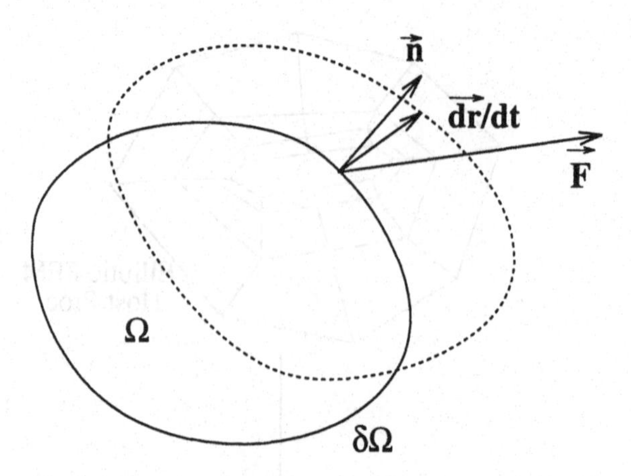

Fig. 3. Integral form of the governing equations.

In a classic multigrid fashion we now formulate a coarse grid representation of Equation 4 based on a restriction operator I_h^H. Thus we define

$$A_H = I_h^H A_h I_H^h \tag{5}$$

$$R_H = I_h^H \left(R_h - A_h \widetilde{\Delta U}_h \right) \tag{6}$$

and solve for the correction to $\widetilde{\Delta U}_h$:

$$A_H \Delta U_H = R_H \tag{7}$$

The solution to Equation 7 is transformed to the fine grid by means of the prolongation operator I_H^h, and finally the fine grid solution is updated by:

$$\widetilde{\widetilde{\Delta U}}_h = \widetilde{\Delta U}_h + I_H^h \Delta U_H \tag{8}$$

As restriction operator we use summation and as prolongation operator injection [7].

To solve the coarse grid system, any suitable parallel sparse matrix solver can be applied. We have used a Jacobi type iterative solver that at each iteration inverts only the diagonal coefficients in the coarse grid system.

4 Computational Results

We here present results both in terms of parallel performance, and the accuracy achieved for a realistic engineering test case of similar size.

N_{Pros}	T_{CFD} (sec.)	T_{FSI} (sec.)	T_{Tot} (sec.)	Overh. (%)	Speed up
2	295	12.3	307	4.0	2.0
4	130	12.3	142	8.6	4.3
8	65.1	11.1	76.2	14	8.0
12	44.4	12.0	56.4	21	10.9
16	38.4	12.0	50.4	24	12.2

Table 1. Ellapsed time used by the CFD code and the rest of the FSI system for a typical problem.

4.1 Benchmark timings

The performance of the software was tested for a typical two-dimensional problem with around 50,000 grid points. The CFD code was running on a Cray T3E, while the coupler and the structure code were both running on a workstation. The timings listed in Table 1 give the CPU time used by the CFD code and the rest of the FSI system, together with the overhead represented by the FSI system and the speed-up for the complete system as a function of the number of processors used by the CFD code. The CPU time for the additional FSI parts, T_{FSI}, include all CPU time associated with FSI, also the TFI routines within the CFD code.

As we can see, acceptable performance is obtained for up to 12 processors. The CPU time for the FSI parts remain almost constant as expected, while the CFD code shows very good scalability. This benchmark is overly pessimistic in terms of the scalability of the complete system, as the FSI part appears as a sequential bottleneck that causes a low parallel efficiency for 16 processors or more. However, if the accuracy of the simulation is to be increased, the need for higher resolution would be on CFD side, calling for a manyfold increase of the number of grid points. Thus, the communication time and sequential CPU time would remain the same, and assuming linear speed-up of the CFD code, the number of processors could grow proportional to the number of grid points without increasing the FSI overhead.

4.2 Validation Test Case

As a validation test case we have simulated wind induced motion of a suspension bridge deck. The bridge profile considered is that of the 2700 m long suspension bridge which is part of East Bridge of the Great Belt Link in Denmark. The bridge profile is of the box girder type, similar to the sketch in Figure 4. Several wind tunnel experiments have been carried out on this bridge prior to construction [8], and the model parameters have been chosen to reproduce these experiments.

Based on the time histories of vertical and torsional motion for several wind velocities, the so-called aerodynamic derivatives, A_i^* and H_i^*, $i = 1, 2, 3, 4$, have

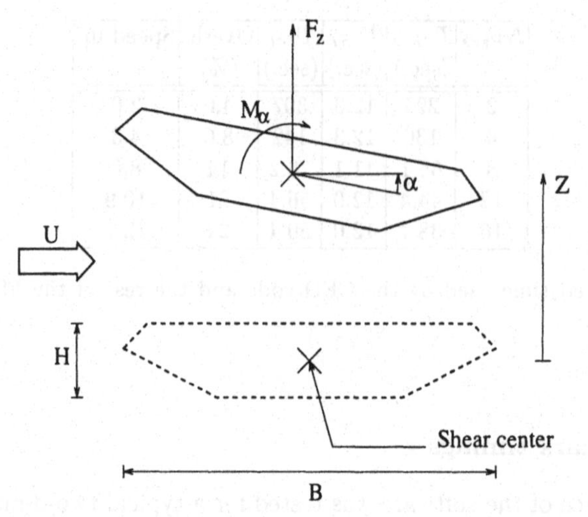

Fig. 4. Typical bridge section profile with forces and vertical and angular displacement.

been estimated. The aerodynamic derivatives express the motion-dependent lift and moment per unit length, $F_{m,z}$ and $M_{m,\alpha}$, as a linear functions of vertical and torsional displacements z and α, and the associated velocities $\dot{z}$ and $\dot{\alpha}$, according to the expressions given in [9]:

$$F_{m,z} = \frac{1}{2}\rho U^2 B \left[kH_1^* \frac{\dot{z}}{U} + kH_2^* \frac{B\dot{\alpha}}{U} + k^2 H_3^* \alpha + k^2 H_4^* \frac{z}{B} \right] \quad (9)$$

$$M_{m,\alpha} = \frac{1}{2}\rho U^2 B^2 \left[kA_1^* \frac{\dot{z}}{U} + kA_2^* \frac{B\dot{\alpha}}{U} + k^2 A_3^* \alpha + k^2 A_4^* \frac{z}{B} \right] \quad (10)$$

where B is the width of the bridge profile and U the wind velocity as seen in Figure 4.

Five time histories of vertical and torosional displacements have been calculated corresponding to wind velocities of 20 m/s, 35 m/s, 50 m/s, 60 m/s, and 65 m/s. Although physically incorrect, the Reynolds number was kept constant at 45000 for all wind velocities because of inadequate turbulence modeling. A typical example of the computed velocity field can be seen in Figure 5, showing the instantaneous pressure distribution around the profile. The length of the time histories was in each case the equivalent of 310 seconds of real time, corre-

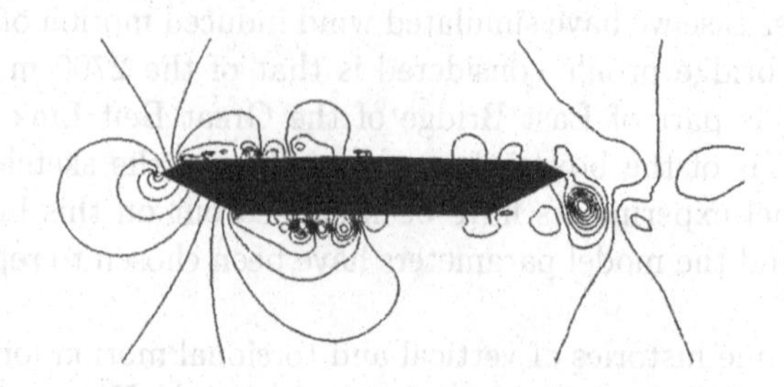

Fig. 5. Computed instantaneous pressure distribution around the bridge profile.

sponding to form $200B/U$ to $650B/U$ depending on the wind velocity. The computations were carried out on a two-dimensional mesh of approximately 40000 grid cells, the time step was set to $0.01B/U$, and three Newton iterations were performed at each time step. The seemingly long time-histories is a result of the vortex shedding frequency being much higher than the oscillation frequency of the bridge profile. This can be seen from Figure 6 showing the displacement time history for 50 m/s, compared to Figure 7 showing the corresponding time history of the lift and torsion coefficients. Note the different scaling of the time-axes in the two graphs, Figure 7 covers only 1/20 of the time shown in Figure 6.

Estimated aerodynamic derivatives are shown in Figure 8 as a function of the non-dimensionalized oscillation frequency. The comparison appears somewhat inconclusive, as some curves agree surprisingly well with the experiments, while other show some discrepancy. However, in almost all cases, the discrepancies are within the deviations between the two different sets of experimental data. The resulting critical flutter speed was thus found to be 69.9 m/s with an associated flutter frequency of 0.214 Hz. This is in good agreement with previous estimates based on wind tunnel experiments [10] giving a value of 74.2 m/s for the critical flutter speed.

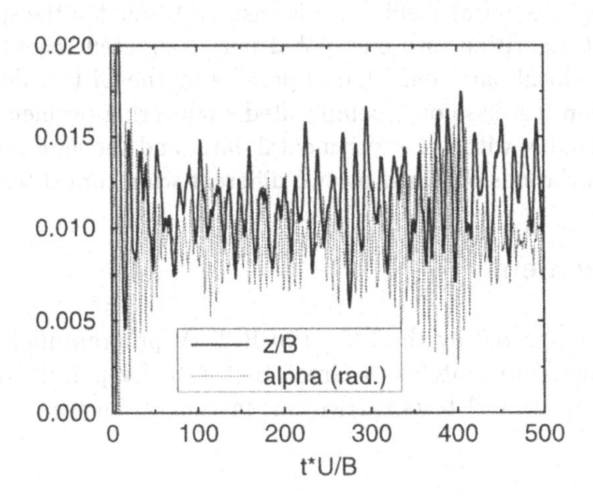

Fig. 6. Computed vertical and angular displacement.

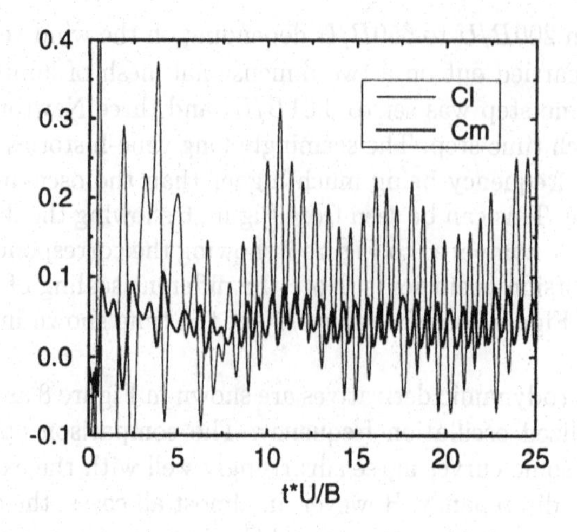

Fig. 7. Computed global forces.

5 Conclusion

We have demonstrated the ability to simulate FSI using two separate existing codes with minor modifications and a coupler running on a separate workstation. The advantages with such an approach is a modular system with interchangeable simulation codes. The parallel efficiency is ensured through a the splitting of the mesh movement algorithm into one global part computed sequentially by the coupler, and one local part computed in parallel by the CFD code.

The validation test case on a complicated engineering problem gives results that agree reasonably with the experimental data, and the agreement is seen to be within the deviations obtained by two different wind tunnel test procedures.

Acknowledgement

This work was supported by the EC - ESPRIT IV program under project no. 2011. The Authors are grateful to Great Belt A.S. for permission to use the results of the wind tunnel tests carried out in connection with the Great Belt Bridge.

References

1. P. Pegon and K. Mehr. Rezoning and remeshing of the fluid domain. Technical Report I.98.26, JRC, ISIS, 21020 Ispra (VA) Italy, 1998.

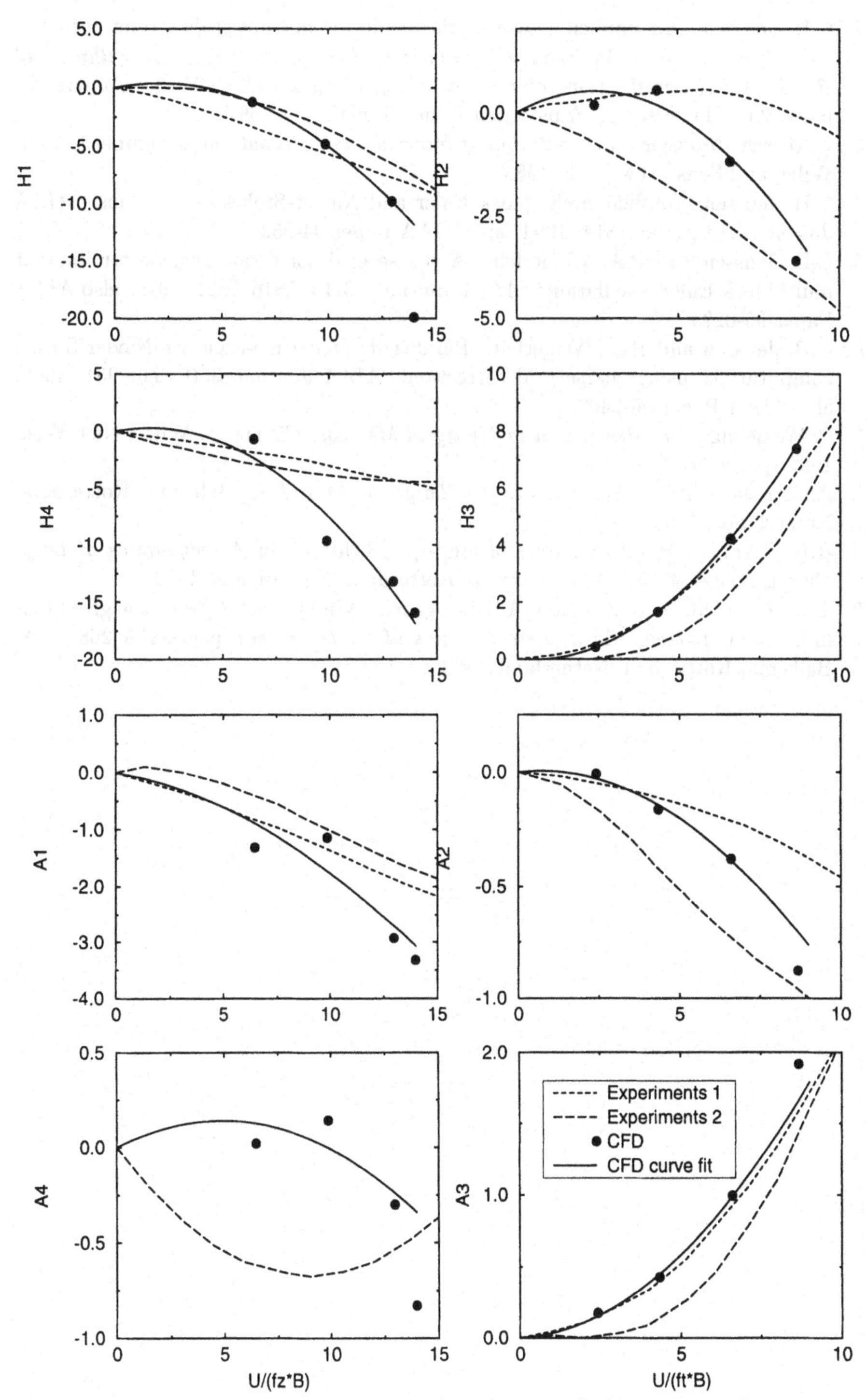

Fig. 8. Esimated aerodynamic derivatives based on simulations compared to estimates based on wind tunnel experiments.

2. C.B. Jenssen. An implicit Navier-Stokes code for moving grids - time accuracy and efficiency issues. In *Parallel Computational Fluid Dynamics: Algorithms and Results Using Parallel Computers, Proceedings of the Parallel CFD'96 Conference*, pages 204–211. Elsevier, Amsterdam, The Netherlands, 1997.
3. C. Hirsch. *Numerical Computation of Internal and External Flows*, volume 1. John Wiley and Sons, New York, 1988.
4. C.B. Jenssen. Implicit multi block Euler and Navier-Stokes calculations. *AIAA Journal*, 32(9):1808–1814, 1994. also AIAA Paper 94-0521.
5. C.B. Jenssen and P.Å. Weinerfelt. A coarse grid correction scheme for implicit multi block Euler calculations. *AIAA Journal*, 33(10):1816–1821, 1995. also AIAA Paper 95-0225.
6. C.B. Jenssen and P.Å. Weinerfelt. Parallel implicit time-accurate Navier-Stokes computations using coarse grid correction. *AIAA Journal*, 36(6):946–951, 1998. also AIAA Paper 96-2402.
7. P. Wesseling. *An Introduction to Multigrid Methods, Chapter 5.* Wiley, New York, 1992. pages 60-78.
8. A. Larsen, editor. *Aerodynamics of Large Bridges.* A.A. Balkema, Rotterdam, Netherlands, 1992.
9. R.H. Scanlan. Wind dynamics of long-span bridges. In *Aerodynamics of Large Bridges*, pages 47–57. A.A. Balkema, Rotterdam, Netherlands, 1992.
10. T.A. Reinhold, M. Brinch, and A. Damsgaard. Wind tunnel tests for the great belt link. In A. Larsen, editor, *Aerodynamics of Large Bridges*, pages 255–268. A.A. Balkema, Rotterdam, Netherlands, 1992.

Parallel Computing Tests on Large-Scale Convex Optimization*

Markku Kallio and Seppo Salo

Helsinki School of Economics
Runeberginkatu 14-16, SF-00100 Helsinki, Finland.
{kallio, salo}@hkkk.fi

Abstract. Large scale optimization models have traditionally been a valuable tool for management within the private sector and government. Their use has ranged from planning of investments over long horizons, to scheduling of day-to-day operations. Applications involving dynamics and uncertainty, e.g. in finance, often result in models of very large scale. Using Cray T3E parallel computer, we test a new solution technique for a general class of large convex optimization models with differentiable objective and constraint functions. The method is based on saddle point computation of the standard Lagrangian. The algorithm possesses a variety of beneficial characteristics, such as a structure that makes it amenable to parallelization, and no requirement that large systems of equations be solved.

We demonstrate the approach using the largest linear programming problem *stocfor3* of the test problem library, as well as several versions of a nonlinear multi-stage stochastic optimization model. The former is a forestry planning model. The latter was developed for risk management in a pension insurance company [1].

1 The Problem

We consider the convex optimization problem specified as follows (see e.g. [2]). Let x be an n-vector, $f(x)$ a convex function, $g(x)$ a convex vector-valued function with m components, and X a closed and convex set of n-vectors. The set X often takes into account simple bounds on primal variables x. Consider the following optimization problem:

$$\min_{x \in X} \{f(x) \mid g(x) \leq 0\}$$

Let y be the m-dimensional dual vector associated with the inequality constraints above, and let Y be the set of nonnegative m-vectors, so that Y accounts for the sign constraints of the dual vector y. Define the Lagrangian

* The authors wish to thank Yrjö Leino for valuable advice. Financial support for this work is gratefully acknowledged from The Foundation for the Helsinki School of Economics.

$L(x,y) = f(x) + y^T g(x)$. A point $(\hat{x},\hat{y})$, with $\hat{x}$ in X and $\hat{y}$ in Y, is a saddle point of L in $X \times Y$ if $L(\hat{x},y) \le L(\hat{x},\hat{y}) \le L(x,\hat{y})$ for all x in X and y in Y. If $(\hat{x},\hat{y})$ is such a saddle point, then $\hat{x}$ is an optimal solution to our convex optimization problem (see e.g. Rockafellar [7]).

2 The Saddle Point Method

The general idea of the saddle point algorithm [3] is, in each iteration, to adjust primal and dual variables in the directions of the gradients of L. The adjusted points obtained are subsequently projected onto feasibility sets X and Y. These gradients are evaluated at perturbed points rather than at the solution at hand at the beginning of the iteration. For the step size, a simple rule is given to guarantee convergence.

The general aim of perturbation is to find ξ in X and η in Y so that the gap $L(x,\eta) - L(\xi,y)$ is positive. Unless (x,y) is a saddle point, such vectors always exist. A practical perturbation is obtained via gradient search. The stepsize to be applied for updating the variables is obtained by dividing this gap by the squared Euclidean norm of the direction vector. To initiate the algorithm, we choose any x in X and y in Y.

3 Implementation on Cray T3E

In the algorithm, only computation of stepsizes for updating is centralized. Other steps have a great potential for parallelization: all components of x and y can be processed in parallel. Besides, it is crucial to exploit sparsity in communication.

For allocating the rows and columns to processors, we think of the Jacobian of g as a block triangular matrix (after possible row and column permutation). Blocks of rows and columns sharing a diagonal block are assigned to the same processor. The overall allocation problem is to assign diagonal blocks to processors. For gradient evaluations in the perturbation phase each processor needs parts of the vectors x and y. In the direction finding phase parts of perturbed vectors ξ and η are needed as well. In both cases, possible need of communication among processors is associated with non-zero offdiagonal blocks of the Jacobian. To minimize communication, we wish the diagonal blocks to be dense matrices so that there are as few nonzeros in the off-diagonal blocks as possible. We employ *shmem_put* subroutines for such communication.

For synchronization, we may use the *barrier* subroutine to ensure completion of communication before gradient evaluations begin. Alternatively, we may employ bilateral synchronization: in each processor gradient evaluation begins if its put-tasks have been completed and all necessary information from other processors has been received. In each iteration, the only necessary *barrier* synchronization precedes the stepsize computation. The latter is done employing the *shmem_real8_sum_to_all* subroutine.

We think of our inequality constraints as a large-scale and sparse system, which does not necessarily possess other structural properties. The (potentially) nonzero data for the gradient of f and the Jacobian of g is stored both column- and row-wise accounting for sparsity.

To accelerate convergence, scaling of primal and dual variables is crucial. The scaling procedure employed by Kallio and Salo [4] is adopted. It defines row and column scaling factors for the Jacobian. Variables are scaled by the inverses of these factors. Scaling is updated in each iteration. Algorithmic steps are applied in the scaled space, while computations are carried out in the unscaled space. Original problem data is not scaled; the scaling factors are employed when needed.

4 Computational Tests

Besides the linear model *stocfor3*, which is the largest one in dimensions in the linear programming test problem library of *Netlib* [5], eight versions of the nonlinear stochastic optimization model [1] are used for tests. Problem names and dimensions are given in Table 1. The number in the *tel*-problem names indicates roughly the number (in thousands) of scenarios considered in the stochastic model. The objective of the *tel*-models is to maximize the expected value of a strictly concave utility function. All constraints are linear.

Problem	Rows (1000)	Columns (1000)	Nonzeros (1000)	Data (MB)	Time (sec)	Iter. (1000)	Speed-up
stocfor3	17	16	74	2	5	6	90
tel-1	8	9	41	2	23	11	86
tel-2	15	17	81	5	31	10	96
tel-4	29	32	154	9	62	11	107
tel-8	47	53	253	17	212	23	111
tel-16	95	105	505	33	470	30	120
tel-32	189	210	1009	66	1384	43	122
tel-96	512	568	2728	187	4727	56	123
tel-192	1023	1137	5456	374	12480	80	124

Table 1. Problem names, number of variables, constraints, non-zeros and size of data, as well as solution time, number of iterations and speed-up. Precision is five digits, except *tel-192* terminating in iterations limit with four digit precision.

The problems were solved to an estimated five digit accuracy in the objective function value - this is expected to be sufficient in practice. Table 1 also shows the run time, number of iterations, and speed-up. All solution times refer to wall-clock seconds (excluding input and output) using 128 of 375 MHz DEC Alpha processors. Speed-up is defined as the sum of time taken by algebraic steps in all 128 processors divided by wall-clock time.

To get an idea of the performance reported in Table 1, *Minos 5.4* [6], a state of the art large-scale optimization package for serial computing, takes 100, 150 and 1400 times longer for *stocfor3*, *tel-1* and *tel-2*, respectively. Problem *tel-4* was unsolvable with *Minos* in one day of computing.

Figures 1 to 4 illustrate the average time distribution of various tasks in an iteration for each individual processor. In each case black refers to algebraic steps, grey to communication and white to waiting due to synchronization. Starting from the left the operations are: communicate information needed for stepsize calculation, update variables, communicate updated variables, wait, perturbation, communicate perturbation, wait, compute direction of update, and wait.

Figures 1 and 2 refer to *stocfor3* with bilateral synchronization and *barrier*-synchronization, respectively. No significant difference in speed-up can be observed in these two cases. Figures 3 and 4 refer to *tel-2* and *tel-192*, respectively. The improvement in speed-up as the problem size increases is quite visible.

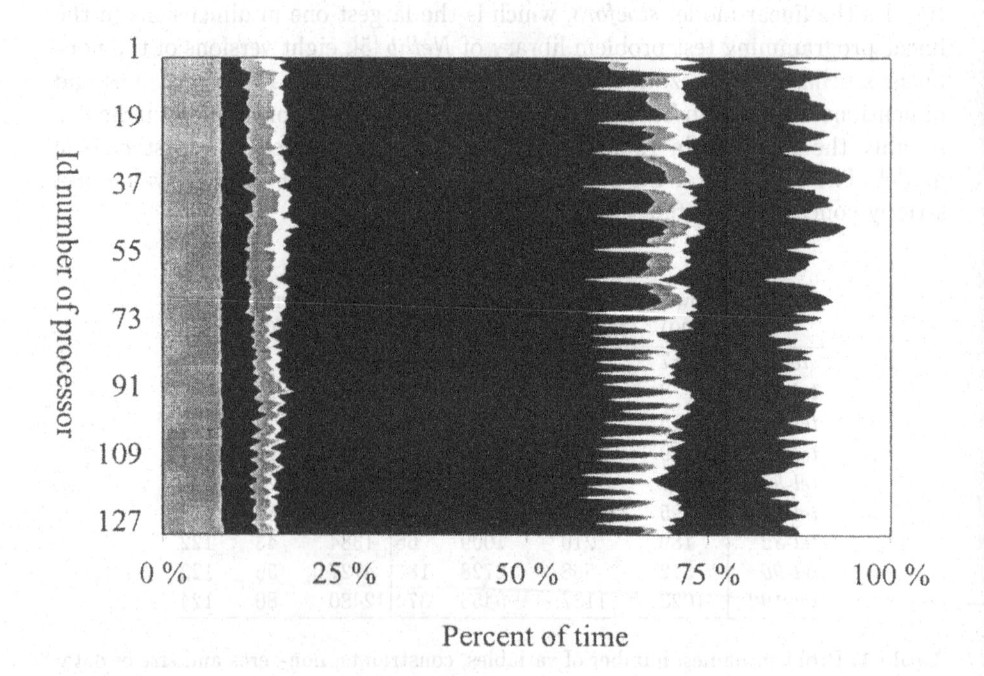

Fig. 1. Time distribution of iterative steps for problem *stocfor3* with bilateral synchronization; computations in black, communication in grey, waiting in white.

References

1. Katja Ainassaari, Markku Kallio and Antero Ranne, "An Asset Management Model for a Pension Insurance Company," Helsinki School of Economics, 1997.
2. M. S. Bazaraa, C. M. Shetty, *Nonlinear Programming*, John Wiley & Sons, 1979.
3. Markku Kallio and Charles H. Rosa, "Large-Scale Optimization via Saddle Point Computation," *Operations Research* (forthcoming).

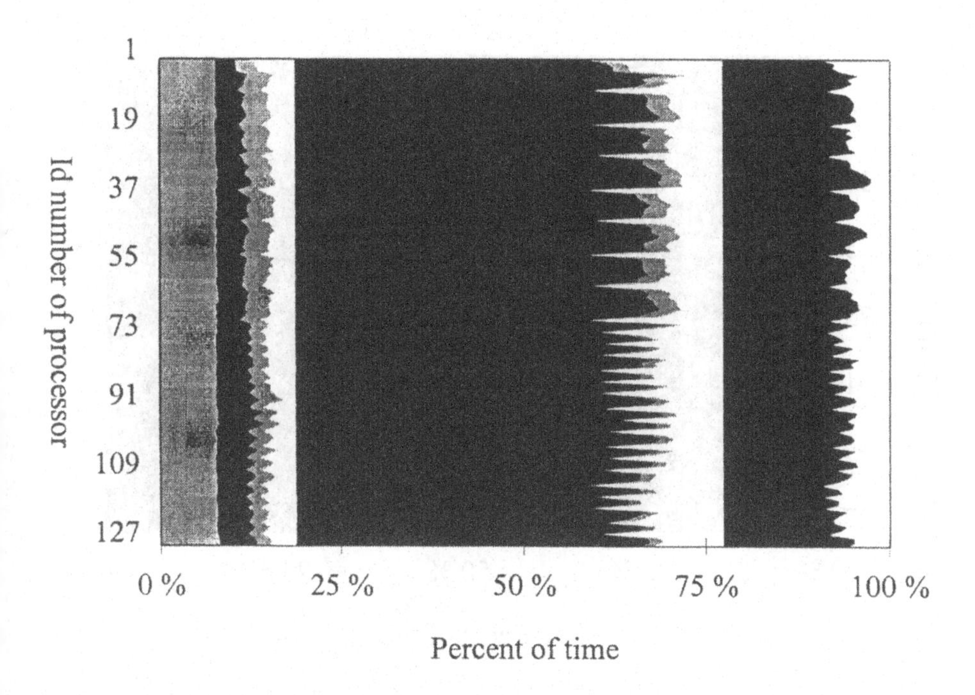

Fig. 2. Time distribution of iterative steps for problem *stocfor3* with barrier synchronization; computations in black, communication in grey, waiting in white.

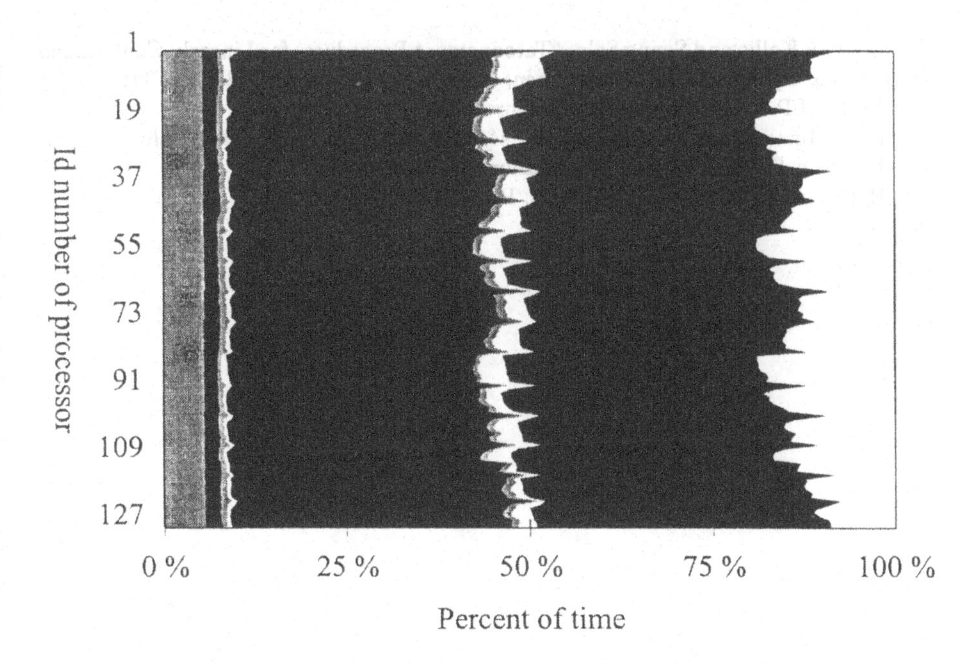

Fig. 3. Time distribution of iterative steps for problem *tel-2* with bilateral synchronization; computations in black, communication in grey, waiting in white.

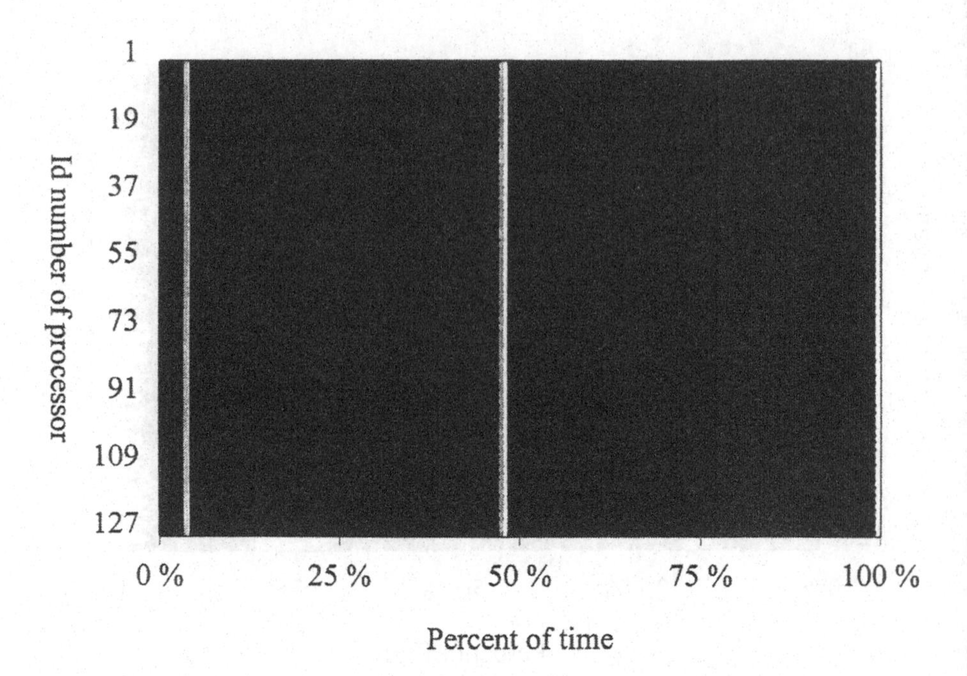

Fig. 4. Time distribution of iterative steps for problem *tel-192* with barrier synchronization; computations in black, communication in grey, waiting in white.

4. Markku Kallio and Seppo Salo, "Tatonnement Procedures for Linearly Constrained Convex Optimization," *Management Science* Vol. 40, pages 788-797, 1994.
5. Netlib, LP Test Problems, Bell Laboratories.
6. B.A. Murtagh and M.A. Saunders, *MINOS - A Modular In-core Nonlinear Optimization System, Version 5.4,* Stanford University, 1993.
7. R.T. Rockafellar, *Convex Analysis*, Princeton University Press, 1970.

Parallel Sparse Matrix Computations in the Industrial Strength PINEAPL Library

Arnold R. Krommer

The Numerical Algorithms Group Ltd,
Wilkinson House, Jordan Hill Road,
Oxford, OX2 8DR, UK
arnoldk@nag.co.uk

Abstract. The Numerical Algorithms Group Ltd is currently participating in the European HPCN Fourth Framework project on *P*arallel *I*ndustrial *N*um*E*rical *A*pplications and *P*ortable *L*ibraries (PINEAPL). One of the main goals of the project is to increase the suitability of the existing NAG Parallel Library for dealing with computationally intensive industrial applications by appropriately extending the range of library routines. Additionally, several industrial applications are being ported onto parallel computers within the PINEAPL project by replacing sequential code sections with calls to appropriate parallel library routines.

A substantial part of the library material being developed is concerned with the solution of PDE problems using parallel sparse linear algebra modules. This paper outlines the scope of these modules and illustrates performance results. Most of the software described in this paper has been incorporated into the recently launched Release 1 of the PINEAPL Library.

1 Introduction

The NAG Parallel Library enables users to take advantage of the increased computing power and memory capacity offered by multiple processors. It provides parallel subroutines in some of the areas covered by traditional numerical libraries (in particular, the NAG Fortran 77, Fortran 90 and C libraries), such as dense and sparse linear algebra, optimization, quadrature and random number generation. Additionally, the NAG Parallel Library supplies support routines for data distribution, input/output and process management purposes. These support routines shield users from having to deal explicitly with the message-passing system – which may be MPI or PVM – on which the library is based. Targeted primarily at distributed memory computers and networks of workstations, the NAG Parallel Library also performs well on shared memory computers whenever efficient implementations of MPI or PVM are available. The fundamental design principles of the existing NAG Parallel Library are described in [5], and information about contents and availability can be found in the library Web page at `http://www.nag.co.uk/numeric/FD.html`.

NAG is currently participating in the HPCN Fourth Framework project on *P*arallel *I*ndustrial *N*um*E*rical *A*pplications and *P*ortable *L*ibraries (PINEAPL). One of the main goals of the project is to increase the suitability of the NAG Parallel Library for dealing with a wide range of computationally intensive industrial applications by appropriately extending the range of library routines. In order to demonstrate the power and the efficiency of the resulting Parallel Library, several application codes from the industrial partners in the PINEAPL project are being ported onto parallel and distributed computer systems by replacing sequential code sections with calls to appropriate parallel library routines. Further details on the project and its progress can be found in the project Web page at `http://www.nag.co.uk/projects/PINEAPL.html`.

This paper focuses on the parallel sparse linear algebra modules being developed by NAG.[1] These modules are used within the PINEAPL project in connection with the solution of PDE problems based on finite difference and finite element discretization techniques. The scope of support for parallel PDE solvers provided by the parallel sparse linear algebra modules is outlined in Section 2. Section 3 provides a study of the performance of core computational routines. Most of the software described has been incorporated into the recently launched Release 1 of the PINEAPL Library.

2 Scope of Library Routines

A number of routines have been and are being developed and incorporated into the PINEAPL and the NAG Parallel Libraries in order to assist users in performing some of the computational steps (as described in [5]) required to solve PDE problems on parallel computers. The routines in the PINEAPL Library belong to one of the following classes:

Mesh Partitioning Routines which decompose a given computational mesh into a number of sub-meshes in such a way that certain objective functions (measuring, for instance, the number of edge cuts) are optimized. They are based on heuristic, *multi-level* algorithms similar to those described in [3].

Sparse Linear Algebra Routines which are required for solving systems of linear equations and eigenvalue problems resulting from discretizing PDE problems. These routines can be classified as follows:

Iterative Schemes which are the preferred method for solving large-scale linear problems. PINEAPL Library routines are based on *Krylov subspace methods*, including the *Conjugate Gradient* method for symmetric positive-definite problems, the *Symmetric LQ* method for symmetric indefinite problems, as well as the *Restarted Generalized Minimum Residual*, the *Conjugate Gradient Squared*, the *Biconjugate Gradient Stabilized*, and the *Transpose-Free Quasi-Minimal Residual* methods for unsymmetric problems.

[1] Other library material being developed by PINEAPL partners includes optimization, Fast Fourier Transform, and Fast Poisson Solver routines (see [1]).

- **Preconditioners** which are used to accelerate the convergence of the basic iterative schemes. PINEAPL Library routines employ a range of preconditioners suitable for parallel execution. These include *domain decomposition*-based, specifically *additive* and *multiplicative Schwarz* preconditioners [7]. Additionally, preconditioners based on classical *matrix splittings*, specifically *Jacobi and Gauss-Seidel* splittings, are provided, and *multicolor orderings* of the unknowns are utilized in the latter case to achieve a satisfactory degree of parallelism [2]. Furthermore, the inclusion of parallel incomplete factorization preconditioners based on the approach taken in [4] is planned.
- **Basic Linear Algebra Routines** which calculate matrix-vector products involving sparse matrices – an operation required by all iterative schemes.
- **Black-Box Routines** which provide easy-to-use interfaces at the price of reduced flexibility.
- **Preprocessing Routines** which perform matrix transformations and generate auxiliary information required to perform the foregoing parallel sparse matrix operations efficiently.

In-Place Generation Routines which generate the non-zero entries of sparse matrices concurrently: Each processor computes those entries which – according to the given data distribution – have to be stored on it. In-place generation routines are also provided for vectors which are distributed conformally to sparse matrices.

Distribution/Assembly Routines which (re)distribute the non-zero entries of sparse matrices or the elements of vectors, stored on a given processor, to other processors according to given distribution schemes. They also assemble distributed parts of dense vectors on a given processor, etc.

Additionally, all routines are available for both real and complex data.

3 Performance of Library Routines

The performance results quoted in this section were obtained for the (diffusion-dominated) *advection-diffusion* problem

$$-\Delta u + \operatorname{div} u = f,$$

on a three-dimensional unit cube $\Omega = [0,1]^3$, discretized on a Cartesian grid using a standard seven-point second-order finite difference scheme. All experiments were carried out for a *small*-size problem configuration, $n_1 = n_2 = n_3 = 30$, resulting in a system of $n = 27\,000$ equations, and for a *medium*-size problem configuration, $n_1 = n_2 = n_3 = 60$, resulting in a system of $n = 216\,000$ equations. All performance tests were carried out on a Fujitsu AP3000 at the Fujitsu European Centre for Information Technology (FECIT). The Fujitsu AP3000 is a distributed-memory parallel system consisting of UltraSPARC processor nodes and a proprietory two-dimensional torus interconnection network.

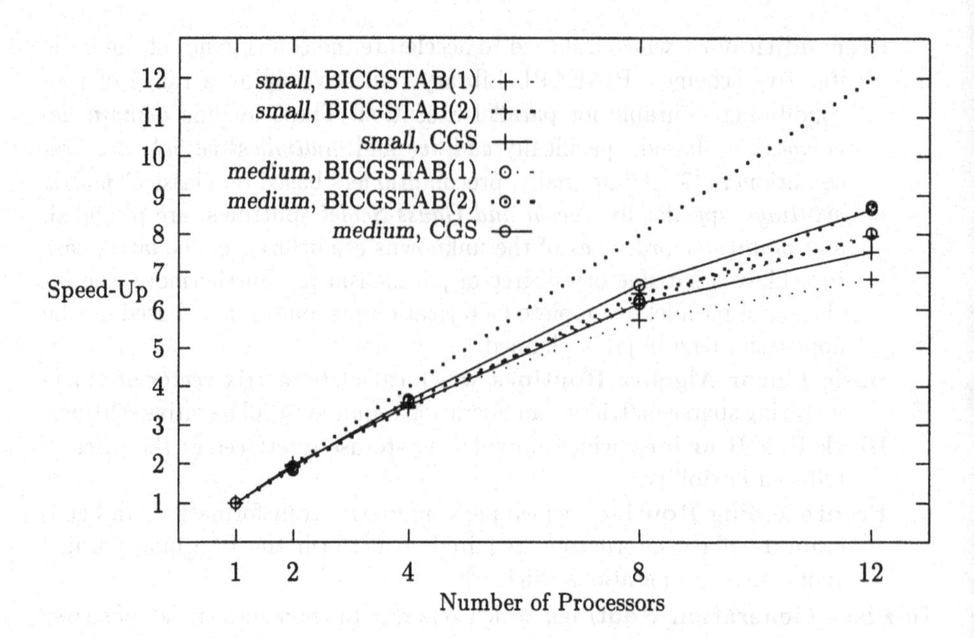

Fig. 1. Speed-Up of Unpreconditioned Iterative Solvers

Performance of Unpreconditioned Iterative Solvers Figure 1 shows the speed-ups of unpreconditioned iterative solvers for the two test problems. The iterative schemes tested – BICGSTAB(1), BICGSTAB(2) and CGS – have been found previously to be the most effective in the unsymmetric solver suite (see [6]). BICGSTAB(2) and CGS can be seen to scale (roughly) equally well, whereas BICGSTAB(1) scales considerably worse (in particular, for the smaller problem). The results are generally better for the larger problem than for the smaller, especially when the maximum number of processors is used. In this case, the improvement in speed-up is between 10 and 20 %, depending on the specific iterative scheme.

Performance of Matrix-Vector Multiplication Figure 2 shows the speed-ups of the sparse matrix-vector multiplication routine in the PINEAPL Library. The results demonstrate a degree of scalability of this routine in terms of both increasing number of processors and increasing problem size. In particular, near-linear speed-up is attained for all numbers of processors for the larger problem.

Acknowledgements I would like to thank Valerie Fraysse and Alan McCoy from Cerfacs, Pasqua D'Ambra from CPS Naples, Lars Soerensen from the Danish Hydraulic Institute, Salvatore Filippone from IBM Italy, Thomas Christensen and Jesper Larsen from Math-Tech ApS, Mishi Derakhshan from NAG Ltd, as well as Eric Ducloux and Frederic Lafon from Thomson-CSF/LCR for useful discussions on the design of the PINEAPL Library routines. I also acknowledge the use of sequential software by Stefano Salvini and Gareth Shaw at NAG Ltd in a number of routines.

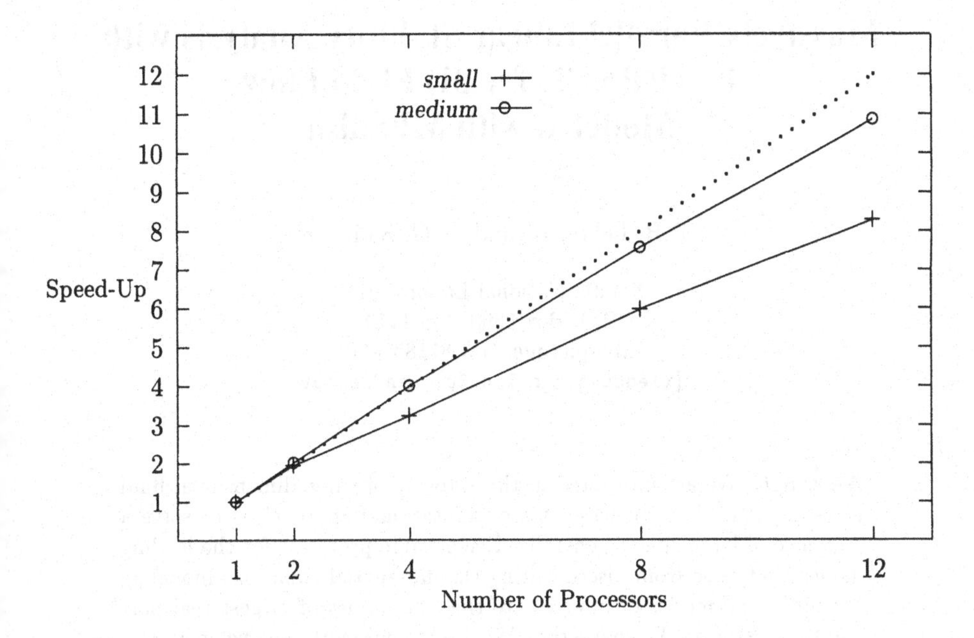

Fig. 2. Speed-Up of Matrix-Vector Multiplication

References

1. D. di Serafino, L. Maddalena, A. Murli, *PINEAPL: A European Project to Develop a Parallel Numerical Library for Industrial Applications*, in "Euro-Par'97 Parallel Processing" (C. Lengauer, M. Griebl, S. Gorlatch, Eds.), Vol. 1300 of *Lecture Notes in Computer Science*, Springer-Verlag, Berlin, 1997, pp. 1333–1339.
2. M. T. Jones, P. E. Plassman, *Scalable Iterative Solution of Sparse Linear Systems*, Parallel Computing 20 (1994), pp. 753–773.
3. G. Karypis, V. Kumar, *Multilevel k-Way Partitioning Scheme for Irregular Graphs*, Technical Report TR-95-064, Department of Computer Science, University of Minnesota, 1995.
4. G. Karypis, V. Kumar, *Parallel Treshold-Based ILU Factorization*, Technical Report TR-96-061, Department of Computer Science, University of Minnesota, 1996.
5. A. Krommer, M. Derakhshan, S. Hammarling, *Solving PDE Problems on Parallel and Distributed Computer Systems Using the NAG Parallel Library*, in "High-Performance Computing and Networking" (B. Hertzberger, P. Sloot, Eds.), Vol. 1225 of *Lecture Notes in Computer Science*, Springer-Verlag, Berlin, 1997, pp. 440–451.
6. S. Salvini, G. Shaw, *An Evaluation of New NAG Library Solvers for Large Sparse Unsymmetric Linear Systems*, Technical Report TR2/96, The Numerical Algorithms Group Ltd, Wilkinson House, Jordan Hill Road, Oxford OX2 8DR, UK, 1996.
7. B. Smith, P. Bjorstad, W. Gropp, *Domain Decomposition: Parallel Multilevel Methods for Elliptic Partial Differential Equations*, Cambridge University Press, Cambridge, 1996.

Massively Parallel Linear Stability Analysis with P_ARPACK for 3D Fluid Flow Modeled with MPSalsa

R.B. Lehoucq[1] and A. G. Salinger[1]

Sandia National Laboratories
P.O. Box 5800, MS 1110
Albuquerque, NM 87185-1110
{rlehoucq,agsalin}@cs.sandia.gov

Abstract. We are interested in the stability of three-dimensional fluid flows to small disturbances. One computational approach is to solve a sequence of large sparse generalized eigenvalue problems for the leading modes that arise from discretizating the differential equations modeling the flow. The modes of interest are the eigenvalues of largest real part and their associated eigenvectors. We discuss our work to develop an efficient and reliable eigensolver for use by the massively parallel simulation code MPSalsa. MPSalsa allows simulation of complex 3D fluid flow, heat transfer, and mass transfer with detailed bulk fluid and surface chemical reaction kinetics.

1 Introduction

Galerkin least squares finite element discretization (GLS FE) of time-dependent Navier–Stokes modeling of incompressible fluid flow (see [8]) produce the matrix system

$$\begin{pmatrix} \mathbf{M} & \mathbf{0} \\ \mathbf{N} & \mathbf{0} \end{pmatrix} \begin{bmatrix} \dot{\mathbf{u}} \\ \dot{\mathbf{p}} \end{bmatrix} + \begin{pmatrix} \mathbf{L} & -\mathbf{C}^T \\ -\mathbf{CR}+\mathbf{G} & \mathbf{K} \end{pmatrix} \begin{bmatrix} \mathbf{u} \\ \mathbf{p} \end{bmatrix} = \begin{bmatrix} \mathbf{d}_1 \\ \mathbf{d}_2 \end{bmatrix}$$

where $\mathbf{u}$ is the fluid velocity and possible temperature, $\mathbf{p}$ is the pressure, $\mathbf{M}$ is the symmetric positive definite matrix of the overlaps of the finite element basis functions, $\mathbf{N}$ is an up-winded mass matrix, $\mathbf{L}$ is the sum of the discretized diffusion, nonlinear convection and any possible reaction operators, $\mathbf{C}$ is the discrete gradient, $\mathbf{C}^T$ is the discrete divergence operator, $\mathbf{R}$ diagonal matrix representing variable density, and $\mathbf{G}$ and $\mathbf{K}$ (pressure Laplacian) are stabilization terms arising from the GLS FE.

Linearizing about the steady state gives rise to the following generalized eigenvalue problem

$$\begin{pmatrix} \mathbf{L} & -\mathbf{C}^T \\ -\mathbf{CR}+\mathbf{G} & \mathbf{K} \end{pmatrix} \begin{bmatrix} \mathbf{u} \\ \mathbf{p} \end{bmatrix} = \begin{pmatrix} \mathbf{M} & \mathbf{0} \\ \mathbf{N} & \mathbf{0} \end{pmatrix} \begin{bmatrix} \mathbf{u} \\ \mathbf{p} \end{bmatrix} \lambda \tag{1}$$

or

$$\mathbf{Jx} = \mathbf{Bx}\lambda.$$

Here, $\mathbf{J}$ is the Jacobian matrix associated with the steady state or base flow and $\mathbf{B}$ is mass matrix, that is singular for incompressible flows. We denote the order of the matrices $\mathbf{J}$ and $\mathbf{B}$ by n.

The steady state is stable if Real$(\lambda) < 0$ for all the eigenvalues of (1). Hence, computing approximations to the right-most eigenvalues determines the stability of the steady state. The goal of our paper is to present preliminary results on providing a massively parallel eigensolver for computing these rightmost eigenvalues.

To compute the right-most eigenvalues, a shift-invert spectral transformation [14] is typically used to transform (1) into the standard eigenvalue problem

$$\mathbf{T}_s\mathbf{x} = (\mathbf{J} - \sigma\mathbf{B})^{-1}\mathbf{B}\mathbf{x} = \nu\mathbf{x}, \quad \nu = \frac{1}{\lambda - \sigma}. \tag{2}$$

The above formulation maps the infinite eigenvalues of (1) (arising from singular $\mathbf{B}$) to zero. By selecting the pole near the imaginary axis, the right-most eigenvalues are mapped by $\mathbf{T}_s$ into those of largest magnitude. However, because $\mathbf{J}$ and $\mathbf{B}$ are real matrices, we only allow a real σ to keep the computation in real arithmetic. Although a natural choice is to select a zero pole, the resulting transformation might miss a Hopf bifurcation (complex conjugate pair of eigenvalues that is in the right half of the complex plane). This occurs, for instance, when the distance to the Hopf bifurcation is greater than the distance to other (perhaps stable) eigenvalues of (1). The paper [3] discusses these issues in some detail.

The computational burden is in solving the linear set of equations with coefficient matrix $(\mathbf{J} - \sigma\mathbf{B})^{-1}\mathbf{B}$. The standard approach is to use a sparse direct solver to factor $\mathbf{J} - \sigma\mathbf{B}$ and then solve linear sets of equations (see [3],[4],[1],[10]). In this paper, we use the related generalized Cayley spectral transformation [14] that is discussed in Section 3.

Because our interest is in three dimensional systems with coupled fluid flow, heat transfer, and mass transfer, we need methods that will work for system sizes of order over $n = 10^5$ and that will scale well up to systems of size $n = 10^8$. Hence, solving linear systems with sparse direct methods is not a viable alternative. They require $\mathcal{O}(n^2)$ operations plus a prohibitive amount of memory. Instead, this paper considers the use of iterative methods for the linear solves on massively parallel machines. Along with a scalable eigensolver, such an approach allows very large system eigenvalue problems to be solved.

2 The Players

In this section, we introduce the massively parallel software for modeling three-dimensional fluid flow, the large scale parallel sparse eigensolver, the iterative linear solver and the massively parallel supercomputer.

2.1 MPSalsa

MPSalsa [20],[18] simulates complex systems with coupled fluid flow, thermal energy transfer, mass transfer and non-equilibrium chemical reactions. Currently, computer simulations of these complex reacting flow problems are usually limited to idealized systems in one or two spatial dimensions when coupled with a detailed, fundamental chemistry model. The goal of the MPSalsa project is to develop, analyze and implement advanced massively parallel numerical algorithms that will allow high resolution three-dimensional simulations with an equal emphasis on fluid flow and chemical kinetics modeling.

MPSalsa uses a GLS FE formulation [8] and has been written for unstructured meshes in two and three dimensions and to run on distributed memory massively parallel computers. A fully-coupled Newton's method is used to solve the nonlinear equations, which requires the inversion of a large sparse matrix at every iteration. With these robust methods, it is often possible to reach steady-state solutions directly from a trivial initial guess without the need of following a time transient. Since the steady-state algorithm does not discriminate between stable and unstable steady states, linear stability analysis is a crucial capability for using MPSalsa as an analysis tool.

MPSalsa has been used to analyze a variety of reacting flow systems, including the analysis of 3D models of Chemical Vapor Deposition (CVD) reactors for the growth of gallium arsenide semi-conducting films [19].

2.2 P_ARPACK

P_ARPACK [12] software is capable of solving large scale symmetric, nonsymmetric, and generalized eigen-problems from significant application areas. The software is designed to compute a few eigenvalues with user specified features such as those of largest real part or largest magnitude. The software implements an implicitly restarted Arnoldi Method [21]. Restarting is used so that the number of Arnoldi iterations (the size of the corresponding Hessenberg matrix) remains fixed. An effective restarting scheme finds ever better starting vectors so that the subsequent Arnoldi run contains the desired eigenvalue approximations. Implicit restarting is a stable and efficient scheme; for more information see [11]. P_ARPACK uses a data parallel model of computation where each processor owns a number of rows of the Arnoldi vectors.

2.3 Aztec

Aztec [9] is a parallel iterative library for solving linear systems of equations. A globally distributed matrix allows a user to specify pieces (different rows for different processors) of the application matrix. Issues such as local numbering, ghost variables, and messages that are crucial for parallelizing an application can be ignored by the user and are instead computed by an automated transformation function. Efficiency is achieved by using standard distributed memory techniques; locally numbered sub-matrices, ghost variables, and message information computed by the transformation function are maintained by each processor so that local calculations and communication of data dependencies is fast. Additionally, Aztec takes advantage of advanced partitioning techniques (Chaco [7]) and utilizes efficient dense matrix algorithms when solving block sparse matrices.

Aztec allows the user to choose between several Krylov-based iterative solution techniques as well as several algebraic preconditioners. In this work, we have exclusively used the GMRES algorithm [17] with a sufficiently large Krylov space (so that no restarts were necessary) that is preconditioned using the ILU(0) domain decomposition preconditioner. The domains for the preconditioning are the same as the partitioning used for parallelizing the MPSalsa calculation, so that the number of domains is the same as the number of processors that the application is run on.

2.4 Computer

The calculations presented here were run on the Sandia-Intel Teraflop massively parallel supercomputer [13] (a.k.a. ASCI Red). This computer has 4,536 nodes, each with two 200Mhz Pentium Pro Processors and 128 MBytes of memory. The theoretical peak is 1.8 TFLOPS and over 1.0 TFLOPS has been achieved. Node to node bandwidth is 800 MBytes/sec. More information can be found at `www.sandia.gov/ASCI/TFLOP/Home_Page.html`.

In 1997, MPSalsa-Aztec was a finalist for the Gordon Bell prize for reaching a sustained rate of 210 Gflops on 3600 nodes of the Sandia-Intel Tflop computer [5].

3 Algorithm

We now briefly describe the algorithm used for computing several (say k) rightmost eigenvalues of (1). As we have already mentioned, the dominant cost of the computation is that of solving the linear systems of equations associated with the shift-invert spectral transformation. Although this transformation maps the eigenvalues near the pole to those of largest magnitude, the transformation also maps the eigenvalues far from the pole to zero. Hence, the spectral condition number (absolute value of the maximum ratio of the eigenvalues) of $\mathbf{T}_s$ can be quite large. The resulting linear systems will be difficult to solve because the rate of convergence with an iterative method [16],[6] depends strongly upon the spectral condition number.

A better conditioned linear set of equations is achieved when using a generalized Cayley [14] transformation resulting in

$$\mathbf{T}_c\mathbf{x} = (\mathbf{J} - \sigma\mathbf{B})^{-1}(\mathbf{J} - \mu\mathbf{B})\mathbf{x} = \nu\mathbf{x}, \quad \nu = \frac{\lambda - \mu}{\lambda - \sigma}. \tag{3}$$

We call μ the zero of the Cayley transform. In contrast to shift-invert transform, the Cayley transform maps any eigenvalues of (1) far from the pole close to one. If we are able to select a pole that is to the right of all the eigenvalues (1) along with $\sigma < \mu$, then the smallest eigenvalue of $\mathbf{T}_c$ is no smaller than one (in magnitude). Moreover, by judiciously choosing the pole, we can approximately bound the largest eigenvalue of $\mathbf{T}_c$ (in magnitude) resulting in a modest (say order ten) spectral condition number.

- Start P_ARPACK with the vector $\mathbf{v} = \mathbf{J}^{-1}\mathbf{B}\mathbf{x}$ where $\mathbf{x}$ is random vector. Select a pole and zero for $\mathbf{T}_c$.
  1. Compute m Arnoldi iterations for $\mathbf{T}_c$ with starting vector $\mathbf{v}$.
  2. Exit if the k eigenpairs satisfy the user specified tolerance.
  3. Implicitly restart the Arnoldi iteration.
  4. Pick a new σ and μ based on the current approximate eigenvalues.

A few remarks are in order. The starting vector is chosen so that it does not contain any components [15] in the null-space of $\mathbf{B}$. The orthogonality of the m Arnoldi vectors is maintained at machine precision. At step 2, the eigenvalues of the m by m Hessenberg matrix are mapped back to the system defined by (1) via the inverse Cayley transformation. The eigen-computation is terminated when these k rightmost approximate eigenvalues satisfy the user specified tolerance. The check that must be satisfied is equivalent to computing $\|\mathbf{J}\hat{\mathbf{x}} - \hat{\lambda}\mathbf{B}\hat{\mathbf{x}}\|$ where $\hat{\mathbf{x}}$ and $\hat{\lambda}$ are the approximate eigenvector and eigenvalue computed. By implicitly restarting the Arnoldi iteration, we compute a new starting vector for the subsequent run (step 1). Finally, at step 4, the new pole and zero for the next Cayley transformation are selected so that the spectral condition number of $\mathbf{T}_c$ is of order ten.

4 Numerical Experiments and Discussion

The flow and heat transfer in a rotating disk CVD reactor is chosen as the targeted test problem for the linear stability analysis. It is an important engineering application where transitions from stable to unstable (and undesirable) flows have been seen experimentally [2]. The flow stability problem is complex because of competing flow effects due to forced convection at the inlet of the reactor, rotationally-driven convection at a spinning disk in the reactor, and buoyancy-driven flow due to large temperature gradients. The system—geometry, equations, and boundary conditions—is axisymmetric, but an understanding of the bifurcations to three-dimensional solutions is crucial for successful design of the next generation CVD reactor.

A three-dimensional, unstructured finite element mesh consisting of 43520 elements is used to discretize the 5 coupled PDEs that govern the flow and heat transfer in this system. The total number of unknowns in the problem, and correspondingly the rank of the eigenvalue problem, is 233925.

For the calculations presented, GMRES was the iterative solver used for all the linear systems along with an overlapping Schwarz domain decomposition preconditioner with ILU(0) on each sub-domain. All computations were done using double precision floating point arithmetic.

Since P_ARPACK is matrix-free (no specified storage format is assumed for the matrices), the VBR (variable block row) storage format provided by Aztec is used for the matrices. A node based distribution of the problem is used hence there are repeated elements (ghost nodes). All the necessary processor communication and associated work is done by Aztec.

The MPSalsa calculation of the steady solution from a trivial initial guess on 200 Processors required 7 Newton iterations and 305 seconds, 90% of which was spent in the linear solver. Each linear solve was required to reduce the residual by 10^{-4} and required Krylov subspaces ranging from 215 to 289, and used 40 seconds on average.

A calculation of 7 leading eigenvalues of this system was performed with the algorithm described above. The total time for the eigenvalue calculation was 1800 seconds. The following discussion presents the times for each step in the algorithm as well as some of the choices in parameters that were made. A tolerance of 10^{-5} used for P_ARPACK.

Only 3 seconds were required to calculate the matrices $\mathbf{J}$ and $\mathbf{B}$. The initial guess of $\mathbf{v} = \mathbf{J}^{-1}\mathbf{B}\mathbf{x}$ for a random vector $\mathbf{x}$ was computed where the linear solver decreased the residual by 10^{-12} in 58 seconds. Using $\sigma = 100$ and $\mu = 1000$ for the initial Cayley transformation, $m = 24$ Arnoldi iterations were computed. The preconditioner was calculated in 10 seconds and and reused in each of linear solves. GMRES required on average 20 seconds and a Krylov subspace size of 175 (without restarts) to reduce the residual by a factor of 10^{-10}.

At this point, none of the computed eigenvalues satisfied the desired tolerance, and a new shift and pole were chosen as $\sigma = -6.86$ and $\mu = 63.5$. Another $m = 24$ Arnoldi iterations were computed and each linear system required 51 seconds and a Krylov subspace size of 345 to reduce the residual by a factor of 10^{-10}. The 7 converged eigenvalues and residuals are listed in Table 1. The numbers listed in the third and fourth columns are $||\mathbf{J}\hat{\mathbf{x}} - \hat{\lambda}\mathbf{B}\hat{\mathbf{x}}||/||\mathbf{B}\hat{\mathbf{x}}||$ (Residual) and $|\hat{\mathbf{x}}^H\mathbf{J}\hat{\mathbf{x}}/\hat{\mathbf{x}}^H\mathbf{B}\hat{\mathbf{x}} - \hat{\lambda}|$ (Rayleigh Quotient Error) where $\hat{\mathbf{x}}$ ($||\hat{\mathbf{x}}|| = 1$) and $\hat{\lambda}$ are the approximate eigenvector and eigenvalue computed. In the case of complex conjugate pairs of eigenvalues, the successive rows list the real and imaginary parts of the vector and scalar quantities.

Table 1. The 7 leading eigenvalues and the residuals are presented. The calculation for a system of rank 233925 required 1800 seconds on 200 Processors. The tolerance used for the eigensolver was 10^{-5}.

Real(λ)	Im(λ)	Residual	RQ errors
-13.867	0.0000	$9.3e-5$	$2.1e-7$
-15.908	10.144	$2.4e-5$	$3.0e-7$
-15.908	-10.144	$3.7e-5$	$1.3e-6$
-23.685	4.513	$4.8e-4$	$3.7e-6$
-23.685	-4.513	$1.7e-4$	$5.2e-6$
-24.137	21.809	$5.8e-5$	$7.5e-6$
-24.137	-21.809	$2.2e-4$	$7.2e-6$

Calculations on this and similar systems showed that the residual reduction in the linear solver had to be near or below 10^{-10} to get accurate eigenvalues. Considering that the linear solves in the steady state calculation required over 200 GMRES iterations (no restarting) just to reach a 10^{-4} residual reduction, it was a concern that requiring 10^{-10} of the linear solver in the eigenvalue calculations was going to severely limit the algorithm. However, solving the Cayley systems proved to be much easier than solving the untransformed system for the steady state. This is seen in the above results, where it took 40 seconds to invert **J** to a tolerance of 10^{-4} in the steady state calculation and only 20 seconds to get 10^{-10} reduction when $\sigma = 100$ and $\mu = 1000$. As explained in Section 3 the Cayley transformed system maps most of the eigenvalues near 1, so that the spectral condition number of the matrix is bounded. It turns out that the spectral condition number of the system $(\mathbf{J} - \sigma\mathbf{B})^{-1}(\mathbf{J} - \mu\mathbf{B})$ is just 8.9 for the initial Cayley transform and 11.0 for the second.

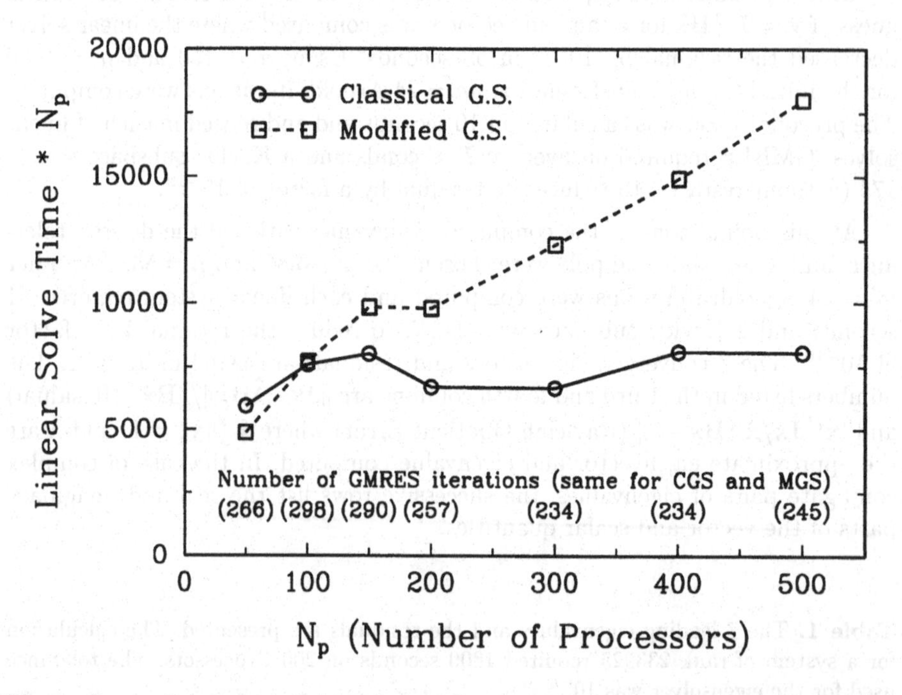

Fig. 1. Comparing the parallel efficiency of orthogonalization schemes used for GMRES. The number of GMRES iterations needed to to solve the linear systems to the specified tolerance using the given number of processors is shown in parenthesis.

Another interesting algorithmic detail that proved important was the choice of orthogonalization used in the GMRES algorithm. Originally, a classical Gram-Schmidt scheme was used, but the lack of numerical stability prevented the GMRES algorithm from reaching the required 10^{-10} tolerance. Two alternative orthogonalization schemes were used successfully: two-step classical Gram-Schmidt

(CGS) and a modified Gram-Schmidt (MGS). The CGS method uses two steps of orthogonalization (the second step is the correction for loss of orthogonality for the Arnoldi basis vectors) but the number of global communication points remains fixed (at two) independent of the GMRES iteration. On the other hand, the MGS scheme requires i communications to orthogonalize i vectors (at GMRES iteration i) but no additional floating point operations (flops). Both schemes reached the 10^{-10} tolerance and provided identical results in terms of the number of GMRES iterations needed.

There was a significant difference in the scalability of the two algorithms as the number of processors was changed. The time required to perform a single linear solve (using a pre-calculated preconditioner) was recorded for the same problem as above for the case of $\sigma = 10$ and $\mu = 50$, with the number of processors being varied from 50 to 500. The results of this calculation are most clear when presenting the total CPU time (calculated as the wall clock time multiplied by the number of processors) as shown in Figure 1. When the problem was run on 50 processors, the extra communications required by MGS were less expensive than the extra flops required by the CGS algorithm. However, as the number of processors is increased up to 500, it is seen that the communication time in MGS starts to dominate, while the total CGS time remains relatively flat. At 500 processors, the CGS routine uses less than half of the time of the MGS method.

The results in Figure 1 show that the two-step classical Gram-Schmidt (CGS) scheme scales much better than the modified Gram-Schmidt (MGS) scheme. It should be pointed out that the inter-processor communication rate of the Sandia-Intel Tflop computer is very fast compared to more loosely coupled parallel machines, where we would expect the crossover point to be at significantly fewer than 100 processors as seen here.

In each case, the number of GMRES iterations is shown. The surprising result is that the number of GMRES iterations does not increase monotonically with the number of processors (and hence domains in the ILU(0) preconditioner). This is currently not understood.

5 Conclusions

The preliminary results presented here show that the algorithms can be used to successfully perform linear stability analysis on reacting flow systems of $\mathcal{O}(10^5)$ on massively parallel computers.

Our results showed that a judicious choice of a Cayley transformation along with a preconditioned GMRES iterative solver for the resulting linear systems allowed the P_ARPACK to compute accurate eigenvalues. Also crucial for the scalability of GMRES was the use of a two-step classical Gram-Schmidt scheme for the necessary orthogonalizations.

Acknowledgments

The authors would like to thank David Day, Louis Romero, Beth Burroughs, John Shadid, and Ray Tuminaro for their contributions to our understanding and the implementation of the algorithms as well as some of the text of this paper.

References

1. N. Alleborn, K. Nandakumar, H. Raszillier, and F. Durst. Further contributions on the two-dimensional flow in a sudden expansion. *Journal of Fluid Mechanics*, 330:169–188, 1997.
2. W. G. Breiland and Greg H. Evans. Design and verification of nearly ideal flow and heat transfer in a rotating disk chemical vapor deposition reactor. *Journal of the Electrochemical Society*, 138:1806–1816, 1991.
3. K. N. Christodoulou and L. E. Scriven. Finding leading modes of a viscous free surface flow: An asymmetric generalized eigenproblem. *Journal of Scientific Computing*, 3:355–406, 1988.
4. K. A. Cliffe, T. J. Garratt, and A. Spence. Eigenvalues of the discretized navier-stokes equation with application to the detection of hopf bifurcations. *Advances in Computational Mathematics*, 1:337–356, 1993.
5. K. D. Devine, G. L. Hennigan, S. A. Hutchinson, A. G. Salinger, J. N. Shadid, and R. S. Tuminaro. High performance mp unstructured finite element simulation of chemically reacting flows. In *Proceedings of SC97*, 1997.
6. Anne Greenbaum. *Iterative Methods for Solving Linear Systems.* SIAM, Philadelphia, PA., 1997.
7. B. Hendrickson and R. Leland. The Chaco user's guide: Version 2.0. Technical Report SAND94–2692, Sandia National Labs, Albuquerque, NM, June 1995.
8. J. R. Hughes, L. P. Franca, and M. Balestra. A new finite element formulation for computational fluid dynamics: Circumventing the Babuska–Brezzi condition: A stable Petrov–Galerkin formulation of the Stokes problem accommodating equal-order interpolation. *Comp. Meth. App. Mech. and Eng.*, 59:85–99, 1986.
9. S. A. Hutchinson, L. V. Prevost, R. S. Tuminaro, and J. N. Shadid. Aztec user's guide: Version 2.0. Technical report, Sandia National Laboratories, Albuquerque, NM, 1998.
10. R. B. Lehoucq and J. A. Scott. Implicitly restarted Arnoldi methods and eigenvalues of the discretized Navier-Stokes equations. Technical Report SAND97-2712J, Sandia National Laboratories, Albuquerque, New Mexico, 1997.
11. R. B. Lehoucq, D. C. Sorensen, and C. Yang. *ARPACK USERS GUIDE: Solution of Large Scale Eigenvalue Problems with Implicitly Restarted Arnoldi Methods.* SIAM, Phildelphia, PA, 1998.
12. K. J. Maschhoff and D. C. Sorensen. P_ARPACK: An efficient portable large scale eigenvalue package for distributed memory parallel architectures. In Jerzy Wasniewski, Jack Dongarra, Kaj Madsen, and Dorte Olesen, editors, *Applied Parallel Computing in Industrial Problems and Optimization*, volume 1184 of *Lecture Notes in Computer Science*, Berlin, 1996. Springer–Verlag.
13. T. G. Mattson and G. Henry. An overview of the Intel TFLOPS supercomputer. *Intel Technology Journal*, 1, 1998.

14. K. Meerbergen, A. Spence, and D. Roose. Shift-invert and Cayley transforms for the detection of rightmost eigenvalues of nonsymmetric matrices. *BIT*, 34:409–423, 1994.
15. Karl Meerbergen and Alastair Spence. Implicitly restarted Arnoldi with purification for the shift–invert transformation. *Mathematics of Computation*, 218:667–689, 1997.
16. Y. Saad. *Iterative Methods for Sparse Linear Systems.* PWS, Boston, MA, 1996.
17. Y. Saad and M. H. Schultz. GMRES: A generalized minimal residual algorithm for solving nonsymmetric linear systems. *SIAM Journal on Scientific and Statistical Computing*, 7(3):856–869, July 1986.
18. A. G. Salinger, K. D. Devine, G. L. Hennigan, H. K. Moffat, S. A. Hutchinson, and J. N. Shadid. Mpsalsa: a finite element computer program for reacting flow problems part 1 - user's guide. Technical Report SAND95–2331, Sandia National Laboratories, Albuquerque, NM, 1996.
19. A. G. Salinger, J. N. Shadid, S. A. Hutchinson, G. L. Hennigan, K. D. Devine, and H. K. Moffat. Analysis of gallium arsenide deposition in a horizontal chemical vapor deposition reactor using massively parallel computations. Technical Report SAND98–0242, Sandia National Laboratories, Albuquerque, NM, 1998.
20. J. N. Shadid, H. K. Moffat, S. A. Hutchinson, G. L. Hennigan, K. D. Devine, and A. G. Salinger. Mpsalsa: a finite element computer program for reacting flow problems part 1 - theoretical development. Technical Report SAND95–2752, Sandia National Laboratories, Albuquerque, NM, 1996.
21. D. C. Sorensen. Implicit application of polynomial filters in a k-step Arnoldi method. *SIAM J. Matrix Analysis and Applications*, 13(1):357–385, January 1992.

Parallel Molecular Dynamics Simulations of Biomolecular Systems

Alexander Lyubartsev and Aatto Laaksonen

Department of Physical Chemistry, Arrhenius Laboratory,
University of Stockholm, S-10691 Stockholm, Sweden

Abstract. We describe a general purpose parallel molecular dynamics code, for simulations of arbitrary mixtures of flexible molecules in solution. The program allows us to simulate molecular systems described by standard force fields like AMBER, GROMOS or CHARMM, containing terms for short-range interactions of the Lennard-Jones type, electrostatic interactions, covalent bonds, covalent angles and torsional angles and a few other optional terms. The state-of-the-art molecular dynamics techniques are implemented: constant-temperature and constant-pressure simulations, optimized Ewald method for treatment of electrostatic forces, double time step algorithm for separate integration of fast and slow motions. The program is written in standard Fortran 77 and uses MPI library for communications between nodes. The scalable properties of the program do not depend on the complexity of the studied system and are determined mainly by the hardware and communication speed. Examples of a few molecular systems differing by the composition will be given: Ionic water solutions, large DNA fragments in water solution with counter ions, a phospholipid membrane system.
Keywords: parallel algorithms, molecular dynamics, computer simulations

1 Introduction

Computer simulation methods, such as, Molecular Dynamics (MD) and Monte Carlo (MC) have now become important techniques to study fluids and solids. These methods provide a link between theory and experiment and they are also the only way to study complex many-body systems when both experimental techniques and analytical theories are unavailable.

The MD method provides a numerical solution of classical (Newton's) equations of motion:

$$m_i(d^2r_i/dt^2) = F_i(r_1, ...r_N) \tag{1}$$

where force $F_i(r_1, ...r_N)$ acting on particle i is defined by the interaction potential (or force field) $U(r_1, ..., r_N)$:

$$F_i(r_1, ...r_N) = -\frac{\partial}{\partial r_i}U(r_1, ..., r_N) \tag{2}$$

The most time-consuming part (sometimes up to 99% of the cpu time) of the MD simulations is the calculations of the forces $F_i(r_1, ...r_N)$. In a typical case of pair interactions the cpu time may scale with number of particles N between $O(N^2)$ and $O(N)$ depending on the algorithm and the type of interaction potential (note, that scaling as $O(N)$ may be reached only at a very large N and for systems with short-range interactions).

Now it is a standard routine procedure to simulate molecular systems consisting of order of 100-1000 particles, which in some cases (e.g. simple liquids) is sufficient to give a good description of corresponding macrosystem. For other systems, a larger number of particles is needed in order to describe them in a realistic way. Complex bio- and organic molecules (e.g. proteins, nucleic acids, membranes, carbohydrates, etc) immersed into a solvent increase the number of involved atoms one to two or more orders of magnitude. Also, larger the molecular systems grow, longer simulations are needed to follow low-amplitude motions and slow conformational transitions. It is clear that the rate of the progress towards more complex molecular models is set, to a large extent, by advances in microprocessor technology and computer architecture as well as by development of appropriate software.

Computer simulations of many-particle systems are well suited for parallel computer systems. The basic reason for this is that the forces acting on each particle can be calculated independently in different processors. However, the most optimal parallel scheme for a particular problem depends both on the hardware in hand and on the system under investigation (size, type of interaction, etc.). Electrostatic interactions, fast intramolecular motions due to explicit modeling of hydrogens, angle and torsional angle forces of macromolecules - all these kinds of forces require a special treatment to create an effective parallel code.

We have developed a general purpose molecular dynamics code (MDynaMix) [1] for simulations of arbitrary mixtures of rigid or flexible molecules employing the most modern simulation techniques: double time step algorithm for fast and slow modes, optimized Ewald method for electrostatic interactions, constant temperature - constant pressure algorithm. The program can be used for simulation of mixtures of molecules interacting with AMBER- or CHARMM-like force fields [2], [3].The code is highly universal and well suitable for simulation of both simple molecules and complex biological macromolecules. The code uses only standard Fortran77 statements, being able to run on any parallel system with MPI-library installed. In the latest version (4.2) additional features were included: separate pressure control in different directions (for simulation of anisotropic systems), generalized reaction field method for electrostatic interactions, truncated octahedron or hexagonal simulation cell, parallel SHAKE algorithm for constrained dynamics, different types of torsional angle potentials and a few other options. Below we describe details of the program organization, parallelization algorithm and performance.

2 Parallel Molecular Dynamics

2.1 General Organization

Complex molecular systems are often described by force fields like AMBER [2] or CHARMM [3]. These force fields contain terms for following interactions:

1) atom-atom short-range interactions (Lennard-Jones potential)

2) atom-atom electrostatic interactions

3) intramolecular interactions: covalent bonds, covalent angles and torsional angles.

4) optional terms (hydrogen bonds, anharmonic bond potential, other specific interactions)

The general functional form of such a force field is:

$$U(r_1,...,r_N) = \sum_{i<j=1}^{N}\left(\frac{A_{ij}}{r_{ij}^6}+\frac{B_{ij}}{r_{ij}^{12}}\right) + \sum_{i<j=1}^{N}\left(\frac{q_i q_j}{r_{ij}}\right) +$$
$$\sum_{bonds} K_{bond}(r_{ij}-r_{ij}^0)^2 + \sum_{angles} K_{ang}(\varphi_a-\varphi_a^0)^2 +$$
$$\sum_{torsions} K_{tors}(1+Cos(m_t\psi_t-\psi_t^0)) + U_{optional} \tag{3}$$

where r_{ij} is the distance between atoms i and j, and other constants define force field parameters for each chemical atom type.

In principle, calculations of pairwise atom-atom interactions (Lennard-Jones and electrostatic forces, i.e the first and the second term in expression (3)) require a double sum over all the atom pairs. Usually this is the most time consuming part of the force calculations. The application of a cut-off radius for these interactions allows one to considerably reduce the cpu time, by effectively setting the interactions between particles separated by distances larger than the cut-off distance to zero.

Two problems, however, emerge here. First, the electrostatic interactions are long-ranged, and strictly, no cut-off without a special treatment can be applied. Attempts to simulate molecular systems with electrostatic interactions using simple spherical cutoff, even of a rather large radius, may lead to serious artifacts [4]. Second, in order to decide whether to calculate the forces between a given atom pair or not (e.g whether the atoms are within or out the cutoff radius), one still should know the distances between all atomic pairs, or at least to have a list of atom pairs with distances less than the cut-off (list of neighbours).

One of the most effective and popular methods of treatment of electrostatic interactions is the Ewald summation method [5]. The Ewald method splits up the total force into a long-range and short range component. The long-range part is calculated in the reciprocal space by the Fourier transform, while the short-range part is treated alongside with the Lennard-Jones forces. The convergence of the two parts of the Ewald sum is regulated by a parameter. The optimal choice of the convergence parameter leads to a scaling of the cpu time as $O(N^{3/2})$. The optimized Ewald method is implemented in the program. Recently a new version of the Ewald method (Particle Mesh Ewald [6]) has appeared, which scales as $NlnN$ for large $N > 10^4$. We are planning to implement the Particle Mesh Ewald method in a future program release.

Creation of the neighbour lists is another problem. In liquids this list should be updated periodically. In the linked-cell method the search of neighboring pairs can be limited only by the current and the touching cells; this leads to an $O(N)$ algorithm. However the true $O(N)$ algorithm is achieved only at very large N; for the "average size" systems of $10^3 - 10^4$ particles periodical (e.g. each 10 MD steps) update of neigbours list by looking through all the atom pairs occurs effective enough. Although the cpu time of this block scales as $O(N^2)$, the coefficient is small, and for example in a test run with 2000 H_2O molecules (6000 atoms) this part of the program consumes about 5% of the cpu time.

Irrespectively to the parallelization, an essential saving of cpu time may be achieved by applying the multiple time scale algorithm. Different kinds of forces in the system fluctuate at different, characteristic time scales. In systems with explicit treatment of hydrogens, covalent bond forces, Lennard-Jones and electrostatic interactions at short distances between atoms require an updated force calculation after 0.2-0.4 fs, whereas long range parts of Lennard-Jones and electrostatic interactions, which consume most of cpu time, may be calculated after 2-4 fs. In the present program the two time scale algorithm is applied [8]. The covalent forces and atom-atom forces for atoms closer than 5Å are calculated at every short time step. Forces between atoms with distance from 5Å to cut-off and reciprocal part of the Ewald sum are calculated at every long time step. Correspondingly, two lists of neighbours are calculated in the program - for fast and for slow forces.

2.2 Parallelization Strategy

There are two main strategies in parallelization of molecular dynamics programs, which called "Replicated data" and "Domain decomposition". In the replicated data method all the nodes know positions of all the particles in the system, while calculation on different contributions to forces goes in parallel. The advantages of this method is a relative simplicity to distribute force calculations over processors: different contributions to the sum (3) have a form of a sum of a large number of similar independent terms (i.e. sums over atom pairs, covalent bond, covalent angles, etc). Such calculations can be done in parallel on different processors equally effective for virtually all kinds of the force fields and molecular structures. The weak point of the replicated data approach is that it requires relatively extensive communications between processors, to collect data on contribution to the forces from all the nodes and to distribute new particle's positions to all the nodes. In the "Domain decomposition" approach particles are distributed between the nodes, usually each node being responsible for the particles in the corresponding subcell. The communication overhead may be lower in this case comparing to the replicated data method, but other sources of overheads arise due to necessity to monitor moving of particles through the subcells, to implement Ewald summation of electrostatic interactions, to get use of the Newton's 3-rd law, etc. An addition of extra functionality to the program in the domain decomposition method may also lead to essential increase of cpu time both for calculations and for communications.

It is believed that the domain decomposition method is more effective for massively parallel computer systems to run simulation programs for particles interacting with simple short range potentials; for parallel simulation of complex biomolecular systems on computers with several or several dozens processors the domain decomposition method is preferable. In our molecular dynamics program, the replicated data method is implemented.

2.3 Parallelization of Specific Parts of the Program

For an effective parallelization of non-bonded interactions (first and second sum in (1))the whole list of atom pairs ($I, J = 1, ...N, I < J$) should be uniformly divided between the available number of nodes. The condition $I < J$ is set to avoid a double count, but it makes difficult to divide equally atom pairs between the nodes. To avoid this, the following scheme was applied. All particles are divided equally between processors (loop $do\ I = TASKID, N, NUMTASK$, where $TASKID$ - node number, $NUMTASK$ - total number of available nodes). For each I we have an inner loop over J with the following condition: $I - J$ is even for $I > J$ and $J - I$ is odd for $I < J$. So the atom pairs occur uniformly distributed between nodes and each one is treated only once. Then for each node lists of closest ($r < 5$Å) and far ($5 < r < R_{cut}$) neighbours are calculated. These lists are local for each node and are recalculated after about 10 long time steps. They are used for the Lennard-Jones and electrostatic (real space part) force calculations. Using of local list of neighbours on each node makes it possible to use the same code both for parallel and single-processor architecture. Moreover, since the list of neighbours is the biggest data structure in the program, distributing it over processors reduce the total memory requirements, or makes it possible to simulate much bigger molecular systems with the same memory limit per processor.

Parallelization of the reciprocal part of the Ewald sum is more or less straightforward. This contribution is expressed as a sum over reciprocal space vectors. Each node calculates force contributions due to a certain (fixed for this node) group of reciprocal vectors.

Calculations of forces due to covalent bonds, angles and torsions can be done independently for each bond, angle, or dihedral angle. At the beginning of simulations all the bonds, covalent angles and torsions are distributed uniformly over available nodes. Each node gets its own list of bonds, angles and torsions and corresponding contributions to the forces and energies are calculated in parallel.

Different forces acting on each atom from other atoms are accumulated first on each node separately. After completing the calculations on each node all the forces acting on a given atom are transfered to the node corresponding to this atom and summed up. The whole operation is made by a single MPI call MPI_REDUCE_SCATTER, which, depending on the implementation, may work essentially faster than consecutive summing up forces (REDUCE) followed by their distribution over the nodes (SCATTER). Then on each node the atoms are moved according to a chosen integration scheme: constant energy (NVE), constant temperature (NVT) or constant pressure and temperature (NPT) molecular dynamics [7]. The double time step algorithm [8] is easily integrated into this scheme. After completing an MD step new positions of atoms are broadcasted to all nodes.

2.4 Portability

The program is written in standard Fortran 77 and highly portable, being able to run on any parallel system with MPI-library installed. It can be also easily made run on a single-processor computer. So far the program was tested on parallel systems IBM SP2, Cray T3E, PC-linux clusters employing MPI calls through an ethernet network, as well as on single processor systems IBM RISC 6000, DEC Alpha, PC-linux. Porting the code to another computer system normally requires only small changes of the Makefile.

3 Examples

The performance of the program was tested in details on several molecular systems:

1. A periodic fragment of DNA in ionic aqueous solution: one helical turn of DNA (635 atoms), 1050 water molecules, 30 Na and 10 Cl ions (3825 atoms totally). In other simulations Na ions was substituted by Li or Cs ions.

2. An NaCl ion solution: 20 NaCl ion pairs and 1960 water molecules, 5920 atoms totally.

3. A lipid bilayer. 64 DPPC lipid molecules (50 atoms each) and 1472 water molecules, 7616 atoms totally.

The three molecular systems are qualitatively different on the composition: the first one contains one macromolecule in solution, the second is a mixture of small molecules, the third one is an inhomogeneous system consisting of a number of middle size molecules and waters.

The dependence of the total computation time on the number of nodes in each case is given in **Table 1**. In all the cases the double time step algorithm with a short step of 0.2 fs and a long step of 2 fs was used. The long-range electrostatic interactions were taken into account by the Ewald method. The cut-off in the reciprocal space was determined from the conditions that the remaining terms were less then 0.001 relative to the total electrostatic energy.

The calculations were performed on IBM SP2 Strindberg at PDC, KTH, and on CRAY T3E in the National Supercomputer Center, Linköping. The program shows good scaling properties in all cases running on up to 32 processors.

Table 1. CPU time (in min) for 1 ps simulation of 3 different molecular systems depending on number of processors. Box sizes and cutoff distance R_{cut} are given in Å

				nodes						
System	Num.atoms	Box sizes	R_{cut}	1	2	4	8	16	32	64
					IBM	SP2	160	Mhz proc.		
1. DNA	3825	33x33x34	13	78	39	20	11	6	3.8	2.5
2. Ion solution	5920	39x39x39	13	138	69	35	18	9.2	5.	3.6
3. lipid bilayer	7616	44x44x64	14	187	94	48	25	13	7.5	5.5
					Cray		T3E			
1. DNA	3825	33x33x34	13	121	60	31	16	9	5.3	3.5
3. lipid bilayer	7616	44x44x64	14	268	134	70	37	19	11	7
					Pentium	-2	300	Mhz		
3. lipid bilayer	7616	44x44x64	14	250	126	-	-	-		

We have used the MDynaMix program for practical simulations of these molecular systems on the nanosecond time scale. Moreover, for the DNA solution simulations were carried out with several different counterion types. Both structural and dynamical properties of the molecules are studied. The reader is referred to the corresponding articles for details [9], [10].

Acknowledgements

We are grateful to the Center for parallel Computing at the Royal Institute of Technology (Stockholm) and National Supercomputing Center (Linköping) for granting the computer facilities, and the Swedish National Science Foundation (NFR) and the Foundation for Strategic Research (SSF) for financial support.

References

1. http://www.fos.su.se/physical/sasha/md_prog.html
2. Weiner, S.J., Kollman, P.A., Ngyuen, D.T., Case, D.A.: An All Atom Force Field for Simulations of Proteins and Nucleic Acids, J. Comput. Chem. **7** (1986) 230-252.
3. MacKerell, A.D., Wiorkiewicz-Kuczera, J., Karplus, M.: An All-Atom Empirical Energy Function for the Simulation of Nucleic Acids, J.Am.Chem.Soc. **117** (1995) 11946-11970.
4. Tironi, I.G., Sperb, R., Smith, P.E., van Gunsteren, W.F.: A Generalized Reaction Field Method for Molecular Dynamics Simulations. J. Chem. Phys., **102** (1995) 5451 - 5458.
5. Allen, M.P., Tildesley, D.J.: *Computer Simulation of Liquids*, Clarendon: Oxford, 1987
6. Darden, T., York, D., Pedersen, L.: Particle Mesh Ewald: A *NlnN* Method for Ewald Sums in Large Systems, J. Chem. Phys., **98** (1993) 10089-10092.
7. Martyna, G.I., Tuckerman, M.E., Tobais, D.J., Klein, M.L.: Explicit Reversible Integrators for Extended Systems Dynamics, Mol. Phys., **87** (1996) 1117-1157.
8. Tuckerman, M.E., Berne, B.J., Martyna, G.J.: Reversible Multiple Time Scale Molecular Dynamics, J. Chem. Phys., **97** (1992) 1990-2001.

9. Lyubartsev, A.P., Laaksonen, A.: Concentration Effects in Aqueous NaCl Solutions. A Molecular Dynamics Simulations. J. Phys. Chem., **100** (1996) 16410-16418.
10. Lyubartsev, A.P., Laaksonen, A.: Molecular Dynamics Simulations of DNA in Presence of Different Counterions. J. Biomolec. Structure and Dynamics (in press).

A Parallel Solver for Animal Genetics

Per Madsen[1] and Martin Larsen[2]

[1] Danish Institute for Agricultural Sciences, Research Centre Foulum, P.O. Box 50, DK-8830 Tjele, Denmark.

[2] UNI-C, Danish Computing Centre for Research and Education, DTU Building 304, DK-2800 Lyngby, Denmark.

Abstract. The use of linear multivariate mixed models in animal genetics, leads to very large, sparse linear systems of equations. The sparse, symmetric coefficient matrix is too large to be constructed explicitly. We describe a parallel, iterative linear equation solver for large sparse systems, developed by DIAS and UNI-C. The solver takes advantage of the structure of the multivariate mixed model equations, and is based on Gauss-Seidel and second order Jacobi iteration. It is parallelized for distributed memory architectures.

1 Introduction

Breeding values are needed to support selection decisions in animal breeding programs. The true breeding value of an animal is unknown, but Best Linear Unbiased Prediction (BLUP) of estimated breeding values (EBVs) can be calculated (Henderson, [1]). The BLUP method consists of a genetic and a statistical model. The genetic model is based on quantitative genetic principles. If all animals in the population are included, the genetic model is called an Animal Model. The statistical model describes how the factors (environmental as well as genetic) affect the measured traits. Breeding programs often involve simultaneous selection for several, possibly correlated, traits. This leads to multivariate models in order to obtain EBVs consistent with the selection criteria.

The BLUP method involves solving a sparse linear system of equations, called the mixed model equations (MME). The size of the system is the sum of number of levels times number of traits, taken over all factors in the model. Assuming the same model for all traits, the size increases in proportion to the number of traits. Taking Danish Dairy Cattle as an example, the number of equations per trait is approx. 10^7. Simultaneous estimation for 3 milk production traits (milk, fat and protein), leads to a system with approx. $3 \cdot 10^7$ equations. The density of the coefficient matrix is approx. 0.1%. In the Danish Dairy Cattle example, the coefficient matrix for the single trait model, fills approx. 0.5 Terabytes using double precision and a half stored, sparse representation. Therefore, the matrix is never constructed and stored explicitly. Instead, a method called "iteration on data" is used (Schaeffer and Kennedy, [3]). During each round of iteration, the coefficients are recalculated from data. The data fill much less than the coefficient matrix (for the Danish Dairy Cattle example, approx. 2 GB per trait).

The computational challenge is to solve very large sparse systems. This paper describes a parallel implementation of "iteration on data" for a distributed memory architecture computer. The development of the parallel solver started as a joint project between DIAS and UNI-C in 1995, and is now supported by EU as a High Performance Computer Network (HPCN) Technology Transfer Node (TTN) project, under the title Continuous Estimation of Breeding Values Using Supercomputers (CEBUS).

2 Model

A general multivariate linear mixed model can be written as:

$$\mathbf{y} = \mathbf{X}_1\boldsymbol{\beta}_1 + \mathbf{X}_2\boldsymbol{\beta}_2 + \sum_{i=1}^{r} \mathbf{Z}_i\mathbf{u}_i + \mathbf{Z}_a\mathbf{a} + \mathbf{e} \tag{1}$$

where $\mathbf{y}$ is a vector of observations, $\boldsymbol{\beta}_1$ and $\boldsymbol{\beta}_2$ are vectors of fixed effects, $\mathbf{u}_i$, i=1, 2, ..., r, are vectors of random effects for the ith random factor other than the animal factor, $\mathbf{a}$ is a vector of random additive genetic effects (animal effects = breeding values), and $\mathbf{e}$ is a vector of random residuals. The dimension of $\mathbf{y}$ is controlled by the number of data records and the number of traits (t). The distinction between $\boldsymbol{\beta}_1$ and $\boldsymbol{\beta}_2$ is for notational convenience, when describing the solving strategy. The design matrices $\mathbf{X}_1$, $\mathbf{X}_2$, $\mathbf{Z}_i$, i=1, 2, ..., r and $\mathbf{Z}_a$ are known, and without loss of generality assumed to have full column rank. The design matrices are structured. Assume records are ordered by traits and the ith random factor only affects each trait once. Then $\mathbf{Z}_i$ is block diagonal containing p_i blocks, where p_i is the number of traits in the ith random factor. This corresponds to a partition of $\mathbf{u}_i$ as $\mathbf{u}_i' = [\mathbf{u}_{i1}'; \mathbf{u}_{i2}'; \cdots ; \mathbf{u}_{ip_i}']$, where $\mathbf{u}_{ij}$ is the effect of the ith random factor on the jth trait. Under the same assumptions, $\mathbf{Z}_a$ is also block diagonal with p_a diagonal blocks, and $\mathbf{a}$ is partitioned as $\mathbf{a}' = [\mathbf{a}_1'; \mathbf{a}_2'; \cdots ; \mathbf{a}_{p_a}']$. The assumptions on variance-covariance components are:

$$\begin{array}{lll}
E[\mathbf{u}_i] = 0 \quad (\forall\, i), & E[\mathbf{a}] = 0, & E[\mathbf{e}] = 0, \\
Var[\mathbf{u}_i] = \mathbf{G}_i = \mathbf{G}_{0_i} \otimes \mathbf{I}_i \ (\forall\, i), & Var[\mathbf{a}] = \mathbf{G}_a = \mathbf{G}_{0_a} \otimes \mathbf{A}, & \\
Var[\mathbf{e}] = \mathbf{R} = \mathbf{R}_0 \otimes \mathbf{I}, & & \\
Cov[\mathbf{u}_i\ ,\ \mathbf{u}_j'] = 0\ (i \neq j), & Cov[\mathbf{a}\ ,\ \mathbf{u}_i'] = 0 (\forall\, i), & \\
Cov[\mathbf{u}_i, \mathbf{e}'] = 0\ (\forall\, i), & Cov[\mathbf{a}, \mathbf{e}'] = 0 &
\end{array}$$

$\mathbf{G}_{0_i}$ is a $p_i \times p_i$ matrix of variances and covariances among the p_i traits in the ith random factor. $\mathbf{G}_{0_a}$ is a $p_a \times p_a$ matrix of additive genetic variances and covariances for the p_a traits included in the animal factor. The dimension of $\mathbf{G}_{0_i}$ and $\mathbf{G}_{0_a}$ depends on the number of traits that are affected by each of the random factors and on whether several correlated random factors affect the same trait. $\mathbf{A}$ is the numerator relationship matrix among the animals. The residual (co)variance matrix $\mathbf{R}$ is a block diagonal matrix (assuming traits are ordered within animals). The diagonal blocks in $\mathbf{R}$ corresponding to each animal depend on which traits are measured. If all t traits are measured the block is $\mathbf{R_0}$, a $t \times t$ matrix. If some of the traits are missing, the corresponding rows and columns in $\mathbf{R_0}$ must be deleted in order to form the diagonal block.

3 Mixed model equations

$$\text{Let} \qquad \mathbf{G} = \sum_{i=1}^{r}{}^{+}\mathbf{G_i}, \qquad \mathbf{Z} = \sum_{i=1}^{r}{}^{+}\mathbf{Z_i}, \qquad \mathbf{u}' = \left[\mathbf{u_1}'; \mathbf{u_2}'; \cdots ; \mathbf{u_r}'\right]$$

where $\mathbf{\Sigma^+}$ is the direct sum operator. Then the MME for model (1) are:

$$\begin{bmatrix} \mathbf{X_1'R^{-1}X_1} & \mathbf{X_1'R^{-1}X_2} & \mathbf{X_1'R^{-1}Z} & \mathbf{X_1'R^{-1}Z_a} \\ \mathbf{X_2'R^{-1}X_1} & \mathbf{X_2'R^{-1}X_2} & \mathbf{X_2'R^{-1}Z} & \mathbf{X_2'R^{-1}Z_a} \\ \mathbf{Z'R^{-1}X_1} & \mathbf{Z'R^{-1}X_2} & \mathbf{Z'R^{-1}Z{+}G^{-1}} & \mathbf{Z'R^{-1}Z_a} \\ \mathbf{Z_a'R^{-1}X_1} & \mathbf{Z_a'R^{-1}X_2} & \mathbf{Z_a'R^{-1}Z} & \mathbf{Z_a'R^{-1}Z_a{+}G_a^{-1}} \end{bmatrix} \begin{bmatrix} \boldsymbol{\beta}_1 \\ \boldsymbol{\beta}_2 \\ \mathbf{u} \\ \mathbf{a} \end{bmatrix} = \begin{bmatrix} \mathbf{X_1'R^{-1}y} \\ \mathbf{X_2'R^{-1}y} \\ \mathbf{Z'R^{-1}y} \\ \mathbf{Z_a'R^{-1}y} \end{bmatrix} \qquad (2)$$

Notice the overall block structure of the coefficient matrix. The diagonal blocks $\mathbf{X_1'R^{-1}X_1}$ and $\mathbf{Z'R^{-1}Z{+}G^{-1}}$ are block diagonal on a lower level, with squared blocks of dimensions equal to the number of traits that are affected by the respective fixed or random factors. The matrix $\mathbf{G^{-1}}$ is easy to compute due to its simple structure. The diagonal block $\mathbf{X_2'R^{-1}X_2}$ for the remaining fixed factors ($\boldsymbol{\beta}_2$) is relatively dense. The diagonal block for the animal factor $\mathbf{Z_a'R^{-1}Z_a{+}G_a^{-1}}$ is sparse, but has a complex structure coming from $\mathbf{G_a^{-1}}$. The matrix $\mathbf{G_a^{-1}} = \mathbf{G_{0_a}^{-1}} \otimes \mathbf{A^{-1}}$ is easy to calculate, because $\mathbf{G_{0_a}}$ is only of dimension p_a, and $\mathbf{A^{-1}}$ can be formed from pedigree information by simple rules due to Henderson [2]. The non-diagonal blocks shown in (2) are sparse, but can in principle contain nonzeros anywhere.

4 Solving strategies

Write (2) as $\mathbf{Cx} = \mathbf{b}$, and decompose the matrix $\mathbf{C}$ as $\mathbf{C} = \mathbf{L} + \mathbf{D} + \mathbf{U}$ where $\mathbf{L}$ is lower and $\mathbf{U}$ is upper triangular and $\mathbf{D}$ is (block) diagonal. The first order Jacobi method is:

$$\mathbf{Dx^{(k+1)}} = -(\mathbf{L} + \mathbf{U})\mathbf{x^{(k)}} + \mathbf{b} = \mathbf{crhs} \Rightarrow \mathbf{x^{(k+1)}} = \mathbf{D^{-1}crhs} \qquad (3)$$

where $\mathbf{x^{(k+1)}}$ denotes the $(k+1)^{th}$ iterate to the solution, and where we introduce the symbol **crhs** for the corrected right hand side. The first order Gauss-Seidel (GS) method is:

$$\mathbf{x^{(k+1)}} = -\mathbf{D^{-1}}(\mathbf{Lx^{(k+1)}} + \mathbf{Ux^{(k)}}) + \mathbf{D^{-1}b} = \mathbf{D^{-1}crhs} \qquad (4)$$

In multi-trait models, (3) and (4) are used as block solvers, because $\mathbf{D}$ is block diagonal with blocks of size equal to number of traits in the corresponding factor.

Use of the first order Jacobi formula (3) does not always lead to convergence. This problem is solved by means of a second order Jacobi method [4], i.e. by updating $x_i^{(k+1)}$ like this (h is a predetermined number):

$$x_i^{(k+1)} = x_i^{(k+1)} + h(x_i^{(k)} - x_i^{(k-1)}) \quad (5)$$

"Iteration on data" goes as follows: The first fixed factor ($\boldsymbol{\beta}_1$) is chosen as the one containing the largest number of levels. In a preprocessing step, the data records are sorted on the level codes of the first fixed factor, and equation numbers are assigned to all factors affecting the records. The relatively dense diagonal block $\mathbf{X}_2^{'}\mathbf{R}^{-1}\mathbf{X}_2$, is in practical applications of limited size. A LU-factorization of this block is stored. Also, the small diagonal blocks contained in $\mathbf{Z}^{'}\mathbf{R}^{-1}\mathbf{Z}+\mathbf{G}^{-1}$ and $\mathbf{Z}_a^{'}\mathbf{R}^{-1}\mathbf{Z}_a+\mathbf{G}_a^{-1}$ are computed, inverted and stored.

During each round of iteration, the sorted dataset is read sequentially. The group of records for one level code for the first fixed factor, is processed together. These records generate one equation in a single trait and t equations in a multi-trait application for the first fixed factor. A new iterate for this equation (t equations) is calculated using the GS formula (4). The same group of records also contributes to matrix elements for the other factors. The products of these contributions and the appropriate components of either $\mathbf{x}^{(k+1)}$ (if available) or $\mathbf{x}^{(k)}$ are absorbed into **crhs**.

When all data are processed, a new iterate for the complete $\boldsymbol{\beta}_1$ vector has been produced, and we are left with the equations for $\boldsymbol{\beta}_2$, **u** and **a** on block diagonal form. New iterates for $\boldsymbol{\beta}_2$ are obtained by backsolving based on the LU-factorization. New iterates for **u** and **a** are calculated using (3) and (5). The calculation by the Jacobi formula (3) is a multiplication of appropriate components of **crhs** from the left with the small inverse diagonal blocks. In a single trait application this is simply a division with a diagonal element.

5 Parallelization

The parallelization of "iteration on data" makes use of a master-slave concept. The master processor distributes disjoint sets of records to the slave processors. Records contributing to the same equation (block of equations in multi-trait) from the first fixed factor, are never split between two slave processors.

In each round of iteration, each slave reads its part of data sequentially. The slave takes exactly the same actions on these records as described for the serial case, i.e. it forms local contributions to **crhs** by formula (4), stored in a local copy of **crhs** on the slave. Only local nonzero components are formed. In parallel to this, the master processor reads the pedigree file, and calculates the contributions to **crhs** due to animal relationships.

A global reduce operation updates the **crhs** on the master. The master copy of **crhs** now contains the global value of **crhs** for all equations in the system. Only nonzero additive contributions are sent from the slaves. The total amount of communication is therefore much smaller than in a standard recursive doubling.

The master sends the global components of **crhs** to those slaves which need them for their Jacobi iteration. The slaves update their disjoint parts of the solution vector for random and animal factors with the new iterate. At the same time, the master backsolves for a new iterate of the solutions to the equations for the remaining fixed factors. The solution components are then exchanged between the processors, and a test for convergence is made (in 2-norm). In case the test is not met, a new round of iteration is performed.

The implementation is made in Fortran 90 and PVM. Tests are made on an IBM SP. The parallel speedup is evaluated in test runs on a single-trait model for protein yield for Red Danish Cattle. The dimension of the equation system is $1.2 \cdot 10^6$. The results (taken from a single series of program runs) in terms of relative speedup is shown in fig.1. A change from 1 to 2 processors causes a small slowdown, since the equations are basically solved on the slaves. From 2 to 6 processors, the speedup is near to linear, and it continues more slowly at least up to 11 processors. Beyond 11 processors the master becomes a bottleneck, due to its pedigree processing and increased communication with the slaves.

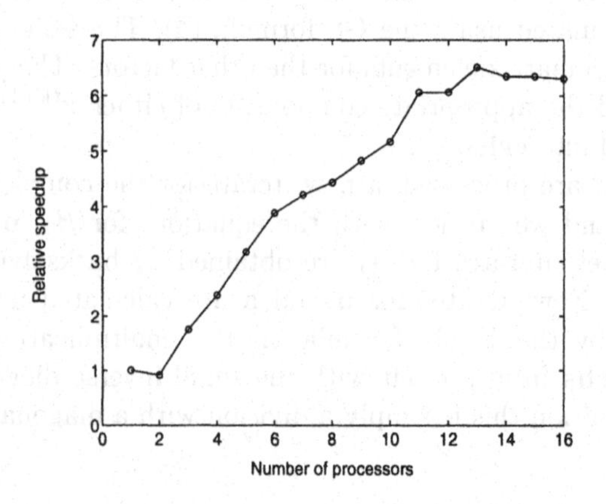

Fig. 1. Speedup in solving the $1.2 \cdot 10^6$ equation system for Red Danish Cattle

References

1. Henderson C.R., 1973. *Sire evaluation and genetic trends.* In Proc. of the Animal Breeding and Genetics Symposium in Honor of Dr. D.L.Lush. ASAS and ADSA, Champaign, Ill.
2. Henderson C.R., 1976. *A simple method for computing the inverse of a numerator relationship matrix used in prediction of breeding values.* Biom. 32:69-83
3. Schaeffer L.R. and Kennedy B.W., 1986. *Computing Solutions to Mixed Model Equations,* Proc. 3.rd. World Congr. on Genetics Applied to Livest. Prod. 12, 382-393. Nebraska.
4. Young D.M., 1971. *Iterative Solution of Large Linear Systems,* Academic Press, chapter 16

Scheduling of a Parallel Workload: Implementation and Use of the Argonne Easy Scheduler at PDC

Lars Malinowsky[1] and Per Öster[1]

Parallelldatorcentrum (PDC),
Royal Institute of Technology, S-100 44 Stockholm, Sweden.
{lama, per}@pdc.kth.se

Abstract. Scheduling of a workload of parallel jobs on a large parallel computer puts specific demands on a scheduling system. Especially if the objective is to give a good service for the largest parallel jobs. In this paper we give an overview of the implementation and extensions of the Argonne EASY scheduler at PDC. We describe necessary developments of EASY to adopt to an inhomogeneous platform, improvements of the scheduling algorithm with e.g.,*recursive backfill*, and added features such as *time-lock*, i.e. reservation and *mixed job resources*.

1 Introduction

To serve the most demanding computational tasks traditional scheduling with variable partitioning is a contradicition. Each job receives a partition of the available processing elements, PEs. At a national computational resource-center, such as PDC, this leads to a fragmentation that prevents large parallel jobs from getting access to the PEs, so called starvation. Hence, less service for the imagined user of a national resource.

After investigating several schedulers we found that our objectives were best met by the Argonne EASY Scheduler[1]. It had fairness, simplicity, and effiency as primary design goals which was well in line of our objectives. Since 1994, when we first started to use EASY, we have incrementally modified the code leading to the version currently in use at PDC.

The original EASY was implemented for a homogenous system. With a system, growing organically such as the the IBM SP, new demands has curraged us to develop new features. The additions and extensions could be categorized as improvements on scheduleing algorithm and supplementary features.

2 Scheduling

The original ANL EASY had following design objectives: fairness, simplicity and effiecent use. It was pointed out and worth to repeat again that these goals are in conflict. Achieving effeicent use will cause starvation. Large jobs, spanning many

PEs, requires many unused PEs ahead of their start which will result in large portions of the machine being unused. In a utilisation perspective it is better to have a steady allocation of small jobs as those easier fit into the machine. Hence, efficient use is in conflict with allocation of large jobs.

The original ANL EASY solution was to serve the jobs on a First In First Out, FIFO, basis. By also requiring the information about the expected run–time, wall–time, of a job when submitted, the scheduler could allow shorter or smaller jobs to leapfrog larger and longer jobs ahead in the FIFO. This is called an optimized FIFO, or back–fill[1].

A job is terminated when it has run for the requested run–time.

The original ANL EASY made certain that optimized job–allocations did not delay the *first* job in the FIFO. We call this a one–step back–fill.

At PDC we developed a version of EASY that moves jobs ahead to allocation only if they do not delay, the predicted start, of *any* other job ahead in the FIFO. This is what we denote recursive back–fill.

Having knowledge of the time of termination of all running jobs and also the requested run–time of all jobs in the FIFO makes it possible to predict the start–time of each job waiting in the FIFO.

The decision of what job to allocate is based on the jobs currently running in the machine and the jobs currently waiting in the FIFO. There is, intentionally, no tracking of earlier predictions. Keeping track of earlier predictions have been suggested by in an algorithm denoted conservative back–fill[3]. This algorithm is less aggressive, optimization–wise, but has the advantage of not delaying the predicted job start.

2.1 Back–Fill

As seen in figure 1 the job is represented by a space that is spanned by the number of processing elements, PEs, and the requested wall–time. One job is waiting in line. For simplicity all jobs in the figures consist of a single space. It is of course possible for a job to consist of several separate spaces, being discontinuous in the PE dimension.

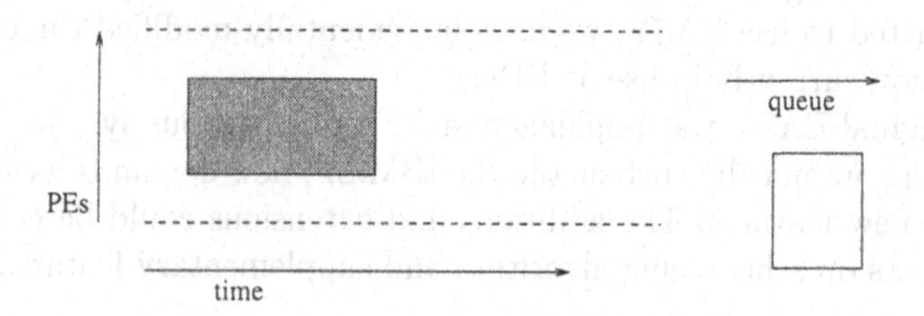

Fig. 1. one runnning and one queued job.

A typical case of back–fill is seen in figure 2. The second job waiting in line could start earlier than the first job without delaying the start of the first job.

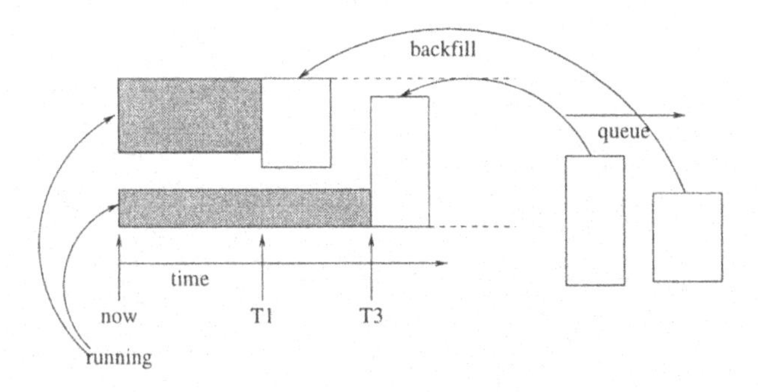

Fig. 2. back–fill.

The second job in figure 3 will not be optimized since that would delay the job ahead of it in line.

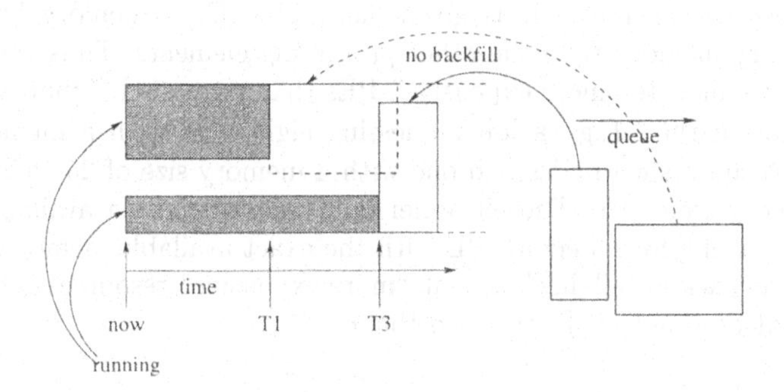

Fig. 3. will not back–fill.

Another scenario is seen by comparing figure 2 and figure 4. The first job in the FIFO is predicted a start–time of T3. The second job is in figure 2 given T1 as a predicted start–time. Since one of the running jobs terminates earlier than expected, figure 4, the estimate in figure 2 will not hold. The FIFO order is always kept and the second job will have to wait until T2, later than T1, as new predicted start–time. This is an example of early termination.

Early termination raises the question of how to keep the users waiting in line happy. Is it better to keep a predicted start–time or should one always choose to start the job that has waited the longest first? That is a policy decision.

2.2 Fall–Back

Given a mixed set of resources, in the sense of primary memory, CPU/memory bandwidth, or CPU capacity there are plenty of choices to be made of out of

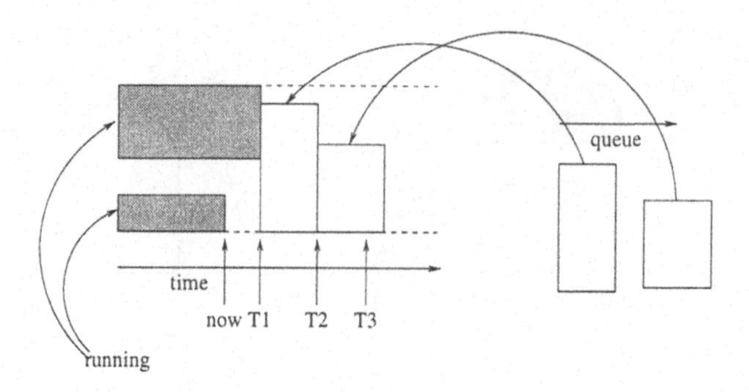

Fig. 4. early termination.

how to schedule them in an effiecent way. The original ANL SP only had one kind of resource while the SP at PDC had PEs with various amount of memory and CPU/memory bandwidth.

It is obvious that a well balanced parallel job does not suffer if one or more of the processing elements it is given has higher CPU/memory bandwith or more primary memory than the other processing elements. Thus we added the possibility to allocate "more expensive" PEs than requested if that would make the job start earlier. E.g., a job requesting eight PEs with a memory size of 128 can get seven such nodes and one with a memory size of 256 and twice the CPU/memory bandwith. The scheduler falls back to another available resource instead of waiting for an eighth PE with the exact available memory.

One drawback of fall–back is that "more expensive" resources gets occupied by jobs that does not explicitly need them.

3 Supplementary Features

3.1 Interactive PEs

The interactive demands of users when developing and debugging code are in contrast to those of batch execution. Also the number of interactive users vary, e.g. there are classes on regular intervals, and research groups tend to be very active during periods of time.

At PDC we did add the concept of interactive PEs. Staff can change the number of interactive PEs as the load changes. Users get access to the interactive PEs on a shared basis with an execution environment similar to the environment a batch job runs in.

3.2 Mixed Job Resources

Certain programming paradigms encourage the use of master/slave PEs. This implies that the scheduler must resolve requests for mixed resources, e.g. a more powerful PE for the master process.

The scheduler accepts a generalized mixed resource request, returning a list of PEs ordered the way the user requests, i.e. a PE having more memory and/or larger CPU/memory–bandwith at the choosen position in a list or any choice of PE/PEs in any order.

3.3 PE Specific Allocation Policy

The variety of demands of users, in terms of minimum consequtive allocation with respect to time, minimum amount of primary memory, and minimum turn–around–time puts restraints on, i.e., the maximum allowed duration of a job.

The original ANL EASY classified jobs as either of day, night or weekend jobs, depending on their requested wall–time. Typically day–jobs ran between 0800 hours and 1800 hours, night–jobs during weekday nights and weekend–jobs during weekends. Allowing for a maximum run–time, wall–clock, of sixty hours during the week–ends.

At PDC we adopted this classification, finding it a good compromise between allowing quite long jobs while still reserving the day–time for shorter jobs, enhancing the turn–around–time during the normal work hours. But we also saw the need for an even more flexible setup and introduced the possibility to tune the allocation policy individually for each PE.

3.4 Absolute and Relative Time–Lock

Several applications are able to checkpoint for a later restart, i.e. store a partial result/state which contains all necessary information for a later continuation. The possibility to chain job–steps in a particular order is necessary or a user must manually submit a new job–step after the completion of the current one. One solution, in a back–filling capable scheduler, is to time–lock job–step $n + 1$ to a known event, i.e. the completion of job–step n. We denote this relative time–lock. The PDC scheduler is also cabable of locking to an absolute event, e.g. time–of–day, which we denote absolute time–lock or reservation.

The absolute time–lock enhances the possibility to achieve successful co–allocation with other schedulers on other sites. We think this feature might become more valuable, see for example the Globus project[4].

3.5 W^3 Graphical Presentation

To give an easy accessible interface for users and system staff we have developed a W^3 application presenting the most necessary information of system and job status.

A graphical interface, a Java applet, has been developed to aid users in utilization of the system. The applet presents running and queued jobs as areas in the time v. PE space. The user can draw own areas or do virtual submissions to get an estimate of when a specified job would start or find out the location and size of future free areas in the time v. PE space.

3.6 Stability

Part from that one expects a scheduler to be robust it could also prevent failures by extensive checking of PE functionality before and after allocating jobs. The amount of faults in the scheduler currently in use are negligible, in the order one fault per year, and malfunctioning PEs are automatically taken out of the allocatable pool. If a PE is unable to access various file–systems, or other important servers, it will also automatically be taken out of the allocatable pool.

3.7 Security

Using Kerberos and the Andrew File System, AFS, posed certain challenges when implementing the EASY scheduler at PDC. Challenges that, when solved, turned into opportunities such as the possibility to schedule resources residing at different sites.

Having two sites, both running their private Kerberos–realm, and if the two of them either use DFS or AFS, it is only a matter of exchanging Kerberos inter–realm–tickets between the kerberos–realms to enable a single user, having an account on each site, running a single job spanning the two sites. If the two sites trust each other one of the sites could schedule jobs on both of the sites as long as one trusts the Kerberos authentication system. This has been succesfully done between the SPs at HPC2N and at PDC.

References

1. Lifka, D.A.: The ANL/IBM SP scheduling system. Job Scheduling Strategies for Parallel Processing. IPPS'95 Workshop Proceedings. Springer-Verlag, Berlin (1995) 295-303
2. Skovira, J., Waiman-Chan, Honbo-Zhou, Lifka, D.: The EASY-LoadLeveler API project. Job Scheduling Strategies for Parallel Processing. IPPS '96 Workshop Proceedings. Springer-Verlag, Berlin (1996) 41-47
3. Feitelson, D., G., Weil, A., M.: Utilization and Predictability in Scheduling the IBM SP2 with Backfilling. Preprint. Institute of Computer Science, The Hebrew University of Jerusalem, 91904 Jerusalem, Israel (1997)
4. Foster, I., Kesselman, C.: Globus: A Metacomputer Infrastructure Toolkit. Intl J. Supercomputer Applications 11(2) (1997) 115-128

An Algorithm to Evaluate Spectral Densities of High-Dimensional Stationary Diffusion Stochastic Processes with Non-linear Coefficients: The General Scheme and Issues on Implementation with PVM

Yevgeny V. Mamontov and Magnus Willander

Laboratory of Physical Electronics and Photonics, Department of Microelectronics and Nanoscience, Chalmers University of Technology and Gothenburg University, S-412 96 Gothenburg, Sweden
{yem, mwi}@fy.chalmers.se

Abstract. The present work deals with spectral-density matrix of the stationary diffusion stochastic process with non-linear coefficients (i.e. drift vector and diffusion matrix) in the case as number of variables of the process is high, i.e. is much greater than a few units. This is an important topic in modern microelectronics and other applied fields associated with non-linear fluctuations in high-dimensional systems. A general scheme for the algorithm to evaluate the above matrix for realistic computing expenses is proposed. This algorithm applies the Monte Carlo method for multifold integrals. An implementation with the Parallel Virtual Machine (PVM) software environment is discussed. To reduce the corresponding message-passing time overheads, the initial-and-final-only-messages (IFOM) technique should be applied. It is shown that the IFOM approach enables to keep the parallelization efficiency at the level of 80-90% even if many tens (or more) processors are involved.

This work deals with diffusion stochastic processes (DSPs) with drift vector $g(x)$ and diffusion matrix $H(x)$ where $x \in \mathbf{R}^n$, $\mathbf{R}=(-\infty,\infty)$, $g \in C^1(\mathbf{R}^n)$, $H \in C^0(\mathbf{R}^n)$, and

$$n \text{ is much greater than a few units.} \tag{1}$$

In many cases, the DSPs with coefficients $g(x)$ and $H(x)$ are granted as solutions of Itô's stochastic differential eqaution (ISDE) system (e.g., [1, (9.3.1)]) $\mathrm{d}x=g(x)\mathrm{d}t+$ $+h(x)\mathrm{d}W(\xi,t)$ where matrix $h(x)$ is coupled with $H(x)$, $H(x)=h(x)[h(x)]^{\mathrm{T}}$, $\xi \in \Xi$ is elementary event, Ξ is the space of elementary events, and $W(\xi,t)$ is the standard Wiener stochastic process of the corresponding dimension. For example, our approach to modeling noise in semiconductor devices and circuits [2]–[6] provides derivation and calculation of drift vector $g(x)$ and diffusion matrix $H(x)$.

The present work assumes that:

- there exists transition probability density $\rho(x,\Delta,y)$ corresponding to coefficients g and H, where $y \in \mathbf{R}^n$, and $\Delta \geq 0$ is the time separation,
- there exists such number $\varepsilon>0$ that $\lim_{\Delta \downarrow 0} (1/\Delta)\int_{\mathbf{R}^n} \|y-x\|^{2+\varepsilon}\rho(x,\Delta,y)dy=0$ for all x,
- and there exists the unique stationary invariant probability density ρ_{inv} corresponding to ρ.

This in particular means that (e.g., [1, (2.5.2), (9.2.14)])

$$\lim_{\Delta\downarrow 0} (1/\Delta)\int_{\mathbf{R}^n}(y-x)\rho(x,\Delta,y)dy=g(x), \quad \lim_{\Delta\downarrow 0} (1/\Delta)\int_{\mathbf{R}^n}(y-x)(y-x)^{\mathrm{T}}\rho(x,\Delta,y)dy=H(x),$$

$$\rho_{inv}(y)=\int_{\mathbf{R}^n}\rho_{inv}(x)\rho(x,\Delta,y)dx, \qquad \text{for all } \Delta>0,\ y.$$

Density ρ_{inv} determines the stationary DSP $\chi:\Xi\times\mathbf{R}\to\mathbf{R}^n$ with drift vector $g(x)$ and diffusion matrix $H(x)$.

In various applied fields, one of the most important characteristics of process χ is its spectral density (e.g., [1, p.24], [7, (86) on p. 76])

$$S(f)=2\int_{-\infty}^{\infty}C(\Delta)\exp(-\iota 2\pi f\Delta)d\Delta, \qquad \text{for all } f\in\mathbf{R}, \tag{2}$$

where f is the frequency, $C(\Delta)$ is covariance matrix of the process (e.g., [1, p. 10]),

$$C(0)=V, \tag{3}$$

$$C(\Delta)=\int_{\mathbf{R}^n\times\mathbf{R}^n}(x-e)(y-e)^{\mathrm{T}}\rho_{inv}(x)\rho(x,\Delta,y)dxdy, \qquad \text{for all } \Delta>0, \tag{4}$$

$$C(\Delta)=[C(-\Delta)]^{\mathrm{T}}, \qquad \text{for all } \Delta<0, \tag{5}$$

e and V are the expectation and variance of χ. Expression (2) can be applied only if (e.g., [1, p. 24], [7, Remark on p. 76]) function C is absolutely integrable over $\mathbf{R}$.

Both densities ρ and ρ_{inv} are associated with the Kolmogorov forward (or Fokker-Planck) equation corresponding to the DSPs with coefficients $g(x)$ and $H(x)$: ρ is its fundumental solution (e.g., [7, pp. 121-122] or [8]) and ρ_{inv} is the proper solution of its stationary version (e.g., [8]). For the n-dimensional DSPs, the above equation is a non-stationary partial differential equation in n-dimensional Euclidean space $\mathbf{R}^n$. However, number n is high (see (1)). Hence, application of the Kolmogorov-equation description of ρ and ρ_{inv} to (4) presents a task of an abnormal complexity which leads to non-realistic computing expenses (see also [8]). Thus, key expression (4) to calculate $C(\Delta)$ in (2) becomes impractical and hence can not be applied as it is.

A potential solution of this problem is proposed in our work [8] and used below. One of the practical advantages of this solution is that it enables not only to allow for condition (1) but also has the feature that the main computer memory necessary for analysis of $S(f)$ under condition (1) is limited with the amount proportional to n^2.

The *purpose* of the present work is to propose a general algorithmic scheme for the above approach and to analyze efficiency of its implementation with Parallel Virtual Machine (PVM) [9]–[12]. In so doing, we concentrate on real part

$$Z(f)=\mathrm{Re}S(f)=2\int_{0}^{\infty}\{[C(\Delta)]^{\mathrm{T}}+C(\Delta)]\}\cos(2\pi f\Delta)\}d\Delta, \qquad \text{for all } f, \tag{6}$$

of spectral density $S(f)$ since (e.g., [8]) the real part is usually of a more practical importance than the imaginary one.

The main result of work [8] is that it theoretically justifies replacement of (4) (used in (6)) with the following *dterministic-transistion* (DT) approximation

$$C_{dt}(\Delta)=\int_{\mathbf{R}^n}(x-e)(\phi(x,\Delta)-e)^{\mathrm{T}}\rho_{sdb}(x)dx, \qquad \text{for all } \Delta\geq 0,$$

where $\phi(x,\Delta)$ is the unique solution of initial-value problem $dy/d\Delta=g(y)$, $y|_{\Delta=0}=x$, and

$$\rho_{sdb}(x)=\exp(U(x)-w), \qquad \text{for all } x, \qquad (7)$$

$$\exp(w)=\int_{\mathbf{R}^n}\exp(U(x))dx,$$

$$U(x)=-(1/2)(x-e)^T\{[H(e)]^{-1}D_*(x)+[D_*(x)]^T[H(e)]^{-1}\}(x-e), \qquad \text{for all } x,$$

matrix $H(e)$ is assumed to be non-singular, $D_*(x)=-2\int_0^1(1-\kappa)\{\partial g(e+\kappa(x-e))/\partial u\}d\kappa$ for all x, u stands for variable $e+\kappa(x-e)$ of function g. Note that the SDB (*simplified detailed-balance*) approximation (7) turns into exact represenation for ρ_{inv} if diffusion function H is independent of x and drift g is a gradient mapping (e.g., see [13, (4.1.5), (4.1.6)] on notion of gradient mapping). Then (6) is reduced to

$$Z(f)=\int_{\Pi}\{\Phi(x,f)(x-e)^{\mathrm{T}}+(x-e)[\Phi(x,f)]^{\mathrm{T}}\}\rho_{sdb}(x)dx, \qquad \text{for all } f, \qquad (8)$$

where sufficiently large rectangular parallelepiped

$$\Pi=[u_1,v_1]\times[u_2,v_2]\times\ldots\times[u_{n-1},v_{n-1}]\times[u_n,v_n], \qquad (9)$$

($u_i<w_i$, $i=1,\ldots,n$) replaces $\mathbf{R}^n$ and $\Phi(x,f)=2\int_0^\infty(\phi(x,\Delta)-e)\cos(2\pi f\Delta)d\Delta$ for all x, f. In practice, it is usually sufficient to evaluate matrix (8) only on a finite set of values of frequency f, say, at $f_1<f_2<\ldots<f_{n_f}$ where $n_f\geq 1$, typically,

$$n_f\approx 50. \qquad (10)$$

One can note that, if $\rho_{sdb}\equiv\rho_{inv}$, then $C_{dt}(\Delta)$ and $dC_{dt}(\Delta)/d\Delta$ coincide with $C(\Delta)$ and $dC(\Delta)/d\Delta$ respectively in the limit case as $\Delta\downarrow 0$. If, besides, drift function g is *linear*, then $C_{dt}(\Delta)$ coincides with $C(\Delta)$ for *all* Δ. In the latter case, expressions $C_{dt}(\Delta)=V\times \times\exp\{\Delta[\partial g(e)/\partial x]^{\mathrm{T}}\}$ for all $\Delta\geq 0$ and $C_{dt}(\Delta)=[C_{dt}(-\Delta)]^{\mathrm{T}}$ for all $\Delta\leq 0$ are valid.

The key advantage of representation (8) compared to (6) is that it does not involve densities ρ and ρ_{inv} by means of (4) and thereby does not necessitate numerical treatment of the above partial differential equation in n-dimensional Euclidean space $\mathbf{R}^n$ under condition (1). In so doing, evaluation of real part $Z(f)$ of spectral density (2) is reduced to calculation of the n-fold integral in (8) for every f_i, $i=1,\ldots,n_f$.

A reasonable way to calculate it is the Monte Carlo technique. According to this technique, the integral is interpreted as the expectation of the integrand with respect to the random variable uniformly distributed in parallelepiped (9). Since accuracy of the Monte Carlo procedure is proportional to $N^{-1/2}$ where N is the number of the trials, as high values as

$$N \approx 10^3 - 10^5 \quad (11)$$

are not uncommon (e.g., [14, p. s79]). In doing so, the trials are produced with the help of a generator of random numbers uniformly distributed on interval (0,1). Since each trial is n-vector (see (1) for n),

the total amount of the random numbers to be generated is nN. (12)

The RANLUXASM programme for MS-DOS, Windows 95, and Windows NT [15] (see also [16], [17] for discussion), advanced random-number generator, generates one number on the avearge for only 1.8 μsec even with the highest value of the so-called luxry-level switch and even in execution on a common PC with the 133-MHz Pentium processor under Windows 95 [15, Table 5]. Hence RANLUXASM can solve task (12) (see also (1) and (11)) for not more than one or two minutes even if n is on the order of a few hundreds. However, in view of (10) and high complexity of the integrand in (8), the total run-time is usually on the order of many hours.

In this situation, parallel computing is clearly a reasonable choice. Then the total run-time is presented as T_0+T where T_0 and T are the times of execution of the non-parallelizable part of the algorithm and sequential execution of its parallelizable part and

T is typically on the order of many hours. (13)

The most cost-effective way to provide parallel computing is to utilize standard office equipment intended for productivity tools that is the Pentium-based PCs under Windows 95. The recent results in PVM development [18]–[20] enables to do that. A few PVM versions like WPVM 2.0 [21], PVM-W95 [22], [23], or PVM 3.4 [24], [25] are available. The computers can be connected into a network with the common 10Base-T Ethernet cables.

The parallel evaluation of (8) can be implemented within the PVM "master–slave" model (also known as the "manager–worker" model [26, Section 8.2]) by means of the *initial-and-final-only messages* (IFOM) scheme. According to this scheme, calculation of each of the n_f values of the integrand in (8), symmetric $n \times n$-matrix, at the above n_f frequency points applies the same N trials, evaluation of these $n_f N$ matirces are equally distributed among $p \geq 1$ processors, and the "master" process:

- sends to each of the "slave" process exactly one, intital (the same for every "slave" process) message which consists of the data to start the simulation
- and receives from each of the "slave" process exactly one, final message which consists of the simulation results obtained by this process.

One can show that, under conditions (1) and (10) and in double precision, the length of the initial message is proportional to $4n^2$ bytes and the length of every final message is proportional to $4(n_f+1)n^2$ bytes or, totally, to $4(n_f+2)n^2$ bytes. If the computers (or hosts) are connected with the standard Ethernet network, every host has one processor, and the PVM procedures `pvm_psend` and `pvm_precv` (recommended in [27] to send and receive messages) are applied, then the total time T_1 to pass the one initial and one final messages is described as

$$T_1 \sim 1.6(n/100)^2 \text{ sec.} \tag{14}$$

This formula accounts the well-known experimental result [28, Section 3.1] that the PVM message-passing speed under Ethernet is 1.25 MB/sec ≈ 1.3 B/μsec. For p processors, the total message time T_p is p times greater than (14), i.e.

$$T_p = T_1 p. \tag{15}$$

Speed-up s and efficiency E of the parallelization with the above IFOM scheme are determined with Amdahl's law (e.g., [29, (6.1), (6.3)]) generalized to expressions

$$s=(T_0+T)/(T_0+T/p+T_p), \tag{16}$$

$$E=s/p. \tag{17}$$

In view of (15), expression (16) is equivalent to

$$s=(T_0+T)p/(T_1 p^2+T_0 p+T). \tag{18}$$

The following properties of this function (at $T_0 \geq 0$, $T>0$, $T_1>0$) can readily be proved.

- As p increases from 0 up to

$$p_M=(T/T_1)^{1/2}>0, \tag{19}$$

  quantity s monotonously increases from 0 up to $s_M = s|_{p=p_M} = (p_M^2+r)/(2p_M+r)$ where $r=T_0/T_1 \geq 0$.
- As p increases from p_M up to ∞, quantity s monotonously decreases from s_M down to 0. The reason for this decrease is time $T_1>0$ in (18) associated with the message passing (see (14)).
- Quantity s_M is the maximum value of s over $p \geq 0$.

It follows from these properties that, to improve spedd-up s, there are no any needs to make number p of the processors greater than p_M, i.e. relation

$$2 \leq p \leq p_M \tag{20}$$

should hold. In view of (19), (14), and (13), one obtains that typical values of p_M may be on the order of a few tens or greater. A highly remarkable feature of (19) is its independence of T_0.

In view of (18) and (19), equality (17) can be written as

$$E=(r+p_M^2)/(p^2+rp+p_M^2). \tag{21}$$

One can prove that (20) is equivalent to the property of quantity (21) to be a non-increasing function of parameter $r \geq 0$. This means that the efficiency achieves its maximum value at $r=0$. Inequality $r \ll 2p_M$ is typically valid. In this case, expression (21) is reduced to $p=[(1-E)/E]^{1/2} p_M$ and hence the following features hold:

- $s_M \approx p_M/2$ (e.g., 50 if $p_M=100$),
- if number α is such that $0<\alpha<1$ and α is not extremely close to 1, then $E \geq \alpha$ and $s \geq \alpha p$ for any $p \leq [(1-\alpha)/\alpha]^{1/2} p_M$; for example, $E \geq 0.8$ and $s \geq 0.8p$ at any $p \leq p_M/2$ (e.g.,

$p \leq 50$ if $p_M=100$), $E \geq 0.9$ and $s \geq 0.9p$ at any $p \leq p_M/3$ (e.g., $p \leq 33$ if $p_M=100$), $E \geq 0.98$ and $s \geq 0.98p$ at any $p \leq p_M/7$ (e.g., $p \leq 14$ if $p_M=100$).

These examples definitely demonstrate very high speed-up and parallelization efficiency which can be achieved with the IFOM scheme for the PVM implementation of the proposed algorithm.

Summing this work up, one can note the following issues.

The general algorithm to evaluate spectral-density matrices of high-dimensional stationary diffusion stochastic processes with non-linear coefficients is porposed. It is based on the theoretically derived approximations and can be implemented for realistic computing expenses. The algorithm does not confine itself to a specific application. It can be used in various and dissimilar fields. The IFOM parallelization scheme discussed in this work enables to keep the PVM speed-up and efficiency very high even if many tens of processors (hosts) are applied. The obtained results can be helpful to efficiently solve practical problems in microelectronics and other fields of science.

References

1. Arnold, L.: Stochastic Differential Equations: Theory and Applications. John Wiley & Sons, New York (1974)
2. Willander, M., Fu, Y., Karlsteen, M., Nur, O., Mamontov, Y. V., Patel, C., Olsson, H. K.: Silicon based nanoelectronics. In: Proceedings of the Second International Conference on Massively Parallel Computing Systems MPCS'96 (Ischia, Italy). IEEE Computer Society, Los Alamitos, CA, USA (1996) 271-277
3. Mamontov, Y. V., Willander, M.: On reduction of high-frequency noise in low-current silicon BTs. Physica Scripta **T69** (1997) 218-222
4. Mamontov, Y. V., Willander, M.: Long asymptotic correlation time for non-linear autonomous Itô's stochastic differential equation. Nonlinear Dynamics **12** (1997) 399-411
5. Mamontov, Y. V., Willander, M.: Model for thermal noise in semiconductor bipolar transistors at low-current operation as multi-dimensional diffusion stochastic process. IEICE Trans.Electronics **E80-C** (1997) 1025-1042
6. Mamontov, Y. V., Willander, M.: Application of ordinary/partial Itô's stochastic differential equations to modelling noise in semiconductor devices and circuits. In: Proceedings of the 8th annual Workshop on Circutis, Systems and Signal Processing ProRISC'97, IEEE Benelux Circuits and Systems Chapter (Mierlo, The Netherlands). STW Technology Foundation, Utrecht, The Netherlands (1997) 401-409
7. Soize, C.: The Fokker-Planck Equation for Stochastic Dynamical Systems and Its Explicit Steady-State Solutions. Worl Scientific, Singapore (1994)
8. Mamontov, Y. V., Willander, M., Lewin, T.: Modelling of high-dimensional diffusion stochastic pro-cess with non-linear coefficients for engineering applications–Part II: Approximations for covariance and spectral density of stationary process. Submitted.
9. Geist, A., Beguelin, A., Dongarra, J., Jiang, W., Manchek, R., Sunderam, V.: PVM 3 User's Guide and Reference Manual. ORNL/TM-12187. Oak Ridge National Laboratory, Oak Ridge, Tennessee, USA (1994)
10. Geist, A., Beguelin, A., Dongarra, J., Jiang, W., Manchek, R., Sunderam, V.: PVM: Parallel Virtual Machine. A Users' Guide and Tutorial for Networked Parallel Computing. The MIT Press, Cambridge, MA, USA (1994) (`http://www.netlib.org/pvm3/book/pvm-book.html`)
11. Pfister, G. F.: In Search for Clusters: The Comming Battle in Lowly Parallel Computing. Prentice Hall PTR, Upper Saddle River, NJ, USA (1995)
12. Geist, G. A., Kohl, J. A., Papadopoulos, P. M.: PVM and MPI: A comparison of features. Calculateurs Paralleles **8** (1996)

13. Ortega, J. M. and Rheinboldt, W. C.: Iterative Solution of Nonlinear Equations in Several Variables. Academic Press, New York and London (1970)
14. Bhavsar, V. C., Isaac, J. R.: Design and analysis of parallel Monte Carlo algorithms. SIAM J. Sci. Stat. Comput. **8** (1987) s73-s95
15. Hamilton, K. G.: Assembler RANLUX for PCs. Computer Physics Communications, **101** (3, 1997) 249-253
16. Lüscher, M.: A portable high-quality random number generator for lattice field theory simulations. Computer Physics Communications **79** (1, 1994) 100-110
17. Hamilton, K. G., James, F.: Acceleration of RANLUX. Computer Physics Communications **101** (3, 1997) 241-248
18. Alves, A., Silva, L., Carreira, J., Silva, J. G.: WPVM: Parallel computing for the people. In: Hertzber-ger, B., Serazzi, G. (eds.): High-Performance Computing and Networking. International Conference and Exhibition (Milan, Italy). Proceedings. Lecture Notes in Computer Science, Vol. 919. Springer-Verlag, Berlin (1995) 582-587 (ISBN 3-540-59393-4).
19. Alves, A.: Parallel computing – Windows style. BYTE **21** (5, 1996) 169-170
20. Dongarra, J., Fischer, M.: Another architecture: PVM on Windows 95/NT. In: Proceedings of the CCC'97 (Atlanta, Georgia).
21. WPVM Home Page: `http://dsg.dei.uc.pt/wpvm/`
22. Santana, M. J., Souza, P. S., Santana, R. C., Souzza, S. S.: Parallel virtual machine for Windows95. In: Parallel Virtual Machine - EuroPVM '96. Third European PVM Conference Proceedings. Springer-Verlag, Berlin (1996) 288-295
23. `http://www.icmsc.sc.usp.br/Grupos/LASD/pvmw95.html`
24. PVM Home Page: `http://www.epm.ornl.gov/pvm/`
25. `http://www.epm.ornl.gov/pvm/EuroPVM97/`
26. Pham, T. Q., Garg, P. K.: Multithreaded Programming with Windows NT. Prentice Hall PTR, Upper Saddle River, NJ, USA (1996)
27. `http://www.epm.ornl.gov/pvm/perf-graph.html`
28. Sunderam, V. S., Geist, G. A., Dongarra, J. J., Manchek, R. J.: The PVM concurrent computing system: Evolution, experience, and trends. Parallel Computing **20** (4, 1994) 531-546
29. Kuck, D. J.: High Performance Computing: Challenges for Future Systems. Oxford Univ. Press, New York, Oxford (1996)

High-Performance Simulation of Evolutionary Aspects of Epidemics*

William Maniatty[1], Boleslaw K. Szymanski[1], and Thomas Caraco[2]

[1] Department of Computer Science
Rensselaer Polytechnic Institute, Troy, NY 12180
http://www.cs.rpi.edu/~{maniattb, szymansk}
[2] Department of Biological Sciences
SUNY Albany, Albany, NY 12192
http://www.albany.edu/biology/caraco/tcaraco.htm

Abstract. Local interactions between individual organisms influence the population dynamics of species and impact their evolution. We describe high-performance simulation of evolutionary aspects of epidemics in spatially explicit, individual based models of multi-species habitat. Evolution consists of two processes, selection between genotypes and mutations producing novel genotypes. In this paper we focus on the effects of selection between genotypes in a model with a single host species and two competing pathogens with fixed (i.e. non-evolving) genotypes. We present the foundations of a model that represents two competing host species, a parasite serving as a disease vector, and a vector borne pathogen. The model is implemented as cellular automaton that tracks individual organisms to account for heterogeneity of the habitat. The implementation targets parallel distributed memory machines (including IBM SP-2 and a network of workstations) and NUMA shared memory architectures (SGI Origin 2000). We demonstrate also that this model yields qualitatively new biological results.

1 Introduction

Certain parasites evolve to impair their host's survival and reproduction only minimally. The primate lentiviruses offer an interesting example [16]. In older coevolutionary associations, such as simian immunodeficiency virus and African green monkeys, infection is essentially non-pathogenic, while in the recent association between HIV and humans, infection leads to serious disease. However, a long-standing coevolutionary association does not always imply reduced parasite virulence [7]. In fact, virulence exhibits a great deal of variation among host-parasite associations, and often varies temporally within a particular association [11, 9, 19].

* This work was supported in part by NSF Grant CCR-9527151. The content of this paper does not necessarily reflect the position or policy of the U.S. Government — no official endorsements should be inferred or implied.

Current theory for the evolution of virulence equates virulence with extra host mortality due to parasite infection, and than assumes that selection should increase a parasite's rate of reproduction. As a parasite exploits host resources at a greater rate, it increases the rate of transmission to new hosts. But the host's mortality rate increases as a consequence, decreasing the length of the period during which the parasite can be transmitted. Depending on the functional relationship between parasite transmission rate and virulence (i.e., the trade-off between transmission rate and infectious period), selection may favor low, intermediate or increasing virulence [15, 16, 10].

A number of recent models address variations of the trade- off just described. Important extensions include analyses of (i) interactions between virulence and host recovery rate ("clearance" by the immune system [1, 2]); (ii)competition between different parasite strains infecting the same host individual ("coinfection" [3, 8, 19]); (iii) competition between parasites when a more virulent strain excludes a less virulent strain infecting the same host ("superinfection" [11, 12, 16]); and and (iv) effects of mutation and relatedness among parasites infecting the same and different hosts [10].

This paper presents a series of computational models of increased complexity, with the goal of creating a final model encompassing analysis of all described above interactions (separately or simultaneously). We introduce two important elements into coevolutionary analysis of virulence. Ecologically, our model assumes an individual-based, spatially explicit basis for birth, death and infection processes [6]. Evolutionally, our model employs genetic algorithms to simulate mutation, recombination [13], and selection processes.

2 A Model of Selection between Competing Strains

To verify that the spatial effects are important for evolution in epidemics, we simulated competition between two microparasite strains. The simplest case of such competition arise when there are just two competing strains transmitted via direct contact and a host is infected by at most one strain at a time [11].

For simplicity, we assume the selection on virulence. The reward function for increased virulence is defined by the primary effect of increased pathogen reproductive rate (as is manifested by infecting nearby hosts). The penalty function for increased virulence (i.e., a constraint) is the consequence of the increased mortality rate for hosts in presence of a more virulent pathogen.

Let s_0, and s_1 denote two pathogen strains that differ in virulence. *Superinfection* is a disease preemption during which a host infected with s_0 becomes infected by strain s_1 via exposure from a nearby host. *Coinfection* occurs when multiple strains infect a host and order of infection matters. To denote coinfection when s_0 arrives first, we adopt the notation for the "pseudo-strain" s_{01}, but when s_1 arrives first then we introduce a pseudo-strain s_{10} [1]. Typically, coin-

[1] To model coinfective systems in which order of infection does not matter one can treat states with infection by either strain as equivalent thereby "merging" these two states into one.

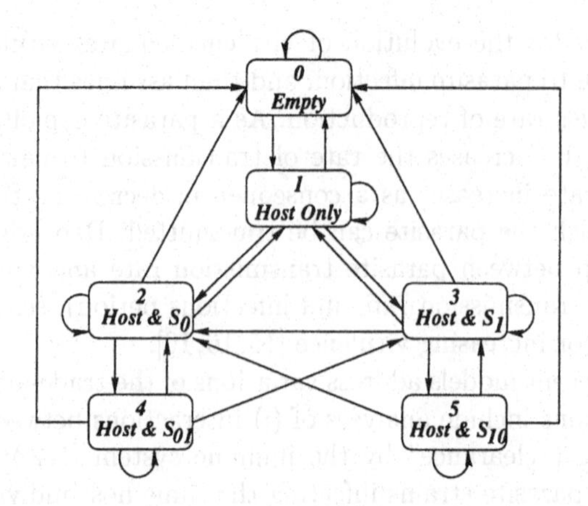

Fig. 1. State Diagram of Coinfection/Competition Model

fection and superinfection are modeled exclusively of each other, so our model provides a mechanism for disabling these events (although, as can be seen below, we permit the user to enable both as well). The state diagram of this simple model is shown in Figure 1.

The parameters for the system are given in Table 1. Let $p(x,y)$ be the probability of going from state x to state y in a single time step.

An empty site will remain unpopulated with probability $p(0,0) = \prod_{i=1}^{5}\left[(1-\rho_i)^{\sigma_i}\right]$, otherwise, it is colonized by an offspring from a nearby populated site. Since congenital immunity is disallowed: $p(0,1) = 1 - p(0,0)$.

Consider next a site occupied by a susceptible (i.e., an uninfected host). It may either die with probability $p(1,0) = \mu_1$ or be infected by a strain from their neighbors. The probability of a host in state 1 avoiding exposure to strain $s_i, i \in 0,1$ is denoted A_i with: $A_0 = (1-\alpha_0)^{\sigma_2}(1-\alpha_{01,0})^{\sigma_4}(1-\alpha_{10,0})^{\sigma_5}$ and $A_1 = (1-\alpha_1)^{\sigma_3}(1-\alpha_{01,1})^{\sigma_4}(1-\alpha_{10,1})^{\sigma_5}$ The Poisson single transition rule implies that if a susceptible is exposed to both strains, only one of them will be successful in infecting the host, so the probability of catching strain 0 is $p(1,2) = (1-\mu_1)(1-A_0)\left[A_1 + \gamma(1-A_1)\right]$ and the probability of catching strain 1 is then: $p(1,3) = (1-\mu_1)(1-A_1)\left[A_0 + (1-\gamma)(1-A_0)\right]$

A host infected with strain s_0 will die (i.e., be removed) with probability $p(2,0) = \mu_2$, and will recover (without immunity) with probability $p(2,1) = (1-\mu_2)\mu_{s_0}$. It will become coinfected with strain s_1 after being infected by $s_{0,1}$ with probability $p(2,4) = (1-\mu_2)(1-\mu_{s_0})(1-A_0)$ and will remain in the same state otherwise.

Finally, consider the case of a host infected with strain s_1. The probability that it dies is $p(3,0) = \mu_3$. It will recover with probability $p(3,1) = (1-\mu_3)\mu_{s_1}$, and superinfection by s_0 will occur with probability $p(3,2) = (1-\mu_3)(1-\mu_{s_1})p_{SI}(1-p_{01})(1-A_1)$. Coinfection by s_0 will occur with probability $p(3,5) = (1-\mu_3)(1-\mu_{s_1})(1-p_{SI})p_{01}(1-A_0)$.

Symbol	Meaning and values
$\alpha_i, i \in \{0,1\}$	Probability of exposure to s_i from an infective in state 2 or 3 within δ.
$A_i, i \in \{0,1\}$	The probability that a susceptible avoids exposure to s_i
$\alpha_{ij,k} i, j, k \in \{0,1\}, i \neq j$	Probability of exposure to s_i from a coinfected host infected by strain s_i then s_j
γ	The competitive advantage of s_0 over s_1 when susceptibles are exposed to both strains
δ_k	The interaction region about site k, $\delta_k \in \{3^2, 11^2, 33^2\}$
$\mu_i, i \in \{1,2,3\}$	Probability of host mortality, $\mu_1 = 0.5, \mu_3 = 0.25$ $\mu_2 \in \{\{0.251, 0.35, 0.45, 0.55, 0.65, 0.75, 0.85, 0.95\}$
$\mu_{s_i}, i \in \{0,1\}$	Probability of recovery from a single strain s_i
$\mu_{ij,k}, i, j, k \in \{0,1\}, i \neq j$	Probability of recovery of a coinfected host from s_k
p_{SI}	Post-Exposure Superinfection probability of s_0 $p_{SI} = 0.3$
p_{01}	The probability that a host infected with s_0 can be coinfected with s_{01} upon exposure to s_1
p_{10}	The probability that a host infected with s_1 can be coinfected with s_{10} upon exposure to s_0
$\rho_i, i \in 1,2,3,4$ a propagule nearby (in δ),	Probability of host in state i placing $(1-\rho_i)^\delta \in \{0.24, 0.2, 0.15, 0.11\}$
$\sigma_i(k), i \in \{0,1,2,3\}$	Number of hosts in state i about site k

Table 1. Symbols Used in Direct Transmission Two Competing Strains Model

Consider a coinfected host, where s_0 infected the host first. The host can die with probability $p(4,0) = \mu_4$, or it can recover from strain s_0 with probability $p(4,3) = (1-\mu_4)(1-\mu_{01,1})\mu_{01,0}$. The host can recover from strain s_1 with probability $p(4,2) = (1-\mu_4)(1-\mu_{01,0})\mu_{01,1}$, otherwise, the host will remain in its current state [2].

Consider the complementary coinfected host, where s_1 infected the host first. The host can die with probability $p(5,0) = \mu_5$. It can recover from strain s_0 with probability $p(5,3) = (1-\mu_5)(1-\mu_{10,1})\mu_{10,0}$. The host could recover from strain s_1 with probability $p(5,2) = (1-\mu_5)(1-\mu_{10,0})\mu_{10,1}$ otherwise the host will remain in its current state [3].

Similar reasoning can be used to extend this model to vector-borne pathogenic strains.

A subset of the parameters presented in Table 1 was selected as a collection of control parameters and based on TEMPEST, we implemented a tool, called

[2] Due to the Poisson single transition rule, we treat the possibility of simultaneous recovery from both strains as an impossible event. However if this event were to be considered possible with non-negligible probability, the formulation might look like $p(4,1) = (1-\mu_4)\mu_{01,0}\mu_{01,1}$

[3] Again, due to the Poisson single transition rule, we treat the possibility of simultaneous recovery from both strains as an impossible event. However if this event were to be considered possible with non-negligible probability, the formulation might look like $p(5,1) = (1-\mu_5)\mu_{10,0}\mu_{10,1}$.

STORM, to simulate the simplified model. In the simulation runs, we varied the spatial parameters governing host fecundity, $\rho_i, i \in \{1,2,3\}$ and rate of disease exposure from local infectives $\alpha_j, j \in \{0,1\}$ so that the intensity of these processes would not be unduly impacted by variations in the area of the ecological stencil of each site.

The space was toroidally wrapped to avoid impact of boundaries on the simulation results. Initially 25% of the sites were populated by susceptible hosts in a spatially uniform density. Small clusters of infective hosts (0.25% of the sites) were placed in the initial environment, with maximum spatial separation to avoid premature extinctions induced by local superinfection events. The experiments presented use a region of 100×100 sites, however STORM has capacity for much larger simulations. STORM was run using the fixed spatial configuration for 2000 generations for each parameter combination.

The results of the experiment have shown a remarkable richness in the range of outcomes generated. First consider a case with two competing strains which are equal except for (i) a small difference in their virulence induced mortality and (ii) the possibility of superinfection by the more virulent strain, s_0. We observe that the the more virulent strain, s_0 can drive the less virulent strain, s_1 to extinction via superinfection as seen in Figure 2(a) and 2(b). The rise and decline of s_1 is governed by the rate at which the epidemic can spread through the environment, which is in turn correlated to the stencil's area, δ. With larger δ values, both s_0 and s_1 infection disperse more quickly so that the superinfection induced extinction occurs earlier.

Now consider a similar case, except that the virulence induced mortality difference is large. In this case, even the competitive advantage of superinfection is not sufficient to prevent extinction of s_0, as shown in Figure 2(e) and 2(f). The rise and decline of s_0 in this case reflects expansion of s_0 infection from its original small cluster and later a denial of access to susceptibles occurs due to the prevalence of s_1 infection.

Finally consider an intermediate level of virulence induced mortality, as seen in Figure 2(c) and 2(d). For a small stencil (e.g., the 3×3 case), the more virulent strain, s_0, drives the less virulent strain, s_1 extinct by gradually excluding it from susceptible hosts. For intermediate sized interaction neighborhoods (e.g., the 11×11 case), the virulent less strain, s_1, drives the more virulent strain, s_0 extinct because of rapid death of hosts infected by s_0. For large stencils (e.g., 33×33) s_0 and s_1 coexist, the first thanks to superinfection and the second because of increased access to susceptibles. This range of results can only be found in spatially explicit models and cannot be generated by spatially homogeneous models.

3 Implementation and Performance of STORM

The STORM model was implemented using C++ and MPI much like the TEMPEST model described in [14]. Prior performance analysis of TEMPEST [17,14] was done on a SIMD mesh architecture (a MasPar MP-1), a network of worksta-

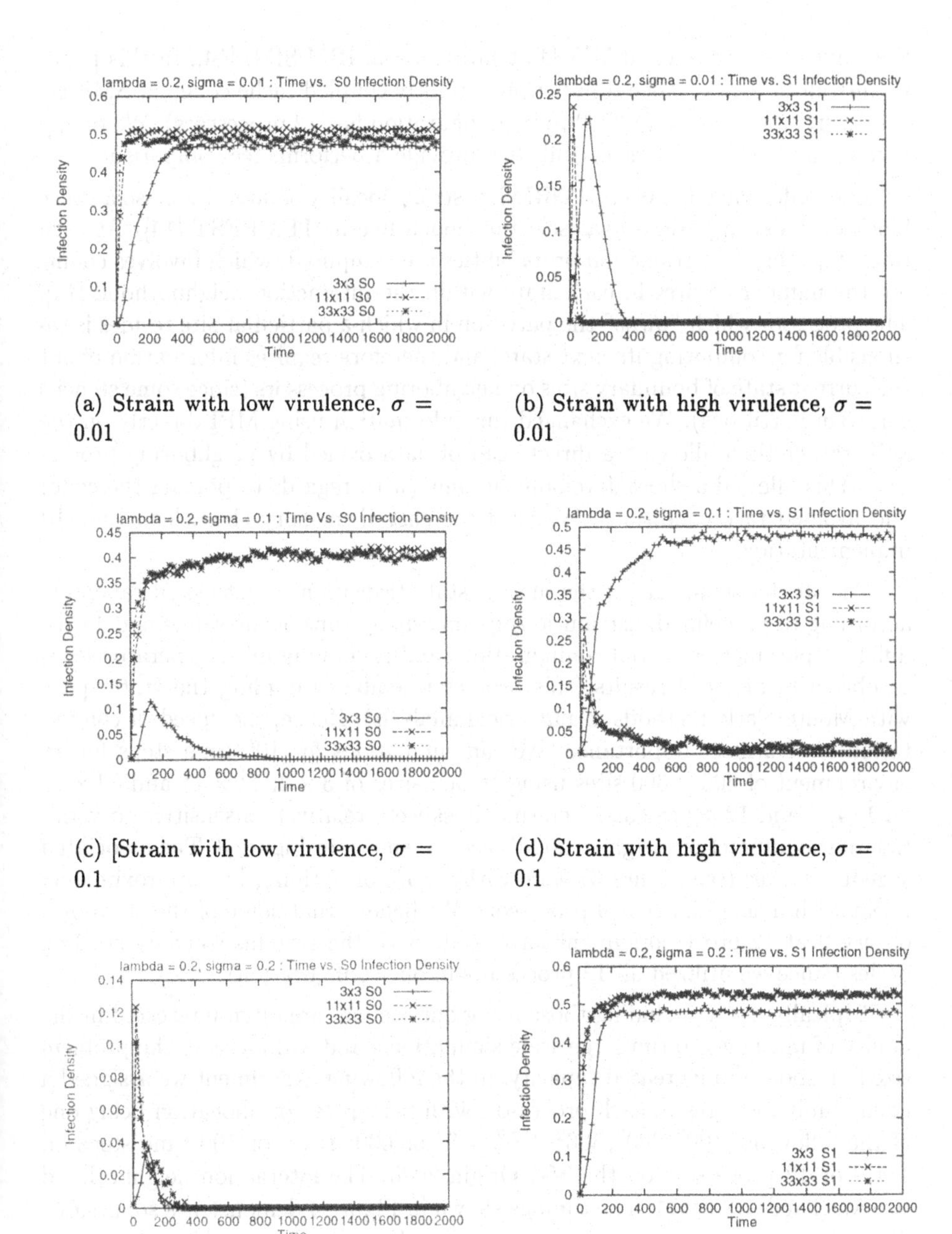

(a) Strain with low virulence, $\sigma = 0.01$

(b) Strain with high virulence, $\sigma = 0.01$

(c) [Strain with low virulence, $\sigma = 0.1$

(d) Strain with high virulence, $\sigma = 0.1$

(e) Strain with low virulence, $\sigma = 0.2$

(f) Strain with high virulence, $\sigma = 0.2$

Fig. 2. The impact of stencil area, δ, and relative virulence levels, $\sigma = \mu_1 - \mu_2$, on Infection densities

tions and a coarse grained MIMD architecture(an IBM SP2) [14]. In this paper we introduce performance results from a tightly coupled cache coherent SMP architecture, an SGI Origin 2000 (our configuration has 12 processors). We briefly review implementation issues and then provide a performance comparison.

The underlying model of STORM has strong locality of interaction, so a static block data decomposition was selected (much like in TEMPEST [14]) At each time step, the state transition probabilities are computed, which involves counting the number of sites in each state within the interaction neighborhood [14]. The processor which "owns" the partition in which a particular site resides is responsible for computing its next state, and therefore requires information about the current state of boundary sites on neighboring processors (since some stencils can span partitions). We exchanged this information using MPI directly on the SGI, rather than allowing a direct read of data owned by neighboring processors. This allowed a short development time (with regards to porting the code) and avoided race conditions. We used the shared memory chameleon (mpich) implementation.

The stochastic nature of the model (state transitions are selected randomly according to predefined distribution) require many runs of the same model with different parameters, initial configurations and random number generator seeds to obtain meaningful results (this process resembles sampling the state space with Monte Carlo methods in numerical analysis). Hence, the speed of computation is of utmost importance. We ran simulations for 100 time steps for an environment of 600×600 sites using stencil sizes of 3×3, 11×11 and 33×33 on 1, 4, 9 and 12 processors. The run times were relatively insensitive to variation of the interaction neighborhood's area as shown in Figure 3. The associated speedup curves (see Figure 3) were within 85% of optimal, but approximately linear with a small knee at 9 processors. We believe that some of the slowdown observed at 12 processors might have to do with the systems software stealing cycles (since we utilized all 12 processors in our configuration).

Typically users, including those using our system, are often interested in the ability of machines to run large scale simulations, and will increase the problem size in response to increased capacity. In the following experiment we assigned a fixed number of sites to each processor, with per processor allocation being one of the following: $100^2, 200^2, 300^2, 400^2, 500^2$ or 600^2 sites for 100 time steps on $1, 4, 9$ or 12 processors on the SGI Origin 2000. The interaction neighborhood size was fixed at 11×11. The timings shown in Figure 4 demonstrate some slowdown when going from single processor to multiprocessor runs (due to copying of boundary information) and another slowdown when all 12 processors were utilized. The slowdown when 12 processors were used was most pronounced for large per processor allocations. This might reflect increased contention of the cache coherent memory, and the overhead of systems software stealing resources to run during the simulations. The 600×600 problem size had a repeatable increase in run time when 4 processors were used over the 9 processor runs of about 10%. We suspect that this increase is due to contention of processors for memory access in the interconnection hub, but more research is required to check this hy-

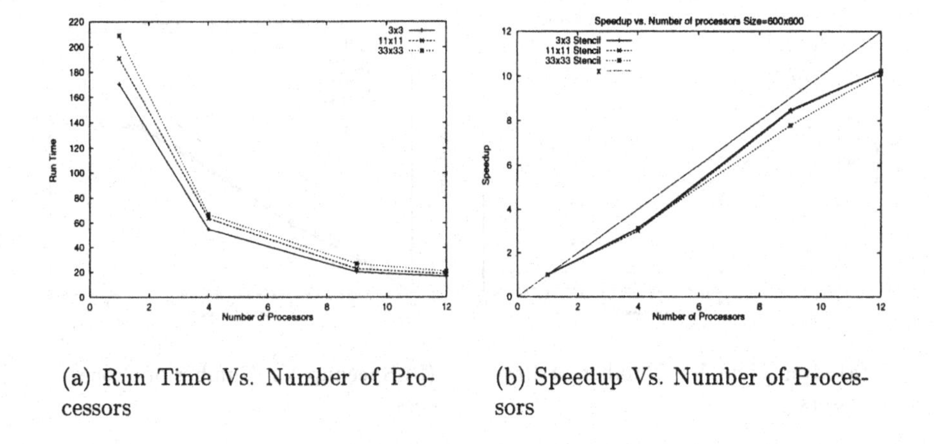

(a) Run Time Vs. Number of Processors

(b) Speedup Vs. Number of Processors

Fig. 3. Performance of Fixed Size Workload vs. Number of Processors on an SGI Origin 2000

pothesis. We noted that the scaled speedup curves looked remarkably similar to the shape of the speedup curves for 9 or less processors, but had a performance degradation when all 12 processors were used (achieving a speedup of about 8). For smaller problem sizes, the system finished within 85% of optimal speedup, but for larger problem sizes, the parallel efficiency was limited to 66%.

Measuring the per processor throughput of the SGI Origin 2000 (denoted T_{SGI}) for the largest problem size yields:

$$T_{SGI} = \frac{100\text{timesteps} \times 4.32 \times 10^6 \frac{\text{transitions}}{\text{timestep}}}{12\text{processors} \times 276.7\text{sec.}} \approx \frac{1.56 \times 10^6 \text{transitions}}{\text{sec.}} \tag{1}$$

Similar computations for TEMPEST presented in [14] showed that the per processor throughput of a MasPar MP-1 was $T_{MP-1} \approx 44 \frac{transitions}{sec.}$, while the per processor throughput of an SP2 was $T_{SP2} = 1.2 \times 10^5 \frac{transitions}{sec.}$. The departmental network of workstations achieved $T_{NOW} = 2.7 \times 10^4 \frac{transitions}{sec.}$. Assuming that the two simulations do roughly equivalent per site computation (which is reasonable), the SGI Origin 2000's processors can be thought of as capable of doing the work of over 35000 MasPar MP-1 processors, 13 SP2 processors and 57 NOW processors.

4 Conclusions and Future Directions

Evolution is the result of the combined effects of selection and mutation. In this paper we presented a model that analyzes only one these processes. By isolating and modeling the impact of selection on simple non-evolving systems, we provided a platform for validating and interpreting results when the complexity

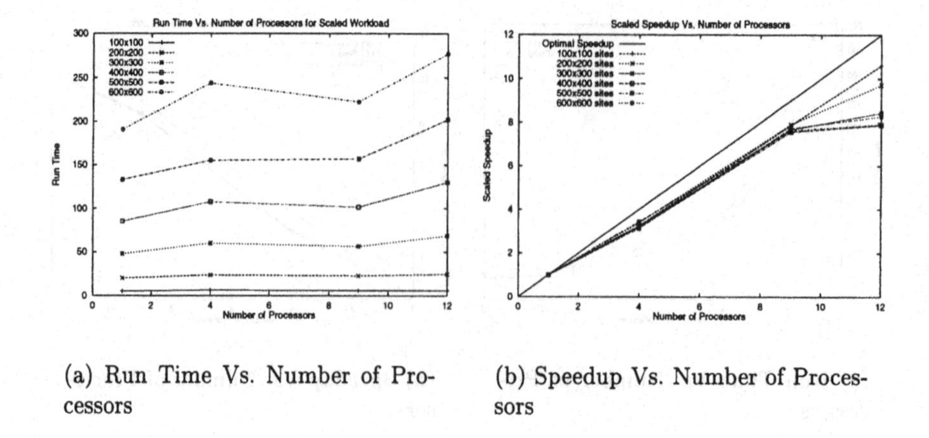

(a) Run Time Vs. Number of Processors

(b) Speedup Vs. Number of Processors

Fig. 4. Performance of Scaled Size Workload vs. Number of Processors on an SGI Origin 2000

of evolution is added. We have observed good performance characteristics on a surprisingly wide range of architectures due to their localized interactions and algorithm selection [18, 14] for frequently performed operations.

Work is in progress to develop a cellular automaton model of genetic variation (and mutation) of two competing host species, one macro-parasite spread via direct contact transmission, and a vector-borne micro-parasite. The evolution of the species is accomplished using the genetic algorithms on the genetic information, copying, mutation, and (where applicable) crossover. Hosts are modeled individually and sites are small enough to contain at most one host, in the same fashion as the model presented here and the Szymanski-Caraco model [17]. The host species are assumed to have sexual reproduction (and hence crossover of genetic material) while the parasitic species are treated as asexual. Such a model poses novel computational challenges, including: computing the lineage of each organism, measuring the simulation trajectory given within site diversity, modeling intrahost competition of parasites. This work also addresses specific defenses against infection (i.e. acquired immunity).

Efficiency and fidelity concerns motivate the consideration of whether a synchronous cellular automata or discrete event simulation engine is better for the simulation kernel. Our experience with Lyme disease simulation (without evolutionary effects, cf. [5, 4]), indicates that in the case of complex species interactions or large diversity of time scales between species an optimistic approach with rollback for correcting potential causality errors may be very efficient. Hence, we plan to implement the presented system using PDES approach and compare efficiency of the resulting system with the current implementation.

5 References

1. R. M. Anderson and R. M. May. *Infectious Diseases of Humans: Dynamics and Control.* Oxford University Press, Oxford, 1991.
2. R. Antia, B. R. Levin, and R. M. May. Within host population dynamics and the evolution and maintenance of microparasite virulence. *American Naturalist*, 144:457–472, 1994.
3. H. J. Bremermann and J. Pickering. A game-theoretical model of parasite virulence. *J. Theoretical Biology*, 100:400–426, 1983.
4. E. Deelman, T. Caraco, and B. Szymanski. Breadth-first rollback in spatially explicit simulations. In *Proc. PADS97, 11th Workshop on Parallel and DistributedSimulation*, Los Alamitos, CA, 1997. IEEE Computer Society.
5. E Deelman, T. Caraco, and B. K. Szymanski. Parallel discrete event simulation of lyme disease. In *Proceedings of the Conference on Biocomputing*, Hawaii, 1996. World Scientific Publishing Co, Singapore.
6. M. Duryea, J. Gardner, T. Caraco, B. K. Szymanski, and W. A. Maniatty. Host spatial heterogeneity and extinction of an SIS epidemics. *Journal of Theoretical Biology*, 1998. To appear.
7. P. W. Ewald. *Evolution of Infectious Disease.* Oxford University Press, New York City, NY USA, 1994.
8. S. A. Frank. A kin selection model for the evolution of virulence. *Proc. Royal Society London*, B250:195–197, 1992.
9. S. A. Frank. Coevolutionary genetics of plants and pathogens. *Evolutionary Ecology*, 7(1):45–75, 1993.
10. S. A. Frank. Models of parasite virulence. *Quarterly Review of Biology*, 71:37–78, 1996.
11. S. Levin and D. Pimentel. Selection of intermediate rates of virulence in parasite-host systems. *American Naturalist*, 117:308–315, 1981.
12. S. A. Levin. Some approaches to the modelling of coevolutionary interactions. In M. H. Nitecki, editor, *Coevolution*, pages 21–65. University of Chicago Press, Chicago, 1983.
13. Hastings I. M. and B. Wedgwood-Oppenheim. Sex, strains and virulence. *Parasitology Today*, 13:375–383, 1997.
14. W. A. Maniatty, B. K. Szymanski, and T. Caraco. Parallel computing with generalized cellular automata. *Parallel and Distributed Programming Practices*, 1(1):85–104, January 1998. Also Technical Report 97-3 Department of Computer Science, Rensselaer Polytenchic Institute, Troy, NY 12180.
15. R. M. May and Anderson R. M. Parasite-host coevolution. *Parastiology*, 100:S89–S100, 1990.
16. M. A. Nowak and R. M. May. Superinfection and the evolution of parasite virulence. *Proc. Royal Soc. London*, B 255:81–89, 1994.
17. B. K. Szymanski and T. Caraco. Spatial analysis of vector-borne disease: A four species model. *Evolutionary Ecology*, 8:299–314, 1994. Online information available at ftp://ftp.cs.rpi.edu/pub/szymansk/eec.ps.
18. B.K. Szymanski, W. Maniatty, and B. Sinharoy. Simultaneous parallel reduction. *Parallel Processing Letters*, 5(1), 1995. RPI Dept. of Computer Science Tech Report CS 92-31.
19. M. Van Baalen and M. W. Sabelis. The dynamics of multiple infection and the evolution of virulence. *American Naturalist*, 146:881–910, December 1995.

A Parallel Algorithm for Computing the Extremal Eigenvalues of Very Large Sparse Matrices*

Fredrik Manne

Department of Informatics, University of Bergen, N-5020 Bergen, Norway
Fredrik.Manne@ii.uib.no

Abstract. Quantum mechanics often give rise to problems where one needs to find a few eigenvalues of very large sparse matrices. The size of the matrices is such that it is not possible to store them in main memory but instead they must be generated on the fly.
In this paper the method of coordinate relaxation is applied to one class of such problems. A parallel algorithm based on graph coloring is proposed. Experimental results on a Cray Origin 2000 super computer show that the algorithm converges fast and that it also scales well as more processors are applied.

1 Introduction

Frequently problems in quantum mechanics lead to the computation of a small number of extremal eigenvalues and associated eigenvectors of

$$Ax = \lambda Bx$$

where A and B are real symmetric and sparse matrices of very high order. The matrices are often of such a magnitude that it is neither practical nor feasible to store them in memory. Instead the elements of the matrices are generated as needed, commonly by combining elements from smaller tables.

A number of methods have been proposed for solving such systems. See Davidson [1] for a survey. More recent methods include among others the implicitly re-started Arnoldi iteration [4] by Lehoucq et. al. Common to most of these methods is that they only require the storage of a few dense vectors. In this paper we consider one of the early methods, the method of coordinate relaxation [2, 7] used for computing the smallest eigenvalue of a large sparse symmetric system. The coordinate relaxation method selects each coordinate of the approximate eigenvector and alters it so that the Rayleigh quotient is minimized. This is repeated until a converged solution is obtained. The advantage of the method is that the amount of memory is restricted and, as noted by Shavitt et. al [8], it is also possible to neglect coordinates if their contribution to the overall solution is insignificant.

* For full paper, see http://www.ii.uib.no/~fredrikm

However, the convergence of the method is only guaranteed if the smallest eigenvalue is simple and has a good separation. Moreover, a good approximation of the eigenvalue must exist. If these conditions are met and the matrix is strictly diagonal dominant, the convergence is usually fast [3].

We present an efficient parallel version of this method applicable for sparse matrices. Designing such an algorithm is a non-trivial task since the computation of the contribution from each coordinate in the algorithm depends directly on the previous computations.

We first show that it is possible to perform the selection of which coordinates to use in parallel. The selected coordinates define a sparse graph. By performing a graph coloring on the vertices of this graph it is possible to divide the updating of the necessary variables into parallel tasks. Since the sparse graph changes from each iteration of the algorithm the graph coloring has to be performed repeatedly.

The presented algorithm has been implemented and tested on a 128 processor Cray Origin 2000 computer using diagonal dominant matrices from molecular quantum mechanics. The results show the scalability of the algorithm.

2 The Coordinate Relaxation Method

Here we briefly review the coordinate relaxation (CR) method. For simplicity we assume that $B = I$ such that we are considering the simplified problem $Ax = \lambda x$.

Given a symmetric matrix $A \in \Re^{n \times n}$. We wish to compute the smallest eigenvalue λ and its corresponding eigenvector x.

Given an initial approximation x of the eigenvector and a search direction y we find a new eigenvector $x' = x + \alpha y$ where the scalar α is determined so that the Rayleigh quotient

$$R(x') = \frac{x'^T A x'}{x'^T x'}$$

is minimized.

In the CR method we take y equal to a unit vectors e_i. Combined with the notation

$$F = Ax \tag{1}$$
$$p = x^T A x \tag{2}$$
$$q = x^T x \tag{3}$$

we get the following quadratic equation for determining α

$$\alpha^2(f_i - a_{ii}x_i) + \alpha(p - a_{ii}q) + px_i - f_i q = 0. \tag{4}$$

When α has been determined one must update the values of p, q, x, F, and λ accordingly:

$$p' = p + 2\alpha f_i + \alpha^2 a_{ii} \quad (5)$$
$$q' = q + 2\alpha x_i + \alpha^2 \quad (6)$$
$$x' = x + \alpha e_i \quad (7)$$
$$F' = Ax' = Ax + \alpha A e_i = F + \alpha A_i \quad (8)$$
$$\lambda' = \frac{p'}{q'} \quad (9)$$

Updating p, q, x, and λ involves only a few scalar operations. The time consuming part of the algorithm involves computing F'. This involves not only $nonz(A_i)$ scalar operations but each element of A_i must also be generated. If A is to large to store in memory this must be done on the fly.

In one complete iteration of the CR method one cycles through every coordinate of x. A threshold that is successively lowered is used to determine if a coordinate should be used or not.

3 A Parallel Algorithm

We now present a parallel version of the CR method for sparse matrices. The algorithm operates in three stages. First we consider which coordinates should be used to update the eigenvector. Then we consider how the calculations can be ordered to allow for parallel execution, and finally how the actual computations are performed. Our computational model is a parallel computer with distributed memory. Communication is done by message passing.

The computation of the different values of α is inherently sequential with each step of the algorithm depending on the previous ones. As described in Section 2 the main work of the algorithm is in updating F according to (8).

Finding Candidates We use the initial values of λ, p, q, and F at the start of the iteration when testing each coordinate to see if it contributes enough to the solution. If so, the coordinate is added to the set of candidates that will be used to update the solution. Thus we postpone the updating of the solution until after we have determined which coordinates to use. To ensure that all significant contributions to the solution are acquired we make repeated passes over the matrix before lowering the threshold value.

By dividing the coordinates evenly among the processors we can now determine the candidates in parallel without the need of communication except for distributing the initial values.

Updating the Solution We show how it is possible to postpone and thus accumulate the updating of F. Let K the set of chosen coordinates. To be able to compute α_j the element f_j must be updated by each α_i where $i < j, i \in K$, and $a_{ij} \neq 0$. Let $K = \{C_1, C_2, ..., C_r\}$, $1 \leq r \leq |K|$ be a partitioning of K such that $a_{ij} = 0$ for $i, j \in C_k$ and $1 \leq k \leq r$. If the coordinates in C_1 are applied first, we can compute each α_i, $i \in C_1$, without performing any update on F. This follows from the fact that $a_{ij} = 0$ for $i, j \in C_1$. Thus the updating of F can be postponed until each α_i, $i \in C_1$ has been computed. Note that the computation of the values of α is sequential. But since this only involves a few scalar operations for each α it can be performed relatively fast. Before we can compute the values of α corresponding to the coordinates in C_2 we must perform an update of F. This can be done in several ways. We choose to immediately perform the complete update of F:

$$F = F + \sum_{i \in C_1} (\alpha_i * A_i) \tag{10}$$

From a parallel point of view we have now restructured the algorithm to consist of fast sequential parts, each one followed by some communication and a potential larger parallel update of F.

To perform the updates on F in parallel we associate the work related to one row of A with one processor making it responsible for updating f_i for each row assigned to it. This requires that each processor has access to the necessary values of α and to the corresponding rows of A.

With this scheme the only communication required is the distribution of the αs. This can be done in one broadcast operation before the parallel update of F. The load balance now depends on how F is distributed and the structure of the rows of A corresponding to coordinates in each C_j. If we distribute x in the same way as F we must gather both x_i and f_i, $i \in C_j$ from each processor before the sequential computation of the α's. We do this on processor 0 which then computes the values of α.

In order to obtain the desired partitioning of K we perform a graph coloring on the adjacency graph $G(K)$. The set C_i now consists of the coordinates whose corresponding vertices are colored with color i. The complete parallel algorithm is as follows:

Parallel Coordinate Relaxation
Calculate initial value of λ and x
Repeat
  Do s times
    Find a set of candidates K
    Perform a graph coloring on $G(K)$
    For each color i:
      Gather f_j and x_j $j \in C_i$ on processor 0
      Processor 0: **For** each $e_j \in C_i$
          Calculate p, q, and α
      Broadcast the values of α
      Update F and x with the coordinates in C_i
  End do
  Lower threshold
  Processor 0: Distribute p and q
Until convergence

Here s is the number of passes we make over the matrix before lowering the threshold.

4 Results

We have performed experiments on a Cray Origin 2000 computer with 128 processors. Here we present results for a matrix from quantum mechanical calculation [5]. The matrix is of order 5189284, and contain on the order of 1.1×10^{11} non-zero elements. The elements are generated on the fly by combining elements from several smaller tables.

Table 1 displays the timings and speedups for one matrix when increasing the number of processors. All times are given in seconds. The quality of the solutions are as good as for the sequential algorithm, and the number of candidates found differ by less than 2%. Finding candidates takes 2.5 seconds on 16 processors

Proc	10	16	26	31	41	51	75	100
Time	1113	662	407	336	260	209	156	128
Speedup	1.0	1.7	2.7	3.3	4.3	5.3	7.1	8.7

Table 1. Execution times and speedup for $m = 38$.

and scales appropriately. The graph coloring takes between 12 and 14 seconds independent of the number of processors used.

To conclude we note that the presented algorithm gives a good speedup on realistic problems but also note that one should be careful to only use the CR method when the convergence criteria are met. Comparisons with the ARPACK parallel package for computing extremal eigenvalues [4] show that for our particular problems the coordinate relaxation method gives an order of magnitude faster convergence.

References

1. E. R. DAVIDSON, *Super-matrix methods*, Computer Physics Communications, (1988), pp. 49–60.
2. D. K. FADDEEV AND V. N. FADDEEVA, *Computational Methods of Linear Algebra*, W. H. Freeman and Co., San Francisco, CA., 1963.
3. G. H. GOLUB AND C. F. V. LOAN, *Matrix Computations*, North Oxford Academic, 2 ed., 1989.
4. R. B. LEHOUCQ, D. SORENSEN, AND P. VU, *ARPACK: An implementation of the implicitly re-started Arnoldi iteration that computes some of the eigenvalues and eigenvectors of a large sparse matrix.* Available from netlib@ornl.gov under the directory scalapack, 1996.
5. I. RØEGGEN. Private communications.
6. A. RUHE, *SOR-methods for the eigenvalue problem with large sparse matrices*, Math. Comp., 28 (1974), pp. 695–710.
7. H. R. SCHWARZ, *The method of coordinate overrelaxation for* $(A - \lambda B)x = 0$, Numer. Math., (1974), pp. 135–151.
8. I. SHAVITT, C. F. BENDER, A. PIPANO, AND R. P. HOSTENY, *The iterative calculation of several of the lowest or highest eigenvalues and corresponding eigenvectors of very large symmetric matrices*, Journal of Computational Physics, (1973), pp. 90–108.

Technologies for Teracomputing: A European Option

Agostino Mathis

Director, ENEA's High Performance Computing and Networking (HPCN) Project,
ENEA - Italian National Agency for New Technology, Energy and the Environment;
Lungotevere Thaon di Revel, 76 - 00196 Roma (Italy);
mathis@enea.it

Abstract. A hardware and software environment with performance above 1 Teraflops (teracomputing) is presently required to face the leading computational challenges not only in fundamental sciences, but also in an increasing number of fields related to applied sciences and engineering. A short survey of the top level computing platforms is made, with a special attention to United States and Japan. The technology options adopted for teracomputing are evaluated. In this perspective, a European line of supercomputers is described, which was originated by the Italian Institute for Nuclear Physics and is now in use also in Germany and France: this line of massively parallel platforms is based on the use of CMOS VLSI custom chips, carefully designed to match the computing needs, so that only the silicon strictly necessary to the operations to be performed is used.

1 Teracomputing: the international perspective

Computer simulation is becoming a powerful means of understanding the structure and behaviour of very complex systems (commonly called "grand challenges"): e.g., the "ab initio" solution of the fundamental equations of a system, to perform self-consistent "computer experiments" to be compared and eventually substituted to "physical experiments". Such a computational environment was originally developed and exploited for specific challenges of fundamental sciences, but is currently requested in an increasing number of fields related to applied sciences and engineering, as: climate and weather forecasting; air and water pollution management; molecular design for new materials and new drugs; biochemistry and bioengineering; numerical wind tunnels for aerodynamics optimisation of cars and planes; electromagnetic design for telecommunications and control systems; optimal design of engines (both for energy saving and pollution control); etc. [1–4]. The "ab initio" solution of this kind of computational challenges typically requires limitless computing power (the higher is the power, the bigger can be the problem domain to be solved, or the better the resolution attained): practically, with the present technology, the target is the availability of platforms offering a computing power above 1 Teraflops (i.e. 1000 billion floating point operations per second) and an almost perfect scalability.

In the worldwide arena, two countries have already demonstrated an explicit awareness of the strategic role of top level supercomputing platforms : the two countries, not strangely, are the United States and Japan.

As regards the United States, in 1995 President Clinton declared an end to all nuclear testing, and offered the three weapons laboratories (Livermore, Los Alamos, Sandia) of the Department of Energy (DOE) a $3 billion package of scientific projects to replace testing. Among the new DOE's Projects, the Accelerated Strategic Computing Initiative (ASCI) has been launched, with a request of $910 million by 2002 in pursuit of computers that will surpass today's systems by a factor of many decades in processing speed and in storage capacity. The contrats awarded by DOE for acquisition of ASCI platforms are reported in Table 1.

Site	Option	Supplier	Power (TFlops)	Cost (M$)	Year of operation
Sandia	Red	Intel	1.8	45	1997
LLNL	Blue Ocean	IBM	3.2	93	1998
Los Alamos	Blue Mountain	SGI/Cray	3+1	110,5	1999
LLNL	White	IBM	10	85	2000

Table 1. *Contracts for ASCI computing platforms awarded by USA's Department of Energy. LLNL stands for Lawrence Livermore National Laboratory.*

Evidently, the ASCI Project has been set up also to spur on technological development, by shaping cutting-edge computer technologies in collaboration with universities as well as industry. Non-defense users will have the possibility of exploiting the computing power, developed under ASCI, for their most challenging problems.

From a technological viewpoint, it has to be observed that, as far as known, all the ASCI platforms by now committed appear built up by assembling a very large number of "commodity processors", and result in huge requirements of cabinets and electrical power supply. The cost per Gigaflops looks in the order of $30,000 for "Option Red" and "Option Blue", and obviously lower for the extension to "Option White".

Originally, five generations of high-performance computers were foreseen over the lifetime of the program (1995-2002), with the following steps in peak computing power (Teraflops): 1.8; 3; 10; 30; 100. In fact, the peak power trend seems slower than planned.

For the steps above 10 Teraflops, the Department of Energy has recently defined a different approach, the "PathForward Program", which shall kick off a series of contracts concerning the missing pieces of the teraflops challenge, in particular the scaling and integration technologies that should tie together clusters of supercomputers, to create platforms far more powerful than currently possible.

Clearly, USA is ready for another leap in computing technologies and computational methodologies, with respect to both Europe and Japan. However, Japan's industry is well established as a supplier of more traditional supercomputing platforms, in particular of the scalable parallel vector architecture. Relevant examples in the field of meteoforecasting are the contracts that Fujitsu has secured to supply the European Center for Medium-range Weather Forecasting (ECMWF) and, more recently, Mto-France, with parallel vector supercomputing systems, replacing the existing USA made supercomputers.

Recently, NEC announced plans to develop "the world fastest ultracomputer with the maximum performance of over 32 Teraflops": this project is part of the Earth Simulator Program promoted by the Science and Technology Agency of Japan, with a budget of the order of $500 million in the period 1998-2002. The Earth Simulator is intended to make a contribution to resolving the problems of the global environment and to establishing countermeasures against natural disasters and catastrophes. The NEC "ultracomputer", expected to be completed in 2002, should achieve a maximum performance of over 32 Teraflops and a main memory capacity of over 4 Terabytes by connecting in parallel thousands of vector type CPUs, each one with a performance capability several times that of the existing ones.

Besides the traditional supercomputers, Japan is building and using some of the most performing, in absolute, massively parallel platforms. It is to be noted that, in the list of TOP500 Supercomputer Sites, several places among the first ten are occupied by Japanese machines. Moreover, the community of the Japanese cosmologists has developed a line of special purpose computing platforms, named GRAPE: the last version of this line, although not classifiable according to the TOP500 benchmarking standard, is considered capable of an effective computing power above 1 Teraflops. A new version of GRAPE-type machine for molecular dynamics calculations, planned for 1999, should attain 100 Teraflops.

A peculiar character of all the Japanese high performance computers is the use of custom designed processors, which exploit the maximum specific performance from a rather conservative silicon technology (CMOS, clock frequency in the order of $100 \div 200$ Mhz).

2 A "petaflops" machine with today's technology?

In recent years the evolution of the business environment of High Performance Computing manufacturers seems to point at a definite trend toward the use of massively produced, commodity processors (e.g.: IBM's SP1 and SP2, INTEL's ASCIred for USA Department of Energy, etc.).

In fact, the increasing performance and availability of general-purpose microprocessors has fostered the spread of clusters of processors, either with shared- or distributed-memory, as an alternative to customized parallel computing platforms. The Intel Pentium II, together with its Microsoft NT operating system (the Wintel model), provides enough overall capacity to displace also the RISC/UNIX workstation in the engineering and scientific marketplace. In the last months every major supplier of RISC based systems indicated they were moving towards utilizing an Intel engine in their systems. The success of the "Wintel" model means that both the microprocessor and the operating system become "commoditized", that is they simply become the brick and mortar for every general purpose computing architecture.

However, this trend by no means shall imply the unification of the vast domain of electronic data processing. On the contrary, it will possibly enlarge the number and the importance of the application fields where the design and fabrication of Application Specific Integrated Circuits (ASIC) shall be competitive on the base of the price/performance ratio. It may be argued that also the field of High Performance Computing and Networking (HPCN) shall be increasingly based on special purpose electronics.

It is easy to see that the clustering of today's commodity processors (CMOS technology) is not viable to obtain the performances that could be expected in 10 years from now [5,6]. In the last ten years, mainly thanks to the performance jump in commodity microprocessors, the peak performance of the fastest computers used in science and engineering skyrocketed by a factor of 500, compared to only a factor of $10 \div 15$ in the preceding decade: it increased from 160 to 2000 Mflops during the 1977-1987 period, and from 2000 to 1,000,000 Mflops (1 Tflops) in the 1987-1997 period. Based on the progress of the last decade, petaflops (1 Pflops $= 10^{15}$ floating point operations per second) computers should become available in another decade or so.

However, if built with today components, a petaflops system would require about one million of microprocessors, each one performing 1 Gflops. Also if we limit, for cost reasons, the main memory to 128 Mbyte per processor (less than the usual balance), the petaflops system would have anyway 128 Tbyte of memory. Such a system would cost at least $20 billion in today's dollars and require perhaps 2 GW of electrical power!

On the other hand, costs and power requirements have decreased steadily in the past, according to Moore's Law (doubling of performances at the same cost every 18 months): along this trend, to reduce the cost of a petacomputer at an acceptable level of $20 million, it would take 10 doubling periods ($2^{10} = 1024$), that is 15 years.

The problem is, the Moore's Law is reaching its limits: due to physical constraints and exploding investment costs, the CMOS technology should keep the past pace of improvement for no more than two devices generations; alternative devices and fabrication technologies usually take 10 to 15 years to be put in commercial production, and no one is yet fully demonstrated in the laboratory and on the road to industry. With CMOS technology, between now and the year 2010, also if the on-chip local clock frequency could reach 6000 MHz, the chip-to-board clock frequency is forecast to increase only from 500 MHz to 1200 MHz, less than a factor of three [7].

Consequently, to maintain the Moore's Law until 2010, that is a 256-fold increase with respect to 1998, the architecture will have to contribute a factor of almost 100, by introducing and exploiting two more orders of magnitude in parallelism inside the microprocessors. Another factor of four in high level parallelism will then be needed to build a petaflops computer with CMOS microprocessors in 2010.

In practice, experience shows that communication and coordination between independent computational processes leads generally to a rapid fall in efficiency as the number of processors grows. The basic problem is latency, which is aggravated dramatically as the size of the system increases. Besides, a general-purpose processor requires considerable space and power, but would only use a very limited fraction of its silicon to perform the more common parallel computing tasks.

Many different scenarios are now under scrutiny to overcome the above described difficulties. Among them [5,6,8]:

- CMOS components, but revolutionary architectures that improve the interaction between memory and processor (in fact, the speed increases in processors 60% per year, in DRAM memories only 7% per year): e.g., multithreaded approaches, Process in Memory (PIM), ecc.;
- components that use new technologies that are much faster than CMOS (like heterojunction bipolar transistors, quantum devices, superconducting electronics), as well as new memory structures;
- special-purpose processors tailored for particular classes of applications: they can be considered as "computational engines" with tremendous advantages in price/performance ratio (e.g., GRAPE machines for N-body astrophysical simulations, several QCD lattice computing platforms [9]).

The only scenario already practically demonstrated, is the last one, the special-purpose "computational engines". A custom design for these "engines" can be conceived for optimal, very lean, architectures , which minimize the needs of silicon and electrical supply (and the related heat dissipation). These features, evidently, are mandatory, if we want a computing structure that shall be easily scalable with limited investment and operation costs. Obviously, the main weakness of the custom approach is the loss of cost reductions due to mass fabrication. However we have to consider the decreasing costs of designing and manufacturing the Application Specific Integrated Circuits (ASIC), owing to the general adoption of Computer Aided Design and Manufacturing techniques.

3 An example of custom-silicon massively parallel architecture

Examples of special-purpose "homebrew" machines that in the last ten years pursued the goals of easy scalability and limited costs, sometimes with very notable results, are the platforms designed and built by several teams of high energy physicists for lattice Quantum Chromo-Dynamics (QCD) simulations. Among the QCD machines, performances of excellence are being obtained by the "APE collaboration", a project of the Italian National Institute for Nuclear Physics (INFN). The APE100 machine is built from 1993, with the name Quadrics for its industrial version. A new generation of APE machines is now built by INFN: it is known as APEmille, and is much more flexible and powerful than APE100 (in its full configuration, the peak power should exceed 1 Teraflops). At present, more than 50 Quadrics systems (APE100 and APEmille technology) have been sold in Italy, Germany (more than 20 systems) and France.

The APE platforms are an instance of distributed memory, single instruction stream, multiple data stream (SIMD) parallel computers. They are based on a set of CMOS VLSI custom chips, carefully designed to match the computing needs, so that only the silicon strictly necessary to the operations to be performed is used. For a given computing power, the volume of circuitry and the electrical power required for circuitry and cooling systems are an order of magnitude lower with respect to platforms built up from commodity processors. The APE platforms are structurally capable to process large data arrays where they are stored, so that the Von Neumann bottleneck between processor and memory can be circumvented. In this sense, they can be considered as smart or active memories, otherwise known as processor in memory (PIM) structures.

The APE100 line was produced, already four years ago, at a cost of the order of $30,000 per Gigaflops (that is of the same order of the first series of ASCI platforms under procurement c/o USA's Department of Energy). The APEmille line, which should afford a ten-fold improvement in peak performance in the same hardware volume, should be produced at a cost of the order of $8,000 per Gigaflops [10]. It is interesting a comparison between APE platforms and some other top level machines in the field of QCD computations, performed by a German scientist, as shown in Table 2 [11].

Computer	Maximum Number of Processors	Peak Perform. (TFlops)	Year of Operation	Efficiency in QCD	Cost/Effective Performance ($/Gflps)	Elect.Power/ Peak Perf. (kW/Tflops)
APE100	1048	0.10	1993	60%	50,000	330
CRAY T3E900	512	0.45	1997	25%	240,000	940
CP-PACS	2048	0.60	1996	60%	170,000	750
ASCIred	9000	1.80	1996	[56%]	54,000	[400]
QCDSP	12288	0.60	1997	30%	15,000	200
APEmille	2048	1.60	1998/9	60%	17,000	<50

Table 2. *Comparison between APE platforms and some other top level machines. CRAY T3E900 is the parallel computer built by SGI-CRAY; CP-PACS is built by Hitachi; QCDSP is built by Columbia University with the use of Digital Signal Processors.*

In Table 2, CRAY T3E900 is the parallel computer built by SGI-CRAY; CP-PACS is built by Hitachi; QCDSP is built by Columbia University with the use of Digital Signal Processors.

Considering the natural scalability of the SIMD architectures, there is a clear opportunity for Europe to invest on the APE line to compete in many computational grand challenges at the same level as USA's ASCI or Japan's Earth Simulator Program, but at a sensibly lower cost [12].

INFN (Italy) and the Desy Laboratory of Hamburg (Germany) are already collaborating in the production and the utilization of the APEmille supercomputer. ENEA (Italy) and HLRZ (Germany) are already collaborating in the exploitation of large APE100 platforms, with a specific effort to explore new fields of applications, outside the particle physics. In particular, they have developed new effective algorithms for N^2-loop computations, which can be of interest in many different fields, as: molecular dynamics, protein folding, neural nets, cosmic clusters, linear algebra, etc. A joint effort is now under way to develop a set of basic linear algebra subprograms (BLAS) for the APE platforms, with reference to the many APE100 already installed both in Germany and in Italy, and obviously with full portability to APEmille.

The new industrial company Quadrics Supercomputers World (70% Finmeccanica, 30% Meiko) is working to complement the SIMD capabilities of APE with the multiple instruction stream, multiple data stream (MIMD) well known capabilities of Meiko's technology [13]. The first bench-marking tool along this direction (known as PQE1) is operating at the ENEA's Casaccia Center (near Rome), and it is used to develop the system software environment and to test the new hybrid (MIMD/SIMD) platform in a wide range of applications [14].

The experience already done in ENEA on the APE100 platforms has demonstrated their high effectiveness and almost perfect scalability in all the problems with nearest-neighbour interactions and with heavy computational requirements: if the problem complexity can grow with the computing platform size, so that its arithmetic capability is always fully exploited, the best level of performance can be mantained also at the maximum single configuration (peak performance of 25 Gflops).

The new hybrid MIMD/SIMD configuration allows an optimal task distribution between the "commodity" RISC processors of the MIMD side and the powerful SIMD "computational engines". Typical examples of problems suitable for such a configuration are: a multi-domain simulation of complex environments (e.g., electromagnetic or fluidodynamics fields); the search of minimal energy configurations in huge molecular structures (in this case the dynamics of the structure is simulated on the SIMD sections, while the optimization algorithm is implemented on the MIMD side); etc. In the Casaccia Center, besides, the hybrid configuration allows, if necessary, the concentration of all the available SIMD platforms on the same large problem (for a total peak performance of about 85 Gflops).

A regularly up-dated status report on the ENEA's activities in these fields can be found at the web site : " http://www.enea.it/hpcn".

The preliminary guide-lines for the next generation of the APE line (the follow-up of APEmille) are being defined by INFN, in collaboration with Desy Laboratory: the new platform, targeted to a peak power of tens of Teraflops, shall be designed with special reference to particle physics requirements, but with an open attitude to take into account any other computational challenge suitable for a SIMD engine.

References

1. Cabibbo N., Mathis A.: "Whatever happened to Europe's high performance computing?". Physics World, January 1995.
2. Griffith V.: "Pharmaceutical supercomputers - Mighty dose of power". Financial Times, May 21, 1997.
3. Taubes G.: "Redefining the Supercomputer". Science, Vol.273, 20 September 1996.
4. Markoff J.: "A Tug-of-War for Elite Supercomputers". International Herald Tribune, March 24, 1998.
5. Messina P.: "High-Performance Computers: The Next Generation (Part I)". Computers in Physics, Vol.11, No.5, Sep./Oct. 1997.
6. Messina P.: "High-Performance Computers: The Next Generation (Part II)". Computers in Physics, Vol.11, No.6, Nov./Dec. 1997.
7. SIA (Semiconductors Industry Association): "The National Technology Roadmap for Semiconductors". 1997 Edition, published by SEMATECH, Inc.
8. Bilardi G., Preparata F. P.: "Horizons of Parallel Computation". Journal of Parallel and Distributed Computing **27** (1995) 172-182.
9. Makino J.: "Supercomputing Symposium Addresses Astrophysics and Related Topics". Computers in Physics, Vol.11, No.6, Nov./Dec. 1997.
10. Maiani L.: "APEmille verso il traguardo". Le Scienze, numero 343, marzo 1997.
11. Schilling K.: "TERAcomputing in Europa: Quo Vamus?". Phys. Bl. 53 (1997) Nr.10.
12. Mathis A.: "High Performance Computing: European Efforts". The IPTS Report, Issue 06, 1996 (European Commission - Joint Research Center).
13. Mathis A.: "How to use HPCN for Creating New Industries". RCI's European Member Management Symposium IX, Noordoostpolder, The Netherlands, 5-6-7 June 1996.
14. Vanneschi M.: "Variable grain architectures for MPP computation and structured parallel programming". Programming Models for Massively Parallel Computers, IEEE Press, March 1998.

High Performance Fortran: Status and Prospects*

Piyush Mehrotra[1], John Van Rosendale[1], and Hans Zima[2]

[1] ICASE, MS 132C, NASA Langley Research Center, Hampton VA. 23681 USA
{pm,jvr}@icase.edu
[2] Institute for Software Technology and Parallel Systems,
University of Vienna, Liechtensteinstr. 22, A-1090 Vienna, Austria
zima@par.univie.ac.at

Abstract. High Performance Fortran (HPF) is a data-parallel language that was designed to provide the user with a high-level interface for programming scientific applications, while delegating to the compiler the task of generating an explicitly parallel message-passing program. In this paper, we give an outline of developments that led to HPF, shortly explain its major features, and illustrate its use for irregular applications. The final part of the paper points out some classes of problems that are difficult to deal with efficiently within the HPF paradigm.

1 Introduction

For distributed-memory architectures two issues are of central importance: the necessity of controlling locality and the complexity of parallel programming. One approach to the control of locality is to use explicit parallelism, e.g., C or Fortran coupled with message passing. However, it became clear that a higher level approach was also possible. One could use a high-level "data parallel" language with a single thread of control coupled with user-specified annotations for distribution of data and computation across processors. In effect, the user can leave some of the low-level details to the compiler and runtime system.

High Performance Fortran (HPF) is at the forefront of such languages, and is important as a milestone on the way to truly expressive parallel languages which will satisfy users' needs more fully. The key concept of HPF – sharing responsibility for exploiting parallelism between the user and the compiler/runtime system – is based on research done by several groups over the years. In this introduction, we first describe some of the most significant language and compiler efforts which facilitated the development of HPF and then give a brief history of HPF itself.

* The work described in this paper was partially supported by the ESPRIT IV Long Term Research Project 21033 "HPF+" of the European Commission, by the Austrian Ministry for Science and Transportation under contract GZ 613.580/2-IV/9/95, and by the National Aeronautics and Space Administration under NASA Contract No. NAS1-19480, while the authors were in residence at ICASE, NASA Langley Research Center, Hampton, VA 23681.

In the context of MIMD machines, Kali (and its predecessor BLAZE) [18, 20] were the first languages to introduce distribution declarations for arrays, using regular as well as irregular distributions and permitting dynamic redistribution. The language also introduced features for controlling the distribution of parallel loop iterations to processors. The Kali compiler [16] was the first compiler to integrate both a static and a runtime communication strategy using the *inspector-executor* paradigm [19].

In the same time period, Thinking Machines, a supercomputer manufacturer, together with COMPASS, a compiler software company, incorporated static layout directives, including alignment of arrays, for a subset of Fortran-8x on the Connection Machine [1].

SUPERB [33] was an interactive restructuring tool, developed at the University of Bonn, which translated Fortran 77 programs into message-passing Fortran for a range of architectures. SUPERB performed coarse-grain parallelization for a distributed-memory machine and was also able to vectorize the resulting code for the individual nodes of the machine.

The Fortran D project [14] followed a slightly different approach to specifying distributions. The distribution of data is specified by first aligning data arrays to virtual arrays known as *decompositions*. The decompositions are then distributed across an implicit set of processors using relative weights for the different dimensions similar to the approach used in CM-Fortran. The language allows an extensive set of alignments along with simple regular and irregular distributions. All mapping statements are considered executable statements, thus blurring the distinction between static and dynamic distributions.

Vienna Fortran [6, 34] was the first language to provide a complete specification of mapping constructs in the context of Fortran. In addition to simple regular and irregular distributions, Vienna Fortran defines a generalized block distribution which allows unequal sized contiguous segments of the data to be mapped to the processors. The language maintains a clear distinction between distributions that remain static during the execution of a procedure and those which can change dynamically, allowing compilers to optimize the code for the two situations. It also defines multiple methods of passing distributed data across procedure boundaries including inheriting the distribution of the actual arguments. The Vienna Fortran Compilation System (VFCS) [3, 35] extended the Superb compiler to HPF and HPF+ [7], targeting a wide range of distributed-memory machines.

Several other projects have also contributed to the understanding necessary for the development of HPF and the compiling technology required for such a language [2, 10, 17, 22, 27–30] including commercial efforts such as [24, 25].

The HPF Forum, a group of about 40 researchers, was formed in 1992, with the objective of developing a unified high-level approach for programming distributed-memory architectures. After extensive discussions, including input from the wider high performance community, the Forum released Version 1.0 of HPF in May 1993, followed by Version 1.1 (November 1994), which mainly incorporated corrections to the language. Based on a requirements document and further input from the community, the Forum released HPF 2.0 in January 1997. All these documents are available at the HPF Web site at `http://www.crpc.rice.edu/HPFF/home.html`.

There are several commercial compilers available for HPF, most of them implementing the features of HPF 1.1 only. The most widely used ones are from Applied Parallel Research, DEC, IBM and Portland Group Inc. The performance of these compilers for regular single grid codes is comparable to that of hand-written message passing code, however, they do not yet perform as well on more complex codes.

The rest of this paper begins by giving a short overview of the HPF 2.0 language (Section 2). In Section 3, we outline some properties of advanced applications requiring highly flexible methods for data and work distribution, and show how they can be handled using features of HPF-2 and beyond. In the final part of the paper (Section 4), we discuss a number of options for the future development of the language. Concluding remarks can be found in Section 5. An extended version of this paper can be found in [21].

2 HPF 2.0

High Performance Fortran is a set of Fortran extensions designed to allow specification of data parallel algorithms for a wide range of architectures. The underlying programming model provides a global name space and a single thread of control. The user annotates the program with distribution and alignment directives to specify the desired layout of data. Explicitly parallel constructs allow the expression of fairly controlled forms of parallelism, in particular data parallelism. Thus, the code is specified in a high level, portable manner with no explicit tasking or communication statements. The goal is to allow architecture specific compilers to generate efficient code for a variety of architectures including SIMD, MIMD shared and distributed-memory machines.

The HPF 2.0 language is based on the current Fortran standard, Fortran 95. It consists of three parts: a) the Base Language, b) the Approved Extensions, and c) Recognized Extrinsic Interfaces. The base language provides the basic set of HPF features which must be supported by an HPF compiler. The Approved Extensions consist of advanced constructs that meet specific needs but are not likely to be supported by the initial compilers. The Recognized Extrinsic Interfaces are a set of interfaces approved by the HPF Forum but which have been designed by others to provide a service to the HPF community.

3 Formulating Advanced Applications With HPF

Many applications in science and engineering use structured grids for modelling physical phenomena, since efficient discretization on such grids can be easily implemented. However, if geometrically complex objects have to be modelled or algorithmic properties such as adaptivity are to be taken into account, using a single structured grid is no longer adequate.

In this section, we will deal with applications based on the use of unstructured grids. We will discuss a possible representation of such grids, an algorithm performing a sweep, and mechanisms for their distribution.

3.1 Unstructured Grids

Many simulations deal with problem domains characterized by complex geometries and require adaptivity to handle shocks, vortex sheets and other regions of strong gradients. An important method for handling such applications covers the domain with an *unstructured grid.* The logical simplicity of regular grids – where the coordinates of one gridpoint can be used to immediately determine the coordinates of all its neighbors – is then lost: the numbering of the vertices in an unstructured grid reflects properties of the grid generation algorithm, the object geometry, and the refinement strategy. In general, it cannot be assumed that the associated order is correlated with the physical location of gridpoints. As a consequence, the neighborhood relation must be explicitly represented.

Assume that the problem space is a bounded region in s-dimensional space, $\Re^s$, where $\Re$ denotes the set of real numbers. Then we can define a *grid* formally as a system $G = (X, E, C)$, where (1) $X = \{x_1, \ldots, x_n\}$ is a set of *gridpoints*, (2) (X, E) is an undirected graph, where E is the set of edges connecting neighbors, and (3) $C : X \rightarrow \Re^s$ defines the *coordinates* for each vertex in the physical problem space.

The data structures representing an unstructured grid and the associated data access patterns are in general not known at compile time. Many analysis tasks – such as determining the communication pattern for accessing nonlocal data in a parallel loop – must therefore be performed at runtime.

Assume that grid $G = (X, E, C)$ is either generated during program execution or read in, and that it remains invariant thereafter. We represent, allocate and initialize G in a Fortran program as illustrated in Figure 1. *GRID* is a one-dimensional, allocatable array, whose elements represent the gridpoints and collectively store all the information pertaining to the grid. The derived type *NODE* represents the element type of *GRID*; it contains the following pieces of information related to each vertex: (1) its coordinates x, y (representing C, under the assumption $s = 2$), (2) linkage information (representing E), specifying the number of neighbors (*NNB*) of the vertex, and their grid indices (*NB*), and (3) a weight, W, for each neighbor, used in the relaxation algorithm, and a flow variable, V. During the sweep over the grid, the neighbors of each vertex are accessed indirectly, using *NB*.

In the remainder of this section we discuss the related issues of data and work distribution in HPF (Section 3.2) and methods for controlling the computation of communication schedules for loops with irregular accesses (Section 3.4).

```
PARAMETER (MAXNNB = ...) ! maximum number of neighbors for a gridpoint
INTEGER :: N, I , IT                 ! number of vertices
REAL        :: T                     ! temporary variable
TYPE NODE
  REAL :: x, y                       ! coordinates
  REAL ::  V                         ! flow variable
  INTEGER :: NNB                     ! number of neighbors
  INTEGER :: NB(MAXNNB)              ! list of neighbors
  REAL    :: W(MAXNNB)               ! weights of neighbors
END TYPE NODE

TYPE (NODE), ALLOCATABLE :: GRID(:)
REAL , ALLOCATABLE :: TMP(:)
      ...
! Read gridsize and allocate GRID and TMP
READ (*, *) N
ALLOCATE (GRID(N), TMP(N))
CALL INIT_GRID(GRID)                ! initialize grid data structure
! Grid processing
DO IT = 1, maxtime                  ! timestep loop
    TMP =  GRID%V
! Sweep over the grid
   DO I = 1, N
      T = 0.0
      DO  J = 1, GRID(I)%NNB
         T = T + GRID(I)%W(J) * TMP(GRID(I)%NB(J))
      ENDDO
      GRID(I)%V = GRID(I)%V + T
   ENDDO
ENDDO
```

Fig. 1. Relaxation sweep over an unstructured grid in Fortran 90

3.2 Grid Partitioning

Grid partitioning subdivides a grid into contiguous, mutually disjoint regions which are mapped to the processors of a parallel machine, thus specifying for each processor its *component* – the portion of the grid which is local to that processor.

The overall criterion for partitioning is the minimization of the total execution time, which depends on many parameters, including the total processing load, the degree of parallelism in the algorithm, the amount of communication, the amount of processing in each node, and the load balance. Different components in a partition may have different sizes.

How can a grid partition be reflected in the data distribution of an associated array? Consider the array *GRID* in the code shown in Figure 1. In general, the size of this array will be very large. As a consequence, replication of the array across all processors is impractical. Using a block or cyclic distribution would balance the load by allocating distribution segments of approximately the same size, if the amount of work per gridpoint is about equal. However, since no assumption can be made about the order of gridpoints, such a distribution – which subdivides the index domain in a regular way – may be highly irregular in terms of the physical locations of gridpoints.

This problem can be dealt with in two ways:

1. *Reordering.* With this method, the elements of the array *GRID* are permuted in such a way that each component of a grid partition is mapped to a contiguous subset of the index domain. After this permutation has been applied, *GRID* can be distributed by *general block.* This method will not be further discussed here.
2. *Indirect Distribution.* An alternative strategy distributes *GRID* in such a way that the distribution segments are directly modeled after the components of the partition. Such sets are specified by an explicit mapping from indices to processors. This will be discussed below in some more detail.

3.3 Indirect Distributions

Indirect distributions can be used to model arbitrarily structured grid partitions by setting up a *mapping array.* Let A, of shape $[1:n]$, be given as above, and let $\delta^G : [1:n] \rightarrow \mathbf{P}$ denote the grid partition. The mapping array, say *MAP*, with the same shape as A, is defined by setting $MAP(i) := \delta^G(i)$. An indirect distribution for A via mapping array *MAP* is represented in the program by *INDIRECT(MAP)*; the effect is to map each index i of A to processor $\mathbf{P}(MAP(i))$.

The application of this method to the example code of Figure 1 yields the following HPF program:

```
!HPF$ PROCESSORS P(NUMBER_OF_PROCESSORS())
      TYPE (NODE), ALLOCATABLE :: GRID(:)
      REAL, ALLOCATABLE :: TMP(:)
```

```
        INTEGER, ALLOCATABLE :: MAP(:)          ! mapping array
!HPF$  DYNAMIC, DISTRIBUTE(BLOCK) :: GRID
!HPF$  DYNAMIC, ALIGN WITH GRID   :: TMP
!HPF$  DISTRIBUTE(BLOCK) :: MAP
        ...
! Read grid size and allocate GRID, TMP, and MAP
       READ(*, *) N
       ALLOCATE(GRID(N), TMP(N), MAP(N))
        ...
       CALL PART(GRID,MAP) ! partition the grid and define MAP
!HPF$ REDISTRIBUTE GRID(INDIRECT(MAP)) ! redistribute GRID,TMP
! Grid processing
```

3.4 Control of Communication Schedules

The analysis of data access patterns involving indirect references must be performed at runtime. Since an HPF parallel loop (annotated by **INDEPENDENT**) does not contain loop-carried dependences, all the communication required for the loop can be done immediately before and/or immediately after its execution. The standard approach to translating such a loop has been the *inspector/executor* paradigm [19]. In this paradigm, the inspector determines the work distribution of the loop and analyzes the access patterns to arrays, resulting in the computation of communication schedules for nonlocal accesses. Schedules may either be *gather schedules* – for nonlocal read accesses, or *scatter schedules* – for nonlocal writes. The executor performs the actual communication according to the schedules determined by the inspector and executes the iterations of the loop. The PARTI and CHAOS libraries [26] provide a set of primitives that support the inspector/executor paradigm.

The inspector phase may be very expensive and represents a significant runtime overhead, which in some cases may dominate the overall execution time [26, 4]. It is therefore important to be able to deal with situations in which the schedule computed by the inspector remains invariant over multiple executions of the parallel loop (for example, if the parallel loop is enclosed in a time-step loop). If this is the case, the inspector needs to be executed only when the loop is entered for the first time, suppressing subsequent computations of the communication schedule. In recent years, a number of studies were performed with the objective of identifying and deleting redundant inspector computations by compile-time or runtime analysis [8, 26]. We have taken an alternative approach by allowing *explicit schedule control*, giving the programmer linguistic mechanisms to control the computation and application of communication schedules. These language extensions, described in detail in [4], have been included in HPF+ and implemented in VFCS. They are illustrated by the example discussed below.

Explicit Schedule Control The code fragment below illustrates the use of *schedule variables* to control the computation of a communication schedule.

```
!HPF+ SCHEDULE :: S
      ...
      DO t = 1, max_time
      ...
!HPF+      INDEPENDENT, ON HOME(C(I)), GATHER (B::S)
        L: DO I = 1, N
             ...
             A(I) = B(IX(I)) + C(I)
             ...
           END DO
           ...
           IF RECONFIGURE(...) THEN CALL RECOMPUTE(IX)
!HPF+         RESET S
           END IF
      END DO
```

Here, S is an explicitly declared schedule variable which is initially undefined. Therefore, the first time the parallel loop is encountered, the inspector is fully executed and determines a communication schedule as discussed above. This schedule is assigned to S. Upon subsequent executions of the parallel loop, this schedule is reused, as long as the call to *RECONFIGURE* yields **false**. Once this call yields **true**, S is *reset*, resulting in the disassociation of S from the previously computed schedule, and setting its value to *undefined*. With the next execution of the parallel loop, the inspector has to be run again, and the above cycle is restarted.

We show the effect of redundant inspector elimination for a kernel based on PAM-CRASH [11], a commercial Finite Element code, with a grid containing 35571 nodes and 35000 elements. Table 1 shows the timings for the kernel as parallelized by VFCS and executed on the Meiko CS-2, without and with explicit schedule reuse.

4 Beyond HPF and Its Paradigm

HPF was primarily designed to support data parallel programming on architectures for which memory locality is important. Consequently, the language is not necessarily well suited to other classes of applications. In this section, we will discuss a few aspects of this issue.

HPF does best on simple, regular problems, such as dense linear algebra and partial differential equations on regular meshes. It is also effective for many irregular problems, if the full scope of the language is used. This includes the important class of multiblock problems and sweeps over unstructured meshes (Section 3); other work in this area includes adaptive N-body simulations [15] and sparse matrix computations [32]. However, there are data parallel applications, for which it is not obvious that they can be handled well in HPF. One

Times without schedule reuse

Processors	Total (No Reuse))	Inspector	Gather	Scatter	Speedup
2	598.98	347.63 (58%)	2.31	2.38	0.9
4	304.75	172.48 (57%)	4.33	6.22	1.8
8	160.19	89.60 (56%)	5.33	7.81	3.4
16	100.04	50.66 (51%)	8.68	10.78	5.5
32	80.99	36.34 (45%)	10.78	11.65	6.7

Times with schedule reuse

Processors	Total (Schedule Reuse)	Inspector	Speedup
2	285.22	1.47	1.9
4	153.00	0.81	3.6
8	79.10	0.47	6.9
16	47.47	0.31	11.5
32	35.31	0.20	15.4

Sequential Time: 545.45 secs

Table 1. Times for Finite Element kernel.

example of such a method is Discrete Simulation Monte Carlo (DSMC) [23].

For some scientific applications, the discipline enforced by the HPF model is too rigid: they need a capability to express parallelism in a less structured way. Examples of such task parallel applications include real-time signal processing, branch and bound problems, and multidisciplinary applications. They can be generally characterized by the fact that tasks may be created dynamically in an unstructured way, different tasks may have different resource requirements and priorities, and that the structure and volume of the communication between a pair of tasks may vary dramatically, from an exchange of boolean signals to data transfers involving millions of numbers.

A number of methods have been proposed to address this issue in the context of HPF. The HPF **INDEPENDENT** directive and the **TASK_REGION** provide limited support for fine-grain tasking. On the other hand, the use of coarse-grain tasks, each comprising an entire HPF program, is both easy and natural. One is, in effect, wrapping HPF in a coordination language. There have been a number of efforts along this line. One such approach is the authors' language Opus [5]. Opus encapsulates HPF programs as object-oriented modules, passing data between them by accessing *shared abstractions (SDAs)* which are monitor-like constructs. Opus was explicitly designed to support multidisciplinary analysis and design, and related applications with significant coarse-grain parallelism.

5 Conclusion

HPF is a well-designed language which can handle most data parallel scientific applications with reasonable facility. However, as architectures evolve and scientific programming becomes more sophisticated, some limitations of the language are becoming apparent. Actual research proceeds in a number of different directions. For example, HPC++ [12] is an effort to design an HPF-style language using C++ as a base. On the other hand, F - - [9] is an attempt to provide a lower-level data-parallel language than HPF. Like HPF, F - - provides a single thread of flow control. But unlike HPF, F - - requires all communication to be explicit using "get" and "put" primitives.

While it is difficult to predict where languages will head, the coming generation of SMP-cluster architectures may induce new families of languages which will take advantage of the hardware support for shared-memory semantics with an SMP, while covering the limited global communication capability of the architectures. In this effort the experience gained in the development and implementation of HPF will surely serve us well.

Acknowledgment The authors would like to thank Kevin Roe (ICASE), Siegfried Benkner and Viera Sipkova (University of Vienna), Guy Lonsdale (NEC) and George Mozdzynski (ECMWF) for many fruitful discussions on this subject. The performance figures were provided by S. Benkner and V. Sipkova.

References

1. E. Albert, K. Knobe, J. D. Lukas, and Jr. G. L. Steele. Compiling Fortran 8x array features for the Connection Machine Computer System. In *Proceedings of the Symposium on Parallel Programming: Experience with Applications, Languages, and Systems (PPEALS)*, pages 42–56, New Haven, CT, July 1988.
2. F. André, J.-L. Pazat, and H. Thomas. PANDORE: A system to manage data distribution. In *International Conference on Supercomputing*, pages 380–388, Amsterdam, The Netherlands, June 1990.
3. S. Benkner et al. *Vienna Fortran Compilation System - Version 1.2 - User's Guide.* Institute for Software Technology and Parallel Systems, University of Vienna, October 1995.
4. S. Benkner, P. Mehrotra, J. Van Rosendale and H. P. Zima. High-Level Management of Communication Schedules in HPF-like Languages. Technical Report, TR 97-5, Institute for Software Technology and Parallel Systems, University of Vienna, April 1997.
5. B. Chapman, M. Haines, P. Mehrotra, J. Van Rosendale, and H. Zima. Opus: A coordination language for multidisciplinary applications. *Scientific Programming (to appear)*, 1998.
6. B. Chapman, P. Mehrotra, and H. Zima. Programming in Vienna Fortran. *Scientific Programming*, 1(1):31–50, Fall 1992.
7. B. Chapman, P. Mehrotra, and H. Zima. Extending HPF for Advanced Data Parallel Applications. *IEEE Parallel and Distributed Technology*, Fall 1994, pp. 59-70.

8. R. Das, J. Saltz, R. von Hanxleden. Slicing Analysis and Indirect Access to Distributed Arrays. In: *Proc. 6th Workshop on Languages and Compilers for Parallel Computing*, 152-168, Springer Verlag (August 1993).
9. B. Numrich. A Parallel Extension to Fortran 90. In: *Proc. Spring'96 Cray User Group Conference*, Barcelona, March 1996.
10. P. Hatcher, A. Lapadula, R. Jones, M. Quinn, and J. Anderson. A production quality C* compiler for hypercube machines. In *3rd ACM SIGPLAN Symposium on Principles Practice of Parallel Programming*, pages 73–82, April 1991.
11. E. Haug, J. Dubois, J. Clinckemaillie, S. Vlachoutsis, G. Lonsdale. Transport Vehicle Crash, Safety and Manufacturing Simulation in the Perspective of High Performance Computing and Networking. *Future Generation Computer Systems*, Vol.10, pp. 173-181, 1994.
12. High Performance C++. *Http://www.extreme.indiana.edu/hpc++/index.html.*
13. High Performance FORTRAN Forum. *High Performance FORTRAN Language Specification, Version 2.0*, January 1997.
14. S. Hiranandani, K. Kennedy, and C. Tseng. Compiling Fortran D for MIMD Distributed Memory Machines. *Communications of the ACM*, 35(8):66–80, August 1992.
15. Y.C.Hu, S.Lennart Johnsson, and S.-H.Teng. A Data-Parallel Adaptive N-body Method. Proc.Eight SIAM Conference on Parallel Processing for Scientific Computing, March 14 – 17, 1997.
16. C. Koelbel and P. Mehrotra. Compiling global name-space parallel loops for distributed execution. *IEEE Transactions on Parallel and Distributed Systems*, 2(4):440–451, October 1991.
17. J. Li and M. Chen. Generating explicit communication from shared-memory program references. In *Proceedings of Supercomputing '90*, pages 865–876, New York, NY, November 1990.
18. P. Mehrotra. Programming parallel architectures: The BLAZE family of languages. In *Proceedings of the Third SIAM Conference on Parallel Processing for Scientific Computing*, pages 289–299, Los Angeles, CA, December 1988.
19. P. Mehrotra and J. Van Rosendale. Compiling high level constructs to distributed memory architectures. In *Proceedings of the Fourth Conference on Hypercube Concurrent Computers and Applications*, March 1989.
20. P. Mehrotra and J. Van Rosendale. Programming distributed memory architectures using Kali. In A. Nicolau, D. Gelernter, T. Gross, and D. Padua, editors, *Advances in Languages and Compilers for Parallel Processing*, pages 364–384. Pitman/MIT-Press, 1991.
21. P. Mehrotra, J. Van Rosendale, and H.Zima. High Performance Fortran: History, Status, and Future. *Parallel Computing*, 1998 (in print).
22. J. H. Merlin. Adapting fortran 90 array programs for distributed memory architectures. In H. P. Zima, editor, *Proc. First International ACPC Conference, Salzburg, Austria*, pages 184–200. Lecture Notes in Computer Science 591, Springer Verlag, 1991.
23. D. Middleton, P. Mehrotra, and J. Van Rosendale. Expressing Direct Simulation Monte Carlo Code in High Performance Fortran. In *Proceedings of the Seventh SIAM Conference on Parallel Processing for Scientific Computing*, pages 698–703, February 1995.
24. *MIMDizer User's Guide, Version 7.02.* Placerville, CA., 1991.
25. D. Pase. MPP Fortran programming model. In High Performance Fortran Forum, January 1992.

26. R. Ponnusamy, J. Saltz, A. Choudhary. Runtime Compilation Techniques for Data Partitioning and Communication Schedule Reuse. Technical Report, UMIACS-TR-93-32, University of Maryland, April 1993.
27. A. P. Reeves and C. M. Chase. The Paragon programming paradigm and distributed memory multicomputers. In *Compilers and Runtime Software for Scalable Multiprocessors, J. Saltz and P. Mehrotra Editors*, Amsterdam, The Netherlands, Elsevier, 1991.
28. A. Rogers and K. Pingali. Process decomposition through locality of reference. In *Conference on Programming Language Design and Implementation*, pages 69–80, Portland, OR, June 1989.
29. M. Rosing, R. W. Schnabel, and R. P. Weaver. Expressing complex parallel algorithms in DINO. In *Proceedings of the 4th Conference on Hypercubes, Concurrent Computers, and Applications*, pages 553–560, 1989.
30. R. Rühl and M. Annaratone. Parallelization of Fortran code on distributed-memory parallel processors. In *Proceedings of the International Conference on Supercomputing*. ACM Press, June 1990.
31. J. Saltz, K. Crowley, R. Mirchandaney, and H. Berryman. Run-time scheduling and execution of loops on message passing machines. *Journal of Parallel and Distributed Computing*, 8(2):303–312, 1990.
32. M.Ujaldon, E.L.Zapata, B.Chapman, and H.Zima. Vienna Fortran/HPF Extensions for Sparse and Irregular Problems and Their Compilation. *IEEE Transactions on Parallel and Distributed Systems*, 8(11), November 1997.
33. H. Zima, H. Bast, and M. Gerndt. Superb: A tool for semi-automatic MIMD/SIMD parallelization. *Parallel Computing*, 6:1–18, 1988.
34. H. Zima, P. Brezany, B. Chapman, P. Mehrotra, and A. Schwald. Vienna Fortran – a language specification. Internal Report 21, ICASE, Hampton, VA, March 1992.
35. H. Zima and B. Chapman. Compiling for Distributed Memory Systems. *Proceedings of the IEEE*, Special Section on Languages and Compilers for Parallel Machines, pp. 264-287, February 1993.

PAVOR - Parallel Adaptive Volume Rendering System

Michael Meißner

WSI/GRIS
University of Tübingen
Auf der Morgenstelle 10, C9
D-72076 Tübingen
Germany
meissner@gris.uni-tuebingen.de

Abstract. Volume rendering is becoming a key technology in the field of scientific computing. The process of generating images from 3D volumetric data is highly demanding and pushes systems to their computational limit.
PAVOR is a parallel adaptive volume rendering system which uses available resources (workstations, PCs, etc). Besides well-known software optimizations used in volume rendering algorithms, we introduce adaptive rendering. This process provides the capability of rendering tiles of the view plane at different sampling rates even allowing different transfer functions per tile. As a result, the frame rate increases by simultaneously increasing the quality of perception.

1 Introduction

Volume rendering puts high demands on systems which claim to be capable of rendering data sets at reasonable frame-rates. Typical sizes of volume data sets range from 128^3 to 512^3 voxel elements each holding values from one to four bytes. Hence, the overall size of such data sets can be up to 512MBytes. It is obvious that the visualization of this data is far beyond nowadays desktop computers. For larger data sets, even supercomputers cannot provide enough power to achieve real-time frame rates. To visualize such data sets, dedicated computer-graphics tools are necessary [8]. The interesting structures hidden in the volume must be identified, visualized and comprehensively understood.

We can classify the previously done work into three categories. The first makes use of supercomputers or massive parallel architectures [2] [11] [20]. These architectures are not widely accessible and not suited for the average user. The second category intends to develop special purpose hardware to provide the required computational power [7] [13] [16] [17] [10] [1]. Many of those architectures are still in experimental stage, except [7], have certain limitations, or will be too expensive for the average user. Further it is not clear how some of these architectures handle oversampling, one of the most important features in volume rendering. Finally, the third strategy uses the widely available computing power on networks and aims to make this power usable for volume visualization [9] [6] [14] [15].

Our PAVOR system belongs to the latter strategy. Challenges are to overcome the limited transfer bandwidth of the network, to establish an even workload, to avoid stall conditions in a multiuser environment, and to effectively exploit the limited computational power available on the network. Since the available systems belong to a multi-user environment, load balancing has to take this into account and stall conditions have to be prevented. We will introduce an improved minimal redundant self scheduling scheme that perfectly fulfills those requirements.

Our approach addresses high quality volume rendering of large data sets providing the feature of oversampling. Furthermore, we introduce adaptive sampling: Only a few areas of the view plane are sampled at cost intense high sampling rate while other areas are rendered at a much lower sampling rate. This is precisely the way, the perception of the human eye works. Tightly coupling an eye-tracking mechanism would greatly improve the system and the usefulness of adaptive sampling rates but has yet not been included.

2 Visualization Algorithm

As mentioned in Section 1, our algorithm is image space oriented. The view plane position is calculated using given rotation and zoom parameters. Once the view plane is located, rays are casted from the eye point through each pixel of the view plane into the view frustum. Figure 1 illustrates this procedure. To

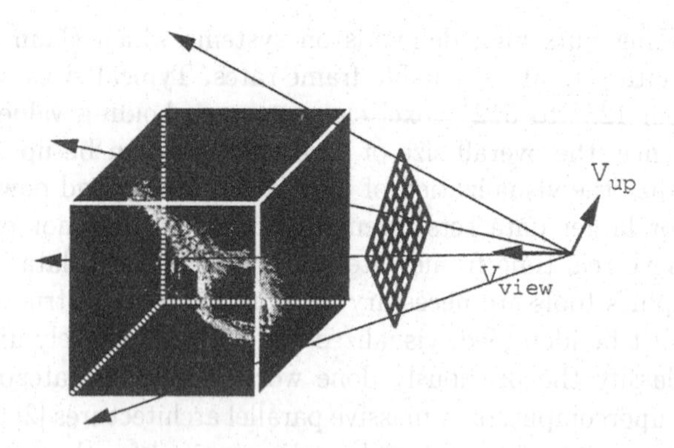

Fig. 1. Rays are casted from the view plane into the view frustum.

detect whether a ray hits the view volume or not, we use a three dimensional extension of the Cohen-Sutherland algorithm. Rays which miss the volume can be discarded. Otherwise samples along the ray will be taken by interpolating values of the neighboring voxels. Additionally, for each sample a gradient is computed out of the gradients at neighboring voxel locations.

Each sample has to be classified. Classification is the stage of assigning a RGBA component to a sample [3] [12]. This is done using a look-up table containing the RGBA values for all possible voxel values. After classification, each colored sample has to be shaded corresponding to its gradient. Currently, we apply Phong shading [4]. This is a very expensive shading model and has to be replaced by a look-up based Phong model which even allows multiple light-sources without loss of performance [19]. The look-up table for this shading requires 1.5KByte of memory and has to be re-computed each time one of the light sources changes its location. Changing light sources is a rare operation compared to changing the view point or the classification functions.

To obtain a final pixel on the view plane, shaded samples along each ray have to be composed to a single pixel value. This is done by composing the values along a ray in a front-to-back or back-to-front order.

3 Parallelism

Our system uses the PVM 3.4 message passing system [5]. It is portable along other platforms as long as PVM supports those platforms. On each machine of the network a PVM daemon must be running, clustering the single systems to a virtual single machine. We were running PAVOR on a dedicated network of 6 R3000 SGI Indigo which is shown in Figure 2. Furthermore we used the machines available at our department. These are SGI machine ranging from R4000 to R1000.

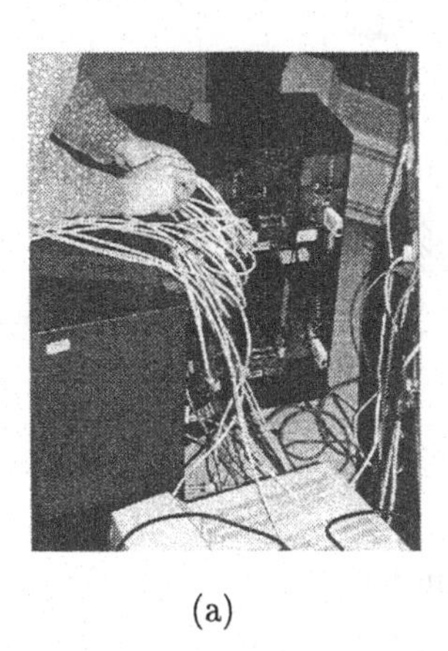

(a)

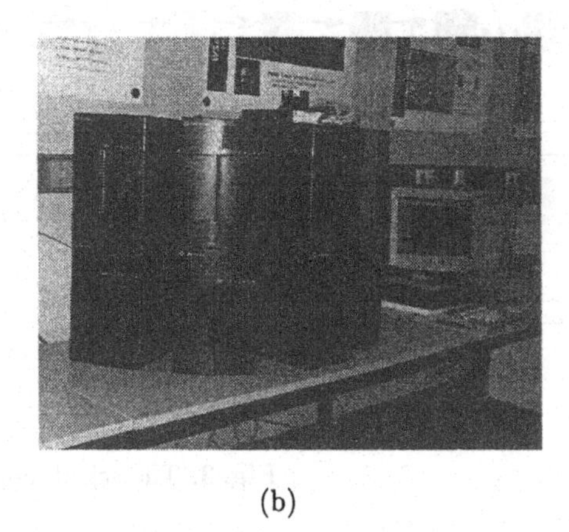

(b)

Fig. 2. Parallel network of eight SGI Indigo (R3000) connected by 10MBit to a switch which again has two 100MBit uplink ports to a SGI Indy workstation which is the master.

3.1 Parallelization of the Algorithm

In our system, the view plane is split into smaller, non-redundant tiles which can be rendered individually. In contrast to other approaches [9] our tiles are not restricted to quadratic shapes.

3.2 Tile Scheduling

The primary intension of our system is to use available resources on an existing network. Therefore, we have to take into account that the available machines will not be dedicated to our purpose but to a multiuser environment. This can lead to unexpected situations and to an extremely unbalanced availability of the machines over time. We solve this problem by applying a self-scheduling scheme [18], which has already been used in other parallel volume rendering implementations [6] [9]. We will briefly explain the overall principle and our improvements of it:

Tasks are organized in a list and spread over the workers one at a time. Once a task is completed, the results are transmitted from the worker to the master, and a new task is sent to the worker. At the same time the completed task is removed from the task list. Figure 3 illustrates this mechanism. Unfortunately,

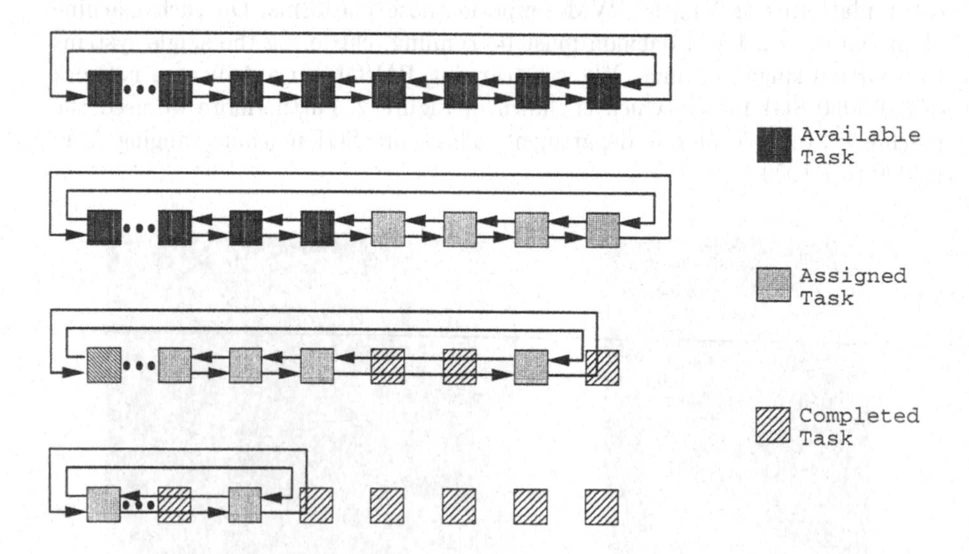

Fig. 3. Tile scheduling principle.

a single machine is able to stall the entire process. Therefore, [9] proposed a *redundant self-scheduling (RSS)* which eliminates this problem as long as the master process does not stall. The idea is to distribute tasks multiple times once fewer tasks are remaining than there are workers. Unfortunately, each tile can be redistributed to all available workers within the network. This leads to a high redundancy which we reduce by introducing a maximum redundancy (MR). How often a tile can be redistributed is limited by the MR value. Depending on the reliability of the network, the MR value has to be set appropriately. Hence, stall conditions can still be prevented by simultaneously allowing controllable redundancy. A value of three turned out to be a good trade-off for up to 10 workers and a moderate work-load on all machines.

Each of the workers will perform the generation of a tile until its completion. Due to the redundant redistribution of tasks, multiple workers can generate the same tile and once one returns the results the other workers will continue their redundant work. This happens very often for the very last tiles of a frame. As a result, during the generation of the first tiles of a frame many workers will still be redundantly working on the last tiles of the previous frame. The described situation is highly undesirable and to further reduce the redundancy, we notify the other workers to stop their work and to grab another task.

Figure 4 gives an overview of the network communication necessary between master and workers.

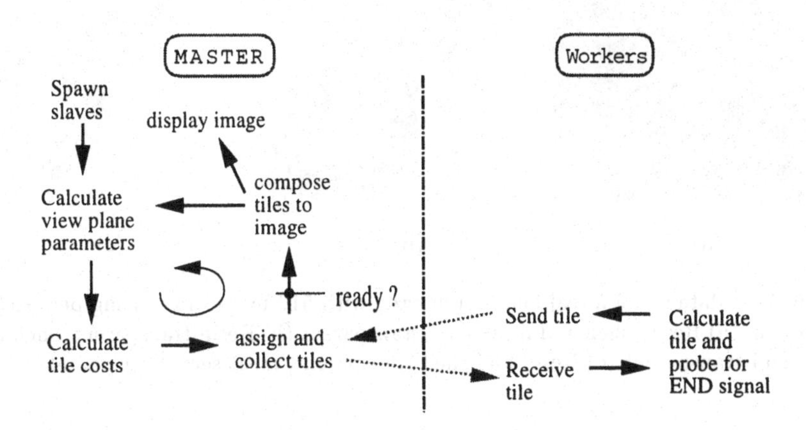

Fig. 4. Communication between master and workers.

3.3 Load Balancing

Due to the possible different sampling rates of the individual tiles, the workload can be quite different for each of them. To ensure a good load balance, we use the following cost function for each tile i:

$$Cost_i = Sampling_I * Sampling_J * Sampling_K \quad (1)$$

Truly, this is a fairly general measure and has to be combined with the time of tile i needed during the generation of the last frame. Even for incremental view point changes this approved to suffice as an accurate measure.

In case the cost functions of all tiles vary too much, i.e., by a factor of more than three, we split tiles into smaller ones and respectively, merge smaller tiles into larger ones. To achieve a good load balance, tiles with high computational cost are distributed first and if possible to the most powerful workers.

4 Results

The data set chosen is a 128^3 down-sampled lobster residing inside a block of resin. Density values of resin range from zero through 40, for the shell from 40 through 80, and for the skin over 80. Figure 5 shows a high quality images of the lobster using different classification functions.

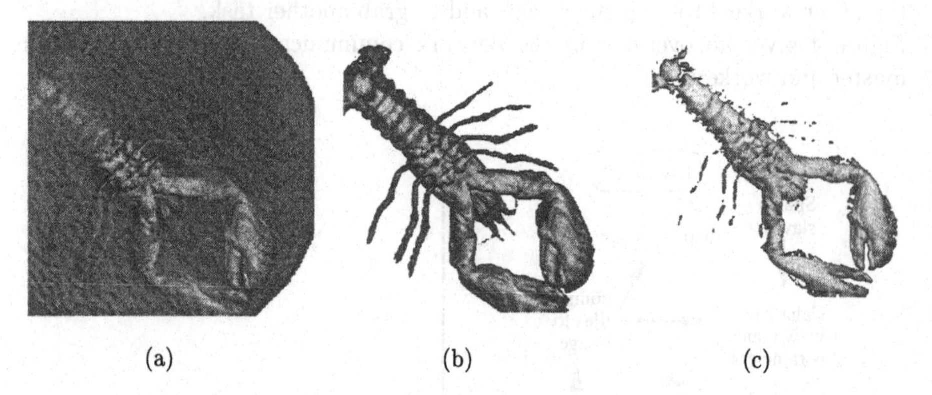

(a) (b) (c)

Fig. 5. 128^3 data set of a real lobster rendered with 512 by 512 rays using perspective projection. (a) Resin, shell and meat semi-transparent (b) Resin transparent, shell and meat semi-transparent (c) Resin and shell transparent, meat semi-transparent

4.1 Redundantly Calculated Tiles

The amount of redundantly calculated tiles depends on the distribution scheme chosen. It stays zero in case each tile is distributed once and increases with the maximum number of times a tile can be redistributed. This dependency is shown in Figure 6. As the two figures illustrate, the redundancy does not mainly depend on the amount of tiles but on the amount of workers being involved in the frame generation.

The overhead spent on generating redundant tiles entire can be reduced by increasing the amount of tiles. Unfortunately, this cannot be done arbitrary since each tile requires communication over the network including setup costs. For frame rates close to interactivity a maximum of 64 tiles per frame turned out to be a good trade-off between overhead, number of tiles, and frame time.

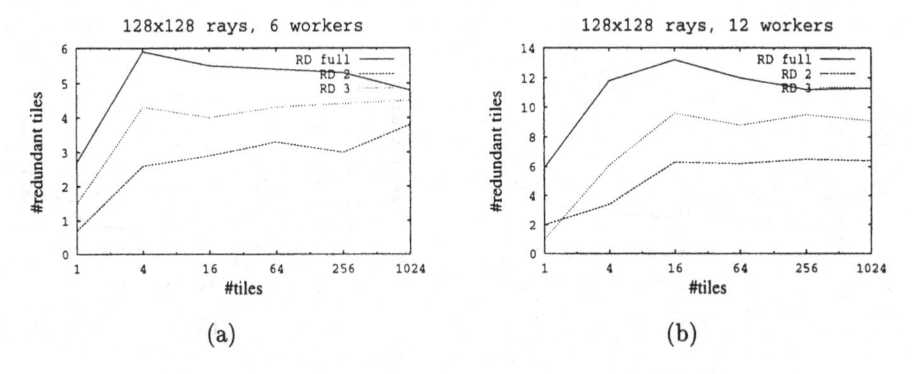

(a) (b)

Fig. 6. Amount of redundantly calculated tiles (a) for six workers (b) for 12 workers.

4.2 Parallelization Speed-up

Generating one image using one-fold sampling ratio takes approximately 15 seconds on an SGI Indigo (R3000). Applying oversampling of two or four will increase the frame time by a factor of 8 or 64. For such tasks, the achieved speed-up, exploiting the parallelism can be considered to be linear.
Figure 7(a) and (b) show the time per frame depending on the strategy chosen. *RD n* denotes that each tile is redistributed up to n times. Additionally, *(k)* denotes that once a tile is returned, all other workers generating this tile will be notified to stop the computation and to grab a new task. The values of Figure 7(a) were measured on an optimal network with no load on the machines and no load on the network. As it can be seen, the strategies do not differ for number of tiles larger than 64. This will change once we get close to interactivity where the number of tiles has to be kept small otherwise network traffic will reduce the frame rate.
Figure 7(b) was generated using 10 workers in a real world situation within a multiuser environment. As it can be observed, distributing each tile only one time does not lead to the best results since a single stalled machine can delay the generation of a single tile, stalling the entire process. Even worse, single frames can be completed in different order than they should appear. This might be acceptable for an off-line generated animation but not for an interactive tool. The scheduling using a RD value larger than zero performed better for a reasonable number of tiles.

5 Conclusions

We presented PAVOR, a parallel adaptive volume rendering system capable of rendering frames with varying sampling-rates and changing classification functions within each frame. Furthermore, we introduced a minimal redundant self-scheduling scheme preventing stall conditions and reducing redundant calculations to a minimum, generally occurring during the computation of the last

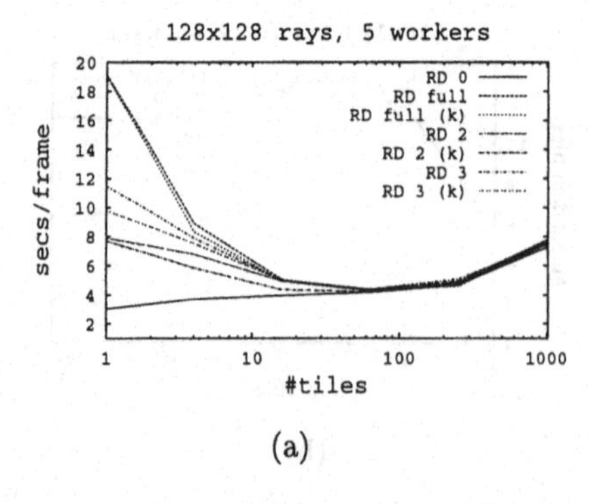

(a)

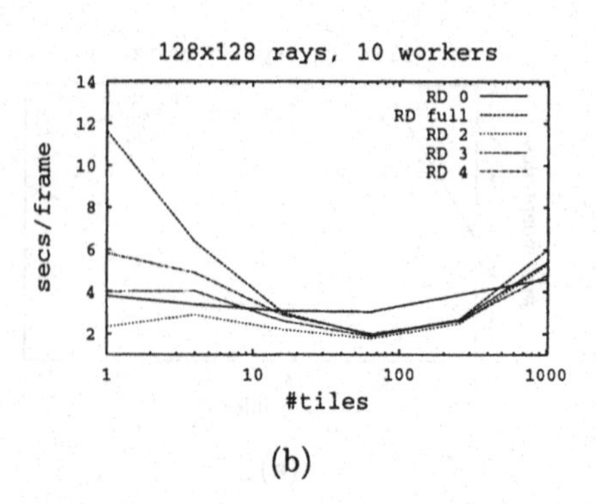

(b)

Fig. 7. Time per frame using different strategies (a) 6 workers, no load on machines and network (b) 10 workers, multi-user environment.

remaining tiles. Thus, we achieve a more efficient usage of the available resources. Additionally, we obtain a good load balance of the tasks by using a cost measure for each tile.

6 Future Work

The current system is not yet capable of delivering interactive frame-rates. Therefore, we will speed-up the implementation by using look-up based gradient generation and shading. Additionally, we intend to compress data before sending it over the network to reduce network traffic.
To enable the visualization of data sets exceeding the main memory size of a single machine, the algorithm has to be extended. So far data is replicated in each worker but this will require splitting the data over the workers.

References

1. I. Bitter and A. Kaufman. A ray-slice-sweep volume rendering engine. In *Proceedings of the 1997 SIGGRAPH/Eurographics Hardware Workshop*, Los Angeles, CA, 1997.
2. G. Cameron and P. E. Underill. Rendering volumetric medical image data on a simd architecture computer. *Proceedings of the Third Eurographics Workshop on Rendering*, 1992.
3. R. A. Drebin, L. Carpenter, and P. Harahan. Volume rendering. *Computer Graphics*, 22:65–74, aug 1988.
4. J. D. Foley, A. van Dam, S. K. Feiner, and J. F. Hueghes. *Computer Graphics: Principles and Practice.* Addison-Wesley, Reading, MA, 1990.
5. A. Geist, A. Beguelin, J. Dongarra, W. Jiang, R. Manchek, and V. Sunderman. *PVM: Parallel Virtual Machine - A User's Guide and Tutorial for Networked Parallel Computing.* MIT Press, Cambridge, MA, 1994.
6. C. Giertsen and J. Petersen. Parallel voluem rendering on a network of workstations. *IEEE Computer Graphics & Applications*, 13(6):16–23, November 1993.

7. T. Guenther, C. Poliwoda, C. Reinhard, J. Hesser, R. Maenner, H.-P. Meinzer, and H.-J. Baur. VIRIM: A massively parallel processor for real-time volume visualization in medicine. In *Proceedings of the 9th Eurographics Hardware Workshop*, pages 103–108, Oslo, Norway, September 1994.
8. A. Kaufman. Volume visualization. *IEEE CS Press Tutorial*, 1991.
9. G. Knittel. A parallel algorithm for scientific visualization. In *Proceedings of the 1996 International Conference on Parallel Processing*, Bloomingdale, USA, aug 1996.
10. G. Knittel and W. Straßer. Vizard - visualization accelerator for realtime display. In *Proceedings of the 1997 SIGGRAPH/Eurographics Hardware Workshop*, Los Angeles, CA, 1997.
11. P. Lacroute. Real-time volume rendering on shared memory multiprocessors using the shear-warp factorization. *Proceedings of the ACM/IEEE '95 Symposium on Parallel Rendering*, pages 15–22, oct 1995.
12. M. Levoy. Display of surfaces from volume data. *IEEE Computer Graphics & Applications*, 8(3):29–37, May 1988.
13. J. Lichtermann. Design of a fast voxel processor for parallel volume visualization. In *Proceedings of the 10th Eurographics Hardware Workshop*, pages 83–92, Maastricht, The Netherlands, 1995.
14. K.-L. Ma and J. S. Painter. Parallel volume visualization on workstations. *IEEE Computer Graphics*, 17(1), 1993.
15. K.-L. Ma, J. S. Painter, C. D. Hansen, and M. F. Krogh. A data distributed, parallel algorithm for ray-traced volume rendering. In *Proceedings of the 1993 Parallel Rendering Symposium*, pages 15–22, San Jose, CA, October 1993.
16. R. Osborne, H. Pfister, H. Lauer, N. McKenzie, S. Gibson, W. Hiatt, and T. Ohkami. Em-cube: An architecture for low-cost real-time volume rendering. In *Proceedings of the 1997 SIGGRAPH/Eurographics Hardware Workshop*, Los Angeles, CA, 1997.
17. H. Pfister and A. Kaufman. Cube-4 - a scalable architecture for real-time volume rendering. *1996 Symposium on Volume Visualization*, October 1996.
18. C. D. Polychronopoulos. *Parallel Programming and Compilers.* Kluwer Academic Publishers, Boston, MA, 1988.
19. J. Terwisscha van Scheltinga, J. Smit, and M. Bosma. Design of an on chip reflectance map. In *Proceedings of the 10th Eurographics Hardware Workshop*, pages 51–55, Maastricht, The Netherlands, 1995.
20. T. S. Yoo, U. Neumann, H. Fuchs, S. M. Pizer, J. Rhoades T. Cullip, and R. Whitaker. Direct visualization of volume data. *IEEE Computer Graphics & Applications*, 12(4):63–71, 1992.

Simulation Steering with SCIRun in a Distributed Environment

Michelle Miller, Charles D. Hansen, and Christopher R. Johnson

University of Utah, Salt Lake City, UT 84112, USA,
{mmiller, hansen, crj}@cs.utah.edu,
http://www.cs.utah.edu/~sci

Abstract. Building systems that alter program behavior during execution based on user-specified criteria (computational steering systems) has been a recent research topic, particularly among the high-performance computing community [1–5]. To enable a computational steering system with powerful visualization capabilities to run in a distributed computational environment, a distributed infrastructure (or runtime system) is required. This infrastructure permits one to harness a variety of machines to collaborate on an interactive simulation. Building such an infrastructure requires devising strategies for coordinating execution across machines (concurrency control mechanisms), mechanisms for fast data transfer between machines, and mechanisms for user manipulation of remote execution.
We are creating a distributed infrastructure for the SCIRun computational steering system [1]. SCIRun, a scientific problem solving environment (PSE), provides the ability to interactively guide or steer a running computation. Initially designed for a shared memory multiprocessor, SCIRun is a tightly integrated, multi-threaded framework for composing scientific applications from existing or new components. High-performance computing is needed to maintain interactivity for scientists and engineers running simulations. Extending such a performance-sensitive application toolkit to enable pieces of the computation to run on different machine architectures all within the same computation proves useful for some scientists. Not only can many different machines execute this framework, but also several machines can be configured to work synergistically on computations (*e.g.*, farming off compute-intensive pieces to the "big iron").

1 Introduction

In recent years, the scientific computing community has experienced an explosive growth in both the size and the complexity of numeric computations possible. This growth is due, in large part, to both the rapid development of parallel machines and the rapid performance gains of RISC microprocessor technology. One of the significant benefits of this increased computing power is the ability to perform complex three-dimensional simulations. However, such simulations present new challenges for computational scientists. How does one effectively analyze and visualize complex three-dimensional data? How does one solve the problems of working with very large datasets often consisting of tens to hundreds of gigabytes? How does one provide tools that address these computational problems while serving the needs of scientific users?

Data analysis and visualization play critical roles in the scientific process. Unfortunately, these tasks are often performed only as a post-processing step after a large-scale batch job is run. For this reason, errors invalidating the results of the entire simulation may be discovered only during post-processing. Further, the decoupling of simulation and analysis can present serious scientific obstacles to the researcher. An analysis package may provide only a limited data analysis capability and may be poorly matched to the underlying physical models used in the simulation code. To add to the problem, many of these packages are simply unable to effectively manage the vast amounts of data being generated by many simulations–especially those generated on large parallel systems. As a result, the researcher may expend significant effort trying to use a data analysis package only to walk away frustrated.

Given the limitations of the batch/post processing cycle, a better approach might be to break the cycle and improve the integration of simulation and analysis. Scientists want more interaction than is currently present in most simulation codes — a desire the scientific computing community is trying to find better ways to address. Clearly, however, the solutions will not be simple ones, since the problems encountered by computational scientists encompass a wide range of issues. To address some of these needs, we have created the SCIRun problem solving environment. In this chapter, we focus on the SCIRun system [1], addressing the issues of interaction and integration of scientific simulation and visualization in a distributed computing environment.

2 Interactivity and Computational Steering

Interactive scientific visualization and computational steering require low-latency and high bandwidth computation in the form of model generation, solvers, and visualization. Latency is particularly a problem when analyzing large datasets, constructing and rendering three-dimensional models/meshes, and allowing a scientist to alter the parameters of the computation interactively (thus "steering" the computation). However, large-scale computational models often exceed the system resources (memory and storage) of a single machine, motivating closer investigation of meeting these same needs with a distributed computational environment comprised of many machines.

2.1 The SCIRun Problem Solving Environment

To achieve execution speeds needed for interactive three-dimensional problem solving and visualization, a multi-threaded problem solving environment and computational steering system, called SCIRun [6, 7], was built to make use of a shared memory multiprocessor, notably the SGI Power Challenge and SGI Origin 200/2000. The SCIRun scientific problem solving environment is a *computational steering* system [1] that allows the interactive construction, debugging, and steering of large-scale scientific computations. SCIRun can be conceptualized as a *computational workbench*, in which a scientist can design via a dataflow programming model and modify simulations interactively. SCIRun enables scientists, for example, to interactively modify geometric models and change numerical parameters and boundary conditions, as well as to modify the level of mesh adaptation needed for an accurate numerical solution, while at the same time visualizing intermediate (and final) simulation results.

An example of the SCIRun system interface is shown in Fig. 1. The center of the picture displays a graphical representation of the dataflow network. The boxes represent computational algorithms (modules), with the lines representing data connections between the modules. Each module may have a separate user interface, such as the matrix solver interface at the left, that allows the user to control various parameters, including, for example, the maximum number of iterations, error tolerance, preconditioner, and iterative solution method. The window on the right is an interactive three-dimensional viewer that combines visualization output with data probes for exploring the data and model. When

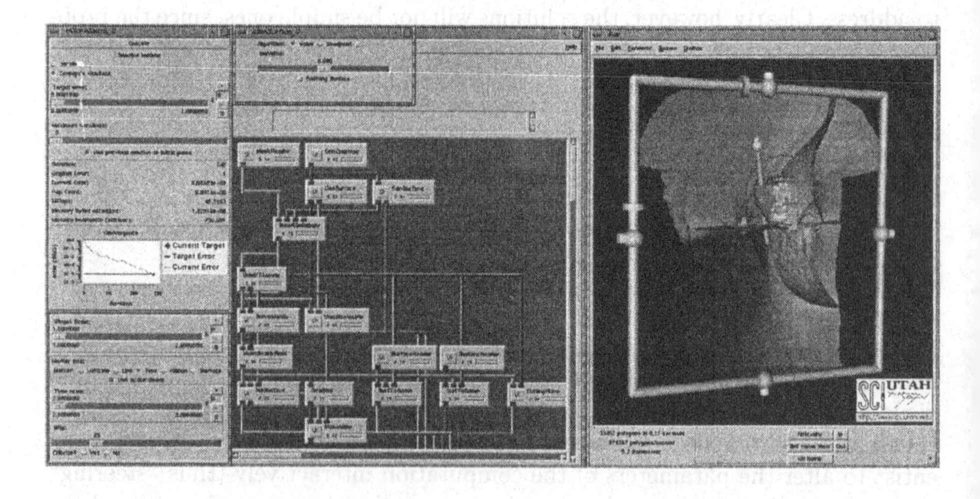

Fig. 1. An example SCIRun network, showing the dataflow programming interface, user interfaces for controlling simulation parameters, and results from an large finite element model.

the user changes a parameter in any of the module user interfaces, the module is re-executed, and all changes are automatically propagated to all downstream modules. The user is freed from worrying about details of data dependencies and data file formats. The user can make changes without stopping the computation, thus "steering" the computational process. As opposed to the typical "off-line" simulation mode - in which the scientist manually sets input parameters, computes results, visualizes the results via a separate visualization package, then starts again at the beginning - SCIRun "closes the loop" and allows interactive steering of the design, computation, and visualization phases of a simulation.

The ability to steer a large-scale simulation provides many advantages to the scientific programmer. As changes in parameters become more instantaneous, the cause-effect relationships within the simulation become more evident, allowing the scientist to develop more intuition about the effect of problem parameters, to detect program bugs, to develop insight into the operation of an algorithm, or to deepen an understanding of the physics of the problem(s) being studied.

Currently, SCIRun manages machine resources to maximize computational efficiency. In a sophisticated simulation, each of the individual components (modeling, mesh generation, nonlinear/linear solvers, visualization, *etc.*) typically consumes a large amount of memory and CPU resources. When all of these pieces are connected into a single program, the potential computational load is enormous. In order to use the resources effectively, SCIRun adopts a role similar to an operating system in managing these resources. SCIRun manages scheduling and prioritization of threads, mapping of threads to processors, inter-thread communication, thread stack growth, memory allocation policies, and memory exception signals (such as segmentation violations). These become important issues as we distribute part of the computation to another machine.

3 Related Work

The concept of steering distributed scientific computation is a relatively new endeavor. However, there are some systems that have investigated aspects of computational steering. We briefly discuss several of these below. The CUMULVS system [3, 8] allows computational scientists with distributed PVM applications to modify a fixed set of simulation parameters while using AVS to visualize the underlying computational model. The Progress and Falcon systems [2, 9] allow application programmers to develop distributed steerable applications through source code modification with the steering being assisted by a runtime system. Basically, these systems explore computational steering by exposing portions of the simulation through their interfaces. The SCIRun system provides a more holistic approach where the entire computation modeling, simulation and visualization are built and executed within a problem solving environment. Such a complete system provides greater flexibility by allowing the user to adapt computational steering at runtime rather than at compile time.

3.1 CUMULVS

The CUMULVS library [3, 8], developed at Oak Ridge National Laboratory, acts as a middle layer between PVM applications and the commercially-provided AVS visualization package [10]. CUMULVS is implemented as a user-programmed

AVS module. It has been used in distributed computational fluid dynamics applications. One of the properties of CUMULVS is that viewers can dynamically plug and unplug from running distributed simulations and visualization packages. After initializing a viewer, the application programmer can provide a list of parameters to be adjusted on-the-fly in a CUMULVS steering initialization procedure call. Separate procedure calls are used for altering scalar or vector parameters from within the application. CUMULVS supports collaborative visualization and simulation by permitting multiple viewers to view and steer a running application. Viewers can attach and detach dynamically during a simulation. An interesting checkpointing capability for rolling back and restarting a failed program run has the potential to allow cross-platform migration and heterogeneous restart of an application.

3.2 Progress and Magellan

The Progress Toolkit [2], developed at the Georgia Institute of Technology, assists application programmers in developing steerable applications. Programmers "instrument" their applications with library calls, using *steerable objects*, which can be altered at runtime through the use of the Progress runtime system. Steerable objects include sensors, actuators, probes, function hooks, complex actions, and synchronization points. Progress uses a client/server program model.

Developed by the same group, the Magellan Steering System [11] is derived from the Progress system, and extends the steering clients and steering servers model used in the initial system. This system uses a specialized specification language, ACSL, which provides commands for monitoring and steering using probes, sensors and actuators. However, application codes still must be instrumented with these commands in order to utilize the steering capabilities of this system. These systems have been used for Molecular Dynamics simulations.

Both systems are layered on top of the Falcon runtime system for on-line monitoring of running simulations. The Falcon system [9], also developed at the Georgia Institute of Technology, monitors a running program, capturing information ranging from a single program variable, much as a debugger would, to complex expressions. It also permits the monitoring of performance data, with interfaces to visualization systems, such as Iris Explorer [12]. Applications can be observed or modified through the use of a probe or actuator, respectively. However, decisions are made about which steering actions to take based on previously encoded routines stored in a steering event database located on a steering server. Clearly, codes must be instrumented to use these facilities via procedure calls and, possibly, the coercion of existing scientific codes to fit a client/server model.

4 Distributing SCIRun

Large-scale scientific computations require multiprocessors of some form due to the vast amount of data inherent in these applications/simulations. However, the computer architecture of multiprocessors varies greatly, differing in the number of processors, communication mechanism, operating system, and memory model, etc. Using abstraction to hide the underlying details of the memory model (*i.e.*, shared memory versus distributed memory) from the application allows us to take advantage of the range of machine types, from shared memory multiprocessors to massively parallel processors. However, such an abstraction creates extra layers in the software infrastructure that generally result in increased software overhead. One must strike a balance of exposing enough details to efficiently tune the application without losing portability.

To achieve our goal of computational steering of distributed simulations, we first created a mechanism for using SCIRun across multiple machines, then we can solve the problem of running a scientific code on a distributed memory architecture. This discussion focuses on the infrastructure mechanisms needed to run one SCIRun computation across multiple machines while maintaining control of the computation from the user's local workstation. The goal of SCIRun has been to maintain interactive rates for users constructing and manipulating models and simulations. To be successful with a distributed version of SCIRun, we must not impair interactive speeds by incurring excessive network latency.

To meet the demand for growing computational problem sizes, we have extended the existing communications infrastructure within SCIRun to include cross-machine communications and execution. In this way, we can preserve existing system functionality for shared-memory executions, as well as providing more architectural flexibility. In constructing a distributed system, our goal is to provide SCIRun users with the illusion of running on a single computer, while they are really utilizing several computers with potentially very different characteristics. Distributing SCIRun has been largely motivated by the large-scale computational needs and the desire to link remote computing facilities for the Utah C-SAFE ASCI Alliance Center and the NCSA PACI Alliance.

4.1 Software Architecture for Distributed SCIRun

The model for running SCIRun on multiple machines calls for the scientist to initiate the computation (*i.e.*, start SCIRun) on a graphics-capable machine, to compose an application from existing SCIRun components (*modules*), to specify which of these should run remotely, and then to start execution of the computation (*dataflow network*). Interacting with modules, viewing the rendered visualization, and interacting with three-dimensional data probes (*widgets*) within the visualization all occur on the initiating machine, which we denote as the Master. The compute-intensive dataflow subnetwork executes on the remote machine, which could be a multiprocessor machine.

We abstract away the differences between running on one or more than one machine by modifying the scheduling, inter-module communication, user interface mechanisms, and other "internal" portions of SCIRun that an application writer would typically never see. Making modifications to the infrastructure allows programs that have been designed for the SCIRun problem solving environment to function in either a multi-machine, networked environment or a single-machine, shared memory environment.

The software architecture, shown in Fig. 2, depicts this solution for a computation distributed across two machines. This architecture is based on a master/slave model. Steering and visualization takes place on the local machine, the user's computational workbench, which we call the Master SCIRun instance. The remote instance of SCIRun will be the Slave. The application writer configures a program by connecting together existing modules, choosing a subset of modules to execute remotely, and choosing the target remote machine (slave). In Fig. 2, the application program (network) consists of reading in data using a `Reader` module, then executing `RMod1`, `RMod2`, and finally the `Salmon` module to render the visualization, in that order. Modules to be run remotely have a proxy module (copy) that remains on the Master side. This proxy keeps a copy of the state and runtime information necessary for the scheduler to perform its task, while the module that performs the actual computation is instantiated on the Slave. The architecture of the Slave consists of a daemon, the *Slave Controller*, which listens on a communication channel for directives from the Master while controlling the modules on the remote side. These directives follow:

- Instantiate a module.
- Destroy a module.
- Instantiate a connection between two modules.
- Destroy a connection between two modules.
- Execute a module.
- Re-execute some portion of the remote network.

This set of directives was based on control commands in the current shared-memory version of SCIRun and completely provides the necessary control for the management of remote computation. By simply extending existing messages and functionality of the scheduling code it is possible to provide all required support for a remote machine.

Control communications is largely uni-directional, moving from the Master to the Slave, using a fire and forget communications strategy. The need for control messages from the Slave to Master is mitigated by having all user interactions initiated on the Master, with the Master then sending notification to the Slave of any changes needed.

In addition, most module-to-module communications are one-way communications. For example, Fig. 2 shows module `Reader` connected to module `RMod1`. Data flows from the output port of `Reader` to the input port of `RMod1`. In most cases, this communication happens after the module, `Reader`, has finished executing, when resulting data is passed out its output port. For remote communications, then, the communications channel must be set up before `Reader` has started to execute.

On the other hand, user interaction with remotely running modules requires bi-directional communication between the Master and Slave. Modules obtain the most current parameter values before executing in SCIRun. If modules are remote, they must request from the Master the current value. In response, the

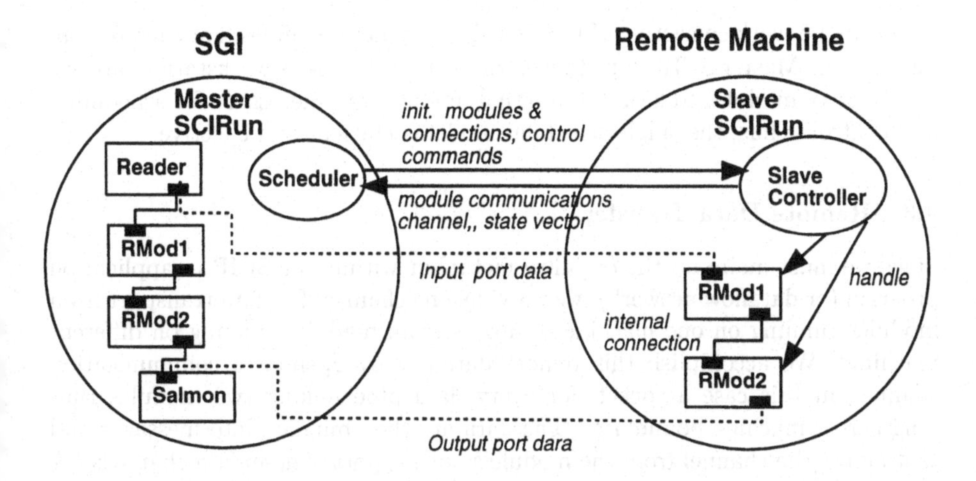

Fig. 2. Software architecture for partitioning a SCIRun dataflow network across two machines. Communications paths are shown. The Slave Controller acts as a daemon listening on a communications channel for commands from the Master SCIRun scheduler. It creates modules, connections between modules, and destroys both. It also forwards control messages, such as `execute()` and `resend()`, to the proper module. Data flows in and out of modules through the typical input and output port mechanism, although a special communication channel will be established for remote module-to-module communication.

Master sends these values to the Slave. Additionally, various GUI updates, displayed on the Master, are sent by the Slave modules running remotely, such as updating the progress bar or runtime indicator for a remote module.

Designing and implementing support for distributed computational environments in the existing SCIRun architecture can be broken down into three separate phases.

1. Design and implement distributed control infrastructure to be integrated within existing SCIRun control mechanisms (specifically the scheduler).
2. Design and implement module-to-module data communication that spans machines.
3. Design and implement user interaction and steering of remotely running modules.

4.2 Distributed Control Infrastructure

The SCIRun scheduler controls the execution order of modules and tracks and signals modules that need to run again. SCIRun is based on a hybrid form of a dataflow network where data passes from upstream modules to downstream ones. To ensure concurrency control, we use one scheduler for all machines working collaboratively on a SCIRun computation. A master-slave configuration yields a single point of control from which module scheduling decisions can be made for the entire network consisting of both local and remote modules. For each remote module, the Master SCIRun instance, running on the user's workstation, instantiates proxy modules to keep the internal module state necessary for scheduling and control. Thus, the single scheduler model is efficient and effective.

4.3 Remote Data Transfer

Since we must maintain the dataflow metaphor within one SCIRun application program (or dataflow network), we provide a mechanism for data transport from modules running on one machine to downstream modules running on different machines. We accomplish this remote dataflow by opening a communication channel, in this case a socket, for every data pipe linking two modules running across machine boundaries. The data are then bundled into messages and sent across the channel from the module's output port. These data channels are instantiated and torn-down when the user makes, or destroys, a connection between modules which cross machine boundaries. Connected modules running on the same machine, even though remote, use the current SCIRun shared memory data communication.

4.4 Remote User Interaction and Steering

SCIRun succeeds in allowing a range of intuitive user interactions with parameters of the model, choice of simulation techniques and their parameters, and the resulting visualization. When we move to a distributed environment, careful attention must be paid to maintaining close coupling between user action and system response. Further, providing visual feedback to the user about the run-time performance traits helps the user choose appropriate modules to execute remotely.

We have built a remote communication mechanism into our Tcl wrapper calls used by the GUI. These are based on a combined *push* and *pull* strategy. Modules always check the most current user-selected parameters before using them, and remote modules do the same. Similarly, a progress bar, which indicates the percentage of the computation completed, is updated on the proxy module residing on the Master by the corresponding Slave module. Finally, if a user changes a parameter in one of the steering windows, the module is rescheduled in both the single-machine and distributed versions of SCIRun.

5 Conclusions and Future Work

We have demonstrated an effective architecture for steering computations with SCIRun by combining machines that have some concept of a global address space. While this work is preliminary, we have shown that by extending the underlying control, data transfer, and user interaction mechanisms of SCIRun, the user need not be concerned with the details of inter-machine communication. Users can choose which modules in the dataflow network are to be run remotely and the target machines on which they should execute. Future work consists of detailed experiments with a variety of platforms and building in support for parallelism in a distributed memory machine. We will also investigate other communication mechanisms, as well as automatic scheduling algorithms for the distribution of the various application modules given specific network and computing infrastructure.

6 Acknowledgments

This work was supported by the NCSA PACI Alliance and the Department of Energy. Equipment was funded by the SGI Visual Supercomputing Center and the National Science Foundation.

References

1. S.G. Parker, M. Miller, C.D. Hansen, and C.R. Johnson. An integrated problem solving environment: the SCIRun computational steering system. In Hesham El-Rewini, editor, *Proceedings of the 31st Hawaii International Conference on System Sciences (HICSS-31)*, volume VII, pages 147–156. IEEE Computer Society, Jan. 1998.
2. J. Vetter and K. Schwan. Progress: A toolkit for interactive program steering. In *Proceedings of the 24th Annual Conference of International Conference on Parallel Processing*, 1995.
3. G.A. Geist, J.A. Kohl, and P.M. Papadopoulos. Cumulvs: Providing fault-tolerance, visualization and steering of parallel applications. *SIAM*, Aug. 1996.
4. D.J. Jablonowski, J.D. Bruner, B. Bliss, and R.B. Haber. Vase: The visualization and application steering environment. In *Proceedings of Supercomputing '93*, pages 560–569. IEEE Computer Society Press, 1993.
5. D.A. Reed, C.L. Elford, T.M. Madhyastha, E. Smirni, and S.E. Lamm. The next frontier: Interactive and closed loop performance steering. In *Proceedings of the 25th Annual Conference of International Conference on Parallel Processing*, 1996.
6. S.G. Parker, D.M. Weinstein, and C.R. Johnson. The SCIRun computational steering software system. In E. Arge, A.M. Bruaset, and H.P. Langtangen, editors, *Modern Software Tools in Scientific Computing*, pages 1–44. Birkhauser Press, 1997.
7. S.G. Parker and C.R. Johnson. SCIRun: A scientific programming environment for computational steering. In *Supercomputing '95*. IEEE Press, 1995.
8. J.A. Kohl, P.M. Papadopoulos, and G.A. Geist. Cumulvs: Collaborative infrastructure for developing distributed simulations. In *Proceedings of the 8th SIAM Conference on Parallel Processing for Scientific Computing*. Minneapolis, Mar. 1997.
9. W. Gu, G. Eisenhauer, E. Kramer, K. Schwan, J. Stasko, and J. Vetter. Falcon: On-line monitoring and steering of large-scale parallel programs. In *Proceedings of the 5th Symposium of the Frontiers of Massively Parallel Computing*, pages 422–429, Feb. 1995.
10. C. Upson and et al. The application visualization system: A computational environment for scientific visualization. *IEEE Computer Graphics & Applications*, 9(4):30–42, July 1989.

11. J. Vetter and K. Schwan. High performance computational steering of physical simulations. In *Proceedings of the 11th International Parallel Processing Symposium.* Geneva, Switzerland, Apr. 1997.
12. Iris Explorer - http://www.nag.co.uk/Welcome_IEC.html.

Addressing the Requirements of ASCI-class Systems

Jamshed H. Mirza

Server Group Architecture, IBM Corporation, 522 South Road, Puoghkeepsie, NY 12601, USA.
mirza@us.ibm.com

Abstract. The talk will discuss IBM's strategy for ultra-large scale systems such as those required for the US Department of Energy's ASCI (Accelerated Strategic Computing Initiative) program. We will give a brief overview of the systems that will address performance requirements of up to 10 TFLOPS over the next two years, and discuss the future evolution of the RS6000/SP system architecture.

The US Department of Energy program called ASCI (Accelerated Strategic Computing Initiative [http://www.llnl.gov/asci/]) has an aggressive plan for acquiring systems with capacity and capability substantially beyond what can be expected merely as a result of market forces. Starting with a one TFLOPS system in 1997, the program aims to acquire a 100 TFLOPS system by 2003.

Right from the begining, one of the key goals for the RS/6000 SP was to design a system that was capable of scaling ultimately to these ultra-large sizes. But as we designed the system, it was also clear to us that market dynamics will not support substantial development that is *unique* to ultra-large systems of the class represented by ASCI. Instead, design of such systems must depend heavily on leveraging mainstream technology and mainstream hardware and software components. Further, these systems must be designed so that smaller clones of the ASCI- class systems can be used effectively for mainstream technical and commercial applications.

This philosophy is reflected in the RS/6000 SP today [http://www.rs6000.ibm.com/sp.html]. The nodes of the SP are standard RS/6000 SMP servers, each running a full-function standard AIX operating system. By preserving all standard distributed UNIX interfaces and programming and execution environment, we can leverage all third party hardware and software developed for the SMP nodes and for clusters of SMPs. This greatly improves the applicability of the SP in diverse environments and application areas. The distributed message passing architecture allows all key hardware and software components to *scale globally* and provides key system availability advantages because the different system components are not tightly coupled and therefore provide good fault isolation. The erector-set like structure provides investment protection and easy migration, and allows maximum flexibility in how a system is configured and operated.

IBM will be delivering a 3+ TFLOPS ASCI system in the second half of 1998. This will be followed in 2000 with a 10 TFLOPS system based on high-performance SMP nodes using the new Power3 64-bit processors. The talk will briefly describe the characteristics of these systems, both hardware and software, and end with a discussion of future technology trends that we believe will get us to 100 TFLOPS and beyond. Given current technology projections, we believe that 100+ TFLOPS systems are feasible with a conventional scaleable parallel architecture. But stress points will emerge throughout the memory/storage hierarchy, which will place greater importance on hardware/software system solutions and programming techniques and algorithms to exploit the available raw performance of these systems.

A Parallel Genetic Algorithm for the Graphs Mapping Problem*

O. G. Monakhov, E. B. Grosbein

Institute of Computational Mathematics and Mathematical Geophysics,
Siberian Division of Russian Academy of Sciences,
Pr. Lavrentieva, 6, Novosibirsk, 630090, Russia
monakhov@rav.sscc.ru

Abstract. The application of generic algorithm to the mapping of the graph of a parallel program on the structure of multiprocessor system is considered. The parallel algorithm based on simulation of evolutionary process and heuristics accelerating of usual genetic algorithm based on the local search on neighbourhood are offered. The results of some experiments of the algorithm conducted on multiprocessor system MBC-100 are presented. The proposed algorithms are included in the library of algorithms for Java-based system TOPAS for investigation, visualization and animation of mapping algorithms: (*http : //rav.sscc.ru/ ~ monakhov/topas.html*).

1 The mathematical model

Let we have the parallel program consisting of separated modules (processes) executed and communicating asynchronously. The modules exchange data via message passing using logic channels between any two interacting modules.

Additionally we have the multiprocessor computing system consisting of processor elements. Each element, in turn, has the direct access only to its local memory and is connected with other elements through physical channels. The processors can exchange data using message passing and interprocessor network is not assumed to be full connected.

Suppose, the modules are attributed of weights describing complexity of tasks performed by them. Similarly, processors have weights describing their computing capacity. Physical channels of interprocessor communication are also attributed of weights describing their traffic capacity. For each pair of modules it is known, whether they exchange data during computation, or not and, if exchange, the volume of transferred data.

Is required to map the graph of the modules into the graph of the interprocessor network to reduce the general time of execution.

Let $H = (P, C)$ be the graph of interprocessor network where P is the set of nodes corresponding to processors,

* This work is supported by RFBR grant N96-01-01632

P_k is weight of node k (in operations per second) corresponding to computing capacity of processor $k \in [0, n]$, $|P| = n + 1$,

C is a matrix of weights of physical channels,

$$C_{kl} = \begin{cases} \infty, & \text{if processor } k \text{ and processor } l \text{ are not directly connected,} \\ & \text{else traffic capacity, in seconds per byte.} \end{cases}$$

Let $G = (M, K)$ be the graph of parallel program where M is the set of nodes corresponding to the modules,

M_i is weight of node i corresponding to number of operations performed by module $i \in [0, m]$, $|M| = m + 1$,

K is the matrix of weights of logical channels,

$$K_{ij} = \begin{cases} 0, & \text{if module } i \text{ and module } j \text{ do not exchange information,} \\ & \text{else volume of communication, in bytes} \end{cases}$$

Let $\pi : M \to P$ be the mapping function, $\mathcal{F}(\pi)$ be the functional evaluating the quality of mapping.

It is necessary to find such mapping π at which $\mathcal{F}(\pi)$ attains the optimum value.

2 The main notions of genetic model

The generic algorithm is based on simulation of the survival of the fittest in the population of entities each of which presents the point in space of solutions. The entities are presented by strings of integers. The elements of strings are named genes. The function named fitness function evaluates the degree of fitness of entity.

The purpose of generic algorithm is the search of global minimum (maximum) of fitness function when the initial structure of population is given through applying to it generic operators: selection, crossover, mutation [1], [2].

2.1 The mutation

The mutation is applied to one entity and produces new one. The mutation consists of the cycle turning over genes. The contents of each gene is replaced on randomly the chosen number with probability p_m that corresponds moving of module from one processor to another one.

The mutation is dedicated for keeping the population non-uniform. This is to protect algorithm from falling in local minimum.

2.2 The crossover

We apply the crossover to two entities and get two new ones. The crossover is realized by cycle turning over corresponding pairs of genes. We exchange the contents of corresponding genes with probability p_u.

2.3 The selection

The selection realizes the principle of the survival of the fittest. It is applied to the old population as a whole. The selection consists in sorting of population on significance of fitness function and copying of several best entities to the new population which then is filled up at using crossover and mutation.

3 The structure of data

We use two populations while searching of optimum, old and new one. Old population is produced on previous iteration (for the first iteration it is filled with randomly chosen entities) and is used for filling new population which becomes old on following iteration.

Each population is the set of entities presenting the various mapping functions π. Each entity is represented with an array *place* of length $m+1$. The genes contain the integers from 0 to n. Thus, $place[i]$ equals to number of processor on which the module with number i is mapped. The generic algorithm does not impose restrictions on fitness function. For this problem we use

$$\mathcal{F}(place) = \max_{k=0,\ldots,n} \sum_{i:\, place[i]=k} \left(\frac{M_i}{P_k} + \sum_{j \in \mathcal{L}(i)} K_{ij} d_{k\,place[j]} \right), \tag{1}$$

where $\mathcal{L}(i)$ is set of modules exchanging data with module i, d_{kl} is distance between processors k and l on graph H computed using C_{kl}.

First term here describes the general time of execution of modules mapped to processor k, second describes the time of interprocessor interaction of these modules with the directly connected modules in graph G. Thus, minimization $\mathcal{F}$ means the minimization of operating time of the most loaded processor. This functional is modification of another one offered in [3]. Another evolutionary approach for parallel mapping problem are found in [4].

4 The local optimization

The local optimization is the attempt to change the solution by replacement of it on close one. The generic algorithms and heuristics of similar grade have the complementary strong and weak distinctive. The generic algorithms are good for search of perspective areas in spaces of solutions but are not so good for exact search in these areas. Heuristics, on the other hand, are good for refinement of approximate solutions but do not take into account the global perspective. The practice shows that the hybrid algorithm uniting genetic one with heuristic optimization frequently produces better results from these both methods used separately.

The local optimization for our problem consists in following: one of modules located on the most loaded processor is moved on processors on which the modules close given on graph G are located. Best entity (solution) of considered replaces initial one in population. We note that after changing of only one gene it should not anew to recalculate functional (1) completely.

Proposition. Let $\mathcal{F} = \max_k g_k$,

$$g_k = \sum_{i:\, place[i]=k} \left(\frac{M_i}{P_k} + \sum_{j \in \mathcal{L}(i)} K_{ij} d_{k place[j]} \right).$$

Let the module m has changed the location from p_1 to p_2 producing new vector *place'*, in which

$$place'[i] = \begin{cases} place[i], & \text{if } i \neq m, \\ p2, & \text{if } i = m. \end{cases}$$

Let the new value of $\mathcal{F}$ be denoted by $\mathcal{F}'$. Then $\mathcal{F}' = \max_k g_k{}'$, where

$$g_k' = g_k + \begin{cases} -\frac{M_m}{P_k} + d_{p_1 p_2} \sum_{i \in \mathcal{L}'(m,k)} K_{im} - \sum_{i \in \mathcal{L}''(m,k)} K_{im} d_{k place[i]}, & \text{if } k = p_1, \\ \frac{M_m}{P_k} - d_{p_1 p_2} \sum_{i \in \mathcal{L}'(m,k)} K_{im} + \sum_{i \in \mathcal{L}''(m,k)} K_{im} d_{k place[i]}, & \text{if } k = p_2, \\ \sum_{i \in \mathcal{L}'(m,k)} K_{im}(d_{kp_2} - d_{kp_1}), & \text{otherwise} \end{cases}$$

$\mathcal{L}'(m,k) = \{j \in \mathcal{L}(m) | place[j] = k\}$, $\mathcal{L}''(m,k) = \{j \in \mathcal{L}(m) | place[j] \neq k\}$.

The mentioned property permits to reduce calculations while performing the local optimization.

5 Short description of parallel mapping algorithm

5.1 Algorithm

In the parallel realization of genetic algorithm the population is divided between processors. Each processor executes the iterations independently over its part of population without exchanges information with other ones. After each N iterations the processors exchange of best entities.

1. Initialization of old population with randomly chosen entities.
2. Performing of N iterations.
3. Interprocessor exchange of best entities.
4. The check of the condition of iterations completion: if it is not satisfied, transition to step 2, otherwise the completion of work.

5.2 The iteration

Next steps are executed by the processors independently.

1. Performing selection using the old population. This fills part of new population.
2. The cycle filling remainder of the new population. While the new population is not filled: a) Random choosing of two entities of the old population. b) With probability p_c or they, or results of crossover between them are subjected to the mutation. c) Filling up the new population with the results produced on the last step.
3. The cycle subjecting each entity of the new population to local optimization.

5.3 The condition of iterations completion

The iterations are finished if the best value of fitness function has not changed during certain time.

5.4 Experimental results

The described algorithm was implemented in C extended with message passing library **routelib** that is used on the multiprocessor MVS-100 consisting of 8 nodes of Transech TTM110 boards analog (Intel i860 XR + SGS Thomson T805 transputer).

A number of experiments were conducted with the proposed algorithm and results of these experiments were compared with the results of the sequential genetic algorithm. Some examples are presented in the table 5.4.

NP	G	H	PS	RE	CT	N	p_m	p_c	$\overline{\mathcal{F}}$	$\mathcal{F}_{\min}$	$\mathcal{F}_{\text{opt}}$	A
100	rg24	hc3	104	8	2-point	10	1/16	0.8	7.05	6	5	7.0
100	hc4	hc3	1000	88	uniform	10	0.9	0.8	11.17	8	8	6.4
100	gr3x8	rg24	1000	88	2-point	12	0.8	0.8	12.24	10	9	7.0
100	rg24	rg8	104	8	2-point	10	1/24	0.8	8.20	6	5	7.0

Some abbreviations are used in the table:

NP — number of passes.

G — graph of program, H — graph of system. Abbreviations: **rg24** — ring of 24 nodes, **rg8** — ring of 8 nodes, **hc3** — cube (8 nodes), **hc4** — hypercube of 4th dimension (16 nodes), **gr3x8** — grid 3×8. All weighs are equal 1.

PS — population size.
RE — number of preserved entites when population is updated (see 5.2).
CT — type of crossover.
N, p_m, p_c — see 2, 5.1.
$\mathcal{F}_{\min}$ — best value of $\mathcal{F}$ found in **NP** passes.
$\overline{\mathcal{F}}$ — mean value of $\mathcal{F}$ past **NP** passes.
$\mathcal{F}_{\text{opt}}$ — optimum for the task.
A — speed-up reached on 8 processors.

Local optimization was used for each task.

6 Conclusion

Results of ongoing research on the development of parallel genetic algorithm for graph mapping problem using local optimization are presented in this paper. The results indicate that the proposed algorithm is able to effectively perform a search for a optimal (suboptimal) solution of the graph mapping problem. The proposed algorithms are included in the library of algorithms for Java-based system TOPAS for investigation, visualization and animation of mapping algorithms: (*http : //rav.sscc.ru/ ~ monakhov/topas.html*).

References

1. Goldberg D.E. *Genetic Algorithms in Search, Optimization and Machine Learning.* Addison-Wesley, 1989.
2. Levine D. Users Guide to PGA Pack Parallel Genetic Algorithm Library. Tech. Rep. Argonne National Lab., available from *http://www.mcs.anl.gov/pgapack.html*, 1996.
3. Monakhov O.G. On parallel recursive mapping algorithm for pyramidal multiprocessor systems. *Bulletin of the Novosibirsk Computing Center* **3** *(1995), 69–75*
4. Monakhov O.G., Grosbein E.B. A parallel evolution algorithm for graph mapping problem, in:*Proc. Inter. Workshop on Parallel Computation and Scheduling (PCS'97)*, CICESE, Ensenada, Baja California, Mexico, 1997, 17-21.

Parallel Wavelet Transforms

Ole Møller Nielsen

UNI•C, Danish Computing Centre for Research and Education, Lyngby, Denmark
uniomni@uni-c.dk,
http://www.imm.dtu.dk/~omni

1 Introduction

Wavelets have generated a tremendous interest in both theoretical and applied mathematics over the past few years, and the fast wavelet transform (FWT) in particular has proven to be an effective tool for e.g. numerical analysis [1] and image processing [2]. Problems from these areas are typically large and the FWT can be quite time-consuming although the algorithmic complexity is proportional to the problem size. The use of parallel computers is one way of speeding up the FWT.

2 The Fast Wavelet Transform

The 1D FWT is a linear mapping of $\boldsymbol{R}^N$ onto $\boldsymbol{R}^N$ defined recursively by a sequence of fundamental linear mappings and it is used to obtain the wavelet coefficients. Many wavelet coefficients are often small enough to be disregarded without introducing a large error; hence the FWT forms the basis of efficient data compression algorithms. See [8, 3, 5] for good expositions of wavelet analysis.

Let $N = 2^J$ and $\boldsymbol{c}^J$ be a given vector with elements $\{c_n^J\}_{n=0,1,\dots,N-1}$. The recurrence formulas for the FWT are

$$c_n^{j-1} = \sum_{l=0}^{D-1} a_l c^j_{\langle l+2n \rangle_{2^j}} \tag{1}$$

$$d_n^{j-1} = \sum_{l=0}^{D-1} b_l c^j_{\langle l+2n \rangle_{2^j}} \tag{2}$$

for $n = 0, 1, \dots, 2^{j-1} - 1$ and $j = J, J-1, \dots, J-\lambda+1$. The expression $\langle l + 2n \rangle_{2^j}$ denotes the modulus function defined such that $\langle l + 2n \rangle_{2^j} \in [0, 2^j - 1]$ for all values of $l \in \boldsymbol{Z}$. λ is an integer between 0 and J that determines how many recursion steps to take, D is an even positive integer known as the wavelet genus, and the numbers a_l, b_l for $l = 0, 1, \dots, D-1$ are the wavelet filter coefficients. The latter are widely available (e.g. in [3]) and they are often chosen such that the mapping $\{\boldsymbol{c}^J\} \to \{\boldsymbol{c}^{J-1}, \boldsymbol{d}^{J-1}\}$ is orthogonal. The number of floating point operations needed to perform the 1D FWT of a vector with N elements is bounded by $4DN$ operations. Since the wavelet transform is a linear mapping of $\boldsymbol{R}^N$ onto

$\boldsymbol{R}^N$, it can be represented by an $N \times N$ matrix $\boldsymbol{W}_N$. Let $\boldsymbol{x} = \boldsymbol{c}^J$ and $\boldsymbol{y}$ be the result of the recursion after λ steps, i.e.

$$\boldsymbol{y} = \begin{bmatrix} \boldsymbol{c}^{J-\lambda} \\ \boldsymbol{d}^{J-\lambda} \\ \boldsymbol{d}^{J-\lambda+1} \\ \vdots \\ \boldsymbol{d}^{J-1} \end{bmatrix}$$

then the wavelet transform can be expressed compactly as

$$\boldsymbol{y} = \boldsymbol{W}_N \boldsymbol{x}$$

The 2D FWT of a matrix $\boldsymbol{X} \in \boldsymbol{R}^{M,N}$ can be defined by combining row transforms with transforms of columns as follows

$$\boldsymbol{Y} = \boldsymbol{W}_M \boldsymbol{X} \boldsymbol{W}_N^T$$

Assuming that data are stored columnwise the expression $\boldsymbol{X}\boldsymbol{W}_N^T$ leads to operations on vectors of length M and stride-one data access which is the most efficient mode on most computer architectures. This is not the case for the expression $\boldsymbol{W}_M \boldsymbol{X}$, because it consists of a collection of columnwise 1D transforms which do not access the memory as efficiently [6]. We therefore rewrite the matrix product as

$$\boldsymbol{Y}^T = \left(\boldsymbol{X}\boldsymbol{W}_N^T\right)^T \boldsymbol{W}_M^T$$

yielding efficient memory access on each processor at the cost of one transpose step. This algorithm can be implemented on a parallel computer similar to parallel 2D FFTs [4].

3 The replicated transform

The most straightforward approach to parallelize the 2D FWT is to distribute the rows of $\boldsymbol{X}$ evenly on P processors (shown here for $P = 4$).

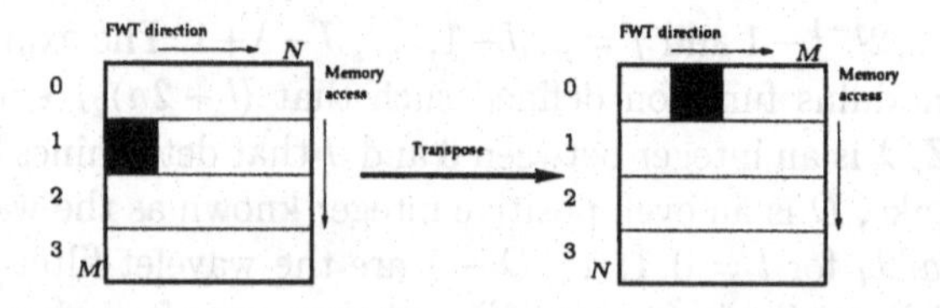

Thus all the wavelet row transforms are done without communication. However, the transposition step causes a substantial communication overhead which depends on the number of processors as well as the problem size.

4 The communication efficient transform

The parallel transpose step can be avoided altogether by parallelizing along the *second* dimension in the first step of the 2D FWT as illustrated below.

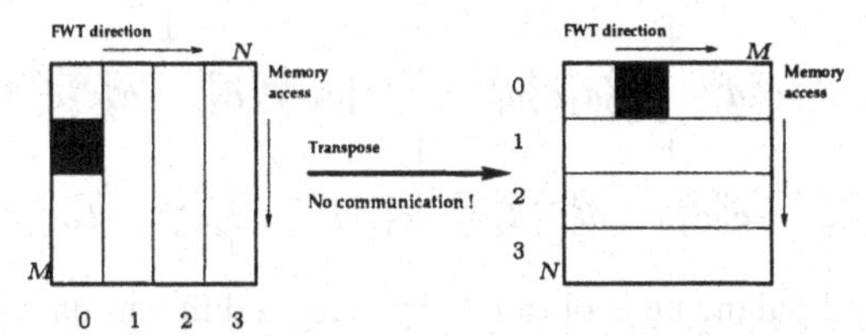

This is no longer the trivial parallelization because the rowwise wavelet transforms now require communication.

4.1 Parallelization of the rowwise 1D wavelet transform

We now describe the parallelization of one row transform which is needed for the first step of the communication efficient transform. Consider the data layout suggested by (1) and (2) when distributed across two processors. This is shown in Table 1. It is seen that distributing the results of each transform step evenly

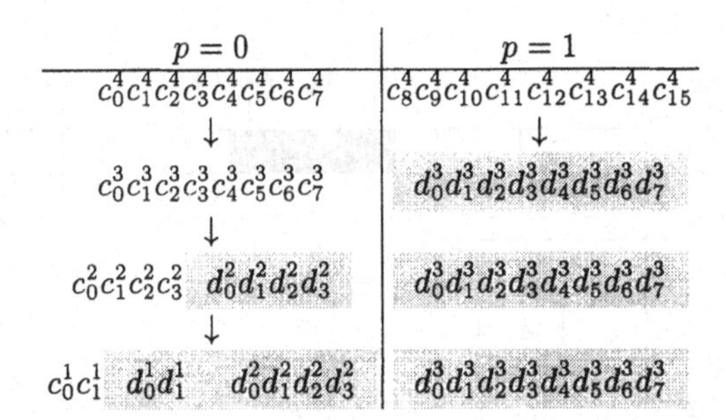

$p = 0$	$p = 1$
$c^4_0 c^4_1 c^4_2 c^4_3 c^4_4 c^4_5 c^4_6 c^4_7$	$c^4_8 c^4_9 c^4_{10} c^4_{11} c^4_{12} c^4_{13} c^4_{14} c^4_{15}$
$\downarrow$	$\downarrow$
$c^3_0 c^3_1 c^3_2 c^3_3 c^3_4 c^3_5 c^3_6 c^3_7$	$d^3_0 d^3_1 d^3_2 d^3_3 d^3_4 d^3_5 d^3_6 d^3_7$
$\downarrow$	
$c^2_0 c^2_1 c^2_2 c^2_3$ $d^2_0 d^2_1 d^2_2 d^2_3$	$d^3_0 d^3_1 d^3_2 d^3_3 d^3_4 d^3_5 d^3_6 d^3_7$
$\downarrow$	
$c^1_0 c^1_1$ $d^1_0 d^1_1$ $d^2_0 d^2_1 d^2_2 d^2_3$	$d^3_0 d^3_1 d^3_2 d^3_3 d^3_4 d^3_5 d^3_6 d^3_7$

Table 1. Standard data layout results in poor load balancing. The shaded sub-vectors are those parts which do not require further processing. Here $P = 2$, $N = 16$, and $\lambda = 3$.

across the processors results in a poor load balancing, because each step works with the lower half of the previous vector only. Processors containing parts that are finished early sit idle in the subsequent steps. In addition, global communication is required in the first step because every processor must know the values on every other processor in order to compute its own part of the wavelet transform. In subsequent steps this communication will take place among the active processors only.

However, we can obtain perfect load balancing and avoid global communication by introducing another ordering of the intermediate and resulting vectors.

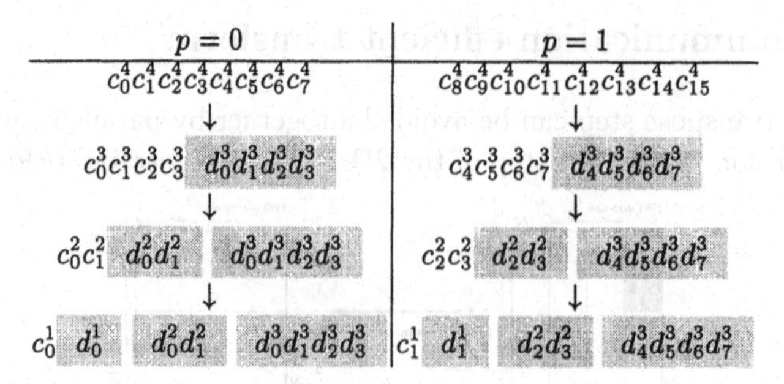

$p = 0$	$p = 1$
$c_0^4 c_1^4 c_2^4 c_3^4 c_4^4 c_5^4 c_6^4 c_7^4$	$c_8^4 c_9^4 c_{10}^4 c_{11}^4 c_{12}^4 c_{13}^4 c_{14}^4 c_{15}^4$
↓	↓
$c_0^3 c_1^3 c_2^3 c_3^3$ $d_0^3 d_1^3 d_2^3 d_3^3$	$c_4^3 c_5^3 c_6^3 c_7^3$ $d_4^3 d_5^3 d_6^3 d_7^3$
↓	↓
$c_0^2 c_1^2$ $d_0^2 d_1^2$ $d_0^3 d_1^3 d_2^3 d_3^3$	$c_2^2 c_3^2$ $d_2^2 d_3^2$ $d_4^3 d_5^3 d_6^3 d_7^3$
↓	↓
c_0^1 d_0^1 $d_0^2 d_1^2$ $d_0^3 d_1^3 d_2^3 d_3^3$	c_1^1 d_1^1 $d_2^2 d_3^2$ $d_4^3 d_5^3 d_6^3 d_7^3$

Table 2. Good load balancing is obtained by using a different data layout. The shaded sub-vectors are those parts which do not require further processing. Again, we have $P = 2$, $N = 16$, and $\lambda = 3$.

This is shown in Table 2. Note that the result is a permutation of the result from the sequential FWT because each processor essentially performs a *local* wavelet transform of its data.

4.2 Communication

Consider the computations done by processor p on a row vector as indicated below

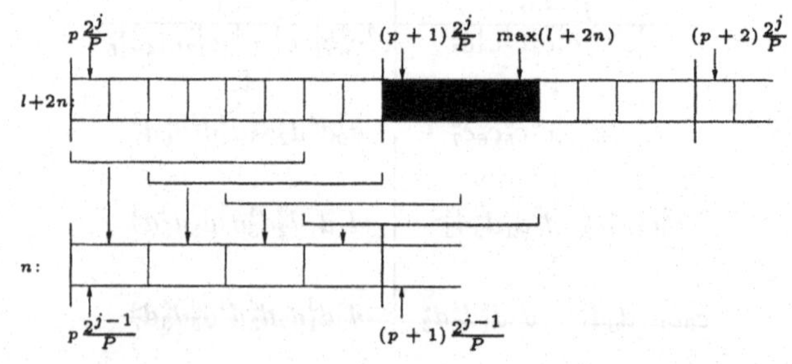

The quantities in equation (1) and (2) can be computed without communication provided that

$$n \leq (p+1)\frac{2^{j-1}}{P} - \frac{D}{2}$$

For a given $n > (p+1)\frac{2^{j-1}}{P} - \frac{D}{2}$ computations are still local as long as

$$l \leq (p+1)\frac{2^j}{P} - 2n - 1$$

However, when l becomes larger than this the index $l + 2n$ will point to elements residing on the processor located to the right of processor p. Exactly $D - 2$ elements per row must be communicated at each step of the FWT, and, if the hardware permits, communication can be overlapped with computations. In [7] a performance model reveals that the efficiency of the this algorithm is independent of the number of processors and that it approaches perfect speedup as the problem size increases.

5 Conclusion

We have developed and tested the performance of different algorithms for computing the 2D wavelet transform in parallel on the VPP300. Some performance results comparing the replicated transform and the communication efficient transform are given below for $N = 512$ and $D = 10$.

		Gflops/s	
P	M	Replicated	Comm. efficient
1	512	1.3	1.3
2	1024	1.6	2.6
4	2048	3.0	5.1
8	4096	3.9	10.2

It is seen that the new communication efficient approach performs significantly better than the old replicated algorithm. This is due to the fact that the new approach avoids the usage of a distributed matrix transpose and that it maintains perfect load balancing. Moreover, the new algorithm is arranged such that communication can be overlapped with computations provided that the underlying hardware permits that. This is the case for the VPP300 and it is seen that almost perfect speedup is achieved.

References

[1] G. Beylkin, R. Coifman, and V. Rohklin. Wavelets in numerical analysis. In Mary Beth Ruskai et al., editors, *Wavelets and their applications*, pages 181–210. Jones and Bartlett, 1992.

[2] Charles K. Chui. *Wavelets: A Mathematical Tool for Signal Analysis.* SIAM Monographs on Mathematical Modeling and Computation. SIAM, 1997.

[3] Ingrid Daubechies. *Ten Lectures on Wavelets.* SIAM, 1992.

[4] Markus Hegland. *Real and Complex Fast Fourier Transforms on the Fujitsu VPP500.* Parallel Computing, 22, pp. 539-553, 1996.

[5] Yves Meyer. *Wavelets: Algorithms and Applications.* SIAM, 1993.

[6] Ole Møller Nielsen and Markus Hegland. *A Two-Dimensional Fast Wavelet Transform for the Fujitsu VP2200.* In preparation

[7] Ole Møller Nielsen Wavelets in Scientific Computing, PhD Thesis. Technical University of Denmark, Department of Mathematical Modelling. March 1998. Avilable at `http://www.imm.dtu.dk/~omni/thesis.html`

[8] Gilbert Strang and Truong Nguyen. *Wavelets and Filter Banks.* Wellesley-Cambridge Press, 1996.

Writing a Multigrid Solver Using Co-array Fortran

Robert W. Numrich[1], John Reid[2], and Kieun Kim[3]

[1] Silicon Graphics, Inc., Eagan, MN 55121, USA
[2] Rutherford Appleton Laboratory, UK
[3] Northeast Parallel Architectures Center, Syracuse University, USA.

Abstract. Co-Array Fortran, known previously as F^{--}, is an extension to Fortran 95 that combines elegance of expression with simplicity of implementation to give an efficient parallel programming language. We illustrate its power and efficiency with a multigrid solver for Poisson's equation in three dimensions. We examine the effect of latency and bandwidth for the co-array version compared with an equivalent MPI version on two machines, the SGI/CRAY ORIGIN 2000 and the CRAY-T3E.

1 Introduction

This paper outlines a method for combining Co-Array Fortran syntax with existing features of Fortran 95 to write efficient and elegant parallel codes. We hope to stimulate discussion about future directions for parallel programming models using the Fortran language. We illustrate our points with a three-dimensional multigrid solver. We picked the multigrid algorithm, not because we want to discuss the best way to implement a multigrid solver, but because it contains many difficult issues that must be addressed in typical parallel codes.

Multigrid methods [1] are a popular and effective technique for solving large, sparse systems of linear equations. Combined with domain decomposition techniques [9], they are also an effective approach to parallel processing. Fortran 95 is a powerful modern language [3] that allows the programmer to express the complicated data structures required to write efficient multigrid algorithms. Co-Array Fortran is a parallel extension to Fortran 95 [5–8] that allows the programmer to combine a simple syntax with Fortran 95 derived types to obtain a clear, concise implementation of the multigrid method.

A Co-Array Fortran program executes as if it were replicated a fixed number of times. The programmer may retrieve this number at run time through the intrinsic procedure `num_images()`. Each copy is called an image and executes asynchronously usually, but not necessarily, on a separate physical processor. Each image has its own index, which the programmer may retrieve at run time through the intrinsic procedure `this_image()`. Each image has its own set of data objects all of which may be accessed in the normal Fortran way. Some objects are declared with additional dimensions, called co-dimensions, in square brackets. Co-dimensions have a different name because they are different from additional normal dimensions. The syntax for co-dimensions, however, is designed such that a programmer uses it very much like normal Fortran syntax to access data across memory images.

2 Data Structures

The multigrid method defines multiple coverings of a geometric domain with different levels of resolution. A parallel implementation adds the complication that each level must be cut into pieces that can be treated independently by parallel processors in a way that balances the work load. The programmer must define mappings not only between levels but also between subdomains within each level. We assume that our problem is defined by a three-dimensional rectangular grid and that each image performs computations on a subgrid defined by upper and lower bounds in each dimension. At the edges, we let each subgrid overlap with the neighboring subgrids so that each image needs to obtain ghost values from neighboring images at specific times during the calculation.

We outline a way to write a multigrid solver that can adapt easily to differing numbers of images, differing numbers of levels, and differing ways of cutting each level into subdomains. To accomplish our goal, we use Co-Array Fortran syntax combined with the existing features of Fortran 95. We combine the Fortran 95 data structure, called a derived type, to define subgrids at each multigrid level. We use co-array syntax to define how the subgrids are related to the full grid and to define mappings among subgrids. For example, we define a `blocked_grid` data type for holding data on a single image for each level as follows:

```
type blocked_grid
  real,pointer,dimension(:,:,:) :: v
  integer,dimension(3,2)        :: bounds
end type blocked_grid
```

The pointer component represents an array that holds information relevant to the particular problem at hand, for example, velocities at each grid point. We use a pointer component in our structure because Fortran 95 does not at the moment allow allocatable components. Other information, such as temperature, density, material properties, might also be included in the structure. We subdivide the grid regularly at each level into rectangular subgrids, one for each image, allocate space for each, and record information about the subdivision in the array component `bounds`. For more difficult problems, we would, of course, define more complicated data structures akin to what object oriented languages would call classes [4].

We need such objects for each level so we declare them as allocatable co-arrays, for example,

```
type(blocked_grid),allocatable,dimension(:)[:,:,:]::u
```

The deferred dimension in the rounded brackets will be filled in at run time by the number of levels we decide to use and the deferred co-dimensions in square brackets will be filled in to specify the decomposition of the grid, which will depend on the number of images we decide to use. If we have `nlev` levels and we want to cut the grid at each level into `np` parts in the `x`-direction, `nq` in the `y`-direction, and `nr` in the `z`-direction, then we allocate objects such as

```
allocate(u(nlev)[np,nq,*])
```

Since the co-size is always equal to the number of images, the final dimension is not required so we represent it with assumed-size notation. The programmer is responsible for making sure that the product `np`×`nq`×`nr` equals the number of images. With 64 images, for example, we might select `np=nq=nr=4`, while with 16 images we might select `np=nq=4` and `nr=1` or perhaps `np=nr=4` and `nq=1`. These parameters might be input parameters to the program at run time, in which case the user specifies how to subdivide the problem, or the program itself might have an algorithm for deciding the best way to subdivide the problem based on its size and the number of images at run time.

At this point we have nothing but empty data structures. We define a procedure, something like a constructor [4], that builds the actual data structure. To allocate space for level `k`, for example, we might call the procedure

```
call new(u(k),nx,ny,nz,np,nq,nr)
```

where `nx,ny,nz` are the lengths of the edges of the full grid at level `k`, and `np,nq,nr` are the parameters that how to subdivide the grid. The constructor, which might look something like the following, uses a predetermined method to subdivide the grid.

```
subroutine new(a,nx,ny,nz,np,nq,nr)
  type(blocked_grid)  :: a[np,nq,*]
  integer, intent(in) :: nx,ny,nz,np,nq,nr
  integer :: xlo,xhi,ylo,yhi,zlo,zhi
  a%bounds(:,:) = cutup(nx,ny,nz,np,nq,nr)
  xlo=a%bounds(1,1)
  xhi=a%bounds(1,2)
  ...
  zhi=a%bounds(3,2)
  allocate(a%v(xlo-1:xhi+1,ylo-1:yhi+1,zlo-1:zhi+1))
contains
  function cutup (nx,ny,nz,np,nq,nr) result(bounds)
    integer, intent(in) :: nx,ny,nz,np,nq,nr
    integer             :: bounds(3,2)
  ! Choose upper and lower bounds in each dimension to give
  ! np blocks in the x-direction, nq blocks in the y-direction
  ! and nr blocks in the z-direction.
    ...
    bounds(:,:) =
  end function cutup
end subroutine new
```

Function cutup subdivides the full grid into subgrids according to some predetermined method based on the arguments passed into the procedure. There are, of course, several options that might be used to subdivide a grid. It makes sense to pick a method that simplifies communication among subgrids. Once we pick a method, we document it as part of the definition of the **blocked_grid** type so that the programmer knows the method used. The work required to do the blocking is hidden away in the constructor where it is done once for all. The cutup procedure returns the lower and upper bounds in each direction as an array, which is recorded as a component of the data structure. Finally, the constructor allocates space for a subgrid on each image with a single layer of ghost cells around each local subgrid, which will be used to communicate overlapping data from one memory image to another.

3 Expressing Ghost Cell Updates with Co-Array Syntax

A big headache with message-passing models for parallel programming is the requirement to subdivide a data set explicitly into independent pieces in a general way that accounts for all the end cases and maintains maps from the global problem to the local problems and back again. When the size of the problem does not divide evenly, end cases arise where some images have more or less data than their neighbors. The differing sizes and differing maps between local and global indices can lead to unreadable and unmaintainable code.

This problem is a holdover from Fortran 77, which does not support allocatable arrays. In Fortran 95, we can use allocatable arrays in such a way that each image allocates its own data retaining global indices but adjusts the upper and lower bounds to match its subsection of the global problem. In that way, each image retains global indices and problems related to end cases disappear because they are never created in the first place.

The benefits of this construction become apparent when it is time to perform ghost cell updates. Updating the ghosts in the x-axis direction requires just two lines of code:

```
if(p > plo) u(k)%v(xlo-1,:,:) = u(k)[p-1,q,r]%v(xlo-1,:,:)
if(p < phi) u(k)%v(xhi+1,:,:) = u(k)[p+1,q,r]%v(xhi+1,:,:)
```

In this example, p,q,r are the image indices for the invoking image relative to the structure u. The programmer may obtain these indices using the following intrinsic procedures:

```
p=this_image(u,1); q=this_image(u,2); r=this_image(u,3)
```

The image to the left has indices [p-1,q,r] and the image to the right has indices [p+1,q,r]. The invoking image needs the yz-plane from its left neighbor, which is just (xlo-1,:,:) in the global indices, and the yz-plane (xhi+1,:,:) from the right. Because we have used global indices, each local image has all the information it needs to move data from one image to another with no worry about end cases.

The symmetry and simplicity of these statements are pleasing to the eye, easy to understand, and easy to maintain. Performing the updates in the other directions is as easy as copying the code, *mutatis mutandis*, making the appropriate changes:

```
if(q > qlo) u(k)%v(:,ylo-1,:) = u(k)[p,q-1,r]%v(:,ylo-1,:)
if(q < qhi) u(k)%v(:,yhi+1,:) = u(k)[p,q+1,r]%v(:,yhi+1,:)
if(r > rlo) u(k)%v(:,:,zlo-1) = u(k)[p,q,r-1]%v(:,:,zlo-1)
if(r < rhi) u(k)%v(:,:,zhi+1) = u(k)[p,q,r+1]%v(:,:,zhi+1)
```

A few lines of code, which look, feel, and behave like Fortran, may replace an entire library of complicated communication procedures. Co-array syntax puts the programmer in control of the code while libraries retain control in the hands of the library developer.

The simplicity of expression holds not just for regular problems like the one described here. For irregular grids, such as those encountered in finite element codes, overlapping data may need to be gathered from other images. But the operation can still be written with a few lines of code. To obtain node information from overlapping elements, for example, we might write

```
elem(index(:,p))%node(:) = elem(jndex(:,p))[p]%node(:)
```

In this case, the array `jndex(:,p)` holds a list of elements that belong to image `p` and need to be held as ghost copies on the invoking image, and the array `index(:,p)` is a list of where the invoking image wishes to hold them in its local memory.

4 Some Results

The limiting factor in our code is the time required to perform ghost cell updates during the relaxation phase at each level. At various times during the relaxation, the six faces of the cube owned by each image must be updated with fresh information from neighboring images. How often and at what points in the relaxation phase this data transfer happens depends on the particular relaxation method used. We used a simple red-black Gauss-Seidel relaxation method with an update between the red and black phases. The communication time shown in our results is the total time to perform these updates.

Some communication time is spent computing residuals and errors and to broadcast data to the coarsest levels where the relaxation is done in serial mode. We did the lowest two levels, $(2^1+1)^3$ and $(2^2+1)^3$, in serial mode. But the time spent for this communication is a small part of the total time and is not the limiting factor for scalability.

Figure 1 shows some timing results. We ran the code on two different machines, the SGI/CRAY ORIGIN 2000 and the CRAY-T3E. We ran two different problem sizes, one with $(2^6+1)^3$ points on the finest level and one with $(2^7+1)^3$ points on the finest level. We ran 7 V-cycle iterations for each problem size, which was sufficient to obtain convergence. We solved each problem twice, once using

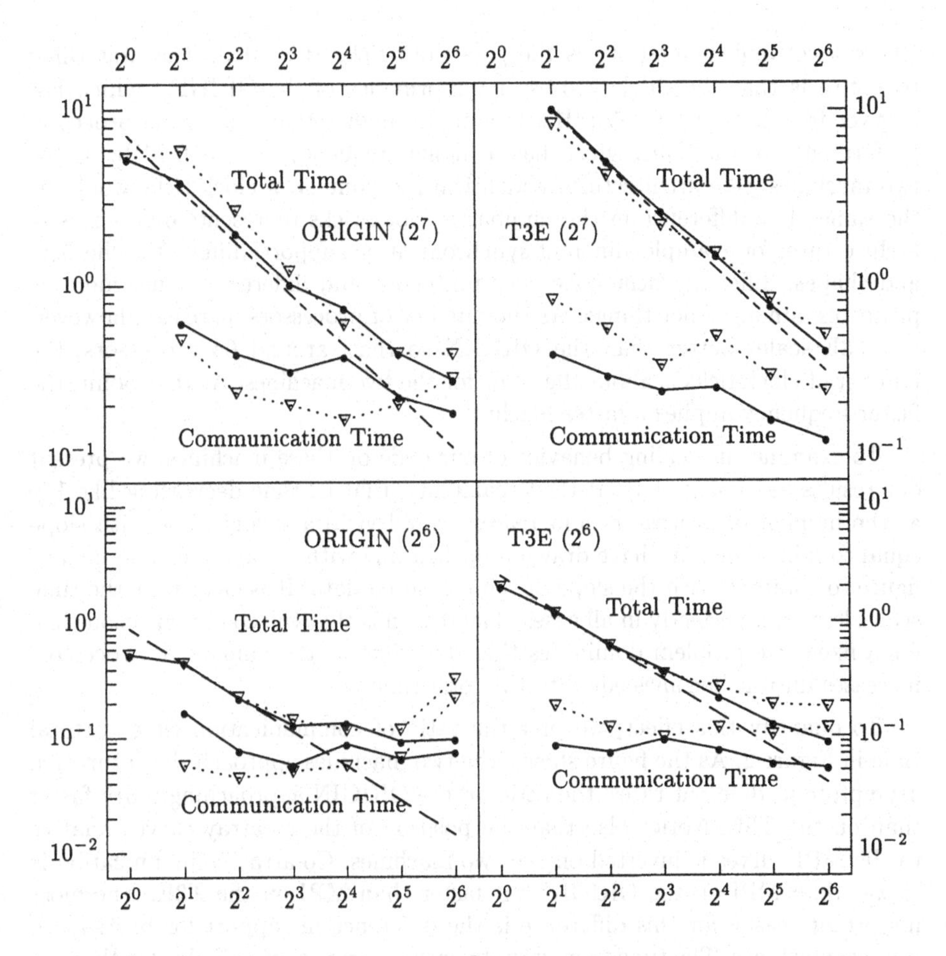

Fig. 1. Time in giga-clock-ticks, $\nu t \times 10^{-9}$, as a function of p, the number of processors. MPI results are marked with dotted lines and co-array results are marked with solid lines. The frequency of the CRAY-T3E is $\nu = 300$ MHz; the frequency of the SGI/CRAY ORIGIN 2000 is $\nu = 195$ MHz. Results are shown for two problem sizes for each machine: seven levels, $(2^7+1)^3$ points on the finest grid, and six levels, $(2^6+1)^3$ points on the finest grid.

co-array syntax and again using MPI [2], while varying the number of processors from one to sixty-four with one memory image per physical processor.

Our results are not reproducible because we used nondedicated machines with varying work loads and we used pre-released software. Rather than trying to explain every wobble in the curves, we concentrate on the overall scalability of the code as a function of the two different programming models and the two different machines.

We multiplied the measured times, t, by the frequency of the machine, ν, in an attempt to reveal architectural differences that might be masked by a higher or lower frequency. It also allows us to present our results in dimensionless form where a convenient quantity is the giga-clock-tick, $\nu t \times 10^{-9}$. The first thing to notice is that the single processor performance of the ORIGIN, which has a lower frequency, $\nu = 195$ MHz, is actually higher than the single processor performance of the T3E, which has a higher frequency, $\nu = 300$ MHz. If the two machines were architecturally identical the number of clock ticks would be the same. The difference in the number of clock ticks represents differences in architecture, for example, different synchronization support, different cache sizes and shapes, different cache coherence protocols, and different virtual memory protocols, among other things. As the number of processors increases, however, the T3E scales better than the ORIGIN so that, around 64 processors, the number of clock ticks is about the same for the two machines. At that point, the faster frequency implies a faster machine.

To examine the scaling behavior of our code on these machines, we present our results as log-log plots. Perfect scalability, that is, time decreasing like $1/p$ as the number of processors p increases, would yield a straight line with slope equal to minus one. We have drawn a dashed line with slope minus one in each figure to compare with the slope of the measured data. It is clear that the time scales less than perfectly in all cases. The reason is that the communication time for a fixed-size problem dominates the total time as the number of processors increases until it becomes equal to the total time.

To quantify this effect, we plot the ratio of communication time to total time in Figure 2. As the figure shows, the two machines approach the saturation asymptote at different rates, the ratio on the ORIGIN approaching unity faster than on the T3E. Notice also that the position of the co-array curves relative to the MPI curves is inverted on the two machines. Co-array communication is slower than MPI on the ORIGIN but faster than MPI on the T3E. The most important reason for this difference is the difference in support for barriers on the two machines. The time to move data with co-array syntax is shorter than for MPI, but the co-array model depends on explicit barrier synchronization, which is supported in hardware on the T3E but not on the ORIGIN. MPI depends on implicit synchronization through the send/receive protocol. Although this protocol often requires extra memory copies from buffer to buffer, it may perform more efficiently on a machine with no support for hardware barriers especially in a nondedicated environment such as we have here.

Another important difference is that the prototype co-array compiler that we used on the ORIGIN has no optimization in it yet to achieve high bandwidth. The MPI library has been tuned over several years to give the highest bandwidth for a long message at the expense of high latency for a short message. Co-array syntax, on the other hand, supports low latency for short messages because the compiler can generate in-line code, which incurs very little overhead. For small numbers of processors, where the messages are fairly large, the MPI version has shorter communication time. But as the number of processors increases, the messages get shorter, the overhead from long startup times becomes a larger fraction of the total time, and eventually the low latency co-array model wins. The co-array compiler on the T3E contains some optimization to achieve high

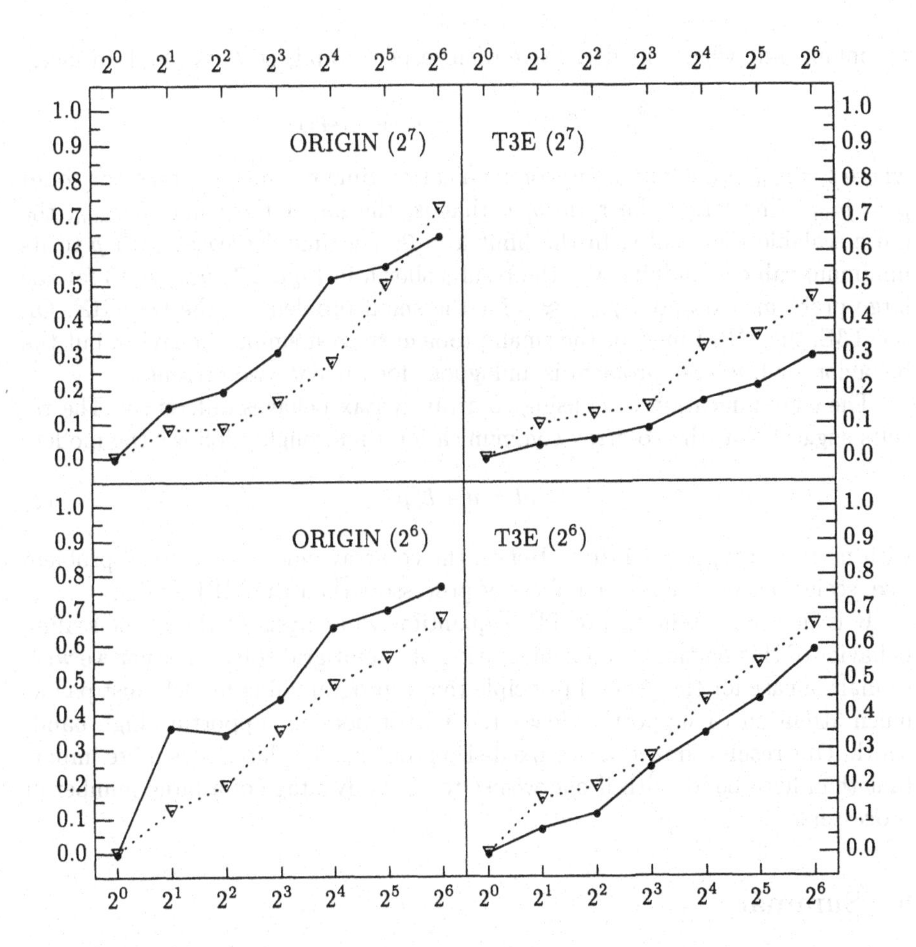

Fig. 2. Ratio of communication time to total time as a function of the number of processors. MPI is marked with dotted lines and co-array syntax is marked with solid lines. The four cases are the same as in Figure 1.

bandwidth. Combining the advantages of hardware support for barriers, low latency, and high bandwidth, the co-array communication time is always lower than MPI on the T3E.

We might try to find an analytical model to describe the differences in the behavior of communication time for the two machines and for the two programming models. The MPI communication times seem roughly to follow the model

$$\nu t = ap + b/p \tag{1}$$

where p is the number of processors and ν is the frequency of the machine. The term proportional to p suggests that there is some resource shared by all the processors and that contention for that resource eventually swamps the calculation. We may estimate the dimensionless parameters a and b by finding the minimum time where the derivative with respect to p is zero. We find, in fact,

$$p_{\min}^2 = b/a \ , \qquad a = \nu t_{\min}/(2p_{\min}) \tag{2}$$

where $t = t_{\min}$ is the minimum communication time reached at processor count $p = p_{\min}$. The larger the ratio b/a, that is, the larger the value of $p_{\min}$, the more scalable the results. In the limit $a = 0$, the time decreases with p to its minimum value at infinity. For the results shown in Figure 1, $p_{\min} \approx 13$ for the large problem and about $p_{\min} \approx 8$ for the small problem on the ORIGIN. On the T3E, the MPI times for the small problem suggest similar behavior, but the behavior for the large problem is ambiguous for the data we present.

The communication time using co-array syntax behaves differently. The results suggest that the co-array communication time roughly follows the model

$$\nu t = a + b/p \tag{3}$$

with no term proportional to p. Hence, the co-array model, for a fixed problem size, scales better to larger numbers of processors than the MPI model.

It is not our intent to provide a quantitative analysis of the exact scaling behavior of this particular implementation of a multigrid solver. Rather we wish to make a case for the general principle that a programming model must pay as much attention to supporting low latency as it does to supporting high bandwidth. Our results show that for fixed-size problems, low latency is more important than high bandwidth if one wishes to take advantage of a large number of processors.

5 Summary

Our results are typical of our early experience with co-array syntax over a number of different applications. The performance of co-array syntax compared with MPI is almost always as good for small numbers of processors but invariably scales better to large numbers of processors. More importantly, the simplicity of the model and its elegant expression of complicated communication patterns make it the model of choice for writing clear, concise code.

In addition, our results demonstrate several important points about the advantages of implementing a parallel programming model as an extension to a language rather than embedding it into a large, cumbersome library. First, since it is part of the language, it understands all the features of the language and can use them to advantage. There is no confusion, for example, about the size of a floating number and there is no difficulty handling arbitrarily complicated Fortran 95 data structures. Second, because the model is expressed as syntax, the compiler can generate in-line code to take advantage of specific architectural features and to optimize for them as well as it can. The result is a high bandwidth model without compromising low latency. Third, by using the object-based techniques available in Fortran 95 combined with co-array syntax, we have shown how to lighten the burden and to remove some of the drudgery associated with message-passing models by encapsulating it within constructors for well defined data structures. Fourth, domain decomposition is often the best way to think about a physical or numerical problem. Once the problem has been decomposed, there should be a natural way to express the decomposition in a syntax that looks like familiar Fortran. The co-array extension is designed to give the programmer precisely that means of expression.

Finally, many programming models and many architectures work well for programs limited by bandwidth. Only a few programming models and a few architectures work well for programs limited by latency. Turning a latency-bound program into a bandwidth-bound program is hard. The situation is similar to the one that existed when high-bandwidth, memory-to-memory, long-vector machines tried to compete with low-latency, register-to-register, short-vector machines. We should underestimate neither the importance of economy of expression nor the importance of a close relationship among the problem to be solved, the programming model used to express it, and the machine used to solve it.

References

1. W. L. BRIGGS, *A Multigrid Tutorial*, SIAM, Philadelphia, 1987.
2. W. GROPP, E. LUSK, AND A. SKJELLUM, *Using MPI, Portable Parallel Programming with the Message-Passing Interface*, The MIT Press, 1994.
3. M. METCALF AND J. REID, *Fortran 90/95 Explained*, Oxford University Press, 1996.
4. C. NORTON, V. DECYK, AND B. SZYMANSKI, *High performance object-oriented scientific programming in Fortran 90*, in Proceedings of the Eighth SIAM Conference on Parallel Processing for Scientific Computing, SIAM Activity Group on Supercomputing, Society for Industrial and Applied Mathematics, March 1997. CD ROM format.
5. R. W. NUMRICH, *F^{--}: A parallel extension to Cray Fortran*, Scientific Programming, 6 (1997), pp. 275–284.
6. R. W. NUMRICH AND J. K. REID, *Co-Array Fortran for parallel programming*. To appear, ACM Fortran Forum, 1998. Draft available from ftp site: jkr.cc.rl.ac.uk/pub/caf.
7. R. W. NUMRICH AND J. L. STEIDEL, *F^{--}: A simple parallel extension to Fortran 90*, SIAM News, 30 (1997).
8. R. W. NUMRICH, J. L. STEIDEL, B. H. JOHNSON, , B. D. DE DINECHIN, G. ELSESSER, G. FISCHER, AND T. MACDONALD, *Definition of the F^{--} extension to Fortran 90*, in Languages and Compilers for Parallel Computing, Z. Li, P. Yew, S. Chatterjee, C. Huang, P. Sadayappan, and D. Sehr, eds., Springer, 1998, pp. 292–306. Lecture Notes in Computer Science #1366.
9. B. F. SMITH, P. E. BJØRSTAD, AND W. D. GROPP, *Domain Decomposition: Parallel Multilevel Methods for Elliptic Partial Differential Equations*, Cambridge University Press, 1996.

Exploiting Visualization and Direct Manipulation to Make Parallel Tools More Communicative

Cherri M. Pancake

Oregon State University, Department of Computer Science, Corvallis, OR 97331, USA
pancake@cs.orst.edu
http://www.cs.orst.edu/~pancake

Abstract. Parallel tools rely on graphical techniques to improve the quality of user interaction. In this paper, we explore how visualization and direct manipulation can be exploited in parallel tools, in order to improve the naturalness with which the user interacts with a parallel tool.
Examples from recent tool research demonstrate that tool displays can be made more communicative and more intuitive to use. Visualization methods can be used to organize complex performance data into layers and perspectives that exploit the user's visual searching capabilities. Direct manipulation techniques allow the user to focus on key elements and then transition smoothly to further levels of detail or interrelated aspects of program behavior. Heuristics derived from studies with parallel users are proposed for when and how the techniques can be applied most effectively.

1 Introduction

The most common complaints about parallel tools have to do with usability. Tools are criticized for being too hard to learn, too complex to use in most programming tasks, and unsuitable for the size and structure of many real-world parallel applications [19, 17, 20]. Part of the problem is that parallel tools can be extremely difficult to implement. They present serious technological challenges, since the tool developer must copy with an inherently unstable execution environment, where it may be impossible to reproduce program events or timing relationships. The fact is that monitoring and other tool activities, intended to observe program behavior, in fact perturb that behavior, sometimes causing errors or performance problems to appear or disappear in unpredictable ways. Further, the notoriously short lifetime of most parallel computers means that there is an extremely small "window of opportunity." Tool development must often begin before the hardware or operating system is stable, but must become available very soon after the computers are first deployed in order to acquire any significant user base [18].

Issues of tool usability can be even more challenging. They can be grouped into three major categories:

- **Data reduction challenges:** The amount of low-level data that can be monitored and recorded during parallel program execution is staggering. For example, the simple recording of key program events during one application run can easily generate gigabytes of data [6]. It is the tool's responsibility to reduce these large volumes of data to a manageable size for presentation to the user. Redundant or extraneous information must be filtered out. Since monitoring data is at the level of machine addresses and stack frames, the remaining data must then be extrapolated or clustered into higher levels of abstraction that can be related to the user's source program.
- **Ergonomic challenges:** The tool developer must also ensure that the interface allows common tasks to be carried out efficiently, in terms of user effort. Not only must the number of user-initiated steps be kept to a reasonable number, but they must be accomplished with a relatively small number of keystrokes or mouse actions. Because the tool serves as an intermediate layer intended to help the programmer understand, debug, or tune the program, care must also be taken to minimize the number of new opportunities for user error that might be introduced by the tool interface.
- **Cognitive challenges:** If the tool is to be of value to the user, it must present information on program behavior in meaningful ways. This is often the most difficult part of tool development. The tool's displays should relate not just to the events that actually transpire during program execution, but also to the user's mental model of program structure and behavior [1]. Information should also be presented in ways that serve to guide the user, in the sense of making further operations obvious.

It is the failure of tools to meet these usability challenges that provokes user dissatisfaction and criticism. The problem is that many tools simply do not convey information effectively to the user. As a result, the user does not understand what the tool is intended to do, or how to apply it to specific programming needs.

The use of graphical techniques can make parallel tools significantly more communicative. Such techniques rely on non-textual attributes such as shape, color, or texture to represent program objects, characteristics, etc., in figurative or symbolic form. In this paper, we examine the two primary graphical techniques available to parallel tool developers: visualization and direct manipulation.

Visualization is the use of graphical representations to portray data or other tool information. As has been demonstrated most notably by [22, 23], graphical displays can be used to make large sets of quantitative data coherent. Effective graphics encourage the eye to compare and contrast elements, revealing patterns or exposing anomalies in the data that would not be discernible if the representations were numeric or textual. Moreover, graphical techniques are capable of revealing information at varying levels or detail, capitalizing on human familiarity with how the appearance of physical objects changes when they are seen from different distances. Finally, graphics are uniquely suited for portraying not just the statistical nature of data, but also its logical characteristics.

Direct manipulation extends the usefulness of visualization by allowing the user to interact with the tool by using a mouse (or some other pointer device, such as a light pen) applied to individual graphical elements. By clicking the mouse button when the cursor is poised over an object, depressing the mouse button and dragging the cursor to a new position, "rubber-banding" a region of the display by clicking-and-dragging to position a rectangular outline over a selected region, etc., the user is able to interact in ways that would otherwise require clumsy sequences of commands.

This paper explores how visualization and direct manipulation can be exploited in parallel tools. Based on observations from field studies with parallel users [20] and tool development efforts that involved user participation [17], heuristics are proposed for when and how the techniques can be applied more effectively. A section on visualization describes how graphical techniques enhance tools' data-reduction and cognitive support. The next section examines how direct manipulation can be introduced in order to further improve ergonomic and cognitive aspects of tool behavior. Examples are drawn from a range of recent tools. (Since direct manipulation in parallel debuggers has been described elsewhere [21], the examples here are drawn from parallel performance analysis tools.) Final sections assess the current "state of the art" and consider what remains to be accomplished if parallel tools are to be truly effective at communicating with the user.

2 Making Visualizations More Communicative

From the standpoint of parallel tools, visualization offers three primary advantages. First, it provides a way to manage the voluminous and complex data associated with parallel program execution. Second, it can capitalize on the user's pattern recognition capabilities. Third, it can facilitate the user explore and "interpret" program behavior by providing alternate views, reflecting different aspects or levels of behavior.

It is not easy to implement graphical displays that exploit these capabilities [7]. Consider the problem of visualizing a large and complex set of performance data. The most straightforward techniques — as evidenced by a number of parallel tools that have been described elsewhere — result in a series of windows, each showing a different portion of the data. This is easily overwhelming, particularly for new or infrequent users. Field studies with actual tool users have shown that they often are confused about what to do with complex visualizations [20, 17]. Common questions include:

- *"What part am I supposed to be looking at?"*
- *"Why can't it show me just the part I need?"*
- *"How do I know what it's* **supposed** *to look like?"*

Here, we propose four heuristics for improving the expressiveness and effectiveness of parallel tools:

1. Visually highlight the most important information

2. Allow the user to change the "perspective" from which data is viewed
3. Use information layering to overcome screen limitations
4. Exploit graphical characteristics to encode information more densely

Each is described below, with examples of how existing parallel tools have implemented similar features.

2.1 Highlight key information visually.

Complex graphics have a potential for overwhelming the eye and distracting attention. The tool designer take conscious pains to ensure that the user's eye will be drawn to the most important display elements. One way of doing this is to consciously reduce the complexity of the display by eliminating information that is of secondary importance; this can be made available through menu choices, display options, or popup displays. Another technique is to organize the information so that key information is scanned first by the eye.

Consider **CXperf** [8], a performance analysis tool developed by Hewlett Packard's Convex Technology Center for use on their Exemplar series of computers. In displaying summary statistics such as elapsed CPU time or cache misses for program functions or code loops, the blocks are sorted so that timings appear in descending order, and values are shown only for the top 15 blocks. The user's eye is thus drawn immediately to the blocks accounting for the highest values. Data on additional blocks are available on-demand; the user can also choose among several sorting orders and determine how many blocks are visible at one time. This example demonstrates that with a little forethought, it is possible to reduce possible distractions without sacrificing the precision or completeness of information available to the user.

2.2 Support multiple perspectives.

The notion of *perspectives* provides another basis for reducing the complexity of tool displays. Rather than presenting all data dimensions in a single view or in a uniform way, it is possible to structure different subsets of data so that each provides a unique perspective, or view. The user can move through the different perspectives to obtain contrasting information about program behavior. If the perspectives are chosen carefully, they can improve the user's ability to recognize important factors or find anomalies.

For example, in addition to showing program metrics plotted against code locations, **CXperf** also creates three-dimensional images that allow the user to see how two different metrics compared for the same code blocks, or to view one metric in relation to two indicators of code location. Figure 1 shows CPU time plotted against both source code functions and threads of execution. A click of the mouse allows the user to change the display to show, say, CPU time and cache misses compared with threads, or compared with source code functions. The sorting and filtering operations described earlier still apply, so the user can adjust the data along each axis as well as the axes themselves.

The advantage of this type of flexibility is that it allows the user to see a quick overview of behavior from a variety of viewpoints; once a pattern (or anomaly) is recognized, the display can be adjusted to show more detail.

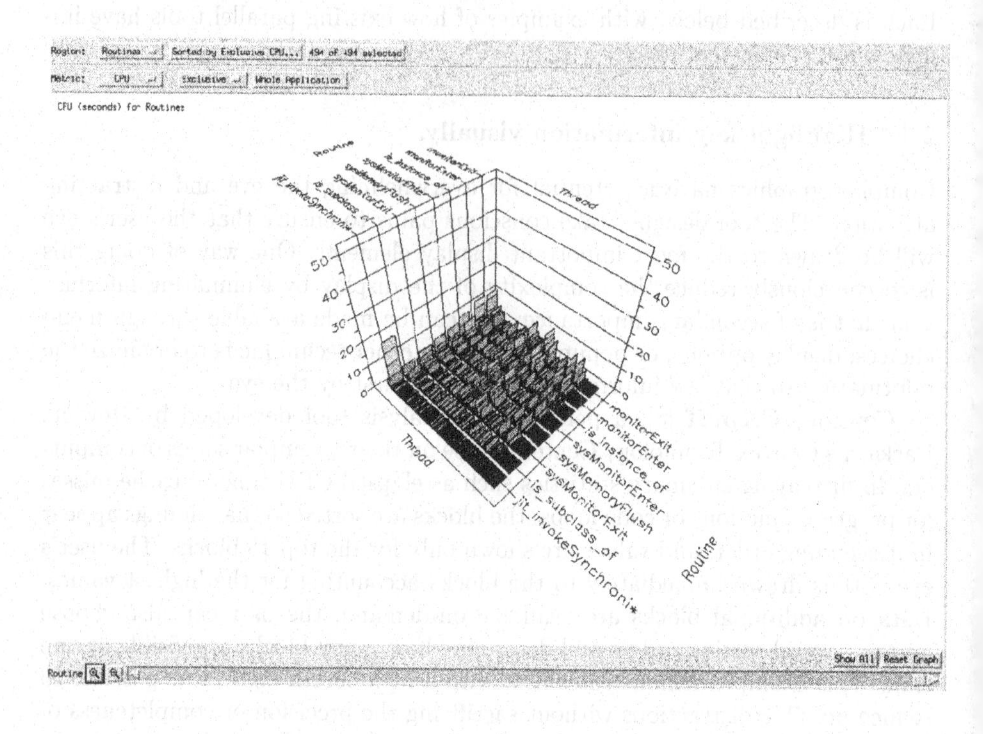

Fig. 1. CXperf [8] display plotting CPU time x thread x code region; the axes may be altered to show not just different metrics, but different combinations of metric/location information.

It is important to note that the user is not having to learn and understand different types of displays. Rather, this is a single display that supports different perspectives into the performance data. The distinction is important, as it allows the user to compare and contrast what is being seen as he/she explores the data space.

A simpler form of perspectives is typified by **Nupshot** [12], an event tracing tool developed at Argonne National Laboratory (derived from an earlier tool called upshot). Like other event-based tools, it portrays the occurrence of program events during execution by projecting them along horizontal "time lines," one for each process involved in the parallel program. Where **Nupshot** differs is that it makes very obvious to the user how events are selected for inclusion. A series of buttons at the top of the window show what color is associated with each type of event (e.g., red for `MPI_Send` and pink for `MPI_Isend` events in a message-passing program), and also can be clicked to remove/reinstate events of each class from the display. While this is not nearly as flexible as **CXperf's** mechanisms, it does allow the user to "ignore" certain types of data and concentrate on others.

2.3 Present information in layers.

Given the complexity of most parallel applications, it is important that the user be able to maintain a sense of context as he/she navigates through the tool's displays. A particularly powerful technique for this is the organization of information into *layers* that reveal different levels of detail. At the highest level, the display may portray information from the whole application (i.e., all processes and/or all regions of the code) in a single, summary visualization. As the user "drills down" into the data, successively more information is portrayed about successively smaller portions of the data space. A so-called thumb-nail image — a compact, low-resolution version of the whole-application image — can help the user maintain a sense of context or direction, indicating which portion of the overall data space is being shown in detail.

Xprofiler [9], a tool developed for IBM's SP/2 computers, provides three levels of timing information. Figure 2 shows the most detailed view. The thumb-nail sketch is highlighted to show which portion of the overall program graph is being viewed. **Xprofiler** also allows the user to collapse areas of the display, hiding them from view so that attention can be focused on particular portions of the program execution. Intel's **SPV** [10], which runs on the Paragon series of computers, uses similar techniques to provide dynamic information about the system-wide use of CPU, memory, I/O, and other resources.

Note that layered information is not simply a matter of magnifying the display so that the user can see a particular area more easily. In fact, different data is presented at each level, so that the finest details are not visible until the user has drilled down to the lowest level.

2.4 Exploit visual encoding properties

A fourth technique that can significantly improve the communicativeness of visualizations is the use of graphical attributes in order to *encode* information. Color, shape, pattern, and other graphical properties can be associated with both quantitative aspects of the data (e.g., magnitude of measured values) and qualitative aspects (e.g., type of event being measured). If the coding is kept consistent across displays, the user can draw inferences about the relationships between different metrics or program behavior.

The **P3T** toolset [5] developed at the University of Vienna, for example, uses colors in its event timeline to encode whether time is being spent by user, MPI, or other system code. The same colors are used in pie-chart summaries of how time is spent overall on different CPUs, allowing the user to determine quickly which sequences of execution state contributed to a particular total. The same tools exploit familiar associations of green and red with stop/go. In a summary of performance for individual source code sections (Figure 3), bars whose length

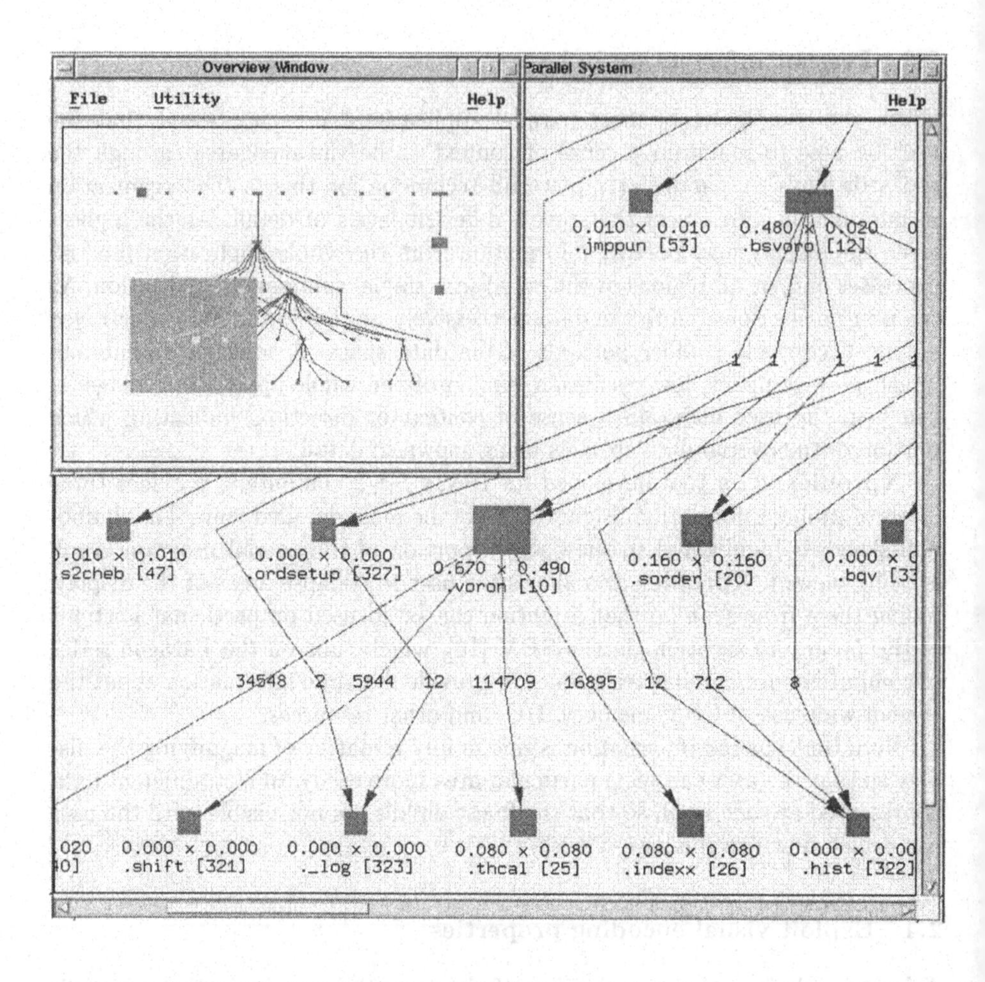

Fig. 2. Xprofiler [9] uses thumb-nail images and differing levels of detail to help the user navigate through large amounts of performance data.

indicate performance time are also colored to reinforce the fact that long bars are "bad" (colored red) and short ones "good" (colored green).

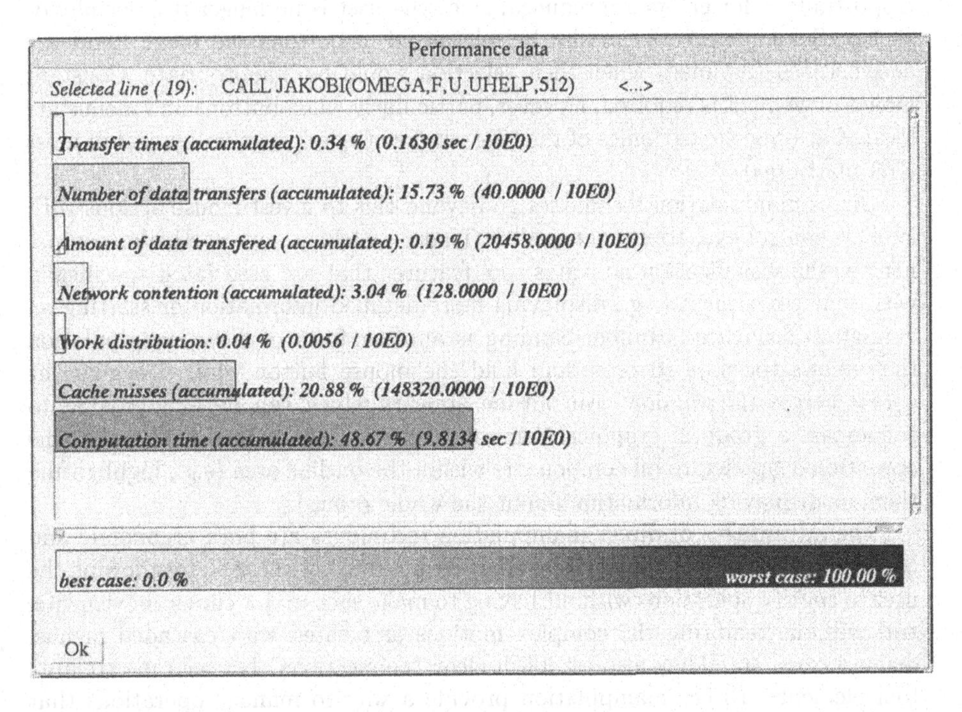

Fig. 3. P3T [5] uses green/red coding to associate rankings of goodness/badness with particular source code sequences.

It is important that the colors or other attributes used as codings be selected carefully. If they conflict with known associations or if the colors are arbitrarily assigned — for example, using yellow/green/red to represent different types of message-passing overheads — the coding can actually hinder usability [4, 24], causing the user to infer associations that are not actually there.

3 Exploiting Direct Manipulation

Once visualization has been implemented by a parallel tool, the graphical techniques can be extended so that the user manipulates the images on the screen. Unlike more traditional ways of supporting user actions via selectable buttons and menus which make use of textual labels direct manipulation allows the user to control the tool without the need to mentally interpret and apply arbitrary word sequences, typed codes, etc.

In windowing interfaces, the user moves the cursor over a button or menu item and clicks the mouse button to select it. These are not good examples of direct manipulation *per se*, since the user is actually manipulating the tool indirectly, by choosing an object that effects some action unrelated to it. Buttons and menus do offer the advantage that user interaction is more ergonomic than typed commends, in terms of the number of physical movements required. Opportunities for errors are reduced, since the user is no longer responsible for syntax. Semantic errors can also be minimized, if buttons and menu items are de-sensitized (dimmed) when their selection would be inappropriate. Constant mouse movement is required, however, particularly when buttons and menus are located at opposite extremes of the tool window (a design policy typical of most GUI platforms).

Direct manipulation techniques go beyond this to invest mouse actions with explicit control over tool functionality. That is, clicking on a particular component of the visualization activates tool features that are associated specifically with that component (e.g., displaying more detailed information or starting an animation sequence). Rubber-banding is another form of direct manipulation that allows the user to press and hold the mouse button while dragging the cursor across the window. An outline appears which can be manipulated to encompass a group of graphical components; when the button is released, the operation is applied to all components within the outline area (e.g., highlighting them or displaying information about the whole group).

The advantages of direct manipulation techniques are both ergonomic and cognitive. They reduce the physical effort required to use the tool, by allowing the user to specify operations without having to make such major cursor movements, and without requiring the complex motions associated with cascaded menus, dialog boxes, etc. They also establish clear "connections" between interrelated tool elements. Direct manipulation provide a way to manage operations that affect multiple windows or graphical regions, through a single user action. More importantly, direct manipulation can make tool operations more intuitive, since the user no longer needs to make a conscious correlation between information displayed graphically and arbitrary textual strings. There is some evidence that this reduction in cognitive load reduces the number of user errors [14, 25].

Again, there are tradeoffs in applying these techniques to parallel tools, as evidenced by field studies involving software tools with direct manipulation [20]. It is difficult to make graphical controls obvious to a new (or infrequent) user. As more controls are added, the tool becomes more flexible, but complexity can quickly get out of hand, bewildering the user. One of the more frustrating aspects for the user is that it can be hard to maintain a sense of context when the displays change in response to each user action. These pitfalls are obvious in the types of questions that emerged during the field studies:

- *"What makes you think I'd click on* **that***?"*
- *"Why does it keep popping up windows?"*
- *"What's this got to do with what I was looking at before?"*

Three heuristics are proposed for applying direct manipulation in order to enhance the communicativeness of parallel tool visualizations:

1. Provide cues to the user indicating when direct manipulation is available

2. Organize the operations into logical pairs so that they can be "undone"
3. Use distinctive mechanisms to differentiate between orthogonal operations

Each is described below, with examples of how existing parallel tools have implemented related features.

3.1 Cue the user when direct manipulation is possible

One of the more interesting observations from the field studies of parallel tools was that while tool developers may "expect" visualizations to allow direct manipulation of some sort, parallel application developers do not [17, 20]. That is, users are not likely to position the cursor over areas of a visualization and experiment with click, shift-click, or other mouse operations. The clear implication for tool developers is that it is necessary to provide some sort of visual cue to the user when such manipulation is possible. Obvious types of cue include textual messages, changes in coloration or background in the areas where manipulation will have effect, and changes in cursor shape.

LCB [15], a tool developed collaboratively by the Parallel Tools Consortium, is intended to provide platform-independent support for viewing information on where a parallel program crashed (or wrote a checkpoint file). It relies heavily on direct manipulation for accessing information that has been structured into hierarchical levels of detail. Because users participated actively in tool design, it was clear early in the design process that visual cues were needed. The tool's status line, which provides brief information on the nodes in the display (which represents the dynamic call graph of the entire parallel program), also indicates when the user can click on the display in order to obtain more information. In addition, the cursor changes to a "pointing hand" whenever it is over an area where direct manipulation is available.

Note that in the example, two types of visual cues are used. User feedback indicated that this redundancy was positive, ensuring that the user was aware of the operations regardless of whether he/she was actively reading the status line [15]. This parallels the results of human factors studies on other visual encoding, which indicate that redundant cues (e.g., the use of both color change and shape change) improve user performance in both searching and tracking activities [4].

3.2 Pair operations logically to support "undo"

It is important to remember that the results of direct manipulation may come as a surprise to the user because some other action was expected, because the user didn't really expect the mouse action to have any effect, because the mouse action was accidental, etc. Whenever the manipulations alter the graphical representation or the target program in some way, it is necessary to provide some form of "undo" operation. A drill-down operation, for example, changes the display to a more detailed view; there must be a corresponding operation that will return the display to the higher level view. (This may not be necessary for direct manipulations that simply pop up additional windows.)

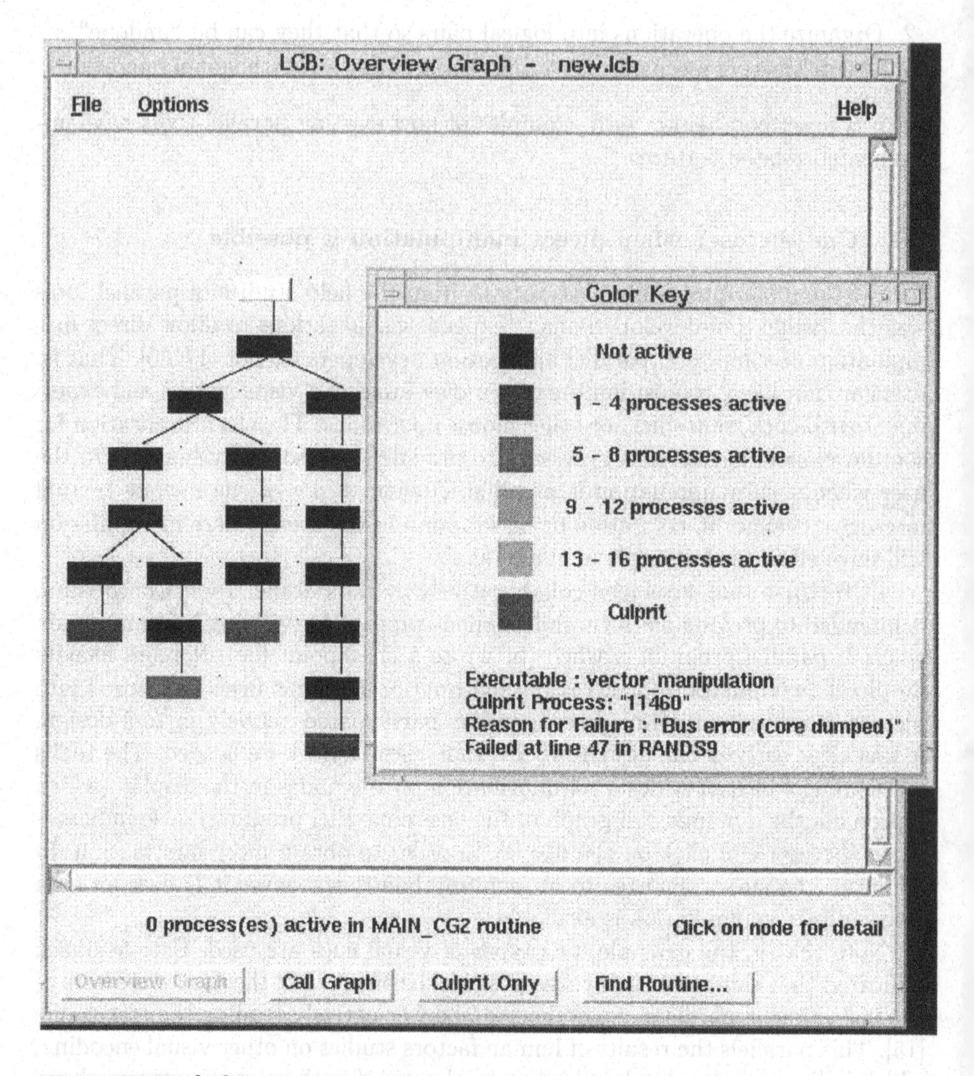

Fig. 4. LCB's [15] displays use hand-shaped cursors when direct manipulation is available, and provide a message reminding the user of its purpose.

As an example, some **CXperf** [8] displays employ a visual representation of the program call graph very similar to that of **LCB**. In this case, however, direct manipulation allows the user to rubber-band selected regions of the graph and "collapse" them. This has the effect of hiding those nodes, which are replaced by a node containing a star to represent the elided portion of the graph. The operation can be performed recursively, so that hidden subgraphs are embedded within others. To undo the operation, the user clicks on the starred node, thereby expanding the subgraph.

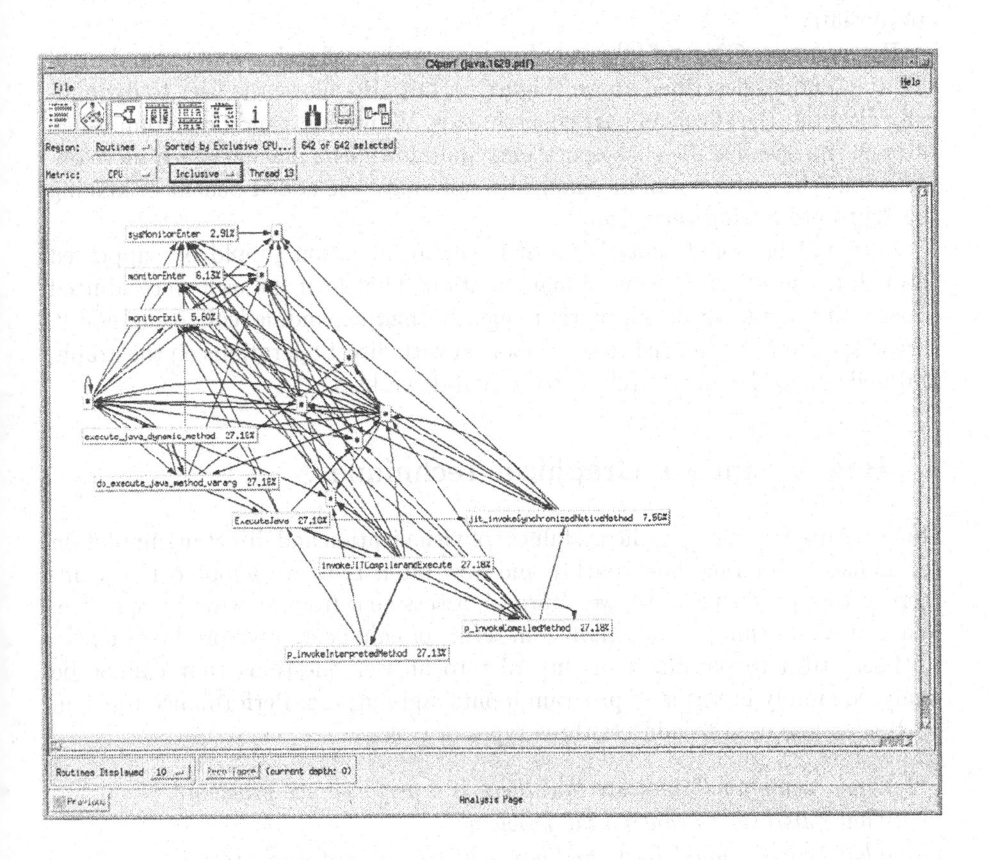

Fig. 5. CXperf allows the user to collapse subgraphs and hide them from view, restoring them later; the nodes with asterisks indicate hidden subgraphs.

Typically, it is not necessary to explain the paired operation used for undoing an action. It would appear that users expect to be able to reverse the effects of graphical changes, since they seem to discover the mechanisms without the need for explicit cues [20].

3.3 Use distinct mechanisms when multiple operations are available

Direct manipulation becomes complex when multiple operations are supported on a single visualization. Since the mouse is a very limited input device, it is difficult to convey to the user that the operations are available, or which operation corresponds to which type of mousing action. While some tools make use of the so-called meta-keys in conjunction with mouse buttons, it may be a source of confusion to users. Generally speaking, users have become accustomed to at most two mousing actions, click-left-button and click-right-button. If more types of operations are needed, radio buttons or option dialogs may be more appropriate.

For example, **CXperf** [8] not only supports the collapsing and expanding of call graph nodes described above (Figure 5), but also allows the user to dynamically filter out the effects of particular threads. While this could be accomplished through the use of shift- click operations, pull-down lists make it much more obvious that the settings can be changed — and have the added benefit of making the "current" setting very clear.

It should be noted that the most common operation should be supported through the most direct form of manipulation. That is, if the user were to need to vary the thread set much more frequently than expanding/collapsing nodes, thread selection would need to be supported with direct operations on the graph, and node control could be relegated to pull-down lists.

4 How Useful are Graphical Techniques?

One measure for assessing the usefulness of visualization and direct manipulation techniques is to track how quickly and how often they are adopted into commercial tool products. Here, we approach assessment from another perspective: What do the techniques accomplish in terms of enhancing user productivity?

Users turn to parallel tools in order to answer questions that cannot be analyzed simply in terms of program inputs and outputs. Performance tools, in particular, are used to address three types of tasks:

- *What "symptoms" indicate that there is a performance problem?*
- *What "disease" is causing the problem?*
- *What "cure" chould I try, and how will I know if it worked?*

The earliest parallel performance tools simply presented information gleaned during program execution, and did not really support any of these user questions directly. This situation is improving with recent tool research.

The **Paradyn** [13] performance monitoring tools are particularly directed at the question of performance symptoms. One component, called the Performance Consultant, searches for evidence of performance bottlenecks, dynamically shifting the focus of monitoring during program execution. By posing hypotheses about the nature and repercussions of bottlenecks, it is able to eliminate some areas of the program from consideration (Figure 6), helping the user to narrow

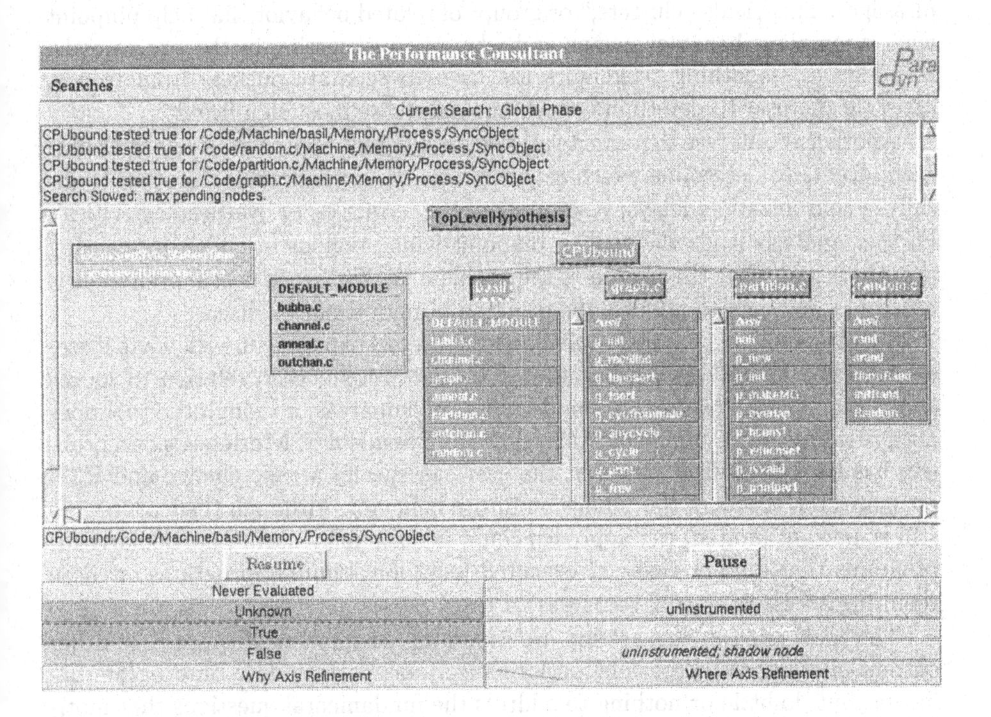

Fig. 6. Paradyn's "experimental" search focuses the user's attention on regions of the program that appear to be causing bottlenecks.

the focus of attention. The tool's emphasis is on identifying problems, however; it cannot really help explain why they are occurring.

Medea [3], goes a step further, applying statistical analysis techniques to identify patterns of program behavior. In a message-passing program, for example, cluster analysis is used to analyze the relative type, frequency, and size of messages. This yields "clusters," or groups of related behavior that help pinpoint what anomalous behavior means and where it is occurring in the source code. In addition, smoothing techniques are used to separate outliers from trends, assisting the user to determine which anomalies are most significant.

Another recent research effort [11] takes a different approach. It analyzes program events to determine which sequences are of unpredictable duration (typically synchronization delays, system use of the resources, or overhead activities). Further analysis leads the tool to pinpoint which regions of codes are causing problems, and what the nature of the problem is. This work is still preliminary, but could do much to address the question of performance "disease."

To date, only one parallel tool allows the user to experiment with what-if scenarios. **P3T** [5] gathers performance data and presents it in relation to source code lines. This helps the user to identify problem areas, although the tool does not provide as much support in this regard as **Paradyn** or **Medea**. Once a problem has been identified, however, the user can specify a code change and **P3T** estimates the effects of the change on future behavior. Although the features are still somewhat limited, the approach could be extremely useful, particularly for programs that are not easily re-executed (e.g., long-running programs or those requiring access to special resources).

The tools cited in this section are somewhat anomalous, compared to other current offerings. Most parallel performance tools portray data on program behavior, but do little or nothing to address the fundamental questions that motivate tool use.

5 Implications for Parallel Tool Research

The heuristics proposed here are intended to improve the naturalness with which the user interacts with a parallel tool. In developing visualizations, the key is to avoid inundating the user with graphical detail. This is done through mechanisms that help focus the user's attention on a relatively small region of the total performance data, and that allow the user to navigate through the data space in order to observe and compare other such regions. The addition of direct manipulation permits the user to interact directly with the visualizations, improving the ergonomics and cognitive directness of the tool interface. Here, the key is to make such interactions both completely obvious and safely reversible.

While the techniques can be highly effective, this will not be true unless elementary usability issues have also been taken into account. Specifically, four criteria are essential for tool usability.

First, the visual displays must be simple to achieve. Users have made it clear that they don't want to spend time preparing their codes for a particular tool [15,

18, 19]. They are particularly reluctant to manually insert instrumentation, since they know from experience that instrumentation can perturb the performance of their programs. All of the tools that have been discussed share the characteristic that no user instrumentation is necessary.

Second, the visualizations must be correlated in some way to the user's source code. At a minimum, performance information must be identified in terms of the corresponding source code function. In fact, users have repeatedly stated that they want to be able to "click back" from a visualization to the lines of source code that caused the behavior they are viewing. As a number of interviewees have stated, "There's no point discovering a bottleneck if you can't find and fix it!" [20]. Most of the tools shown here do not provide such support, however — as users are quick to point out.

Third, tool-related idiosyncracies must be hidden from the user. If the tool's monitoring perturbs the timing relationships of the target program, for example, the tool should calculate what the effects are and compensate for them automatically. What the user wants to see is the behavior of their program, not the effects of the tool. Unfortunately, while techniques for compensating for tool intrusiveness are known (e.g., AIMS [26]), they are employed only rarely. None of the tools reviewed here performs such compensation.

Fourth, machine-induced idiosyncracies should also be hidden. Consider a computer employing clocks that are not synchronized. If timestamps are used by the tool to identify and order the occurrence of program events, lack of clock synchronization may make it appear that impossible sequences took place (such as the arrival of a message before it has been sent). The tool should be able to recognize this type of anomaly and indicate its presence to the user. Again, techniques for this have been known for some time, but only a couple of tools (viz., AIMS [26] and VAMPIR [16]) have actually applied them.

In conclusion, recent developments in parallel tool research have established that tool displays can be made more communicative and more intuitive to use. Visualization methods can be used to organize complex performance data into layers and perspectives that exploit the user's visual searching capabilities. Direct manipulation techniques allow the user to focus on key elements and then transition smoothly to further levels of detail or interrelated aspects of program behavior.

Parallel tools traditionally have applied visualization and direct manipulation to portray program behavior in order to help the user identify potential problems. They have also been used to related program behavior to the source of problems or anomalies. No single tool does both effectively, however. Furthermore, the mechanisms are still confusing and incomplete - they haven't really achieved general levels of usability yet. Nor do current tools approach the most promising application of visual techniques — to help the user identify what can be done to *solve* performance problems. This is the next challenge for parallel tool research.

6 Acknowledgements

We gratefully acknowledge the assistance of the following in obtaining the tool screendumps included here: Hugh M. Caffey, Hewlett-Packard; Jhy-Chun Wang, IBM; Thomas Fahringer, Univ. of Vienna (P3T); Ken Ferschweiler, Oregon State University (LCB); and Bart Miller, Univ. of Wisconsin (Paradyn).

References

1. Baecker, R. M. and W. A. S. Buxton, "Cognition and Human Information Processing," in *Readings in Human-Computer Interaction: A Multidisciplinary Approach*, ed. Baecker and Buxton, Morgan Kaufmann, 1987, pp. 207–218.
2. Browne, S., J. Dongarra and K. London, "Review of Performance Analysis Tools for MPI Parallel Programs," available online at `http://www.cs.utk.edu/ browne/perftools-review/`, 1998.
3. Calzarossa, M. *et al.*, "Medea: A Tool for Workload Characterization of Parallel Systems," *IEEE Parallel & Distributed Technology*, Winter, 1995, pp. 72–80.
4. Christ, R. E., "Review and Analysis of Color Coding Research for Visual Displays," *Human Factors*, 1975, 17(6), pp. 542–570.
5. Fahringer, T., "Estimating and Optimizing Performance for Parallel Programs," *IEEE Computer*, 1995, 28 (11): 47–56.
6. Hansen, G.F., C. A. Linthicum and G. Brooks, "Experience with a Performance analyzer for Multithreaded Applications," in *Proc. Supercomputing '90*, IEEE Computer Society, New York, pp. 123–131.
7. Heath, M. T., A. D. Malony and D. T. Rover, "The Visual Display of Parallel Performance Data," *IEEE Computer*, 1995, 28 (11): 21–28.
8. Hewlett-Packard Corporation, *CXperf User's Guide*, Hewlett-Packard Corporation publication B6323-96001, available online at `http://docs.hp.com:80/dynaweb/hpux11/dtdcen1a/0449/@Generic__BookView`, 1998.
9. IBM Corporation, *IBM AIX Parallel Environment: Operation and Use*, IBM Corporation publication SH26-7231, 1996.
10. Intel Corporation, *System Performance Visualization Tool User's Guide*, Intel Corporation, publication 312889-001, 1993.
11. LeBlanc, T., Meira Jr., W. and V. Almeida, "Using Cause-Effect Analysis to Understand the Performance of Distributed Programs," in *Proc. SIGMETRICS Symposium on Parallel and Distributed Tools*, ACM, 1998, pp. 101–111.
12. Lusk, E., "Visualizing Parallel Program Behavior," Technical Report, Argonne National Laboratory, available online at `http://www-fp.mcs.anl.gov/ lusk/papers/upshot2/paper.html`.
13. Miller, B. P. *et al.*, "The Paradyn Parallel Performance Measurement Tools," *IEEE Computer*, 1995, 28 (11): 37–46.
14. Morgan,m K., R. L. Morris and S. Gibbs, "When Does a Mouse Become a Rat? Or Comparing Performance and Preferences in Direct Manipulation and Command Line Environments," *The Computer Journal*, 1992, 34 (3): 267–271.
15. Muddarangegowda, M. and C. M. Pancake, "Basing Tool Design on User Feedback: The Lightweight Corefile Browser," Technical Report, Oregon State University, available online at `http://www.CS.ORST.EDU/ pancake/papers/lcb/lcb.html`, 1995.

16. Pallas GmdH, "VAMPIR – Visualization and Analysis of MPI Resources" , Pallas GmdH publication available online at `http://www.pallas.de/pages/vampir.htm`, 1998.
17. Pancake, C. M., "Can Users Play an Effective Role in Parallel Tool Research?" in *International Journal of Supercomputing and HPC*, 11 (1), 1997, pp. 84–94.
18. Pancake, C. M., "Establishing Standards for HPC Systems Software and Tools," *NHSE Review*, 2 (1), Fall 1997. Available online at `nhse.cs.rice.edu/NHSEreview`.
19. Pancake, C. M. and C. Cook, "What Users Need in Parallel Tool Support: Survey Results and Analysis," *Proc. Scalable High Performance Computing Conference*, 1994, pp. 40–47.
20. Pancake, C. M. *et al.*, unpublished results of user surveys conducted on behalf of Intel Corporation, IBM Corporation, Hewlett-Packard Corporation, Convex Computer Corporation, DOD HPC Modernization Program, and the Parallel Tools Consortium, 1989–1997.
21. Pancake, C. M. "Direct Manipulation Techniques for Parallel Debuggers," *Proceedings of Supercomputer Debugging Workshop '92*, 1993, pp. 179–208.
22. Tufte, E. R. The Visual Display of Quantitative Information. Graphics Press, 1983.
23. Tufte, E. R. Visual Explanations: Images and Quantities, Evidence and Narrative. Graphics Press, 1997.
24. Ware, C. and J. C. Beatty, "Using Color Dimensions to Display Data Dimensions," *Human Factors*, 1988, 30(2), pp. 127–142.
25. Woods, D. D., "Visual Momentum: A Concept to Improve the Cognitive Coupling of Person and Computer," *International Journal of Man-Machine Studies*, 184, 21: 229–244.
26. Yan, J., S. Sarukhai and P. Mehra, "Performance Measurement, Visualization and Modeling of Parallel and Distributed Programs Using the AIMS Toolkit," *Software — Practice and Experience*, 1995, 25 (4): 429–461.

Deploying Fault-Tolerance and Task Migration with NetSolve*

James S. Plank[1], Henri Casanova[1], Micah Beck[1], and Jack Dongarra[1,2]

[1] Computing Sciences Department, University fo Tennessee, Knoxville, TN, 37996.
{plank, casanova, mbeck, dongarra}@cs.utk.edu.
[2] Computer Science and Mathematics Division, Oak Ridge National Laboratory, Oak Ridge, TN 37831–6367.

Abstract. Computational power grids are computing environments with massive resources for processing and storage. While these resources may be pervasive, harnessing them is a major challenge for the average user. NetSolve is a software environment that addresses this concern. A fundamental feature of NetSolve is its integration of fault-tolerance and task migration in a way that is transparent to the end user. In this paper, we discuss how NetSolve's structure allows for the seamless integration of fault-tolerance and migration in grid applications, and present the specific approaches that have been and are currently being implemented within NetSolve.

1 Introduction

The advances in computer and network technologies that are shaping the global information infrastructure are also producing a new vision of how that infrastructure will be used. The concept of a *Computational Power Grid* has emerged to capture the vision of a network computing system that provides broad access not only to massive information resources, but to massive computational resources as well. Such computational grids will use high-performance network technology to connect hardware, software, instruments, databases, and people into a seamless web that supports a new generation of computation-rich problem solving environments for scientists and engineers.

Grid resources will be ubiquitous. However, for the average scientific user, harnessing their power will present a challenge. Consider such a user. In the world of uniprocessor workstations, his life is relatively simple. Software packages such as MATLAB [21] and Mathematica [41] enable him to solve a wide variety of numerical problems with a convenient and flexible user interface. For less standard problems, he may obtain software solutions from a repository like Netlib. These are typically rather simple to incorporate into his programming platform. However, when his computational needs grow beyond the capability of

* This work was supported by the Applied Mathematical Sciences Research Program, Office of Energy Research, U.S. Department of Energy, under contract DE-AL04-94AL85000 with Lockheed Martin Energy Research Corporation.

a workstation, he must turn to parallel computing platforms. He will be driven to employ computational grids.

While computational grids offer tremendous computing power, they also combine and amplify all the well-known complexities of distributed and parallel computing environments. Moreover, they deprive the programmer of convenient and flexible user interfaces. The obstacles and difficulties standing in the way of routine and effective use of such environments include the following:

- **Distributed ownership** – Often multiple machines are available to perform computation, but each is owned by a different person or group, which may need exclusive use of the machine periodically. Some machines have shared ownership, making the machines available on set schedules. Alternatively, a machine may never available for exclusive use, having the capacity to perform computations in one minute, and to be loaded by other computations in the next minute. Finally, distributed ownership implies that each resource must be considered transient as there is no single administrative authority. All of these characteristics make the task of managing the collection of machines as a single resource very difficult.
- **Platform heterogeneity** – While the processing power of a set of machines may be large, it may be hard to harness because each machine is a different type. This makes the task of partitioning a program among the available machines difficult. Moreover, it makes porting software difficult because different machines may have different compilers and libraries.
- **Network reliability and performance** – When machines are geographically separated, proximity becomes an important issue, because communication times are not uniform between all machines in the collection. Moreover, as larger distances separate machines, the chance of network disconnections increases.
- **Storage availability** – Like CPU capacity, both primary and secondary storage may exist in abundance on a grid, but their availability may be transient and hard to predict.

Thus, while the sheer processing power and storage capacity of computational grids make them attractive, their usability is a major concern. NetSolve, developed at the University of Tennessee and Oak Ridge National Laboratory, is a software environment for network computing that addresses this concern. Its central purpose is to enable the creation of complex applications that can deliver the immense power of computational grids to the desktops of users without being complicated to use or difficult to deploy. To achieve this aim, NetSolve uses a modular, *client-agent-server* architecture that composes a system in which processing platforms and computational software of arbitrary complexity become accessible to users through programming and problem solving interfaces that are already familiar and easy to use.

In this paper, we present the general structure of NetSolve, and how that structure accommodates grid computing. We then elaborate on how NetSolve integrates fault-tolerance and migration into its architecture so that it may harness the power of dynamically changing grid resources in ways that are transparent to the end user. This integration is split into two research directions. The first focuses on how NetSolve's design and implementation can enable fault-tolerance and migration, while the second simply uses NetSolve to explore a variety of approaches to fault-tolerance and migration. We will mention early results and design decisions throughout.

2 NetSolve

2.1 Overview

NetSolve is a software environment for networked computing designed to transform disparate computers and software libraries into a unified, easy-to-access computational service. It aggregates the hardware and software resources of any number of computers that are loosely connected across a network and offers their combined power through client interfaces that are familiar from the world of uniprocessor computing (e.g. MATLAB, simple procedure calls). It uses a *client-agent-server* paradigm (Figure 1) to deliver the power while hiding the complexity of the underlying system.

A NetSolve server may be arbitrarily complex, from a uniprocessor to a MPP (Massively Parallel Processor) or a large networked cluster of machines; moreover there is no upper bound on the number of servers that can be aggregated to form a NetSolve resource pool, and new servers can be added with little effort.

When a user wants a certain computational task to be performed, he/she can use any one of a number of conventional software clients to contact an agent with the request. The agent keeps track of information about all the servers in its resource pool, including their availability, load, network accessibility, and the range of computational tasks they can perform. The agent then selects a server to perform the task, and the server responds to the client's request. The client's computations are performed remotely at the server. The client's data is sent to the server and the server uses its installed software libraries to perform the computation. When the server finishes the computation, it sends the result back to the client and the agent is notified of the completion of the task.

As shown in Figure 1, there may be multiple instances of NetSolve agents on the network, and different clients may contact different agents depending on their locations. The agents exchange information about their different servers, and allow access from any client to any server if desirable. NetSolve can be used either via the Internet or on an intranet, such as inside a department of a university, without participating in any Internet-based computation.

2.2 Strengths of NetSolve

The elements of the NetSolve architecture are for the most part well understood. Depending on your perspective, NetSolve's design can be viewed as an enhanced client-server architecture, or as an enhanced RPC (remote procedure call) environment. However, its three-tiered approach achieves a logical separation that

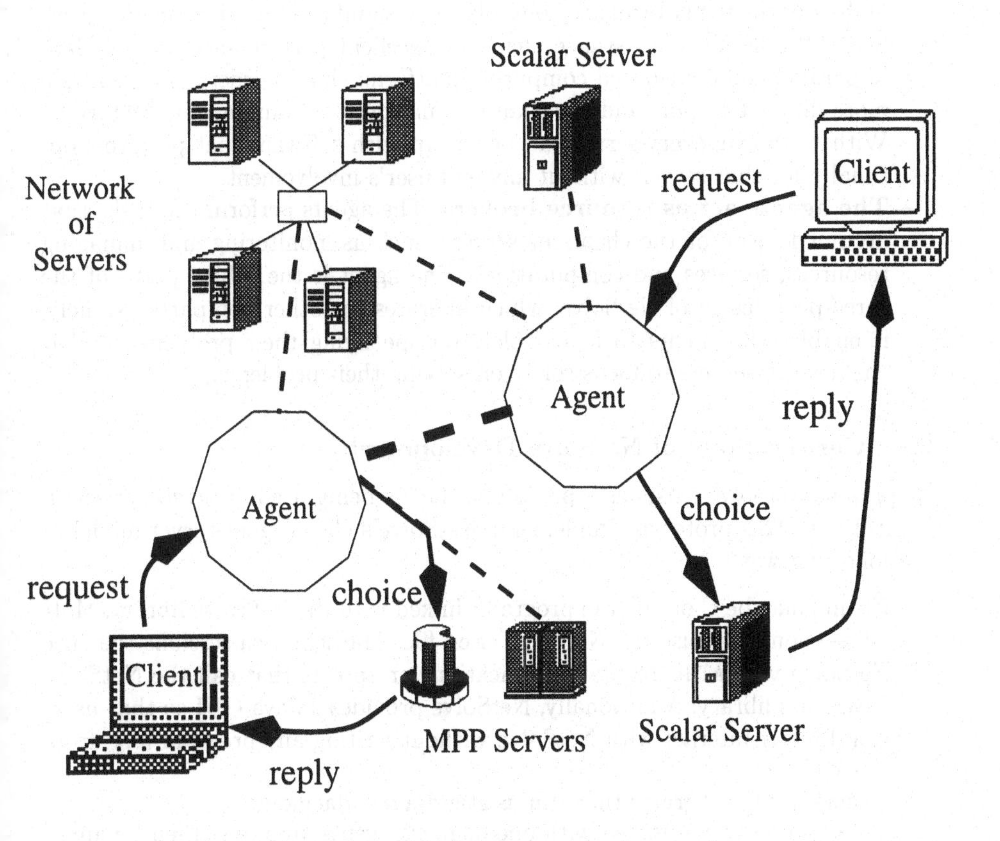

Fig. 1. NetSolve's organization

allows each part to do what it does best without worrying about the details of the other parts. Specifically:

- **The client simply specifies the problem**. This simplicity of use is not only exactly what users typically want, it's what users actually get with uniprocessor packages such as MATLAB or LAPACK [2]. In the parallel world, however, even the simplest software packages burden the user with setting up the parallel environment, distributing data, specifying block sizes and processor grids, and so on. And this does not take into account heterogeneity, failures, or load-balancing. With NetSolve, the user goes back to the uniprocessor model, and lets the NetSolve agent and pool of servers worry about the other details.
- **The servers can optimize computations for their particular architectures**. One of the biggest challenges in writing software for solving problems on parallel systems is making the software general-purpose enough to work on a variety of parallel platforms, yet customizable enough to take advantage of the architectural features of each platform. The typical result is that the software either has sub-optimal performance and is easy to use,

or has optimal performance, but places a significant burden on the user. ScaLAPACK [6], for instance, achieves excellent performance on a variety of parallel and distributed computing platforms, but is orders of magnitude more difficult to port and use than its uniprocessor counterpart, LAPACK. With NetSolve, every server can be set up with software that performs optimally for that server, without any end user's involvement.

– **The agents act as resource brokers**. The agents perform the tasks that are not logical for the clients or servers, such as monitoring and managing resources, services and computations. The agent is the "third part" of the three-part design of NetSolve, which liberates the other two parts. Namely, it enables the clients to focus solely on specifying their problems, and it enables the servers to focus solely on solving their problems.

2.3 Current State of NetSolve Development

At present NetSolve exists as a prototype that is being used in several research institutions. This prototype implements NetSolve's client-agent-server model in the following way.

– Clients may be C or Fortran programs linked with the NetSolve library, Matlab sessions that use the NetSolve `.mex` file, Mathematica sessions that use NetSolve via `MATHLINK`, Java applications or applets that call the NetSolve Java class library. Additionally, NetSolve provides a Java GUI so that users can directly interact with NetSolve without writing any program (e.g. from the Web).
– Agents are C programs that run as stand-alone daemons.
– Servers may be registered with one or more agents, and may then be annotated with the software services that they can perform. Servers may register or unregister at any time.

Currently, the agents and servers run on all variety of Unix machines, and clients may run on Unix and Windows 95/NT.

We have developed a suite of server routines that are very easy to install as part of any uniprocessor server. These correspond to standard computations from different fields, such as linear algebra, FFTs, optimization, curve fitting, iterative methods, etc. We have also developed servers that run any of the ScaLAPACK [6] routines on any parallel processing environment that supports PVM or MPI. It is a straightforward task to embed other software into a server. Finally, we have also developed servers composed of a collection of workstations managed by Condor [25]. These are described below in section 5.1.

Though NetSolve is still a prototype, its potential for providing easy access to large computational resources has already attracted early users within the scientific community. Below are two illustrations of such early deployments:

– **Image processing with NetSolve:** Researchers at the ICG institute at Graz University of Technology, Austria, currently use NetSolve to make sophisticated image processing functions available for remote execution to a large community of users (See [9]).

- **Neuro-science with NetSolve:** At the Computational Neurobiology Lab of the Salk Institute, the NetSolve prototype is being integrated into MCell, which is a software that performs three dimensional Monte Carlo simulations of cellular microphysiology for biologists and biochemists. MCell simulations require solutions to extremely large data parallel problems in multidimensional parameter spaces. NetSolve is ideal for this kind of problem. Because of NetSolve's ability to serve up a vast array of computational resources, including the 256 node Cray T3E located at the San Diego Supercomputer Center, integration of MCell with NetSolve is expected to enable MCell simulations that previously took two weeks to run to finish in an afternoon.

3 Transparent Fault-Tolerance and Load Balancing

As mentioned earlier, the computational resources on a grid may be vast, but due to distributed ownership and the nature of loosely connected systems, their availability may be quite variable. Privately owned workstations may be available for computation one minute, then revoked by the owner the next. Shared machines may exhibit variable load, or may have set schedules for exclusive use. Under these conditions, issues of fault-tolerance and load-balancing become especially important. There has been a vast amount of research on embedding fault-tolerance and load-balancing into parallel and distributed computing platforms. Approaches that have been explored include user-transparent checkpointing and migration libraries (e.g. [10,12,38,25]), programming paradigms that facilitate the task of fault-tolerance or load balancing (e.g. [27,35]), or modified algorithms for performing certain specific computations in a fault-tolerant manner (e.g. [7, 20,30]). While the effectiveness of these techniques has been demonstrated experimentally, none of them have made a large impact on the scientific computing community because using them requires far too much end user involvement.

All the types of fault-tolerance and load balancing listed above fit seamlessly into NetSolve's structure. We divide fault-tolerance and load balancing into **inter-server** and **intra-server** categories. For example, the agent may keep tabs on the progress of a server. If the server is not progressing at a reasonable pace, then the agent (or client) may instruct another server to perform the computation. With a little more effort, the agent may be able to instruct another server to actually continue the computation using the current state of the old server. These are examples of inter-server techniques for fault-tolerance and load balancing. Intra-server techniques are possible as well. For instance, servers of all kinds may employ transparent or non-transparent techniques to render themselves resilient to faults. Servers composed of multiple machines may implement local load balancing to make themselves more efficient.

The important point about the use of all such techniques within the NetSolve framework is that they are totally transparent to the end user. Thus, by employing NetSolve, a user can reap the benefits of research on fault-tolerance and load balancing without actually incorporating it into their code. In fact they may reap these benefits without even knowing it! In this way, NetSolve can make a major contribution in the area of *deploying* support for fault-tolerance.

4 Inter-server Paradigms

If a server fails or is not progressing fast enough, then it is logical for the agent to detect this and move the computation to another server. In this section we discuss different levels of inter-server fault-tolerance that will be explore, and how they fit into the NetSolve framework.

4.1 The Current Approach

The first level of inter-server fault-tolerance is implemented in the current NetSolve prototype. The agents periodically collect information about the servers and their status. If a server that is currently executing a computation becomes unavailable, then the agent selects a new server to take over the computation. Currently, this involves simply having the client contact the new server, and start the computation anew. The agents use a variety of mechanisms for determining server availability, including standard Unix calls such as `uptime` to determine server load. We have also incorporated the Globus Heart Beat Monitor [17] into the agents to perform reliable failure detection. We plan to integrate the efforts of the Network Weather Service [42] into the NetSolve agents as well.

4.2 Integrated Storage Services

A simple way to improve upon the current model is to have the new server query the old server, and if the old server is alive but overloaded, then the old server can send the state of the computation to the new server, which resumes processing from that point. This can be done in a generic manner as in Condor [25], or it can be done in a more application-specific manner, as in the PUL library [35].

This simple method is a good first step, and indeed provides more fault-tolerance than most currently used software for scientific computation. However, it is limited by the fact that the failed server must somehow provide its state to the new server. To address this problem, we introduce a new class of server to the NetSolve framework, called *storage servers*. At a high level, storage servers, just like regular servers, are managed by the agents. However, their job is not to compute, but to hold data. Storage servers will store checkpoints of the computations running on other servers. Very simply, when a computation server decides it should save its state, it selects a storage server with the help of a NetSolve agent, and sends its checkpoint to the storage server. The storage server may store the data in any way that it chooses - in physical memory or on secondary storage. If the storage server needs to make some guarantees about availability, it may replicate the data. Note that the storage server does not need to be concerned with the format of the checkpoint data or with what the data means. If an agent detects that a server has failed or is not performing a computation quickly enough, then it selects a new server to resume the computation from the most recent checkpoint in the associated storage server. This removes any dependency on the old server, which may not be able to respond to queries from the new server. When the computation is finished, the server or the agent may notify the relevant storage servers so that they may discard their checkpoints.

Thus, the concept of inter-server fault-tolerance fits nicely within the framework of NetSolve. One may view storage servers as computation servers that merely hold data and do not perform computations. The management of the all servers is still performed by the agents, which, as in the non-fault-tolerant case, do not concern themselves with the actual computations or checkpoint files. Instead they provide a repository for information about servers and a broker for the resources of servers. As discussed below in Section 5.3, this design of inter-service fault-tolerance allows for interesting interactions of inter and intra-service fault-tolerance.

A first prototype of storage servers has already be developed and will be distributed with the next NetSolve software release.

5 Intra-server Fault-tolerance

NetSolve's logical separation of clients and servers and its dynamic access to heterogeneous resources make it an ideal test-bed that can use real applications to compare the performance of different techniques for fault-tolerance and load balancing. In this section we present different techniques for fault-tolerance and load-balancing that we are considering implementing within NetSolve. These are termed *intra-server* techniques because they are implemented within one server. We also discuss the mixing of intra-server and inter-server techniques for fault-tolerance.

5.1 The Current Approach

Currently, NetSolve allows a pool of workstations managed by Condor [25] to act as a NetSolve server. Condor provides transparent checkpointing and restart so that computations may be moved from loaded/failed machines to idle/available machines. As such, it is able to make use of idle cycles in a shared workstation environment. Employing Condor is a significant first step towards intra-server fault-tolerance. However, it has the following limitations:

- Condor can only migrate jobs between machines of the same architecture.
- Condor can only migrate jobs within its server.
- Condor only works with serial (non-parallel) programs.
- Condor never completely migrates off of its original "host" machine. In particular, system calls are always forwarded to and executed by the original host machine.

In the following sections, we describe the approaches that we take to fault-tolerance and load balancing that attempt to alleviate these limitations.

5.2 Coordinated Checkpointing and Rollback Recovery

The most straightforward scheme for providing fault-tolerance for parallel programs is to employ coordinated checkpointing and rollback recovery. At set intervals (determined by parameters such as program size, storage speed, etc. [40]), the processors involved in a computation save their computation states to stable storage. This is called a coordinated checkpoint. Following a failure of any or all components, an equal number of processors use the states on stable storage to restart the computation from the point of the checkpoint. Several checkpointing libraries have been written for performing coordinated checkpointing on various parallel computing platforms. For example, MIST [10] and CoCheck [38] provide transparent checkpointing for PVM and MPI programs on networks of workstations, CLIP [12] provides semi-transparent checkpointing for Intel Paragon programs, and the PUL-library [35] provides non-transparent checkpointing for a certain class of parallel applications. Here, transparency refers to the amount of programmer involvement necessary to get checkpointing to work.

As a first step, we will add non-transparent coordinated checkpointing as a basic method for intra-server fault-tolerance. The reason for making it non-transparent is because doing so facilitates making the checkpoints platform-independent, and therefore the processor configuration following a failure does not need to be constrained by the configuration at the time of checkpointing. For example, the program may restart on a different number of processors or on processors of differing architectures.

We have already developed a prototype of a NetSolve server for the ScaLAPACK package that performs portable checkpointing and rollback. We will then attack other parallel server packages. The obvious next step is to employ the storage servers described in Section 4.2 to store the checkpoints so that the applications may be restored on any server, not just the server with failed processors.

5.3 Diskless Checkpointing Techniques

One way to improve the performance of coordinated checkpointing is to remove stable storage from the protocol. This is called diskless checkpointing [30]. Here, the coordinated checkpoint is stored in the memory of the application processors and combined with error correcting coding and some extra checkpointing processors so that the loss of a limited number of processors may be tolerated. These techniques have already been embedded into some of the ScaLAPACK routines and have demonstrated excellent performance [30]. There are three interesting research directions for diskless checkpointing in NetSolve.

First, we will incorporate the above code into the ScaLAPACK servers. Note that the structure of NetSolve allows us to embed such application-specific fault-tolerant techniques into the servers so that users may take advantage of them without any knowledge of the underlying fault-tolerant concepts. This would be a breakthrough in bringing complex research concepts into more mainstream use.

Second, there are novel ways in which we can leverage diskless checkpointing to mix intra and inter-server fault-tolerance. For example, instead of storing the error correcting coding in checkpointing processors within a server, we may store them at the storage servers, which may facilitate the migration of the application to another server, or perhaps free the server from having to allocate checkpointing processors altogether.

Finally, by incorporating diskless checkpointing into the server applications, we may discover interesting new fault-tolerant paradigms. This was done with the right-looking matrix factorizations of ScaLAPACK, which did not perform as well as some of the other ScaLAPACK algorithms when diskless checkpointing was employed. The result was a new paradigm for fault-tolerant matrix algorithms called checksum and reverse computation [24]. We anticipate that there will many more opportunities to explore new fault-tolerant paradigms within individual server applications. These may be based on some combination of checkpointing, algorithm-based fault tolerance [20] or backward assertion [7] techniques.

5.4 Distributed Shared Memory and Other Programming Paradigms

There are some problems that lack the regular structure of (for example) dense matrix operations, and map more naturally to the shared memory paradigm of parallel programming [1, 23]. Since NetSolve does not constrain the server, such programs may employ shared memory solutions in the servers without the end user being aware of it. The use of shared memory has implications for fault-tolerance and load balancing. In particular, checkpointing strategies may take advantage of the replication and redundancy inherent in shared memory systems to achieve better performance. This has been explored by researchers for transparent runtime libraries that implement distributed shared memory [8, 11, 22, 39], and for programs that make explicit use of data structures with shared memory semantics [5, 33].

Relatedly, there has been research on fault-tolerant shared tuple spaces [3] and other models of parallel programming such as farming [36], master-slave [4], and coarse-grained dataflow [14] that are more restrictive than general message passing, and facilitate the addition of fault-tolerance and computation migration. A great strength of NetSolve is that if a server functionality maps well to one particular programming paradigm, then the server may implement that paradigm, and thereby allow that functionality to have fault-tolerance and load balancing embedded in it rather seamlessly.

We anticipate implementing a few fault-tolerant servers with some of these programming paradigms. There are three obvious benefits to such implementations. First, they will add to the number of deployable fault-tolerant servers available with NetSolve. Second, they will enable us to compare the performance and tradeoffs of using traditional fault-tolerant strategies, (such as coordinated checkpointing on message-passing architectures) and using fault-tolerant strategies that make use of a specialized programming paradigm. Finally, these implementations may serve as examples for other researchers to implement their own fault-tolerant servers for different applications.

Extending these ideas to inter-server fault-tolerance is also an interesting avenue for research. For example, storage servers may be employed to mirror the state of shared memory, or perhaps to back up a shared tuple space. Similarly, storage servers may implement a remote memory service [16], which may improve the performance of inter-server checkpointing and migration.

5.5 Migration Issues

There are three other research issues in migration that we will explore in the context of NetSolve. First is architecture independent checkpointing. Above, we describe a scheme where checkpointing is embedded directly into an application, giving it the ability to checkpoint and restart on differing computational platforms. There are other ways of duplicating this functionality, such as using a preprocessor to automate the task of embedding type program information into the checkpoint [32] employing restricted languages that may embed architecture independent checkpointing into the compiler and runtime system [37] or employing special checkpointable data structures [5]. While prototype implementations of the above techniques have been promising, the structure of NetSolve allows us to test their use with real applications, compare their performance and deploy them. We anticipate that by exploring these concepts in the framework of NetSolve, we will gain a better understanding of how to write easy-to-deploy and efficient code for architecture independent checkpointing.

Second, the use of privately owned resources as NetSolve servers is an important issue. Specifically, the CPU capacity of most workstations is rarely utilized by their owners. Separate studies have shown that if a computation may be structured so that it only uses idle cycles of privately owned machines, an enormous amount of computation may be performed [13, 26, 28]. The issue of using privately owned resources has been touched in NetSolve upon by the Condor servers, but ideally the brokering of such resources should be performed by the NetSolve agents, and then the migration co-managed by the servers and the agents. We plan to leverage off the similar efforts of CARMI [31] and Globus [18] so that NetSolve can tap into the processing capacity of privately owned workstations.

Finally, The function of the NetSolve agent is closely related to that of a class of systems known as Object Request Brokers, most notably Object Management Group's CORBA [29]. These brokers allow a server to register an interface and facilitate clients in finding and attaching to these interfaces. The objective of such systems are to create an infrastructure for the construction of distributed enterprise level applications.

Because it is designed as a near-universal architecture, CORBA suffers from a combination of generality and structure. Clients and servers can be implemented in any programming language, but all interfaces are defined using object-oriented constructs. Since even OO languages have diverse type systems, CORBA defines its own in the form of the Interface Definition Language (IDL), and requires every language to map IDL to its own constructs. As a result, the CORBA user must make use of extensive mapping tools and invest significant effort just to connect to the system.

The Java world seeks to overcome the generality of CORBA by creating single-language solutions. These solutions make use of the fact that client and server are both implemented in Java to eliminate the need for an IDL. They make use of remote call mechanisms such as Sun's Remote Method Invocation (RMI) to pass complex structures transparently between client and server. Unfortunately, Java is not currently an appropriate standard for world's high performance software.

NetSolve takes a very practical approach: the interface is modeled after the Fortran type system, and mappings are defined for other programming languages. Because Fortran's type system is so simple, these mappings are generally straightforward. Because so much of the software interfacing to Netsolve is written in C and Fortran, there are few software barriers in porting a large body of single-processor code to Netsolve. While this framework does not have the potential to ease the task of brokering complex objects as in CORBA or HORB, it does not enforce a programming paradigm (e.g. object-oriented programming) that would get in the way of porting the large body of code that is implemented in C and Fortran. Simply put, NetSolve is a tool that works without getting in the way. It is an open research area to investigate how more complex brokering paradigms fit into the NetSolve framework.

5.6 Related Grid-oriented Software

As grid-based computing has begun to emerge, approaches to grid-oriented software that are both similar to and different from NetSolve's are under development. Ninf [34], is a project that bears similarities to NetSolve, and a *bridge* has been developed so that NetSolve and Ninf can share resources and clients. The Network-Enabled Optimization Server (NEOS) [15] is in some ways comparable to NetSolve. The differences, however, are significant. First, NEOS addresses a specific field of computational science (optimization) whereas NetSolve, can integrate virtually any processing of user data. Second, NetSolve's software architecture allows it to be deployed on any scale with great flexibility as opposed to NEOS which is centralized at the Argonne National Laboratory. Third, NetSolve provides several more user interfaces than NEOS, including a Matlab and two Java interfaces.

Other than Condor, the two leading middleware systems for creating computational grids are Globus [18] and Legion [19]. They address somewhat different audiences. NetSolve targets any scientist or engineer and provides them with a high level service. By contrast, both Globus and Legion are built on their own lower level directory and communication services, and therefore are significantly more elaborate to deploy. However, both Globus and Legion can be accommodated by the NetSolve model (as agents/servers). As they gain maturity we intend to review the current NetSolve design/implementation and to gradually integrate new components from these systems. As mentioned in Section 4.1, we have already integrated the Globus Heart Beat Monitor into NetSolve. Analogous investigations of the integration of NetSolve with Legion's approach to grid computing are also underway.

6 Conclusion

NetSolve is an environment for networked computing whose goal is to deliver the power of computational grid environments to users who have need of processing power, but are not expert computer scientists. It achieves this goal with its three-part client-agent-server architecture. In order to deliver the full capacity of grid resources, NetSolve must deal with the potential for these resources to be unstable, which means that fault-tolerance and/or computation migration must be employed. We have described how the current version of NetSolve addresses these issues, and how NetSolve will evolve to address them more completely. The most significant impact of this research is in NetSolve's *deployability* of techniques for fault tolerance and migration. Specifically, by incorporating primitives for inter-server fault-tolerance within the NetSolve model, and by developing fault-tolerant software for NetSolve servers, we can deliver fault-tolerance and migration to end-users without any burden on the end user.

References

1. C. Amza, A. L. Cox, S. Dwarkadas, P. Keleher, H. Lu, R. Rajamony, W. Yu, and W. Zwaenepoel. TreadMarks: Shared Memory Computing on Networks of Workstations. *IEEE Computer*, 29(2):18–28, February 1996.
2. E. Anderson, Z. Bai, C. Bischof, J. Demmel, J. Dongarra, J. Du Croz, A. Greenbaum, S. Hammarling, A. McKenney, S. Ostrouchov, and D. Sorensen. *LAPACK Users' Guide, Second Edition.* SIAM, Philadelphia, PA, 1995.
3. D. E. Bakken and R. D. Schilchting. Supporting Fault-Tolerant Parallel Programming in Linda. *IEEE Transactions on Parallel and Distributed Systems*, 6(3):287–302, March 1995.
4. A. Baratloo, P. Dasgupta, and Z. M. Kedem. CALYPSO: A Novel Software System for Fault-Tolerant Parallel Processing on Distributed Platforms. In *4th IEEE International Symposium on High Performance Distributed Computing*, August 1995.
5. A. Beguelin, E. Seligman, and P. Stephan. Application Level Fault Tolerance in Heterogeneous Networks of Workstations. *Journal of Parallel and Distributed Computing*, September 1997.
6. L. S. Blackford, J. Choi, A. Cleary, E. D'Azevedo, J. Demmel, I. Dhillon, J. Dongarra, S. Hammarling, G. Henry, A. Petitet, K. Stanley, D. Walker, and R. C. Whaley. *ScaLAPACK Users' Guide.* Society for Industrial and Applied Mathematics, Philadelphia, PA, 1997.
7. D. Boley, G. H. Golub, S. Makar, N. Saxena, and E. J. McCluskey. Floating Point Fault Tolerance with Backward Error Assertions. *IEEE Transactions on Computers*, 44(2), February 1995.
8. G. Cabillic, G. Muller, and I. Puaut. The Performance of Consistent Checkpointing in Distributed Shared Memory Systems. In *Proceedings of the 1995 European Intel Supercomputer Users' Group Meeting*, 1995.

9. H. Casanova and J. Dongarra. NetSolve's Network Enabled Server: Examples and Applications. *IEEE Computational Science & Engineering*, to appear.
10. J. Casas, D. L. Clark, P. S. Galbiati, R. Konuru, S. W. Otto, R. M. Prouty, and J. Walpole. MIST: PVM with transparent migration and checkpointing. In *3rd Annual PVM Users' Group Meeting*, Pittsburgh, PA, May 1995.
11. M. Castro, P. Guedes, M. Sequeira, and M. Costa. A checkpoint protocol for an entry consistent shared memory system. In *Thirteenth ACM Symposium on Principles of Distributed Computing*, Los Angeles, CA, August 1994.
12. Y. Chen, J. S. Plank, and K. Li. CLIP: A Checkpointing Tool for Message-Passing Parallel Programs. In *SC97: High Performance Networking and Computing*, San Jose, November 1997.
13. P. E. Chung, Y.Huang, S. Yajnik, G. Fowler, K. P. Vo, and Y. M. Wang. Checkpointing in CosMiC: a user-level process migration environment. In *Pacific Rim International Symposium on Fault-Tolerant Systems*, December 1997.
14. D. Cummings and L. Alkalaj. Checkpoint/Rollback in a Distributed System Using Coarse-Grained Dataflow. In *24th International Symposium on Fault-Tolerant Computing*, pages 424–433, Austin, TX, June 1994.
15. J. Czyzyk, M. Mesnier, and J. Moré. NEOS : The Network-Enabled Optimization System. Technical Report MCS-P615-1096, Mathematics and Computer Science Division, Argonne National Laboratory, 1996.
16. M. J. Feeley, W. E. Morgan, F. H. Pighin, A. R. Karlin, and H. M. Levy. Implementing Global Memory Management in a Workstation Cluster. In *15th Symposium on Operating Systems Principles*, pages 201–212. ACM, December 1995.
17. I. Foster, C. Kesselman, C. Lee, G. von Laszewski, and P. Stelling. A Fault Detection Service for Wide Area Distributed Computations. In *Proc. of the High Performance Distributed Computing Conference*, to appear.
18. I. Foster and K Kesselman. Globus: A Metacomputing Infrastructure Toolkit. In *Proc. Workshop on Environments and Tools*. SIAM, to appear.
19. A. Grimshaw, W. Wulf, J. French, A. Weaver, and P. Jr. Reynolds. A Synopsis of the Legion Project. Technical Report CS-94-20, Department of Computer Science, University of Virginia, 1994.
20. K-H. Huang and J. A. Abraham. Algorithm-Based Fault Tolerance for Matrix Operations. *IEEE Transactions on Computers*, C-33(6):518–528, June 1984.
21. The Math Works Inc. *MATLAB Reference Guide*. 1992.
22. G. Janakiraman and Y. Tamir. Coordinated Checkpointing-Rollback Error Recovery for Distributed Shared Memory Multicomputers. In *13th Symposium on Reliable Distributed Systems*, pages 42–51, October 1994.
23. K. L. Johnson, M. F. Kaashoek, and D. A. Wallach. CRL: High-Performance All-Software Distributed Shared Memory. In *15th Symposium on Operating Systems Principles*, pages 213–228. ACM, December 1995.
24. Y. Kim, J. S. Plank, and J. Dongarra. Fault Tolerant Matrix Operations using Checksum and Reverse Computation. In *6th Symposium on the Fontiers of Massively Parallel Computation*, October 1996.
25. M. Litzkow and M. Livny. Experience with the Condor Distributed Batch System. In *Proc. of IEEE Workshop on Experimental Distributed Systems*. Department of Computer Science, University of Winsconsin, Madison, 1990.
26. M. W. Mutka and M. Livny. The available capacity of a privately owned workstation environment. *Perfomance Evaluation*, August 1991.
27. V. K. Naik, S. P. Midkiff, and J. E. Moreira. A Checkpointing Strategy for Scalable Recovery on Distributed Parallel Systems. In *SC97: High Performance Networking and Computing*, San Jose, November 1997.

28. D. A. Nichols. Using Idle Workstations in a Shared Computing Environment. *Operating Systems Review: Proceedings of SOSP-11*, 21(5):5–12, November 1987.
29. R. Orfali and D. Harkey. *Client/Server Programming with Java and CORBA*. John Wiley & Sons, Inc, 1997.
30. J. S. Plank, Y. Kim, and J. Dongarra. Fault Tolerant Matrix Operations for Networks of Workstations Using Diskless Checkpointing. *Journal of Parallel and Distributed Computing*, 43:125–138, September 1997.
31. J. Pruyne and M. Livny. Parallel Processing on Dynamic Resources with CARMI. In *First IPPS Workshop on Job Scheduling Strategies for Parallel Processing*, April 1995.
32. B. Ramkumar and V. Strumpen. Portable Checkpointing and Recovery in Heterogeneous Environments. In *27th International Symposium on Fault-Tolerant Computing*, 1997.
33. D. J. Scales and M. S. Lam. Transparent Fault Tolerance for Parallel Applications on Networks of Workstations. In *Usenix 1996 Technical Conference on UNIX and Advanced Computing Systems*, San Diego, January 1996.
34. S. Sekiguchi, M. Sato, H. Nakada, S. Matsuoka, and U. Nagashima. Ninf : Network based Information Library for Globally High Performance Computing. In *Proc. of Parallel Object-Oriented Methods and Applications (POOMA), Santa Fe*, 1996.
35. L. M. Silva, J. G. Silva, S. Chapple, and L. Clarke. Portable Checkpointing and Recovery. In *Proceedings of the HPDC-4, High-Performance Distributed Computing*, pages 188–195, Washington, DC, August 1995.
36. L. M. Silva, B. Veer, and J. G. Silva. Checkpointing SPMD Applications on Transputer Networks. In *Scalable High Performance Computing Conference*, pages 694–701, Knoxville, TN, May 1994.
37. B. Steensgaard and E. Jul. Object and native code thread mobility among heterogeneous computers. In *15th Symposium on Operating Systems Principles*, pages 68–78. ACM, December 1995.
38. G. Stellner. CoCheck: Checkpointing and Process Migration for MPI. In *10th International Parallel Processing Symposium*, April 1996.
39. G. Suri, B. Janssens, and W. K. Fuchs. Reduced Overhead Logging for Rollback Recovery in Distributed Shared Memory. In *24th International Symposium on Fault-Tolerant Computing*, pages 279–288, June 1994.
40. N. H. Vaidya. Impact of Checkpoint Latency on Overhead Ratio of a Checkpointing Scheme. *IEEE Transactions on Computers*, 46(8):942–947, August 1997.
41. S. Wolfram. *The Mathematica Book, Third Edition.* Wolfram Median, Inc. and Cambridge University Press, 1996.
42. R. Wolski. Dynamically forecasting network performance to support dynamic scheduling using the Network Weather Service. In *6th High-Performance Distributed Computing Conference*, August 1997.

Comparison of Implicit and Explicit Parallel Programming Models for a Finite Element Simulation Algorithm

Joanna Płażek[1], Krzysztof Banaś[1], Jacek Kitowski[2,3]

[1] Section of Applied Mathematics UCK, Cracow University of Technology, ul. Warszawska 24, 31-155 Cracow, Poland
[2] Institute of Computer Science, AGH, al. Mickiewicza 30, 30-059 Cracow, Poland
[3] ACC CYFRONET, ul. Nawojki 11, 30-950 Cracow, Poland

Abstract. In this paper we compare efficiency of two versions of a parallel algorithm for finite element compressible fluid flow simulations on unstructured grids. The first version is based on the explicit model of parallel programming (with message-passing paradigm), while the second incorporates the implicit model (in which data-parallel programming is used). Time discretization of the compressible Euler equations is organized with a linear, implicit version of the Taylor-Galerkin time scheme, while finite elements are employed for space discretization of one step problems. The resulting nonsymmetric system of linear equations is solved iteratively with the preconditioned GMRES method.
The algorithm has been tested on HP Exemplar SPP1600 computer using a benchmark problem of 2D inviscid flow simulations – the ramp problem.

1 Introduction

According to the progress in computer technology many different parallel algorithms for the finite element method (FEM) have been proposed (see e.g. [1], [2]). In the present paper we do not concentrate on isolated problems, like parallel solving systems of linear equations or (parallel) generation (or adaptation) of unstructured meshes, but propose a general approach, taking into account a computational fluid dynamics (CFD) problem to be solved in parallel as a whole.

The most profitable way to exploit full potential of a parallel computer is to construct the algorithm mapping well on the architecture. Due to different characteristics of ccNUMA and sCOMA architectures and slower communication between SMP nodes than within the nodes it would be probably not the best choice to incorporate DSM in a whole program. Since SMP nodes are specially designed for implicit programming the better way is to use implicit programming within the SMP nodes while staying with message-passing paradigm between the nodes.

In the paper we present a parallel algorithm to approximate the Euler equations on unstructured grids. The algorithm incorporates a linear, implicit version of the Taylor-Galerkin [3] scheme for time
discretization. Finite elements are applied for space discretization of one step problems and an overlapping domain decomposition algorithm based on advancing front [4] with the iterative preconditioned GMRES method is used to solve the resulting system of linear equations [5]-[7], Preconditioning is achieved by using block Jacobi iterations to perform matrix-vector multiplications within the GMRES. Two programming paradigms are applied: message-passing and data parallel, which reflect explicit and implicit programming models respectively.

2 Implicit Programming Model

The GMRES algorithm is parallelized using compiler directives and pragmas whenever a loop over patches of elements, nodes or individual degrees of freedom is encountered. In particular the parallelization is applied for loops over blocks of unknowns in the Jacobi method at the steps of blocks construction, computation of element stiffness matrices and assembly into patch stiffness matrices as well as in the Jacobi iterations.

We performed implicit parallelization for HP/Convex SPP 1600 computer using its native compiler and three kinds of directives and pragmas depending upon the nature of dependencies between the data considered [8]. The *loop_parallel* directive and pragma force parallelization of the immediately following loop. The compiler does not check for data dependencies or perform variable privatization. We must synchronize any dependencies manually, and manually privatize loop data as necessary. The *loop_private* directive and pragma declare a list of variables and/or arrays private to the following loop. The *prefer_parallel* directive and pragma cause the compiler to parallelize the loop if it is free of dependencies and other parallelization inhibitors. The compiler automatically privatizes any loop variables that must be privatized. *prefer_parallel* directive and pragma require less manual intervention than *loop_parallel.*

On HP/Convex SPP 1600 system processes run on virtual machines called subcomplexes, which are arbitrary collections of processors. For indicated loops the compiler generates a parallel code that will automatically run on as many processors as are available at runtime. Normally, these are all processors of the subcomplex on which our program is running. We can specify a smaller number of processors via operating system commands (like *mpa,* at runtime) or using some attributes from compiler's directives and pragmas (at compilation stage).

3 Explicit Programming Model

In this model the algorithm is designed for MIMD systems (usually with distributed memory). The practical realization, in our case, uses PVM for message passing between different processing units. There exists one master process that controls the solution procedure and several slave processes performing in parallel most of calculations (master-slave model). The flow diagram of the whole simulation is presented in Table 1 with master process in **boldface** and slave processes in *italic*.

start slave processes on different processors of Parallel Machine
divide the whole mesh into submeshes and send
  subdomain data to corresponding processors (slave processes)
receive subdomain data
for each time step
  create patches of elements corresponding to blocks of unknowns
    in the Gauss-Seidel preconditioner
  compute element stiffness matrices, assemble them into matrices
    of block problems, invert block problems' matrices
  enter GMRES
  perform subsequent steps of the GMRES algorithm:
    the steps involving matrix-vector product consist of loops over all internal
      nodes of the submesh followed by the exchange of the data concerning
      the nodes on the boundary of the submesh
    computations of vector norms and scalar products require selection
      of one process that gather results from submeshes and then broadcast
      the final result to other processes
    *other vector operations (*axpy *products) are performed perfectly in parallel*
  when GMRES converge compute the error in time step and give back control
    to the master process
  either go directly to the next step or adapt the mesh
  if the mesh has been adapted
    divide the whole mesh into submeshes and send
      subdomain data to corresponding processors (slave processes)
    receive subdomain data
    go to the next time step

Table 1. Diagram of the whole simulation (indicating **master** and *slave* processes)

4 Heterogeneous Programming Model

We suggest the following combination of both models for computer architectures based on SMP nodes (hypernodes). Explicit model is applied at the level of hypernodes. A finite element mesh is divided into a number of submeshes equal to the number of hypernodes. Slave programs are executed on different hypernodes and PVM is used for passing messages between hypernodes. Parallel execution on a hypernode is achieved using the implicit model. Compiler pragmas are applied to the loops in the slave algorithm and data distribution among different processors of a hypernode is done by the compiler at run-time.

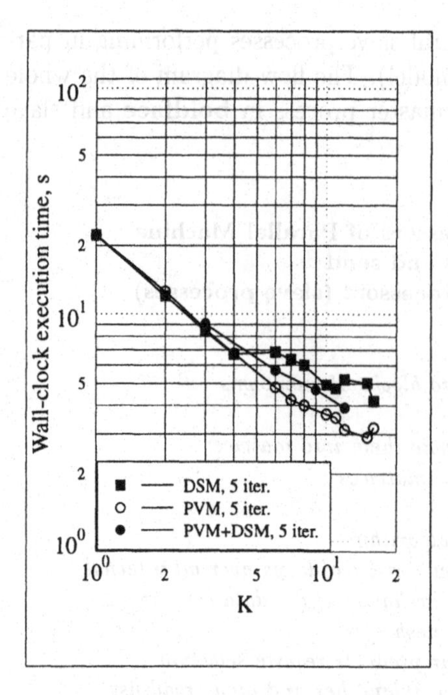

Fig. 1. Wall-clock execution time for one simulation step, three programming models and the mesh with 4474 nodes.

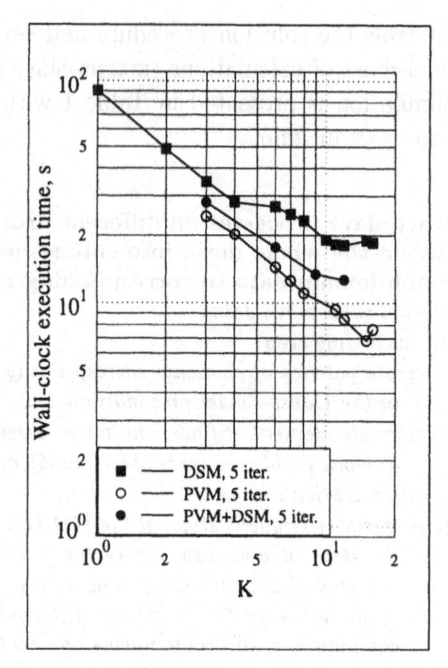

Fig. 2. Wall-clock execution time for one simulation step, three programming models and the mesh with 16858 nodes.

5 Results for explicit and implicit programming models

We tested our parallel algorithm with an example of flow simulation where implicit finite element problems appear within time discretization algorithm. We have chosen a well known transient benchmark problem – the ramp problem [9]. The results refer to one time step (one finite element problem) chosen as a representation for the whole simulation. Two meshes, one with 4474 nodes and the second with 16 858 nodes were employed.

For domain decomposition we used a version of mesh partition algorithm that ensures vertical alignement of subdomain interfaces. The strategy results in the smallest number of the interface nodes minimizing the communication requirements.

We performed simulations on a HP/Convex Exemplar SPP 1600 system with a subcomplex consisting of 16 processors from four SPP hypernodes (4 processors from each hypernode at most).

In Figs. 1 and 2 we present execution wall-clock times for solution of one step problem using three programming models. It follows that for the explicit programming model (with PVM) efficiency is independent of the procesing node organization – with no substantial influence of lower bandwidth between the hypernodes – since it is similar for both subcomplexes. For the implicit programming model worse scalability has been obtained than for the previous case, however the scalability is still within the acceptable range (especially for one hypernode).

The implicit program is much more sensitive to communication speed than the explicit one. Changing from one hypernode execution to multihypernode execution only mildly affects the performance of the explicit code, while greatly deteriorates execution times of the implicit code. Still we hope implicit model is worth testing due to the relative ease of programming and possible future improvements in compiler technology. The explicit model needs programmer's care concerning synchronization between computations done in the domains thus often implementation of load-balancing procedure is necessary. On many-processor machines with hierarchical network layer architecture probably the heterogeneous model would be the choice, since efficiency degradation is not high in comparison with message-passing model with better memory utilization. Moreover, further optimization for SMP programming is still possible.

6 Acknowledgments

The work has been sponsored by Polish Committee for Scientific Research (KBN) Grants No. 8 T11F 21 14, 8 T11F 003 12 and 8 TC11 006 15.

References

1. Carey, G.F., Bose, A., Davis, B., Harle, C., and McLay, M., Parallel computation of viscous flows, in *Proceedings of the 1997 Simulation Multiconference. High Performance Computing '97, 6-10 April 1997, Atlanta, USA.*
2. *Finite Element Method in Large-Scale Computational Fluid Dynamics, International Journal for Numerical Methods in Fluids*, **21**, (1995).
3. Donea, J., A Taylor-Galerkin method for convective transport problems, *International Journal for Numerical Methods in Engineering*, **20**, 101–119, (1984).
4. Banaś, K., and Płażek, J., Dynamic load balancing for the preconditioned GMRES solver in a parallel, adaptive finite element Euler code, in: Desideri, J.-A., et al. (eds.), *Proc.of the Third ECCOMAS Computational Fluid Dynamics Conference*, Sept.9-13, 1996, Paris, J.Wiley & Sons Ltd., 1996, pp. 1025–1031.
5. Saad, Y., and Schultz, M., GMRES: a generalized minimal residual algorithm for solving nonsymmetric linear systems, *SIAM Journal of Scientific and Statistical Computing*, **7**, 856–869, (1986).
6. Saad, Y., Iterative Methods for Sparse Linear Systems, PWS Publ. Company, Boston, 1996.
7. de Sturler, E., and Van der Vorst, H.A., Communication cost reduction for Krylov methods on paralle computers, *Proc. of High Performance Computing and Networking Conference*, April 18-20, 1994, Munich, vol. II, 190-195, Springer Verlag.
8. CONVEX Exemplar Programming Guide, First Edition, October 1994, CONVEX Press, Richardson, Texas, USA.
9. Woodward, P., and Colella, P., The numerical simulation of two dimensional fluid flow with strong shocks, *Journal of Computational Physics*, **54** (1984) 115–173.

Parallel Algorithms for Triangular Sylvester Equations: Design, Scheduling and Scalability Issues

Peter Poromaa

Department of Computing Science, Umeå University
S-901 87 Umeå, Sweden.
peterp@cs.umu.se
http://www.cs.umu.se/~peterp

Abstract. A new scalable algorithm for solving (quasi)triangular Sylvester equations on a logical 2D-toroidal processor grid is presented. Based on a performance model of the algorithm, a static scheduler chosing the optimal processor grid and block sizes for a rectangular block scatter (RBS) mapping of matrices is incorporated in the algorithm.
Scalability properties implying good scalability of the algorithm are also derived from the performance model. Performance results from our ScaLAPACK-like implementations on an 64 processor IBM SP verify the scalability potential of the algorithm.

1 Introduction

The parallel algorithm developed in this paper solves the triangular Sylvester equation

$$AX - XB = C, \tag{1}$$

where the $m \times m$ matrix A and the $n \times n$ matrix B both are in Schur canonical form (i.e., A and B in (1.1) are upper (quasi)triangular). The unknown X and the right hand side C are both $m \times n$ matrices.

We also extend the algorithm for the standard Sylvester equation to solve the generalized Sylvester equation

$$\begin{aligned} AX - YB &= C \\ DX - YE &= F, \end{aligned} \tag{2}$$

where X and Y are unknown $m \times n$ matrices, $(A, D), (B, E)$ and (C, F) are given pairs of $m \times m, n \times n$, and $m \times n$ matrices, respectively. Similar to the standard case, we assume that the matrix pairs (A, D) and (D, E) are in generalized Schur form with A, B upper (quasi)triangular and D, E upper triangular. It is well known that the standard (generalized) Sylvester equation has a unique solution if and only if A and B ($A - \lambda D$ and $B - \lambda E$) have disjoint spectra [20].

The Sylvester equation can, among applications in control theory [4, 6] and in condition estimation of matrix functions [16], be used to compute error bounds for the equation itself and for computed subspaces or eigenvectors [9, 3, 10, 13, 11]. Both the Sylvester equation [2] and the generalized Sylvester equation [10, 13] have been employed in the task of reordering eigenvalues in the real (generalized) Schur form. The algorithms developed in this paper runs on any DMM system able to simulate a 2D-toroidal connectivity.

Any non-standard notation used will be explained in the surrounding context.

2 Blocked Algorithms

We aim at solving triangular Sylvester equations on a logical 2D-toroidal processor grid, where the requirements on the hardware and software are limited to the ability to act as if the logical connectivity exist. In the rest of this paper, we will

```
D_a = ⌈m/M⌉ = number of diagonal blocks in A
D_b = ⌈n/N⌉ = number of diagonal blocks in B
for j = 1 to D_b
    for i = D_a downto 1
        { Solve the (i, j) sub-system }
        A_ii X_ij - X_ij B_jj = C_ij
        for k = 1 to i - 1
            { Update block column j of C }
            C_kj = C_kj - A_ik X_ij
        for k = j + 1 to D_b
            { Update block row i of C }
            C_ik = C_ik - X_ij B_jk
```

Fig. 1. Blocked version of a triangular Sylvester solver.

mainly focus on the standard Sylvester equation and assume the matrices A and B to be upper (quasi)triangular, knowing that the extension to the generalized case (2) is straight forward.

Using Kronecker products, (1) can be rewritten as

$$Zx = y, \tag{3}$$

where

$$Z = I_n \otimes A - B^T \otimes I_m, \quad x = \mathrm{col}(X), \quad y = \mathrm{col}(C). \tag{4}$$

Let M and N be some block sizes such that A and B are partitioned in $M \times M$ and $N \times N$ blocks, respectively. $D_a = \lceil m/M \rceil$ and $D_b = \lceil n/N \rceil$ are the number of diagonal blocks in A and B, respectively. With this partitioning we solve the system (1) by solving for $D_a D_b$ sub-systems:

$$A_{ii}X_{ij} - X_{ij}B_{jj} = C_{ij} - \left(\sum_{k=i+1}^{D_a} A_{ik}X_{kj} - \sum_{k=1}^{j-1} X_{ik}B_{kj}\right) \equiv C_{ij} - G_{ij}, \tag{5}$$

for $j = 1, 2, \ldots, D_b$ and $i = D_a, D_a - 1, \ldots, 1$. Each sub-system, which is indeed a small Sylvester equation, can be written as an $MN \times MN$ linear system

$$Z_{ij}x = y \equiv h - f,$$

where

$$Z_{ij} = I_N \otimes A_{ii} - B_{jj}^T \otimes I_M, \quad x = \mathrm{col}(X_{ij}), \quad h = \mathrm{col}(C_{ij}), \quad f = \mathrm{col}(G_{ij}),$$

and f is the contribution from earlier sub-systems (i.e., the solution to a sub-system is used to update the remaining system). A sequential block algorithm implementing (5) is outlined in Figure 1. At the innermost level, Z_{ij} is dense and $MN = 1, 2$ or 4 ($2, 4$ or 8 in the generalized case), and these small sub-systems are solved by computing an LU factorization of Z_{ij} followed by a forward and backward substitution (for details see [18, 14]).

2.1 A parallel block algorithm PDTRSY

In order to construct a parallel block algorithm for solving the triangular Sylvester equation (1), we examine the data dependency in equation (5). We see that all X_{ij}-blocks appearing on the same block diagonal can be solved for in parallel. As soon as X_{ij} is computed, the updating of the ith block-column and the jth block-row of C can be done in parallel (for details see [12] or [17]). The aim is to develop a scalable parallel block algorithm, taking full account of the inherent parallelism in the problem. We will from now on refer to our parallel block algorithm for solving a triangular Sylvester equation (1) as PDTRSY.

2.2 Data partitioning

Assume that we have p processors labeled $0, 1, \cdots, p-1$ connected in a $P \times Q$ processor grid with P rows and Q columns of processors. These can be referred to as $0, 1, \cdots, P-1$ and $0, 1, \cdots, Q-1$, respectively, in order to identify an individual processor in the grid. Moreover, the neighbors associated with any processor on the grid, is unique and easy to compute.

By partitioning an $m \times n$ matrix C in blocks of size $M \times N$, where $M \leq \lceil m/P \rceil$ and $N \leq \lceil n/Q \rceil$, different mapping strategies can be obtained by regarding $p = PQ$ processors as a $1 \times PQ, PQ \times 1$, or as a $P \times Q$ processor grid. We can define a function $\mathrm{f}(\gamma, \delta) = ((p_r, q_c), (i, j), (\rho, \tau))$ that returns a triple of tuples assigning the matrix entries (γ, δ) of an $m \times n$ matrix, with $\gamma = 0, 1, \cdots, m-1$ and $\delta = 0, 1, \cdots, n-1$ to processor (p_r, q_c), at local block index (i, j), and index in the local array (ρ, τ) as

$$\mathrm{f}(\gamma, \delta) = \left(\left(\lfloor \frac{(\gamma \% T)}{M} \rfloor, \lfloor \frac{(\delta \% R)}{N} \rfloor \right), \left(\lfloor \frac{\gamma}{T} \rfloor, \lfloor \frac{\delta}{R} \rfloor \right), (\gamma \% T \% M, \delta \% R \% N) \right), \quad (6)$$

where $T = MP$, $R = NQ$ and $\%$ is the binary modulus operator. This partitioning strategy is used because of its simplicity to accomplish a block scattered mapping and any needed local or global information (indices) can easily be computed. Moreover it is similar to the strategy used in ScaLAPACK [5, 8].

2.3 Distributed algorithm and some of its characteristics

PDTRSY uses a block scattered mapping of data on a logical 2D-toroidal processor grid. In Figure 2 we show how the matrices A, B and C can initially be mapped onto a 2×4 processor grid. In our implementation the solution X overwrites the right hand side C. That is, the matrices C and X share the same physical memory. Figure 2 also shows the data needed to solve for the block X_{32}, and to update the second block-column and the third block-row of C, with respect to X_{32} and the third block-column and the second block-row from A and B, respectively. At first nothing may look right, but if A and B are shifted three times east and north, respectively, all required data is in the right place at the right time. In general, by shifting the matrices A and B k times east and k times north, respectively, from their initial positions, all matrix blocks needed for all computations regarding the X_{ij} blocks belonging to the kth diagonal of X, will be at the right places when needed. The same result can be obtained by shifting C and A k times east and north-east[1], respectively, or by shifting C and B k times west and south-west, respectively. This implies that we can always let the largest matrix remain at its initial position. The resulting high level algorithm PDTRSY is outlined in Figure 3 and more detailed versions can be found in [19].

Since there are several interesting applications for Sylvester equations where, e.g., $m \gg n$ [3, 2, 14, 13, 18], PDTRSY and PDTGSY (Generalized case) as outlined in [19], efficiently implements that case too.

3 Performance Modeling and Scheduling

The granularity, or how the block sizes M and N, and how P and Q for a given p are chosen, is crucial for the performance of PDTRSY. These parameters should be determined by taking some machine dependent characteristics into account, e.g., the cost for floating point operations in relation to communication cost, the actual cache sizes, vector register length and the number of processors of the target machine. In order to predict the execution time, one can construct a performance model. The model should take m, n, p, P, Q, M and N as input parameters and predict the execution time. This model can then be used in a static scheduling algorithm of PDTRSY for chosing M, N and P, Q. For details on modeling and scheduling read Poromaa [19].

4 Scalability

PDTRSY can, by using similar techniques as Kumar [15], be proven to be scalable due to sustained efficiency. According to the results of the analysis, the soefficiency function equals $p^{3/2}$ for PDTRSY [19]. This means that we can keep efficiency at a level recieved for a certain size of a problem on a given number

[1] A north-east shift in the grid is defined as a north shift followed by an east shift, or vice versa.

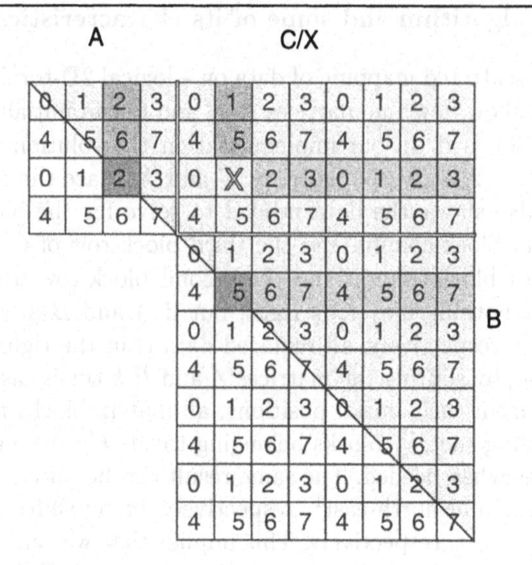

Fig. 2. Initial distribution of data using a 2 × 4 processor grid.

for $k = 1$ **to** number of diagonal blocks in C
  if $(m = n)$
    if $(Q \neq 1)$ **Shift** A East
    if $(P \neq 1)$ **Shift** B North
  else if $(m < n)$
    Shift A South-East
    if $(P \neq 1)$ **Shift** C South
  else
    Shift B North-West
    if $(Q \neq 1)$ **Shift** C West
  endif
  { Solve sub–systems on current block diagonal in parallel }
  if (mynode is owner to C_{ij}), solve for X_{ij} in
    $A_{ii}X_{ij} - X_{ij}B_{jj} = C_{ij}$
    broadcast X_{ij} to processors that need X_{ij}
  elseif (mynode needs X_{ij}) **receive** X_{ij}
    Update block column j of C in parallel
    Update block row i of C in parallel

Fig. 3. High level parallel algorithm for triangular Sylvester equation.

of processors for a larger problem by adding processors but without having to add more memory per processor. Just as the isoefficiency does, scaled speedup reflects the total overhead of a parallel algorithm, e.g., coming from an unbalanced load, inherent sequentiality at a block level (including software overhead) and communication that can not be hidden behind computations. Note that isoefficiency gives an absolute measure and scaled speedup a relative measure of the overhead [19].

We can conclude that a parallel efficiency E_p equal to one, and a parallel speedup S_p of p is according to the analysis in [19] reachable for PDTRSY, if only the problem is big enough. For example, starting with $m = n = 256$, and increasing the work by a factor 10^6, we get matrices of size 25616×25616. This problem could, according to the analysis [19], be solved on 10^6 processors on a 1000×1000 grid with about 33% parallel efficiency.

5 ScaLAPACK Implementations

ScaLAPACK is intended to be a distributed memory version of the LAPACK [1] software package for dense and banded matrix computations. Important goals are scalability and portability. We have developed ScaLAPACK implementations for the standard Sylvester equation (1), as well as for the generalized Sylvester equation (1.3), utilizing as much as possible of BLAS, BLACS [7] and LAPACK code.

5.1 The PDTRSY and PDTGSY routines

The subroutines, named PDTRSY and PDTGSY, have almost [19] the same functionality as the LAPACK subroutines DTRSYL and DTGSYL [14] and calling sequences as specified for ScaLAPACK.

In PDTRSY we use the LAPACK subroutine DTRSYL to solve sub-systems. When we are concerned with the generalized triangular Sylvester equation, PDTGSY [19] is used and the routine used for solving sub-systems of generalized Sylvester equation is DTGSY2 [14]. In the updating-phase of PDTGSY there are two calls to BLAS DGEMM instead of one as in PDTRSY. Also, all messages will be twice as large or twice as many, depending on how things are implemented. Thus, twice the amount of data will be sent and twice as much computational work is done compared to PDTRSY.

5.2 Performance results of PDTRSY and PDTGSY

All test results presented here are from using the Scalable POWERparallel system (SP) at High Performance Computing Center North (HPC2N) which includes 64 Thin Nodes (120 MHz). Single processor peak performance is 480 Mflops/s. System peak performance is around 32 Gflops/s.

Performance results presented in this section was obtained by using block sizes proposed by the scheduling algorithm for PDTRSY, described in [19]. Performance results in Mflops/s from the IBM SP system are shown in Tables 5.1.

Missing figures only appear when the problem could not be solved, due to lack of local memory resources. The number of flops W is estimated as $mn(m+n+1)$ and $2mn(m+n+1)$ for PDTRSY and PDTGSY, respectively.

In Table 5.1 where $m = n$, the performance scales nicely in both directions. Performance increases with increasing p and fixed W, as well as with increasing W and fixed p, except for the sometimes relative loss of performance for the largest problem on a fixed number of processors. This can be explained by the excessive use of virtual memory (i.e., swap from/to disc).

The overall results are good but somewhat unpredictable. Some cases show superlinear speedup, unpredictable drops in performance and so on. Explanations to these exceptions can possibly be found in local memory hierarchy utilization, virtual memory utilization and in the fact that the communication switch may occasionally be shared with other users. For more performance results see [19].

Table 5.1 Performance in Mflops/s for PDTRSY and PDTGSY on IBM SP, $m = n$.

	PDTRSY							PDTGSY						
$m = n$	1p	2p	4p	8p	16p	32p	64p	1p	2p	4p	8p	16p	32p	64p
256	145	87	92	113	131	158	50	43	48	67	72	108	71	91
512	109	133	150	207	259	210	170	54	68	106	119	118	115	156
1024	101	202	225	358	468	510	529	90	101	114	220	387	487	579
2048	103	110	338	491	695	923	1168	109	107	179	327	633	794	1145
2560	188	280	374	555	749	1136	1318			104	367	721	1014	1344
3072	204	151	137	516	871	1275	1473				294	612	1118	1504
4096	219			336	980	1339	1710					659	1329	1704
5120	257				1095	1618	1976						1296	2001
8192						268	2548							2321
10240							2808							

6 Conclusions

We have presented distributed algorithms for triangular Sylvester equations which are scalable and more general but as efficient as existing algorithms [19]. This is accomplished by the fact that the new algorithms allow the largest matrix of a problem to remain fixed after its initial distribution. Furthermore, the RBS mapping of *all* data on a 2D-toroidal grid guarantees that a minimal amount of memory is required per processor, and also supports excellent load balancing.

The scalability properties of the new algorithms are good and verified by our ScaLAPACK-like implementations which show good performance on an 64 processor IBM SP system.

Together with the algorithms, we can use a scheduling technique relying on a performance model. The resulting static scheduler can, given any problem size and number of available processors, choose a configuration, including granularity and logical processor grid dimensions for close to optimal performance. Readers intrested in more deails should read Poromaa [19].

References

1. E. Anderson, Z. Bai, C. Bischof, J. Demmel, J. Dongarra, J. Du Croz, A. Greenbaum, A. Hammarling, A. McKenney, S. Ostrouchov, and D Sorensen. *LAPACK Users Guide*, Release 2.0. SIAM Press, September 1994.
2. Z. Bai and J. Demmel. On Swapping Diagonal Blocks in Real Schur Form. *Linear Alg. Appl.*, 186:73–95, 1993.
3. Z. Bai, J. Demmel, and A. McKenney. On Computing Condition Numbers for the Nonsymmetric Eigenproblem. *ACM Trans. Math. Softw.*, 19:202–223, October 1993.
4. S. P. Bhattacharyya and E. de Souza. Pole Assignment via Sylvester's Equation. *Systems and Control Letters*, 1:261–263, 1982.
5. J. Choi, J. J. Dongarra, R. Pozo, and D. W. Walker. ScaLAPACK : A Scalable Linear Algebra Library for Distributed Memory Concurrent Computers. Progress paper, Oak Ridge National Laboratory and Univerity of Tennesee, 1991.
6. B. N. Datta. Parallel and Large-Scale Matrix Computations in Control: Some Ideas. *Linear Alg. Appl.*, 121:243–264, 1989.
7. J. Dongarra and C. Whaley. A User's Guide to the BLACS v1.0. LAPACK working note 94, Department of Computer Science, University of Tennessee, Knoxville, TN 37996, June 1995.
8. J. J. Dongarra and D. W. Walker. Software libraries for linear algebra computations on high performance computers. *SIAM Review*, 37(2):151–180, June 1995.
9. N. J Higham. Perturbation theory and backward error analysis for $AX - XB = C$. *BIT*, 33(1):124–136, 1993.
10. B. Kågström. A Direct Method for Reordering Eigenvalues in the Generalized Real Schur Form of a Regular Matrix Pair (A, B). In G. H. Golub M. S. Moonen and B.L.R. De Moor, editors, *Linear Algebra for Large Scale and Real-Time Applications*, pages 195–218. Kluwer Academic Publicher, 1993.
11. B. Kågström. A Perturbation Analysis of the Generalized Sylvester Equation. *SIAM J. MATRIX ANAL. APPL.*, 15(4):1045–1060, October 1994.
12. B. Kågström and P. Poromaa. Distributed Block Algorithms for the Triangular Sylvester Equation with Condition Estimator. In F. Andre' and J.P. Verjus, editors, *Hypercube and Distributed Computers*, pages 233–248. North-Holland, 1989.
13. B. Kågström and P. Poromaa. Computing Eigenspaces with Specified Eigenvalues of a Regular Matrix Pair (A, B) and Condition Estimation: Theory Algorithms and Software. *Numerical Algorithms*, 12:369–407, 1996.
14. B. Kågström and P. Poromaa. LAPACK-Style Algorithms and Software for Solving the Generalized Sylvester Equation and Estimating the Separation between Regular Matrix Pairs. *ACM Transactions on Mathematical Software*, 22(1):78–103, 1996.
15. V. Kumar, A. Grama, and G. Karpys. *Introduction to Parallel Computing.* Benjamin/Cummings, September 1994.
16. R. Mathias. Condition Estimation for Matrix Functions via the Schur Decomposition. Technical report, Department of Matematics, College of William and Mary, Williamsburg, USA, October 1994.
17. P. Poromaa. Design, Implementation and Performance for the Triangular Sylvester Equation with Condition Estimators. UMINF-189-90, *Licentiate Thesis*, Institute of Information Processing, Department of Computing Science, University of Umeå, Sweden, November 1990.

18. P. Poromaa. On Efficient and Robust Estimators for the Separation between two Regular Matrix Pairs with Applications in Condition Estimation. UMINF-95-05, Department of Computing Science, Umeå Uneversity, Sweden, August 1995.
19. P. Poromaa. *High Performance Computing: Algorithms and Library Software for Sylvester Equations and Certain Eigenvalue Porblems with Apllications in Condition Estimators, PhD Thesis.* UMINF-97.16, Department of Computing Science, University of Umeå, Sweden, May 1997.
20. G.W. Stewart. Error and perturbation bounds for subspaces associated with certain eigenvalue problems. *SIAM Review*, 15:727–764, 1973.

Fast and Quantitative Analysis of 4D Cardiac Images Using a SMP Architecture

V. Positano, M. F. Santarelli, L. Landini*, A. Benassi

C.N.R. Institute of Clinical Physiology,
Via Savi, 8 - 56126 Pisa, Italy
positano@sunecho.ifc.pi.cnr.it
*Department of Information Engineering: EIT, University of Pisa,
Via Diotisalvi, 2 - 56126 Pisa, Italy

Abstract. In the present research a parallel algorithm for medical image processing has been proposed, which allows 3D quantitative analysis of left ventricular cardiac wall motion in real time. It is a fundamental task in evaluating a lot of indexes useful to perform diagnosis of important diseases. However, such analysis involves expensive tasks in terms of computational time: tridimensional segmentation and an accurate cavity contour detection during the entire cardiac cycle. In this paper an implementation of a dynamic quantitative analysis algorithm on low-cost Shared Memory Processor machine is described. In order to test the developed system in actual environment, a dynamic sequence of 3D data volume, derived from Magnetic Resonance (MR) cardiac images, has been processed.

1 Introduction

Health Care problems solution is an important application area for the new emerging technologies in computer science. In particular, the study of dynamic phenomena in medicine, such as in cardiac field, satisfies a real society's need in the primary aspect of the Health care: with respect to cardiology, ischaemic heart disease is the most widespread pathology in Europe and in the United States, where it represents the first reason of death. Moreover, modern diagnosis mainly relies on results produced by sophisticated techniques able to improve quality and availability of medical data and procedures. The recent progress in computer technology and communication network has increased our ability to generate, collect, and manipulate vast amount of data in different fields in medicine.

One important topic in medical imaging is the contour detection problem aimed to perform quantitative and automatic analysis of different human districts. In fact, from measurements of left ventricular geometry during cardiac cycle, it would be possible to evaluate the presence of an ischaemic zone in the wall, and its degree of severity.

In literature, a lot of work exists about contour detection [1], [2], but all proposed algorithms consist of contour evaluation based on bidimensional images.

In precedent works the authors demonstrated the importance of parallel algorithms to perform real time visualization of dynamic 3D medical images ([3], [4], [5]); such work was also supported by an HPCN-ESPRIT project named PACOMIP (PArallel Computer-based Medical Image Processing). PACOMIP services were exploited in two major departments of cardiology, where diagnosis is based on multimodal imaging approach. During clinical trial, a number of technical improvements were addressed, two of which steered the work herein proposed: 1. possibility to improve software potentialities by including quantitative analysis of left ventricular wall in real time, i.e. measurement of geometry changes during cardiac cycle including dynamic volume evaluation; 2. portability of the developed software on low cost hardware platforms in order to spread HPC technology to health care community.

So that, in the proposed method, the whole algorithm is based on volumetric operations. In fact, both classification and segmentation of the ventricular wall include data volume processing extended to the whole data set, thus exploiting, during endocardial border detection, information relevant to all directions at the same time. It results in a more effective measurement of endocardial borders during the cardiac cycle.

Moreover, the algorithm herein proposed has been implemented for low-cost parallel computer with Unix-based operative system. In particular, preliminary performance results are shown by using a Shared Memory Processors (SMP) machine based on Intel Pentium.

2 Algorithm description

2.1 Hardware and software tools

In the present research the parallel algorithm for volumetric quantitative analysis has been developed in C language, for Unix-based Operative System. In such a way the developed software can be automatically ported on other Unix based multiprocessors that follows the SMP philosophy. For algorithm performance tests. a double Pentium PRO 200, each processor with 512 Kbytes cache, was used. The source program was compiled with 'gcc compiler' under Linux operative system.

2.2 Algorithm phases

The algorithm proposed consists of three fundamental steps: data interpolation, 3D segmentation and quantitative analysis on detected data.

Data interpolation. A pre-processing phase of the algorithm includes interpolation operation. In fact, in the currently available biomedical image data the resolution between slices is often much less than the resolution within each slice. So that, interpolation operation is done in order to make data voxelization, obtaining the same spatial resolution along the three cartesian axes. A weighted threelinear interpolation method has been developed.

3D Segmentation. This operation encodes interpolated data in order to classify the volume in a data structure that contains two fields corresponding to two classification functions: opacity and shading. Opacity values are user-defined, while shading value is function of both the density image value and the gradient module of the relevant voxel. The gradient is calculated by the central differences method [6].

As shown before, classification operation during volume rendering includes an user based definition of the opacity function, that automatically allows the selection of the regions of interest, such as the cardiac cavities. In fact, as a result of a suitable choose of the opacity function, the gradient evaluation allows to easily detect cavity borders. In fact, it is sufficient to automatically select the surface relevant to the higher gradient values borders.

Quantitative analysis on detected left ventricular volumes. After selecting left ventricular internal border, the algorithm allows to process the inner data voxels in order to extract useful indexes for diagnostic aims (as left ventricular wall measurement and cavity volume evaluation). Thus, thanks to HPC technology, it is possible to quantitatively follow the evolution of indexes values during the cardiac cycle.

Moreover, by using volume rendering algorithm, as described in [3], [4], [5], [7], [8], it is also possible to visualize how the left ventricular cavity moves during the cardiac cycle

3 Results

In order to asses algorithm efficiency, a clinical study was based on Magnetic Resonance (MR) images corresponding to a normal heart. The cardiac MR study consists of 20-24 slices (each one of 128x128 pixels) covering the heart.

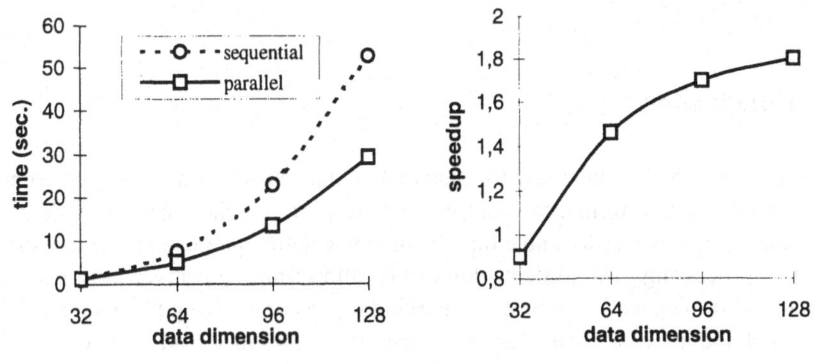

Fig. 1. Processing time vs. data dimension $(.)^3$ **Fig. 2.** Speedup vs. data dimension $(.)^3$

Each slice was acquired several times, typically 12 times during the cardiac cycle. The dimensions of each volume data after voxelization (data interpolation) can be chosen by operator in dependency of minimal error requested in quantitative analysis. The choice of 128x128x128 dimension in interpolated volume preserves all information included in original data, but a decimation of original volume compatible with clinical requirements obviously reduces the processing time.

Figure 1 shows the processing times for both sequential and parallel versions of the algorithm, as function of interpolated data volume dimension. The results are referred to a sequence of 12 volumes. In Figure 2 speedup analysis is shown as function of data volume dimension. From speedup analysis it is evident that the parallel algorithm is very effective for the typical data volume dimension i. e. 128^3.

Moreover, from preliminary results the proposed method appears very effective to detect cavity borders in MR cardiac images, also in presence of low signal to noise ratio, as often occurs in pathologic cases. Figure 3 shows the normalized values of the cardiac volume during the entire cardiac cycle, evaluated by means of the automatic algorithm (continue line) and by manual segmentation (dotted line). Figure 4 shows a image of human hearth extracted from the volume segmented by our algorithm. The left and the right ventricles are clearly defined.

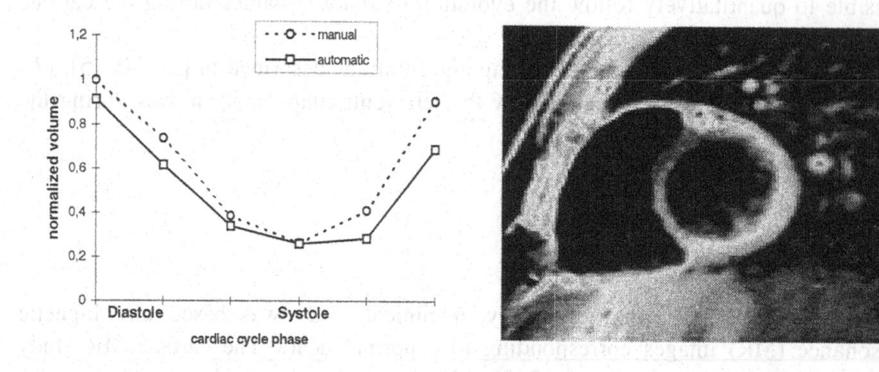

Fig. 3. Volumes evaluation during a cardiac cycle **Fig. 4.** A segmented MR image.

4 Conclusions

The use of the SMP architecture for quantitative analysis of cardiac images seems to be a good cost/performance compromise respect to the use of dedicated supercomputer. It is worth to note that the implementation proposed preserves the on-line data processing. The system, doing this, allows an innovative, easy and rapid evaluation, for diagnostic purposes, of moving organs or evolving phenomena. At the other hand, the use of low-cost hardware platform allows to spread the use of parallel computing technique on the medical community using the resources yet available in a typical medical environment.

In perspective, the use of the message-passing programming technique, based on PVM or MPI libraries, appears to be a good approach to exploit an existing heterogeneous LAN to increase the algorithm performance.

References

1 Kass, M. Witkin, W. Terzopoulos, D. Snakes: Active contour models. Int. Computer vision, 1(4), 321-331 (1988).

2 Leymarie, F. Levine, M. D. Tracking deformable objects in the plane using an active contour model. IEEE Trans. on pattern analysis and Machine intelligence, 15(6), 617-633 (1993).

3 Santarelli, M. F. Positano, V. Landini, L. Parodi, O. Serafini, T. A parallel system for dynamic 3D medical imaging. Lecture Notes in Computer Science 1225, High-Performance Computing and Networking. B. Hertzberger and P. Sloot (Eds.), Springer-Verlag, Berlin, Heidelberg; 989-90 (1997).

4 Santarelli, M. F. Positano, V. Landini, L. A parallel volume rendering algorithm: performance analysis and implications in medicine. Proceedings of the _Int. Conf. on Parallel and Distributed Processing Techniques and Applications (PDPTA '97), H. R. Arabnia (eds.) Las Vegas, Nevada, USA, June 30- July 3, 1997, pp. 1490-1496.

5 Santarelli, M.F. Positano, V. Landini, L. Real time multimodal medical image processing: a dynamic volume rendering application, IEEE Transaction on Information Technology in Biomedicine, Vol 1, n 3, Sept 1997, pp 171-178.

6 Foley, J. D. Introduction to Computer Graphics. Addison-Wesley Publ. Company Inc., Reading, Massachusetts (1994).

7 Lacroute, P. and Levoy, M. Fast Volume Rendering Using a Shear-Warp Factorization of the Viewing Transformation. Proceedings of SIGGRAPH Conference 1994, 451-457 (1994).

8 Lacroute, P. Analysis of a Parallel Volume Rendering System Based on the Shear-Warp Factorization. IEEE Transactions on Visualization and Computer Graphics, 02 (03) (1996).

Ab Initio Electronic Structure Methods in Parallel Computers

Sami Pöykkö

Laboratory of Physics, Helsinki University of Technology,
P.O.Box 1100, FIN-02015 HUT, Finland.
sip@fyslab.hut.fi

Abstract. A short review of different electronic structure methods employing density functional theory is given. The focus is on the plane wave pseudopotential (PWPP) method. For the PWPP method possible parallelization strategies are discussed. It is shown that by combining several data-driven approaches to parallelize calculations excellent performance in massively parallel computer is obtained. The performance is demonstrated by making usual scaling tests and by analyzing scalar performance using performance analyzing tools. Also some future aspects of the parallel computing and electronic structure methods will be given.

1 Introduction

The electronic structure calculations employing the density functional theory (DFT) [1] are widely used tools in computational physics and chemistry. These methods have been successfully applied to calculate accurate structures and energetics for a broad range of systems from molecules and crystalline solids to liquid metals and semiconductors [2].

The computer programs employing DFT minimize the ground state energy

$$E = E_k + E_{ps}^L + E_{ps}^{NL} + E_H + E_{xc} \tag{1}$$

by iterating the Schrödinger equation

$$H\Psi_i(\mathbf{r}) = \epsilon_i \Psi_i(\mathbf{r}), \tag{2}$$

where

$$H = -\frac{1}{2}\nabla^2 + V_{ps}^L(\mathbf{r}) + V_{ps}^{NL}(\mathbf{r}) + V_H(\mathbf{r}) + V_{xc}(\mathbf{r}) \tag{3}$$

using orthonormality constraint

$$\int d\mathbf{r}\Psi_i^*(\mathbf{r})\Psi_j(\mathbf{r}) = \delta_{ij}. \tag{4}$$

Above E_k is kinetic, E_{ps}^L is local pseudopotential, E_{ps}^{NL} is non-local pseudopotential, E_H is electrostatic (hartree) and E_{xc} is exchange-correlation energy. $V_{ps}^L(\mathbf{r})$,

$V_{ps}^{NL}(\mathbf{r})$, $V_H(\mathbf{r})$, and $V_{xc}(\mathbf{r})$ are corresponding potentials and $-\frac{1}{2}\nabla^2$ is kinetic energy operator.

Most widely used method to solve the problem is expand one electron wave functions in a plane wave (PW) basis. A number of other methods have also been developed. In real space methods standard parallelization strategies (like domain decomposition) lead to very good performance in parallel computers [3, 4]. More sophisticated methods like finite element [5], adaptive grids combined both with plane wave basis [6] and with real space method [7] have been introduced, but these methods have not gained great popularity in the scientific community. The orthonormality constraint leads to the $O(N^3)$ scaling (N is the number of atoms in the simulation) for the above mentioned methods. A lot of effort has been directed towards developing a method having linear scaling property [8–10], but these methods are not yet widely used. Plane wave pseudopotential (PWPP) method, which has remained the most popular DFT method, has been considered to be a hard task to implement on parallel computer due to the global nature of the three dimensional fast fourier transformation (3D-FFT).

It has turn out that the electronic structure methods having $O(N^3)$ scaling with the system size are ideally suited for the current massively parallel computers. The latest generation of MPP systems has enabled the use of the system sizes in excess of 200 atoms in a standard scientific simulation. The $O(N^3)$ scaling will however effectively limit the maximum attainable system size in the near future. In order to significantly increase the system sizes methods with better scaling properties (without making further compromises in the accuracy) have to be developed. On the other hand the limitations in the accuracy of electronic structure methods due to the use of DFT in the LDA [1] (or GGA [1]) approximation are becoming more severe problem, since the increased computational power with more effective programs have successfully eliminated many other sources of errors (like finite size effects).

2 Plane wave pseudopotential method on parallel computer

The computational task of a PWPP code is to find the set of coefficients $c(n_{pw}, n_{states}, n_{kpts})$ included in the plane-wave expansion of one-electron wave functions (band index j, $\mathbf{k}$-vector $\mathbf{k}$)

$$\Psi_{j,\mathbf{k}}(\mathbf{r}) = \sum_{\mathbf{G}}^{n_{pw}} c(\mathbf{G}, j, \mathbf{k}) exp[i(\mathbf{k}+\mathbf{G})\cdot\mathbf{r}], \tag{5}$$

which are used to generate the valence electron density

$$\rho(\mathbf{r}) = \sum_{j}^{n_{states}} \sum_{\mathbf{k}}^{n_{kpts}} w_{\mathbf{k}} \Psi_{j,\mathbf{k}}^*(\mathbf{r}) \Psi_{j,\mathbf{k}}(\mathbf{r}) \tag{6}$$

minimizing the DFT total-energy functional. Above n_{pw} is the number of plane-waves included in the expansion, n_{states} is the number of electronic states, n_{kpts}

is the number of **k**-points used in Brillouin zone integration and $w_{\mathbf{k}}$ is a weight factor of the **k**-point. The expansion (5) is in fact a Fourier transformation from the momentum space representation of the wave functions to the real space representation. The use of the plane-wave basis set thus implies the periodic boundary conditions; the periodic replicas of the modeled system fill the infinite three dimensional space. This periodically repeated system is called a supercell. In the expansion (5) the **G** are the supercell reciprocal lattice vectors, defined by $\mathbf{G} \cdot \mathbf{a} = 2\pi m$, where **a** is a lattice vector of the supercell and m is an integer. The number of terms in the expansion (n_{pw}) is restricted by taking only terms corresponding to the kinetic energy smaller than a cut-off energy $E_{cut} = \frac{1}{2}\|\mathbf{k} + \mathbf{G_{cut}}\|^2$ into account $\Leftrightarrow \|\mathbf{G} + \mathbf{k}\| < \mathbf{G_{cut}}$. This truncation sets a limit for the highest oscillations of the wave functions and via the Nyqvist criterion the truncation determines also the size of the 3D-FFT grid. From the computational point of view finding the set $c(n_{pw}, n_{states}, n_{kpts})$ requires Fast Fourier Transformations (FFTs) between the real- and momentum-spaces, vector-vector and vector-matrix multiplications to be performed. The parts of the computations performed in the real and reciprocal spaces are illustrated in the Table 1.

Table 1. A flow chart of a typical PWPP program. $FFT(\Psi)$ and $FFT(\rho)$ are, respectively, FFTs for the valence electron density and one-electron wave functions (see text for explanation), n is the number of the one-electron wave functions in the simulated system.

G-SPACE			**R**-SPACE
$\{c_i^{\mathbf{k}}(\mathbf{g})\}$	$\Longrightarrow$	$\overset{n \cdot FFT(\Psi)}{\Longrightarrow}$	$\Psi_i^{\mathbf{k}}(\mathbf{r})$
V_{ps}^{NL}	E_{ps}^{NL}		$\Downarrow$
$\rho(\mathbf{g})$	$\Longleftarrow$	$\overset{FFT(\rho)}{\Longleftarrow}$	$\rho(\mathbf{r})$
$\Downarrow$			$\Downarrow$
V_H	E_H		$\Downarrow$
V_{ps}^L	E_{ps}^L		$V_{xc} \quad E_{xc}$
$V_{ps}^L + V_H$	$\Longrightarrow$	$\overset{FFT(\rho)}{\Longrightarrow}$	$V_L = V_H + V_{ps}^L + V_{xc}$
$\{c_i^{\mathbf{k}}(\mathbf{g})\}$	$\Longrightarrow$	$\overset{n \cdot FFT(\Psi)}{\Longrightarrow}$	$V_L \Psi_i^{\mathbf{k}}(\mathbf{r})$
$V_L c_i^{\mathbf{k}}(\mathbf{g})$	$\Longleftarrow$	$\overset{n \cdot FFT(\Psi)}{\Longleftarrow}$	$V_L \Psi_i^{\mathbf{k}}(\mathbf{r})$
$+V_{ps}^{NL}$			
$+\sum(\mathbf{k}+\mathbf{g})^2 c_i^{\mathbf{k}}(\mathbf{g})$	E_k		
$(H - \epsilon_i) c_i^{\mathbf{k}}(\mathbf{g})$	$\Longrightarrow$	new $c_i^{\mathbf{k}}(\mathbf{g})$	

In typical calculation the three most time-consuming parts of the program (the FFTs, the orthonormalization of electronic states and calculation of the forces acting on the plane-wave coefficients) use about 90 percent of the total time. The PWPP method is computationally challenging also because of the memory needed to store a huge number of plane-wave coefficients included in the expansion of the one-electron wave functions (the total number of plane-waves included can be of the order of 10^8). In an effective implementation of a PWPP program to a massively parallel system all FFTs as well as all linear algebra have to be performed effectively in a parallel manner and also the use of the memory has to be parallelized.

3 Parallel PWPP code

Because the PWPP method provides computational burden both in the number of calculations to be performed and in the memory usage, a data-driven parallelization method is the most practical way to start the parallelization of a PWPP program. The plane-wave expansion of the wave functions (eq. 5) provides three straightforward ways to parallelize the computation. One can parallelize with respect to A) **k**-points, B) electronic states, or C) **G**-vectors [12]. From these three strategies the first two are useful in small parallel machines, but in the massively parallel environment only the strategy C) is effective enough. The details of each parallelization strategy can be found from Refs. [12] and [13]. Each of these parallelization strategies require substantial communication between all processors participating in the calculation.

3.1 Parallel FFT for PWPP program

The memory usage of PWPP program is dominated by the need to store huge number of the plane waves and in the G-parallel PWPP code each electronic state is distributed over the range of processors. Thus 3D-FFT has to be performed for the data which has been distributed and simple methods used to parallelize FFT can not be used.

The maximum frequency of the wave function oscillations in PW-calculations is fixed by the maximum length of G-vector ($\mathbf{G_{cut}}$) in system. Since ($\rho(\mathbf{r}) = |\Psi(\mathbf{r})|^2$), $2 \cdot \mathbf{G_{cut}}$ has to be used for the density in order to have the same accuracy for Ψ and ρ. During one self-consistency loop of PWPP code $3{\cdot}n_{states}$ FFTs for the wave function coefficients ($FFT(\Psi)$) and 3-5 FFTs for the valence electron density ($FFT(\rho)$) have to be performed. These two FFTs are not computationally equally demanding, because the two maximum lengths for the G-vectors. The 3D-FFT grid has to be chosen so large that a sphere of radius $2 \cdot \mathbf{G_{cut}}$ fits completely in the box. The set of the plane-waves does not extend over the whole grid, as is illustrated in Fig 1 **Z**, since the $(\mathbf{G}+\mathbf{k})$-vectors span only a sphere with radius of $\mathbf{G_{cut}}$. For an effective massively parallel PWPP code it is crucial to take full advantage of the difference between these two FFTs.

The 3D-FFT is calculated as a sum

$$F(\mathbf{r}) = \sum_{j_1=1}^{n_x} W_x \sum_{j_2=1}^{n_y} W_y \sum_{j_3=1}^{n_z} W_z F(\mathbf{G}), \tag{7}$$

where $W_x = \exp(i2\pi x j_1/n_x)$, $W_y = \exp(i2\pi y j_2/n_y)$ and $W_z = \exp(i2\pi z j_3/n_z)$, (n_x, n_y, nz) is the dimension of the 3D-FFT grid. As one can see from Eq. (7) 3D-FFT can be computed by taking 1D-FFTs sequentially on each dimension.

The important difference between $FFT(\Psi)$ and $FFT(\rho)$ is the fact that for $FFT(\rho)$ 1D-FFT has to be performed for each column at each direction, but for the $FFT(\Psi)$ only for the columns where at least one plane wave coefficient resides (gray areas in Figs. 1), since all elements in other columns are zero.

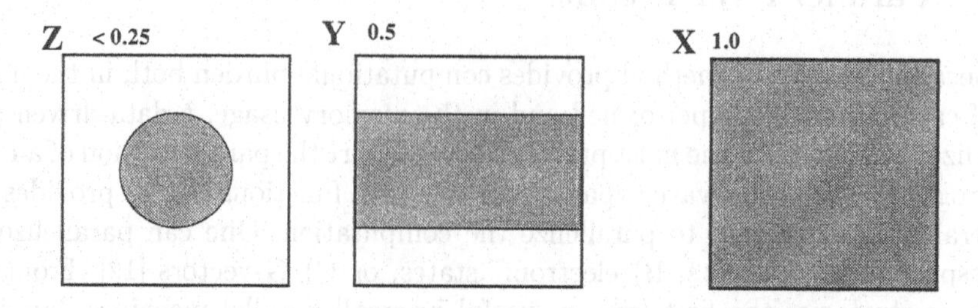

Fig. 1. The gray areas show the columns of the 3D-FFT grid containing at least one plane-wave coefficient (*ie.* non-zero element) in the column. The numbers above each direction show the fraction of the gray area.

In practice the data distribution is done so that the z-columns are distributed in such manner that the plane-wave coefficients are distributed as evenly as possible, since the number of the plane waves in the nodes dictates the load balance. There is no need to distribute data in such way that the areas in each node are continuous.

In a parallel computer 3D-FFT are performed in such way that first 1D-FFTs are performed locally in each processor for their own data and after that data is re-organized in such way that the columns in the next direction are distributed over the processors, and next 1D-FFTs are performed and yet one re-organization and 1D-FFTs are needed to complete 3D-FFT. If the forward and inverse transformation are done in reverse order for dimensions ($x->y->z \Rightarrow z->y->x$) no more re-organizations are needed. The big benefit in the parallel computer comes from the fact that for the $FFT(\Psi)$ only data from (or to, depending on which is smaller) gray areas in Figs. 1 has to be communicated. This reduces the amount of data to communicate by the factor of three.

3.2 Multi-level parallelism

Linking together two or three of the data-driven parallelization strategies leads to a situation where all the calculations are done within small blocks of processors and all interprocessor communication is done within these small blocks instead of communicating over all participating processors (see fig. 2).

In the FINGER (FINnish General Electron Relaxator) program approaches A) and C) are used simultaneously. In the first level of the parallelization, **k**-points are distributed. At this stage four blocks of five processors are created in the case of the example shown in Fig. 2. Calculations in these n_{kpts} blocks are relatively independent of each other and only the total electron density has to be globally

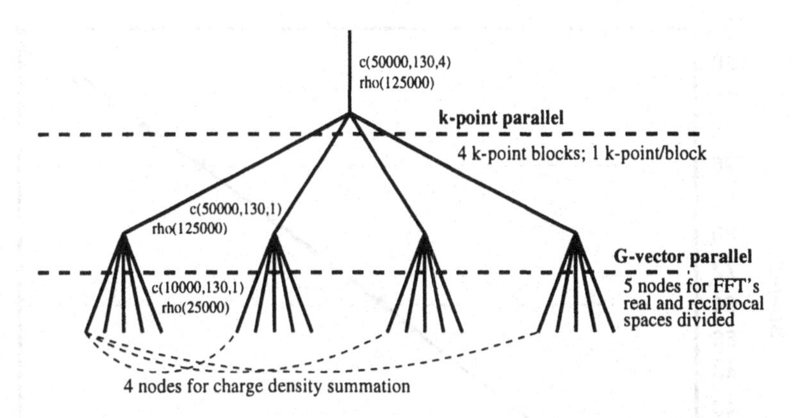

Fig. 2. Example of the data driven parallelization strategy used in the FINGER program. A typical size of a 3D-FFT grid (electronic density in real space) and a table $c(n_{pw}, n_{states}, n_{kpts})$ used to store plane-wave coefficients are also shown.

summed over these blocks. The size of the table used to store the plane-wave coefficients reduces by a factor of n_{kpts} (number of **k**-points), while the 3D-FFT grid remains unaltered.

Inside each of the n_{kpts} blocks parallelization with respect to the **G**-vectors is used in the second level. The most difficult part at this stage is parallelization of the Fast Fourier Transforms (FFTs). In the FINGER program the very effective FFT parallelization introduced originally in the CETEP (Cambridge Edinburgh Total Energy Package) program [11, 12] has been employed. In this scheme the real and reciprocal spaces are distributed inside each block of processors. Thus the number of plane-waves and the fraction of the 3D-FFT grid stored in the local memory reduces by the factor of the number of processors in the block (five in the example shown in Fig. 2). After this data distribution all communications are done within small blocks (in the example shown in Fig. 2 there are blocks of 5 processors for the FFTs and for the orthonormalization of states, and blocks of 4 processors for the construction of the density). Direct communication between all processors is not needed.

The speedup of the FINGER program as the number of processors increases is shown in Fig. 3. The ideal speedup shown in Fig. 3 does not essentially tell anything about the performance of a program in a parallel machine without the knowledge about the speed/processor the program can obtain. Using the FINGER program a sustained speed up to 130 Mflops/processor has been obtained in a Cray-T3E computer system [14]. In the scaling test shown in Fig. 3 sustained speed of about 13 Gflops was obtained in a 128-processor run. The physical system used in the scaling test shown in Fig. 3 was a zinc vacancy in ZnSe described by a 64-atom-site supercell and four **k**-points (the dimensions for the plane-wave coefficient tables are close to the ones shown in Fig. 2). Since the size of cache memory available in the calculation scales linearly with the number of the processors, speedups better than the linear one (as seen in Fig. 3 in the case of 64 processors) may take place. The minimum number of processors in

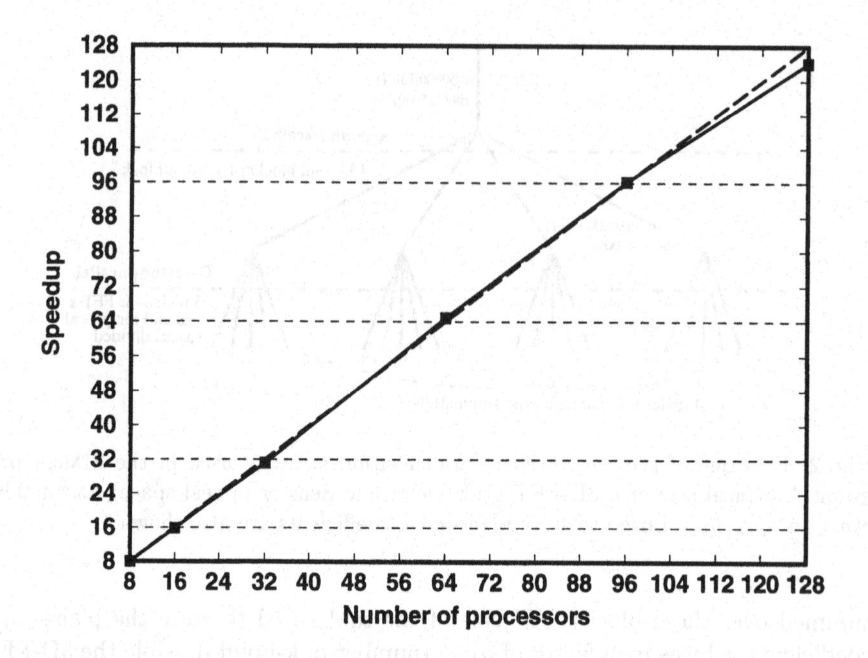

Fig. 3. Speedup versus the number of processors for the FINGER program running in a Cray T3E system. The dashed line corresponds to the ideal speedup.

the scaling test shown in Fig. 3 is limited to 8 due to the maximum memory available in a single node. Scaling tests employing smaller system sizes show linear speedup also in the region for less than 8 processors.

4 Summary

In summary it was shown that by combining several data-driven approaches to parallelize calculations excellent performance in massively parallel computer is obtained using the PWPP method. In order to significantly benefit from the next generation of supercomputers (more than shorter computing times) the need is two-folded; on the other hand methods having better scaling properties are needed to make larger systems sizes attainable and on the other hand more accurate methods capable to treat of the order of 100 atoms are needed to improve the predictive capability of the computational electronic structure methods.

5 Acknowledgments

The author wishes to thank Dr. Stewart Clark and DR. Tomi Mattila for many fruitful discussions.

References

1. See, for example, R. O. Jones and O. Gunnarson, "The density functional formalism, its applications and prospects", Reviews of Modern Physics, **61**, 689 (1989).
2. See, for example, M. Parrinello, "From silicon to RNA: the coming of age of *ab initio* molecular dynamics", Solid State Communications **102**, 107 (1997).
3. E. L. Briggs, D. J. Sullivan, and J. Bernholc, "Real-space multigrid-based approach to large-scale electronic structure calculations", Physical Review B **54**, 14 364 (1996).
4. J. R. Chelikowsky, N. Troullier, and Y. Saad, "Finite-difference-pseudopotential method: electronic structure calculations without a basis", Physical Review Letters **72**, 1240 (1994).
5. E. Tsuchida and M. Tsukada, "Electronic-structure calculations based on the finite-element method" Physical Review B **52**, 5573 (1995).
6. F. Gygi, "*Ab initio* molecular dynamics in adaptive coordinates", Physical Review B **51**, 11 190 (1995).
7. G. Zumbach, N. A. Modine, and E. Kaxiras, "Adaptive coordinate real-space electronic structure calculations on parallel computers", Solid State Communications **99**, 57 (1996).
8. Y. Wang, G. M. Stocks, W. A. Shelton, D. M. C. Nicholson, Z. Szotek, and W. M. Temmerman, "Order-N multiple scattering approach to electronic structure calculations", Physical Review Letters **75**, 2867 (1995).
9. P. Ordejón, E. Artacho, and J. M. Soler, "Self-consistent order-N density functional calculations for very large systems", Physical Review B **53**, 10 441 (1996).
10. E. Hernándes, M. J. Gillan, and C. M. Goringe, "Basis functions for linear scaling first-principles calculations" Physical Review B **55**, 13 485 (1997).
11. M. C. Payne, M. P. Teter, D. C. Allan, T. A. Arias, and J. D. Joannopoulos, "Iterative minimization techniques for *ab initio* total-energy calculations: molecular dynamics and conjugate gradients", Reviews of Modern Physics **64**, 1045, (1992).
12. L. J. Clarke, I. Štich and M. C. Payne, "Large-scale *ab initio* total energy calculations on parallel computers", Computational Physics Communications **72**, 14 (1992).
13. J. S. Nelson, S. J. Plimpton and M. P. Sears, "Plane-wave electronic-structure calculations on a parallel supercomputer", Physical Review B **47**, 1765 (1993).
14. My estimate for the Mflops/processor-rate is obtained using `pat` performance analysis tool in Cray-T3E (375 MHz DEC alpha processors).

Iterative Solution of Dense Linear Systems Arising from Integral Equations

Jussi Rahola[1,2]

[1] CERFACS, 42 av. G. Coriolis, 31057 Toulouse Cedex, France
Jussi.Rahola@cerfacs.fr

[2] Center for Scientific Computing, P.O. Box 405, FIN-02101 Espoo, Finland
Jussi.Rahola@csc.fi

Abstract. In this paper we show how to solve dense linear systems arising from integral equations with iterative solvers. We present three case studies: a volume and a surface integral equation for electromagnetic scattering and a surface integral equation for the electric potential in bioelectromagnetism. For two of the methods, iterative solvers convergence quickly even without preconditioning while for the third an approximate inverse preconditioner is employed. We show how the matrix-vector products can be computed efficiently with special methods that do not form the matrix explicitly.

1 Introduction

Many problems in science and engineering can be cast either as partial differential equations (PDEs) or as integral equations (IEs). Each of these approaches have their benefits and limitations.

PDEs involve only local interactions and they thus give rise to systems of linear equations with sparse coefficient matrices. However, these sparse systems tend to be badly conditioned and the conditioning gets worse when the discretisation is refined. The whole object has to be discretised which can lead to large numbers of unknowns, especially for 3D problems. In computational electromagnetics the boundary conditions are in many cases given at infinity and thus also the space surrounding the object has to be discretised up to a suitable absorbing boundary condition.

IEs by contrast involve local and global interactions, thus giving rise to dense coefficient matrices. In many applications, these dense matrices are well conditioned, making them suitable for iterative solvers. IE methods in electromagnetism need no absorbing boundary conditions because the boundary conditions at infinity are automatically incorporated in the integral equation formalism. In many cases it is also possible to formulate the physical problem as a surface integral equation involving only the unknowns on the surface of the object.

The bottleneck for IEs is the solution of dense linear systems. The solution of dense systems with the state-of-the-art LAPACK library requires $\mathcal{O}(N^3)$ operations and $\mathcal{O}(N^2)$ units of storage, where N is the number of unknowns. In many cases the computational requirements, especially the huge memory space needed, severely limit the applicability of the IE method.

2 Iterative solution of dense systems

In this work we have applied iterative solvers for dense problems arising from three integral equation formulations. For all of these, the general recipe is as follows. First, try to find an efficient iterative method that converges with as few iterations as possible. In many cases, already this step may reduce the computational complexity by an order of magnitude as compared to the direct solution by LAPACK. The second step consists of finding a good preconditioner, if necessary. Because the problem is dense, not all types of preconditioners can be applied directly to these problems. The final step in the reduction of computational complexity is to find a special method for computing the matrix-vector products without forming the matrix explicitly. This can lead to huge saving in both the memory consumption and in the CPU requirements. We will now see how this recipe works for the three example problems.

2.1 Problem A: a volume integral equation for electromagnetic scattering

We first study the application of iterative solvers to linear systems arising from a volume integral equation formalism for electromagnetic scattering. The formalism is described in [1] and an earlier, related formalism in [2]. The volume integral equation is targeted for anisotropic or inhomogeneous scatterers and for this method we are mostly interested in dielectric or weakly conducting scatterers.

For the volume integral equation formalism we have used cubic computational cells and piecewise constant basis functions. The dense linear systems arise from the use of a collocation technique. The resulting matrices are complex symmetric, i.e., non-Hermitian. The iterative solution of these systems has been studied in [3, 1] where the complex symmetric version of QMR [4] was chosen as the iterative solver. Iterative methods converge very quickly for this problems even without preconditioning. Moreover, for a fixed physical problem, the required number of iterations does not depend on the discretisation of the problem.

For this problem we noticed that the eigenvalues of the coefficient matrices lie mostly on a line in the complex plane in the case of a spherical scatterer. For the sphere we can analytically explain this behaviour by finding the spectrum of the volume integral operator [5]. Figure 1 gives an example of the eigenvalue distribution and the related analytically calculated points in the spectrum.

If the computational cells sit on a regular lattice, it is possible to enlarge this to a full cube, for which the matrix-vector product is a convolution. This convolution can be efficiently computed with 3D fast Fourier transforms (FFT). The computational complexity of the matrix-vector product is reduced from $\mathcal{O}(N^2)$ to $\mathcal{O}(N \log N)$ and the memory requirements from $\mathcal{O}(N^2)$ to $\mathcal{O}(N)$. With this technique we have solved dense linear systems with hundreds of thousands of unknowns. We have also parallelised this solver, where most of the work is done by parallel FFTs.

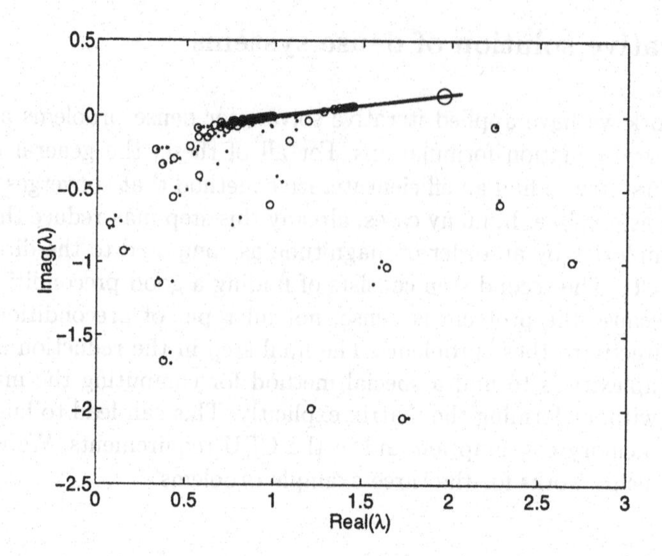

Fig. 1. The eigenvalues of the coefficient matrix (small black dots) and some points in the spectrum of the integral operator due to the resonances (small circles). The spectrum of the integral operator also includes the line segment that extends from the point one to the point marked with a large circle. The eigenvalues were computed from a discretisation of a sphere with 1064 computational cells.

2.2 Problem B: a surface integral equation for electromagnetic scattering

Our second problem is a surface integral equation from electromagnetic scattering calculations. This part is joint work with A. Bendali and M'B. Fares from CERFACS who provided the author with the integral equation code.

The surface integral equation formalism uses triangular surface elements where the degree of liberty is the vectorial flux across the edge of the triangle. The unknowns are the surface currents from which the electric fields and the far-field pattern can be computed. The flux finite elements can describe a vector field whose normal components are continuous across the triangles. A Galerkin procedure yields a complex symmetric coefficient matrix. The chosen surface integral formalism uses an impedance boundary condition can be applied to all types of scatterers [6]. However, the experiments described in the article all relate to the case of perfectly conducting scatterers.

For this problem we chose the complex symmetric version of QMR as the iterative solver. Already without any preconditioner the iterative solver is faster than a direct method for this problem. We have used a sparse approximate inverse preconditioner to speed up the convergence of iterative solvers. The preconditioner M is constructed by first choosing a sparsity pattern for M and then finding the nonzero values of the preconditioner by solving the problem

$$\min_M \|I - AM\|_F^2 , \tag{1}$$

where $\|\cdot\|_F$ denotes the Frobenius norm. Note that this leads to independent least-squares problems for each column of M, *i.e.*

$$\min_{m_j} \|e_j - Am_j\|_F^2,\ j = 1, \ldots, N\ , \tag{2}$$

where m_j and e_j are the jth columns of M and I, respectively.

We pick the sparsity pattern for the preconditioner M starting from the computational mesh. For each column of M (and for the corresponding degree of freedom) we include to the sparsity pattern all unknowns corresponding to triangles that are, say, at most two triangles apart from the triangle containing the original degree of freedom. Note that in solving equations (2) we would in principle need full columns of the coefficient matrix A, which would result in enormous least-squares problems. To speed up the construction of the preconditioner, we use only some entries in each column of A for each least-squares problem. In practice, we typically solve a least-squares problem of size 40 by 25 for each column of M.

Figure 2 shows the convergence of QMR when the sparsity set of the preconditioner M is enlarged. This preconditioner allows us to solve problems with 10 000 unknowns with less than 100 iterations. For problems of this size, the preconditioned iterative solver is about 20 times faster than the direct method from the LAPACK library [7].

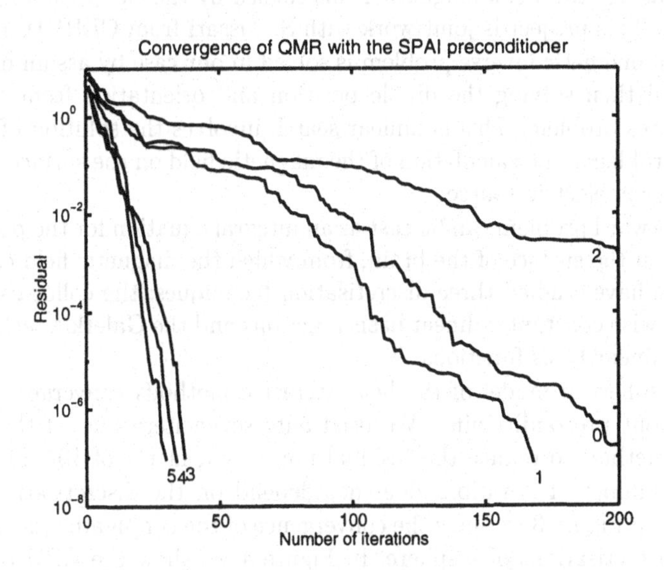

Fig. 2. Convergence of QMR with the sparse approximate inverse preconditioner. The numbers below the curves indicate the level of preconditioner used, with 0 denoting unpreconditioned QMR. Level 1 means that the neighbouring triangles are used in the construction of the preconditioner, for level 2 the neighbours of the neighbours are also used, and so on.

For this integral equation formulation the matrix-vector product can be computed with the fast multipole method (FMM). The FMM was originally developed for computing all pairwise interactions of a collection of particles [8]. It has been used in molecular dynamics simulations and in gravitational simulations. The fast multipole method groups far-away computational cells together and uses truncated potential expansions to represent the combined effect of these cells in an evaluation point. The near-by points must always be computed directly.

We have implemented a hierarchical multi-level version of the FMM using the so called diagonal translation operators [9]. Our preliminary tests show that the FMM becomes faster than the direct method of computing the matrix-vector product for less than two thousand unknowns. The CPU requirements for the FMM scale as $(N \log N)$, while the memory requirements scale only linearly with the problem size. For example, a single matrix-vector product with the FMM takes about 80 seconds of CPU time on a Silicon Graphics Power Challenge for a problem of 100 000 unknowns.

2.3 Problem C: a surface integral equation for potential calculations in bioelectromagnetism

Our third problem is related to the biomagnetic inverse problem: to localise the bioelectric activity in the brain or in the heart with the help of magnetic or electric measurements. We will concentrate on magnetoechephalography (MEG), which measures the weak magnetic fields caused by the electrical activity in the brain [10]. This project is joint work with S. Tissari from CERFACS.

The biomagnetic inverse problem is solved in our case by assuming a dipolar source and then solving the dipole position and orientation from a nonlinear least-squares problem. This nonlinear search involves the solution of a series of forward problems, the calculation of the magnetic field on the surface of the skull from a known electric source.

The forward problem can be cast as an integral equation for the potential distribution on the surface of the brain, from which the magnetic field can be computed. We have studied three discretisation techniques: the collocation method with piecewise constant or linear basis functions and the Galerkin technique with piecewise linear basis functions.

For a spherical model of the head iterative methods converge very quickly even without preconditioning. We used only seven iterations of the conjugate gradient method to reduce the residual norm by a factor of 10^8. Moreover the required number of iterations does not depend on the discretisation for these problems. In Figure 3 we show the convergence of the conjugate gradient method for two discretisations of a sphere. In Figure 4 we show the CPU time for the conjugate gradient method and for the direct solver from LAPACK. For a problem of slightly less than 3000 unknowns the iterative solver is about 40 times faster than the direct one.

For a realistically shaped model of the brain we have used the biconjugate gradient method stabilized (BiCGstab) [11] together with the incomplete LU

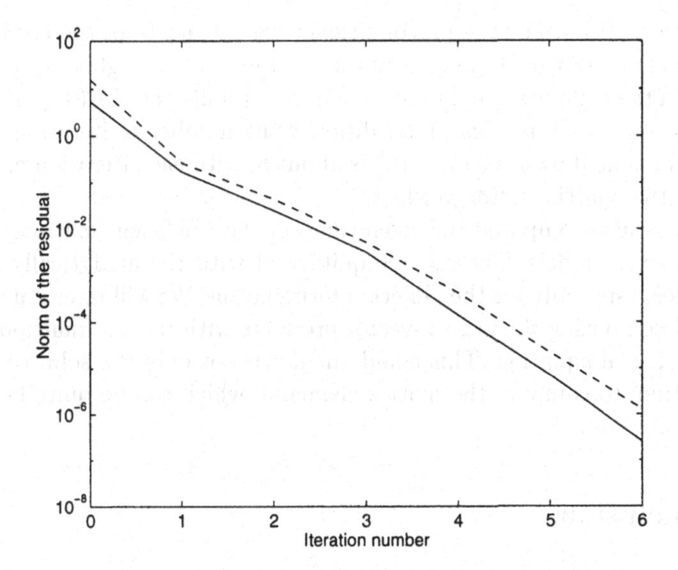

Fig. 3. Convergence of the unpreconditioned conjugate gradient method for the surface integral equations arising from bioelectromagnetism. The convergence is shown for two discretisations of the unit sphere with 80 triangles (solid line) and 720 triangles (dashed line). It can be seen that the convergence speed is the same for both methods. The only difference is that the initial residual for the case of 720 triangles is larger due to the larger number of unknowns.

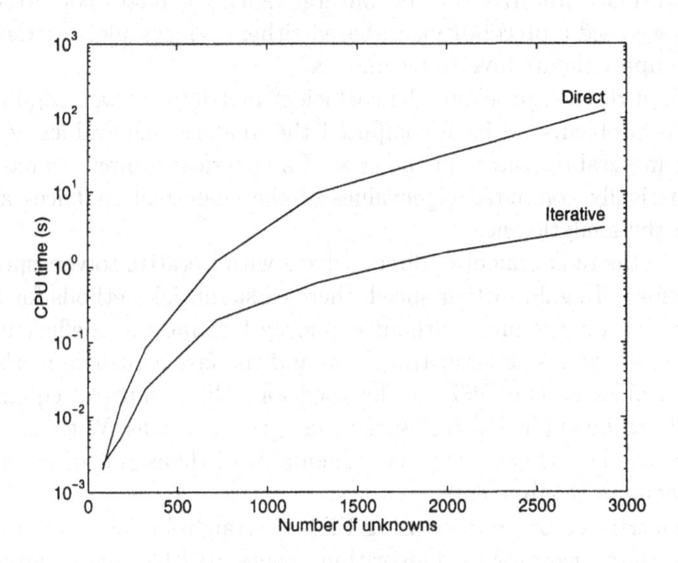

Fig. 4. CPU time for the unpreconditioned conjugate gradient method and direct method (LAPACK) for the surface integral equations arising from bioelectromagnetism when the collocation technique is used together with piecewise constant basis functions.

factorization (ILU). For the preconditioner we extract from the coefficient matrix a sparse matrix containing only the interactions of neighbouring degrees of freedom in the computational mesh. Then we compute the ILU(0) preconditioner from this matrix. With this preconditioner the number of iterations to reduce the initial residual by a factor of 10^8 is about 8, with each iteration of BiCGstab requiring two matrix-vector products.

We have also computed the eigenvalues of the coefficient matrices for spherical conductor models. These agree quite well with the analytically calculated eigenvalues, especially for the Galerkin formulation. We will later study the possibility of computing the matrix-vector products with the fast multipole method for this integral equation. This could cut down not only the solution time, but also the time to compute the matrix elements, which can be quite large for the Galerkin method.

3 Conclusions

We have studied the iterative solution of dense linear systems of linear equations arising from discretisation of integral equations. The example problems are a volume and a surface integral equation of electromagnetic scattering together with a surface integral equation from bioelectromagnetism. For all the three cases presented, iterative solvers can outperform direct solvers from LAPACK. For our example problems, the number of iterations is very low and for some of the problems independent of the number of unknowns. In some cases even unpreconditioned iterative solvers converge quickly enough. For two of the test problems we used a preconditioner, based either on incomplete factorizations or on sparse approximate inverse techniques.

For all of the test problems the coefficient matrices are well conditioned. For two of the problems we have computed the analytic eigenvalues of the corresponding integral operators in the case of a spherical computational geometry. The numerically computed eigenvalues of the coefficient matrices agree quite well with the analytic ones.

Already the replacement of direct solvers with iterative solvers speeds up the computations. To gain further speed, there exist special methods for computing the matrix-vector products without explicitly forming the coefficient matrices. We have used the fast Fourier transform and the fast multipole method for the example problems. The FFT can be used for volume integral equations while the FMM can be applied also to surface integral equations. With these methods one can solve dense linear systems of hundreds of thousands of unknowns with modest memory requirements.

The matrix-vector product using FFT is straightforward to parallelise. We have used the transpose-based algorithm, where 2D FFTs are computed locally in each processor. The computational grid has to be transposed between the processor after which the 1D FFTs in the remaining direction can be computed locally. The FMM is much more challenging to parallelise and we have not yet tackled this problem.

References

1. J. Rahola. *Efficient solution of dense systems of linear equations in electromagnetic scattering calculations.* PhD thesis, Helsinki University of Technology, 1996. CSC Research Reports R06/96, Center for Scientific Computing, 1996.
2. K. Lumme and J. Rahola. Light scattering by porous dust particles in the discrete-dipole approximation. *Astrophys. J.* **425**, 653–667, 1994.
3. J. Rahola. Solution of dense systems of linear equations in the discrete-dipole approximation. *SIAM J. Sci. Comput.* **17**, 78–89, 1996.
4. R. W. Freund. Conjugate gradient-type methods for linear systems with complex symmetric coefficient matrices. *SIAM J. Sci. Stat. Comput.* **13**, 425–448, 1992.
5. J. Rahola. On the eigenvalues of the volume integral operator of electromagnetic scattering. CERFACS Technical Report TR/PA/98/19, CERFACS, 1998.
6. A. Bendali, M.G. Fares, and J. Gay. Finite element solution to impedance boundary integral equation in electromagnetic scattering. CERFACS Technical Report TR/EMC/97/35, CERFACS, 1997. Submitted to IEEE Trans. Antennas Propagat.
7. J. Rahola. Iterative solution of dense linear systems in electromagnetic scattering calculations. In *Proceedings of the 14th Annual Review of Progress in Applied Computational Electromagnetics, Monterey, CA, March 16–20, 1998*, pages 1126–1133, Monterey, California, 1998. The Naval Postgraduate School.
8. L. Greengard and V. Rokhlin. A fast algorithm for particle simulations. *J. Comp. Phys.* **73**, 325–348, 1987.
9. J. Rahola. Diagonal forms of the translation operators in the fast multipole algorithm for scattering problems. *BIT* **36**, 333–358, 1996.
10. M. Hämäläinen, R. Hari, R.J. Ilmoniemi, J. Knuutila, and O.V. Lounasmaa. Magnetoencephalography — theory, instrumentation, and applications to noninvasive studies of the working human brain. *Rev. Mod. Phys.* **65**, 413–497, 1993.
11. H. van der Vorst. Bi-CGSTAB: A fast and smoothly converging variant of Bi-CG for the solution of nonsymmetric linear systems. *SIAM J. Sci. Stat. Comput.* **13**, 631–644, 1992.

Comparison of Partitioning Strategies for PDE Solvers on Multiblock Grids

Jarmo Rantakokko

Department of Scientific Computing, Uppsala University, Sweden
jarmo@tdb.uu.se

Abstract. Different partitioning strategies for multiblock grids have been compared experimentally. The numerical experiments have been performed on a 512 processor Cray T3D using a compressible two dimensional Navier-Stokes solver. Some complementary results are made with an advection equation solver using a Cray T3E-900. The results show that the behavior of the different parallelization strategies depends very much on the number of subgrids and their sizes as well as the number of available processors. In order to get optimal performance for a certain problem and processor configuration, the partitioning strategy must be chosen with regard to these aspects. Our results give guidelines for this.

1 Introduction

The numerical solution of partial differential equations is computationally very intensive, e.g. consider simulation of airflow around an aircraft. Parallel computing is necessary, both to reduce the computational time and to distribute the problem so data can fit into the main memory. An efficient numerical approach is to use finite differences on structured grids. Complex geometries can be modeled with a set of rectangular and structured grids, a composite grid. The computational domain is decomposed into logically rectangular domains. Each subdomain is discretized building a component of the composite grid. The subgrids are connected at the adjoining subdomain boundaries. The connections can be irregular for a complicated geometry. We have then irregularly coupled regular meshes. The different sizes of the subgrids and the irregular data dependencies between the grids makes data partitioning non-trivial. Composite grids exhibit both a coarse-grain parallelism at the subgrid level and a fine-grain parallelism at the mesh point level. Most of the previous work treats either the coarse-grain parallelism, see for example [7], or the fine-grain parallelism [5]. In [2] we have developed additional partitioning strategies that exploit both the fine-grain and the coarse-grain parallelism at the same time. The partitioning strategies are briefly described in Section 2 below.

Multiblock grids constitutes a special case of composite grids where the subgrids or blocks do not overlap. This simplifies the solvers but the partitioning strategies are still the same. Multiblock grids are frequently used in aerodynamical computations. In this paper we will compare how the different partitioning strategies perform for a multiblock solver. In Section 3 we show numerical results from an application where we solve the compressible Navier–Stokes equations simulating airflow in a two-dimensional expanding-contracting tube. We also give some complementary results from a simpler advection equation solver. Finally, in Section 4 we give guidelines for which strategy to choose and when.

2 Partitioning strategies

A multiblock solver can be parallelized by partitioning data from the grids over the processors, i.e. the processors compute the solution in different parts of the domain in parallel. In [2], different partitioning strategies are suggested and described. Here, we will compare four of the most central strategies for parallelizing a multiblock solver. We will refer to the blocks of the multiblock grid as *grids*, *element grids* or *subgrids* without any distinction. Also, we will talk about partitioning or dividing the grids in blocks. The partitions will then consist of rectangular data blocks. The structure of the multiblock grid is *not* changed unless explicitly stated. Briefly explained, the four strategies are:

1. ***Single grid partitioning***: The grids are partitioned two dimensionally in $p \times q = N_p$ rectangular blocks and distributed one by one over all processors (N_p). Each processor then gets one block of each grid. In principle we can use any partitioning strategy for a single grid but the rectangular partitioning is sufficient as we have almost constant workload on the grids. This strategy exploits only the fine-grain parallelism.
2. ***Block partitioning***: The grids are not further divided, they are distributed as entities to the processors. As a consequence, using more processors than element grids gives no extra performance. The extra processors will only be idle. The naming of this strategy can seem to be somewhat confusing. By blocks we refer here to the subgrids. This strategy is based on the inherent coarse-grain parallelism. If the load balance is very important, i.e. we have comparably few subgrids and they are of different sizes, then the blocks should be distributed using e.g. a bin-packing method disregarding the neighbor relations. If we have relatively many blocks or if the blocks are of the same sizes it is better to use a graph partitioning method with the blocks as vertices to minimize the inter-processor communication.
3. ***Cluster partitioning***: A further development of the two previous strategies above is to divide the processors into subsets, one per subgrid, and to partition the grids within the different subsets of processors. Again, we can use any single grid partitioning algorithm within the clusters. Here, we have used the recursive coordinate bisection method. The idea with this strategy is similar to the concept of processor subsets in HPF-2 but here we have developed an optimal clustering algorithm with respect to load balance between the clusters.
4. ***Graph partitioning***: This strategy is new and exploits both the fine-grain and the coarse-grain parallelism, including the couplings between the subgrids, at the same time. Here we divide the element grids into a number

of smaller data blocks, set up a connectivity graph for the blocks over the whole composite grid including inter-grid connections, partition the graph, and finally merge blocks on the same partitions. The smaller data blocks we have the better load balance we get. But, there will then be an overhead in copying data between neighboring blocks. The last step in the algorithm tries to reduce this overhead.

Partitioning strategies for composite grids are non-trivial to implement, especially the fourth strategy above, due to the irregular inter-grid couplings. In addition, existing software are usually integrated in the particular solver environment and can not easily be extracted. The above described strategies are all implemented in a general independent software package for partitioning composite grids. The package has an object-oriented design which yields a new and flexible way to create algorithms by composition of low-level operations [3].

3 Numerical experiments

We have experimentally investigated how the different partitioning strategies, *graph, single, block,* and *cluster*, perform on grids of different sizes and number of subgrids. As a realistic case study we have used a compressible Navier–Stokes solver but we have also done some complementary experiments with an advection equation solver. The experiments were carried out on a Cray T3D and T3E at Edinburgh Parallel Computing Centre. See the Tables 1 and 2 for the characteristics of the computers.

Table 1. Hardware configurations of the computers used in the numerical experiments. The T3E can do two floating point operations per clock cycle, compared to one for the T3D, and has *stream buffers* to enhance the performance. In practice the T3E was five times faster than the T3D on a single node.

Computer model	*CPU/MHz*	*Memory*	*Cache*	*2:nd Cache*
Cray T3D MCN512-8	Alpha 21064/150	64Mb	8Kb	-
Cray T3E-900/LC240	Alpha EV5.6/450	64Mb	8Kb	96Kb

Table 2. Communication characteristics of the computers used in the numerical experiments. Approximate timings of the Cray specific low-level communication library *shmem.* The high-level *mpi* library has then some additional overhead.

Model	*Topology*	*Latency*	*Bandwidth*
Cray T3D	3D torus	1-2μs	120Mb/s
Cray T3E	3D torus	1-2μs	330Mb/s

The implementation of the solvers are made in Cogito, *Composite Grid Software Tools.* An overview, illustrating the ideas and motivating the design, of the Cogito project is given in [6]. The software tools are written in Fortran 90 with an object-oriented design. Cogito is specially designed to solve this kind of applications. It uses MPI with persistent communication and derived data types to handle the irregular communication in an efficient way.

3.1 The Navier–Stokes solver

Our first example is the compressible Navier-Stokes equations simulating airflow in an expanding-contracting tube. The geometry is modeled with a five element multiblock grid, see Figure 1. The equations are solved explicitly until steady-state with a five-stage Runge-Kutta scheme, Figure 2. A fourth order term of artificial viscosity is added for stability. At the block boundaries the solution is interpolated as an average of the grid points in the respective grids. At the inlet, outlet, and the tube walls both physical and numerical boundary conditions are prescribed. For a more detailed description of the numerics see [1].

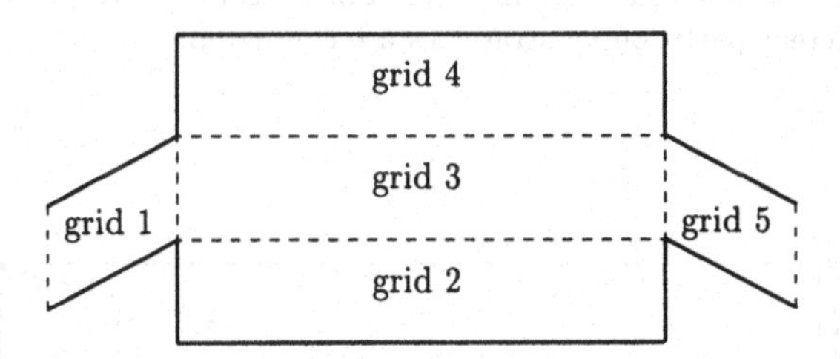

Fig. 1. Schematic view of the multiblock grid used in the airflow simulation. We have one grid for the inlet, one for the outlet, and three grids for the body center.

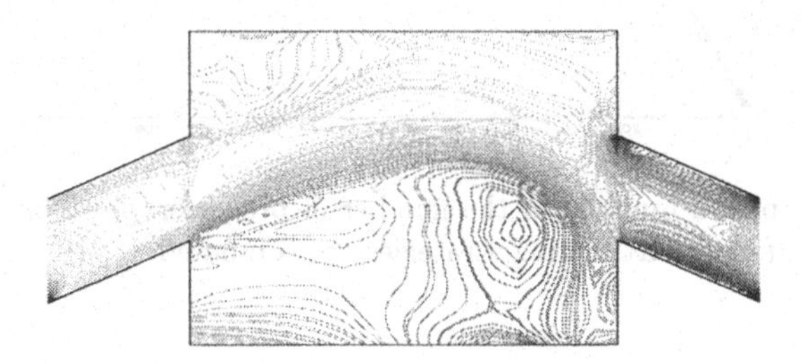

Fig. 2. Density contour-lines of the steady-state solution of airflow in an expansion tube with constant inflow on the left.

We have measured *fixed size speedup* on the Cray T3D for three different grid resolutions, which we will call *small, medium,* and *large.* The *small* grid is very small but fine enough to give a coarse solution. The *large* grid has many grid points for this problem but is still quite small to run on the largest processor configuration, i.e. on all the 512 nodes. The results from the experiments are plotted in the Figures 3, 4, and 5.

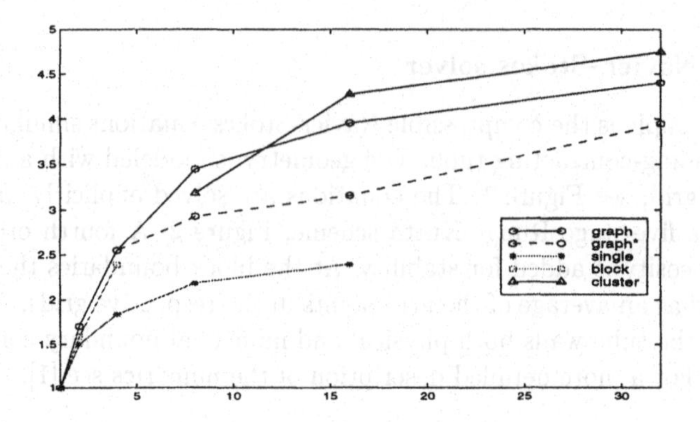

Fig. 3. Speedup as a function of the number of processors, *small grid* with totally 1785 grid points. Different partitioning strategies are compared.

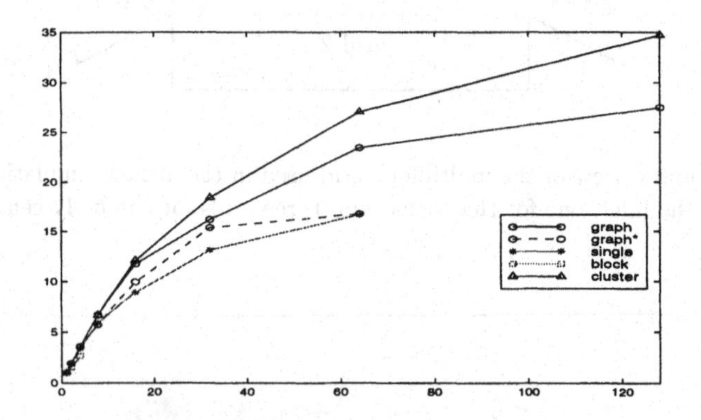

Fig. 4. Speedup as a function of the number of processors, *medium grid* with 27 500 grid points. Different partitioning strategies are compared.

In the figures we have two different curves for the graph strategy, (i) *graph* where we have optimized for performance, i.e. taking into account the communication, arithmetic, and serial overhead, and (ii) *graph** where we have optimized for load balance, i.e. considering only the arithmetic work. The load balance is usually considered as the most important parameter. From the experiments we can see that there is a large overhead in the communication and also that the number of blocks has a considerable significance for the performance. The *graph* variant gives overall good results while the performance drops significantly for

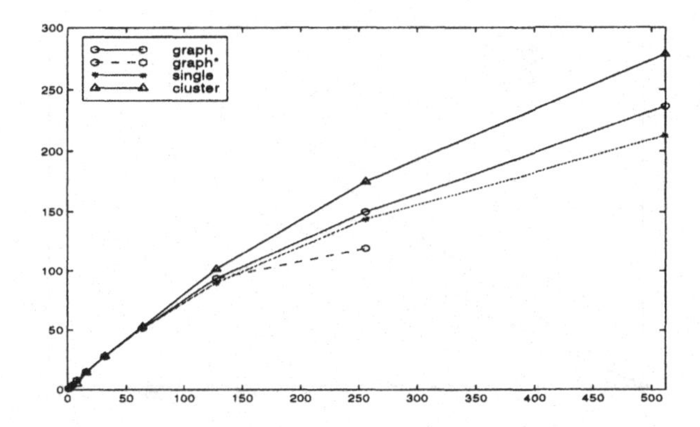

Fig. 5. Speedup as a function of the number of processors, *large grid* with 450 000 grid points. Different partitioning strategies are compared.

the *graph** variant. The *cluster* strategy has the best performance for large processor configurations while for small processor configurations the load balance between the processor clusters has some significance. The *block* strategy has the lowest number of edge-cuts and blocks per processor and is then competitive despite of its poor load balance for small processor configurations. The performance of the *single* strategy is degraded due to a large communication overhead.

As an additional experiment we have measured *sizeup*. This is another way to compare the algorithms. Here we measure how the problem size can be scaled as the number of processors is increased. *Sizeup* is defined as the ratio between the work on p processors divided by the work on a single processor as the run time is kept constant, [4]. The results from the experiments are summarized in Figure 6. We can see that the *single* grid partitioning scales perfectly up to 512 processors, constant problem size per processor, and that for the other two strategies we can even increase the problem size per processor as we increase the number of processors. The explanation is that we have so few grid points per processor that the serial overhead in calling subroutines, starting up loops, etc is relatively high, especially for the complicated boundary conditions. With the *cluster* and *graph* partitioning we distribute the processors over different grids so that the processors are not involved in all the computations and thus the serial overhead is reduced. The processors can then do more other work. This effect is most obvious for the *cluster* strategy where each processor is only involved in computations on one subgrid and the corresponding boundary conditions.

3.2 The advection equation solver

The number of subgrids is very limited in the Navier-Stokes example so one should be very careful of drawing general conclusions from these experiments. We have then made some complementary experiments solving the advection equations in 2D. Again we have used explicit finite differences and modeled the geometry with a multiblock grid. This is a *toy* example with a known solution.

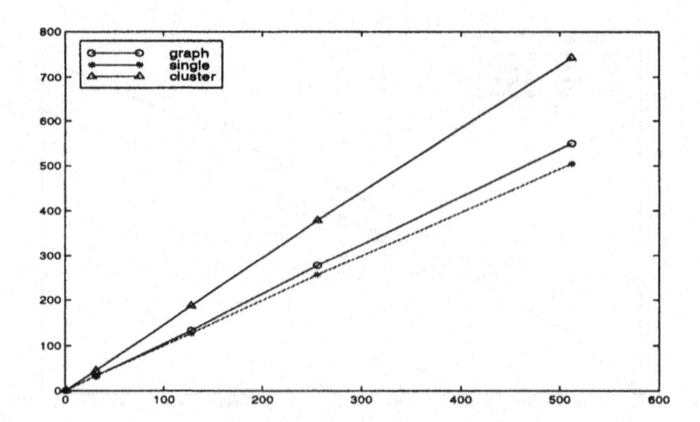

Fig. 6. Sizeup for the Navier-Stokes solver, the largest problem is 450 000 grid points partitioned with the cluster strategy for 512 processors.

The geometry is the unit square but we still use a multiblock grid with 20 elements to simulate a more complex solver. The results from the experiments are shown in Figure 7. Here we have used the Cray T3E-900. As we can expect the load imbalance becomes more significant for small processor configurations in the *cluster* strategy and results in a poor performance below 64 processors. The general tendencies are still the same as for the Navier-Stokes solver.

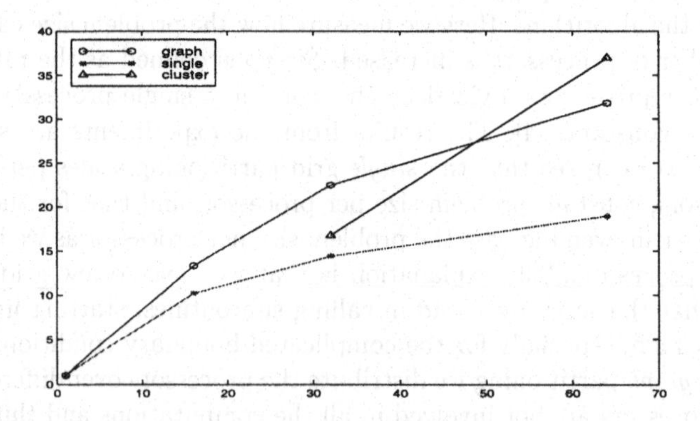

Fig. 7. Fixed size speedup of the advection equation solver on the Cray T3E.

4 Guidelines and conclusions

We have compared four different strategies for partitioning composite grids using two different parallel computers, Cray T3D and Cray T3E, respectively. Experiments were carried out with two different PDE solvers, one for the compressible Navier–Stokes equations and one for the advection equations. Moreover, we have measured fixed size speedup and sizeup. Both these measures are very intuitive and reflect how parallel computers are used. Either the problem size is fixed and the parallel computer is used to compute the solution faster or the execution time is fixed and the parallel computer is used to solve a larger problem. Sizeup is less sensitive to processor speed and memory hierarchy than speedup. The conclusions from the experiments are:

1. *Single grid partitioning*: Gives good performance for small number of processors on large grids. Positive effects on the memory management as less data are active at a time. A drawback is the communication overhead.
2. *Block partitioning*: Obvious parallelization strategy but the load balance is very critical which limits the performance and the scaling.
3. *Cluster partitioning*: Gives the best performance for large processor configurations. For small number of processors the load imbalance between the clusters can be significant reducing the performance. Scales very well.
4. *Graph partitioning*: Very flexible strategy that can be tuned to give good performance for all number of processors on different sets of composite grids.

In summary, the behavior of the different parallelization strategies depends very much on the number of subgrids and their sizes as well as the number of available processors. For a few large subgrids the *block* partitioning is not practically useful while the other strategies work well depending on the number of processors. On the other hand, for many small subgrids the *cluster* and *single grid* partitionings do not work well while the *block* strategy can be a very good alternative. The *graph* strategy gives an overall good performance.

Acknowledgments The numerical experiments were performed at the Edinburgh Parallel Computing Centre supported by the *Training and Research on Advanced Computing Systems* (TRACS) programme.

References

1. R. Enander, *Grid patching and residual smoothing for computations of steady state solutions of first order hyperbolic systems*, Thesis, Dept. of Scientific Computing, Uppsala University, Uppsala, 1991.
2. J. Rantakokko, *Data Partitioning Methods and Parallel Block-Oriented PDE Solvers*, Ph.D. thesis, Uppsala University, Uppsala, Sweden, 1998.
3. J. Rantakokko, *Software tools for partitioning of block-structured applications*, to appear in the proceedings of the ISCOPE '98 conference, Santa Fe, December 1998.
4. X. Sun, J.L. Gustafson, *Toward a better parallel performance metric*, Parallel Computing, 17:1093-1109, 1991.
5. M. Thuné, *Straightforward partitioning of composite grids for explicit difference methods*, Parallel Computing, 17:665-672, 1991.
6. M. Thuné, E. Mossberg, P. Olsson, J. Rantakokko, K. Åhlander, K. Otto, *Object-oriented construction of parallel PDE solvers*, in Modern Software Tools in Scientific Computing, pp 203–226, E. Arge, A.M. Bruaset, H.P. Langtangen editors, Birkhäuser, Boston, 1997.
7. A. Ålund, P. Lötstedt, M. Sillén, *Parallel single and multigrid solution of industrial compressible flow problems*, Computers & Fluids, 26:775-791, 1997.

Ship Design Optimization

Claus Risager and John W. Perram

The Maersk Mc-Kinney Moller Institute for Production Technology, Odense University and IT Department, Odense Steel Shipyard Ltd
cri@oss.dk
jperram@mip.ou.dk

This contribution is devoted to exploiting the analogy between a modern manufacturing plant and a heterogeneous parallel computer to construct a HPCN decision support tool for ship designers. The application is a HPCN one because of the scale of shipbuilding – a large container vessel is constructed by assembling about 1.5 million atomic components in a production hierarchy. The role of the decision support tool is to rapidly evaluate the manufacturing consequences of design changes. The implementation as a distributed multi-agent application running on top of PVM is described

1 Analogies between Manufacturing and HPCN

There are a number of analogies between the manufacture of complex products such as ships, aircraft and cars and the executioin of a parallel program. The manufacture of a ship is carried out according to a production plan which ensures that all the components come together at the right time at the right place. A parallel computer application should ensure that the appropriate data is available on the appropriate processor in a timely fashion.

It is not surprising, therefore, that manufacturing is plagued by indeterminacy exactly as are parallel programs executing on multi-processor hardware. This has caused a number of researchers in production engineering to seek inspiration in other areas where managing complexity and upredictability is important. A number of new paradigms, such as Holonic Manufacturing and Fractal Factories have emerged [1,2] which contain ideas rather reminiscent of those to be found in the field of Multi-Agent Systems [3, 4].

Manufacturing tasks are analogous to operations carried out on data, within the context of planning, scheduling and control. Also, complex products are assembled at physically distributed workshops or production facilities, so the components must be transported between them. This is analogous to communication of data between processors in a parallel computer, which thus also makes clear the analogy between workshops and processors.

The remainder of this paper reports an attempt to exploit this analogy to build a parallel application for optimizing ship design with regard to manufacturing issues.

2 Shipbuilding at Odense Steel Shipyard

Odense Steel Shipyard is situated in the town of Munkebo on the island of Funen. It is recognized as being one of the most modern and highly automated in the world. It specializes in building VLCC's (supertankers) and very large container ships. The yard was the first in the world to build a double hulled supertanker and is currently building an order of 15 of the largest container ships ever built for the Maersk line. These container ships are about 340 metres long and can carry about 7000 containers at a top speed of 28 knots with a crew of 12.

Odense Steel Shipyard is more like a ship factory than a traditional shipyard. The ship design is broken down into manufacturing modules which are assembled and processed in a number of workshops devoted to, for example, cutting, welding and surface treatment. At any one time, up to 3 identical ships are being built and a new ship is launched about every 100 days.

The yard survives in the very competitive world of shipbuilding by extensive application of information technology and robots, so there are currently about 40 robots at the yard engaged in various production activities. The yard has a committment to research as well, so that there are about 10 industrial Ph.D. students working there, who are enrolled at various engineering schools in Denmark.

3 Tomorrow's Manufacturing Systems

The penetration of Information Technology into our lives will also have its effect in manufacturing industry. For example, the Internet is expected to become the dominant trading medium for goods. This means that the customer can come into direct digital contact with the manufacturer.

The direct digital contact with customers will enable them to participate in the design process so that they get a product over which they have some influence. The element of unpredictability introduced by taking into account customer desires increases the need for flexibility in the manufacturing process, especially in the light of the tendency towards globalization of production. Intelligent robot systems, such as AMROSE, rely on the digital CAD model as the primary source of information about the work piece and the work cell [5,6].This information is used to construct task performing, collision avoiding trajectories for the robots, which because of the high precision of the shipbuilding process, can be corrected for small deviations of the actual world from the virtual one using very simple sensor systems. The trajectories are generated by numerically solving the constrained equations of motion for a model of the robot moving in an artificial force field designed to attract the tool centre to the goal and repell it from obstacles, such as the work piece and parts of itself. Finally, there are limits to what one can get a robot to do, so the actual manufacturing will be performed as a collaboration between human and mechatronic agents.

Most industrial products, such as the windmill housing component shown in Fig. 1, are designed electronically in a variety of CAD systems.

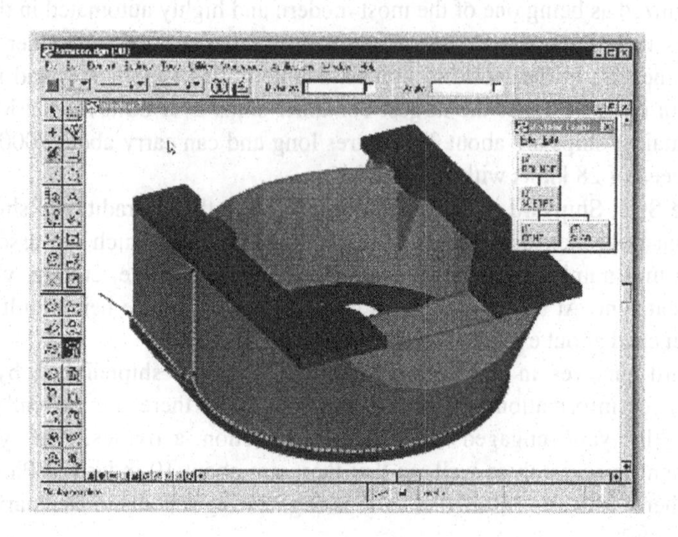

Fig. 1. Showing the CAD model for the housing of a windmill. The model, made using Bentley Microstation, includes both the work-piece and task-curve geometries.

4 Today's Manufacturing Systems

The above scenario should be compared to today's realities enforced by traditional production engineering philosophy based on the ideas of mass production introduced about 100 years ago by Henry Ford. A typical production line has the same structure as a serial computer program, so that the whole process is driven by production requirements. This rigidity is reflected on the types of top-down planning and control systems used in manufacturing industry, which are badly suited to both complexity and unpredictability.

In fact, the manufacturing environment has always been characterized by unpredictability. Today's manufacturing systems are based on idealized models where unpredictability is not taken into account but handled using complex and expensive logistics and buffering systems.

Manufacturers are also becoming aware that one of the results of the top-down serial approach is an alienation of human workers. For example, some of the car manufacturers have experimented with having teams of human workers responsible for a particular car rather than performing repetitive operations in a production line.

This model in fact better reflects the concurrency of the manufacturing process than the assembly line.

5 A Decision Support Tool for Ship Design Optimization

Large ships are, together with aircraft, some of the most complex things ever built. A container ship consists of about 1.5 million atomic components which are assembled in a hierarchy of increasingly complex components. Thus any support tool for the manufacturing process can be expected to be a large HPCN application.

Ships are designed with both functionality and ease of construction in mind, as well as issues such as economy, safety, insurance issues, maintenance and even decommissioning. Once a functional design is in place, a stepwise decomposition of the overall design into a hierarchy of manufacturing components is performed. The manufacturing process then starts with the individual basic building blocks such as steel plates and pipes. These building blocks are put together into ever more complex structures and finally assembled in the dock to form the finished ship.

Thus a very useful thing to know as soon as possible after design time are the manufacturing consequences of design decisions. This includes issues such as whether the intermediate structures can actually be built by the available production facilities, the implications on the use of material and whether or not the production can be efficiently scheduled [7].

Fig. 2. shows schematically how a redesign decision at a point in time during construction implies future costs, only some of which are known at the time. Thus a decision support tool is required to give better estimates of the implied costs as early as possible in the process.

Simulation, both of the feasibility of the manufacturing tasks and the efficiency with which these tasks can be performed using the available equipment, is a very compute-intense application of simulation and optimization. In the next section, we describe how a decision support tool can be designed and implemented as a parallel application by modelling the main actors in the process as agents.

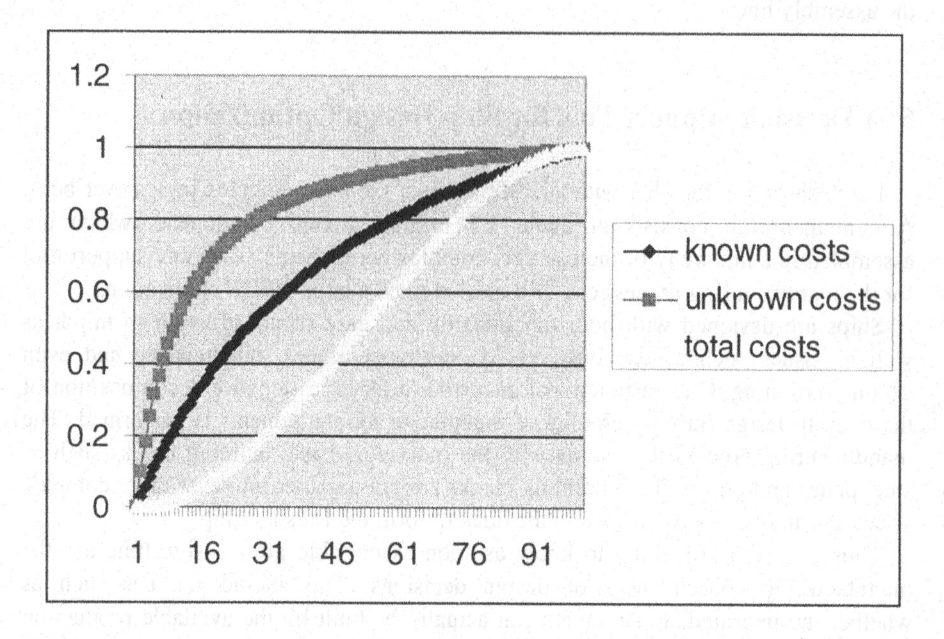

Fig. 2. Economic consequences of design decisions. A design decision implies a future commitment of economic resources which is only partially known at design time.

6 Multi-Agent Systems

The notion of a software agent, a sort of autonomous, dynamic generalization of an object (in the sense of Object Orientation) is probably unfamiliar to the typical HPCN reader in the area of scientific computation. An agent possesses its own beliefs, desires and intentions and is able to reason about and act on its perception of other agents and the environment.

A multi-agent system is a collection of agents which try to cooperate to solve some problem, typically in the areas of control and optimization. A good example is the process of learning to drive a car in traffic. Each driver is an autonomous agent which observes and reasons about the intentions of other drivers. Agents are in fact a very useful tool for modelling a wide range of dynamical processes in the real world, such as the motion of protein molecules [8] or multi-link robots [9]. For other applications, see [4].

One of the interesting properties of multi-agent systems is the way global behaviour of the system emerges from the individual interactions of the agents [10]. The notion of emergence can be thought of as generalizing the concept of evolution in dynamical systems.

Examples of agents present in the system are the assembly network generator agent which encapsulates knowledge about shipbuilding production methods for planning assembly sequences, the robot motion verification agent, which is a simulator capable of generating collision-free trajectories for robots carrying out their tasks, the quantity surveyer agent which possesses knowledge about various costs involved in the manufacturing process and the scheduling agent which designs a schedule for performing the manufacturing tasks using the production resources available.

7 Parallel Implementation

The decision support tool which implements all these agents is a piece of Object-Oriented software targetted at a multi-processor system, in this case, a network of Silicon Graphics workstations in the Design Department at Odense Steel Shipyard. Rather than hand-code all the communication between agents and meta-code for load-balancing the parallel application, abstract interaction mechanisms were developed. These mechanisms are based on a task distribution agent being present on each processor. The society of task distribution agents is responsible for all aspects of communication and migration of tasks in the system.

The overall agent system runs on top of PVM and achieves good speedup and load balancing. To give some idea of the size of the shipbuilding application, it takes 7 hours to evaluate a single design on 25 SGI workstations.

References

1. H. van Brussel, P. Valkenaers, L. Bongaerts and J. Wyns, *Artificial and system design issues in holonic manufacturing systems*, Proc. 3rd. IFAC workshop on Intelligent Manufacturing Systems (IMS'95), Romania 1995.
2. See also articles in the Proceedings of IMS'98, published by EPFL, Lausanne.
3. See the article by S. Bussman in Ref. [2].
4. Recent proceedings of the European Workshops on Multi-Agent Systems (MAAMAW'94, 96, 97), published in vols 1038, 1069 and of LNCS are a good place for the unitiated to find out about MAS.
5. L. Overgaard, H.G. Petersen and J.W. Perram, *A general algorithm for dynamic control of multi-link robots*, Int. J. Robotics Res., [14], 281-294, (1995).
6. J.W. Perram, *Applications of HPCN in manufacturing industry,* Proc. PARA'96, LNCS, [1184], 586-89, Springer Verlag, (1996).
7. Kaare Christiansen, *Concurrent engineering; a knowledge engineering concept for a shipyard,* Ph.D. Thesis, Technical University of Denmark, (1996).

8. L. Baekdal, W. Joosen, T. Larsen, J. Kolafa, J.H. Oversen, J.W. Perram, H.G. Petersen, R. Bywater and M. Ratner, The Object-Oriented development of a parallel application in protein dynamics: why we need software tools for HPCN applications, Comp. Phys. Comm., [97], 124-135, (1996).
9. L. Overgaard, H.G. Petersen and J.W. Perram, Motion planning for an articulated robot: a multi-agent approach, LNCS, [1069], 206-219, Springer Verlag, (1996).
10. J.W. Perram and Y. Demazeau, A multi-agent architecture for distributed constrained optimization and control, Proc. SCAI'97, Frontiers in AI, [40], 162-175, IOS, (1997).

Parallelization Strategies for the VMEC Program

Luis F. Romero[1], Eva M. Ortigosa[1], Emilio L. Zapata[1], Juan A. Jiménez[2]

[1]Dep. Computer Architecture, University of Málaga, 29071 Málaga, Spain
[2] Dep. Nuclear Fusion. CIEMAT, 28040 Madrid, Spain

Abstract. The magnetohydrodynamic equilibrium problem in magnetic confinement fusion devices is solved by using the Variational Moments Equilibrium Code (VMEC). The sequential version of the code is computationally very expensive and may take several days of CPU time. In this work we present two parallelization techniques for the code. First, an automatic tool has been used as a guidance for the parallelization of the critical parts of the code, and some deficiencies of its automatic parallelization techniques have been determined . In a second approach, these shortcomings are overcome to develop manual techniques which offer better results and can be used in future automatic parallelizers.

1 Introduction

VMEC (Variational Moments Equilibrium Code) [1] is used to solve the magnetohydrodynamic equilibrium problem, found in the plasma physics in magnetic confinement fusion devices. It is particularly interesting to remark that this code takes into account the fully 3D nature of some designs (stellarators). This is an iterative variational method, which uses a steepest descent method in order to minimize the MHD energy functional. The inclusion of matrix preconditioning techniques drastically reduces the number of iterations, thus providing a great convergence speedup. The code assumes nested flux surfaces, which are described in terms of Fourier coefficients (moments). The minimization of the MHD forces is done in Fourier space (spectral code); so the transformations from real space to Fourier space and viceversa are very critical steps from the point of view of optimization. Our study centers on the most complex version of the code, which computes the equilibrium using a mobile boundary for the plasma (free-boundary code) [2]. The complexity involved by this boundary condition translates into a very high computational cost, which makes this problem specially suitable for parallel processing.

In this paper we describe two different approaches to parallelize this code for distributed-shared memory machines. A computational description of VMEC can be found in Sect.2. In Sect.3 we describe the automatic parallelization for the most expensive parts of the code and a manual parallelization technique for the whole code. Several comparisons and results are shown in Sect.4. And finally, in Sect.5, some conclusions are presented.

2 Computational Description of VMEC

Execution of VMEC is performed in two main phases. In the first one, a solution to the problem is computed using a coarse grid over the 3-dimensional computational domain (24x36x17 points for a typical problem size); the program iterates several hundreds times until a weak convergence criterion is satisfied. Then, in the second phase, after an interpolation step, the system is solved again, using a finer grid (typically 24x36x140), iterating several thousands times until the final (strong) convergence criterion is found. Each iteration strongly depends on the results of the preceding one and no parallelism can be found at this level.

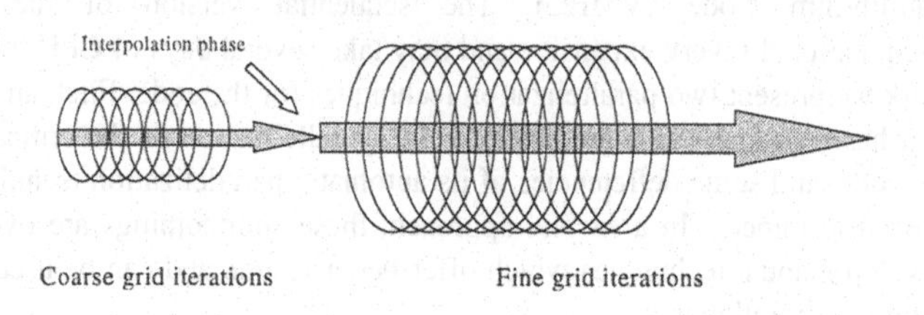

Fig. 1. Coarse and fine grid iterations

In every iteration, a routine (*funct3d*) computes the MHD forces. We can also find a test for convergence and periodical outputs of results, but *funct3d* involves a 99.6% of the execution time, so we will focus just on it. Routine *funct3d* consists on nine consecutive calls to nine different subroutines, four of them involving a 90% of CPU time (*totzsp, bcovar, forces* and *tomnsp*). An additional call to routine *vacuum* is required if free boundary conditions are considered. In this case, *vacuum* consumes a 50% of the time, and it includes four new expensive routines: *bextern, foumat, fouris* and *solver*.

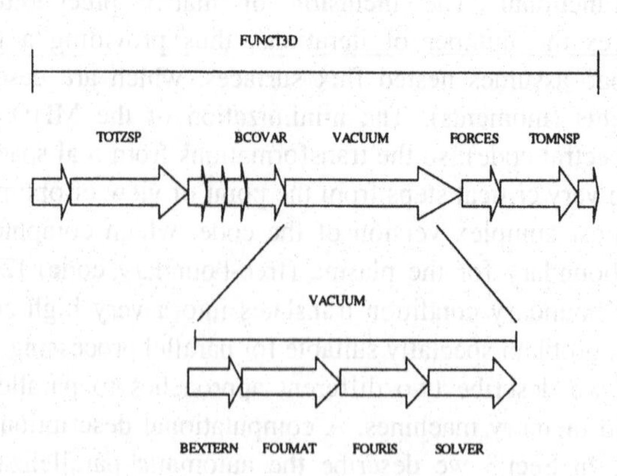

Fig. 2. Structure of routine *funct3d* (only the most expensive subroutines are shown)

Data dependencies between these nine routines (thirteen if *vacuum* is included) are very strong. So, the position of the calls must be preserved and no parallelism seems to be possible at this level. Inside the named subprograms, we find loops, reduction operations or other elementary tasks which will be candidates for further parallelization.

3 Parallelization Strategies

The parallelization of the code has been first performed using an automatic tool: the Silicon Graphics' *Power Fortran Analyzer, PFA* [5]. Later, the outputs from PFA have been strongly modified by hand to obtain a new parallel version of the code. The work has been developed on a shared memory machine (Silicon Graphics Origin 2000). The resulting code was later translated to a message passing version using *shmem,* and finally, it was ported to a distributed memory machine (CRAY T3E)

3.1 Automatic parallelization

In a first approach, we used the PFA tool in order to extract parallelism from VMEC. PFA is an automatic parallelizer for Fortran codes, which locates every loop in the program and determines if it can be parallelized. PFA uses well known techniques as loop interchange, loop fusion, loop unrolling and strip mining, and it introduces parallelization by distributing selected loops across the network of processors (using a `C$DOACROSS` directive).

Our first results using PFA were discouraging. Tested with the most expensive routines in VMEC (*totzsp* and *tomnsp*) PFA only achieved a maximum speedup of two on those routines (for any number of processor). Then, several assertions and directives were included, reporting to the analyzer about typical sizes of variables, absence of memory aliases or other unreal data dependencies, but the results were only slightly better.

3.2 Manual parallelization

Using PFA, the execution of a program is usually performed on a single (master) processor. When a loop is marked as parallel by the Analyzer, using a `C$DOACROSS` directive, the other processors are activated and the iterations of the loops are distributed between them. When the loop ends, all the processors, but the master, are deactivated. The cost of launching and stopping new threads is very expensive and so, PFA only marks a loop as parallel if its size is large enough.

Our first optimization consists on the elimination of these expenses by creating parallel sections, each one composed by many loops. If the loops have been parallelized by PFA, each processor pre-computes the local bounds for that loop as soon as possible (at the beginning of the program in most cases), and each one works on its local iterations. If the loops cannot be parallelized, they can be inserted in a parallel region, although they are computed by just one processor and the other ones wait for the results. Sometimes it is more efficient to replicate the computations in every processor in order to reduce memory traffic. Using this strategy, there is no extra cost in a parallel loop. But the parallel sections also implies overhead, and this overhead can be minimized if we reduce the number of sections by increasing their sizes and connecting them each other. As a consequence, we increase the speedup of routine *totzsp* and *tomnsp* to 7 using 8 processors on a SGI Origin 2000.

Nevertheless, when this technique was applied to the full code, the efficiency still remained very low (speedup lower than three), because of two reasons. First, PFA is very conservative when interprocedural analysis is performed. For example, PFA doesn't regard about memory locality between different parallel loops, and so, there are a lot of cache misses along the execution of VMEC. Second PFA disregard many loops because of their sizes, so increasing load unbalance and memory traffic.

Our next strategy takes into account the weakness of PFA in several aspects. To enlarge the size of a parallel section, any operation can be included in them. In one hand, parallel loops, even those initially rejected by the analyzer because of their sizes which can be now distributed without any expense. On the other hand, fully sequential operations or loops, which can be replicated in all the processors or can be masked in most of them.

The parallelization of each loop is fitted to a pre-established data locality for the matrices, in order to reduce memory traffic. Most matrices will be distributed looking for a good load balance and a low communication overhead, but some matrices will be replicated in every processor if they remain invariant, they are small, or they are fully required for computations by all the processors. Finally, there are matrices stored in only one processor.

The resulting code, with only one parallel region, is performed by all processors, although some operations are masked in most of them. We completely eliminated the overhead of starting and stopping threads, and we parallelized almost every loop, even inner and small ones. As a result, we achieved a speedup of 5.8 for 8 processors.

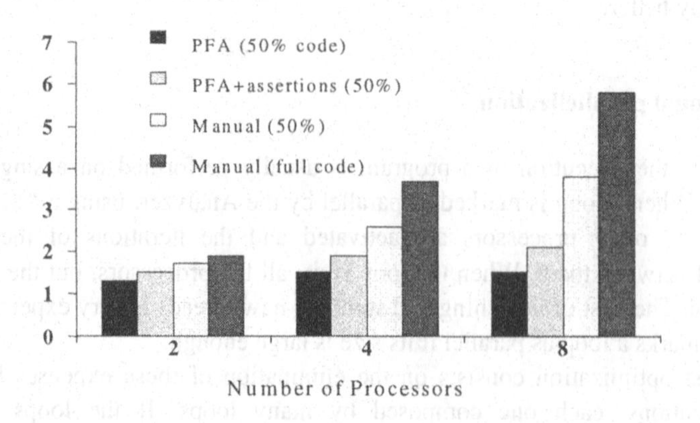

Fig. 3. *Speedup* for different parallel codes

3.3 Manual code analysis and optimizations

But we are still very far from the perfect speedup, specially when the number of processor is 16 or higher. The reasons of the low efficiencies can be concluded from the next graph (Fig. 4), in which the time spent by the different subroutines in all the processors are shown:

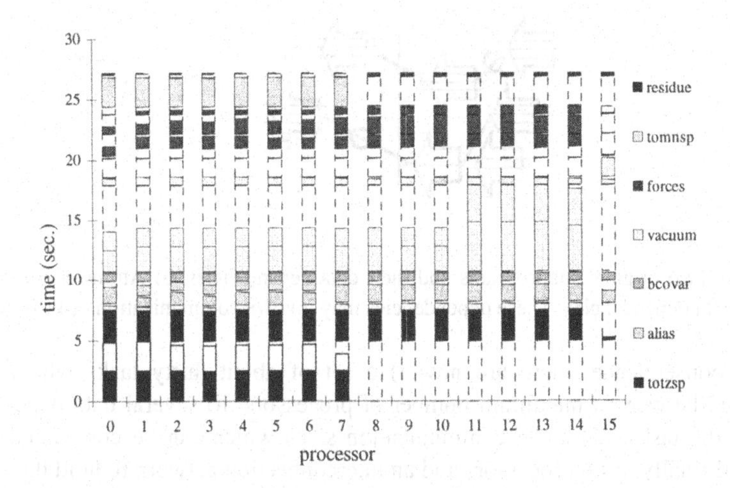

Fig. 4. CPU time of routines in *funct3d.* (16 processors, 50 iterations)

In this figure, white sections in each column indicates communications: *shmem_put* calls (continuous lines) or *shmem_wait* calls (dotted lines). The high density of white sections in the chart indicates that the processors are inactive during a long period, because they are sending a message, waiting for the arrival of a message, or they are expecting the end of an unbalanced loop. For example, in time $t=0$, it can be observed a distributed loop with only eight parallelizable steps.

So, communication overhead and load unbalance are, as expected, the main sources of the inefficiency of the parallel code. In order to increase its efficiency, a reorganization of the routines inside *funct3d* can be done, with two main aims: to achieve a better load balance, and to overlap communications with computations. But data dependencies between the thirteen routines in *funct3d* are very strong, and there are very few possibilities of new arrangements.

To increase the number of possible arrangements, we split the procedures in *funct3d* and created a set of atomic tasks, each one holding a single loop, a sequential operation, a reduction operation, or a small group of these tasks. Every atomic task will be either i) distributed between a maximum number of processors, or ii) fully sequential, or iii) cloned in every processor.

Data dependencies between atomic tasks are much more flexible, and allow us to reorder the computations in multiple ways. Furthermore, whenever this data dependence implies a communication between processor memories, it can be easily overlapped with other computations.

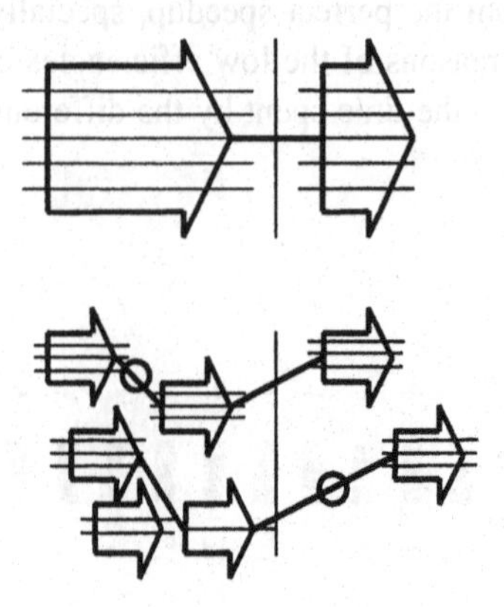

Fig. 5. a) Two original subprograms and their data dependencies. b) Atomized subprograms and new data dependencies. These dependencies may involve communication tasks (circles)

As a consequence, there are now i) a set of about thirty tasks which can be distributed between a maximum number of processors, ii) several data dependencies between the tasks, iii) some communication steps which can be considered as new tasks, and finally, iv) n processors and an interconnection network to hold the tasks.

We have only to find how to order the tasks to reduce the execution time. We first reordered the tasks by hand trying to find the best load balance between the processors, and overlapping every communication step with communications (*manual sort search*). As a result, we achieved a speedup of 6.4 using 8 processors for the whole code in an Origin 2000, and similar results were obtained in a CRAY T3E machine with also 8 processors.

A second approach (*automatic sort search*) tries with all the possible arrangements for the routines satisfying the data dependencies ("topological sorting"). For each one, all the possible loop allocations are tried, and the most efficient is chosen. This strategy finds the best ordering, but it is also an N-P complete problem, and it is very expensive, even when the number of tasks is low (about thirty, in our problem).

A third approach (*semi-automatic sort search*) starts from a manual loop allocation (which globally obtains a good load balance). Then, each processor computes the best topological sorting of the works it has to perform (using the *Dijkstra* algorithm, it can be done very fast (O(n·log(n))). Finally, a poll among the processors determines the best arrangement, considering the proposed from every processor. This technique can be included in our program to select the best order for each set of input parameters, and also considering the number of processors. This program could also consider tasks belonging to the preceding iteration of *funct3d* in a similar way to that presented in [3] for a Biconjugate Gradient Stabilized.

4 Results

The final parallel version of VMEC uses the semi-automatic arrangement proposed in the previous section, although only the most expensive routines has been considered for the *Dijkstra* step.

In the next figure, the time spent for each processor in each of the routines in *funct3d* are shown, now using the new arrangement:

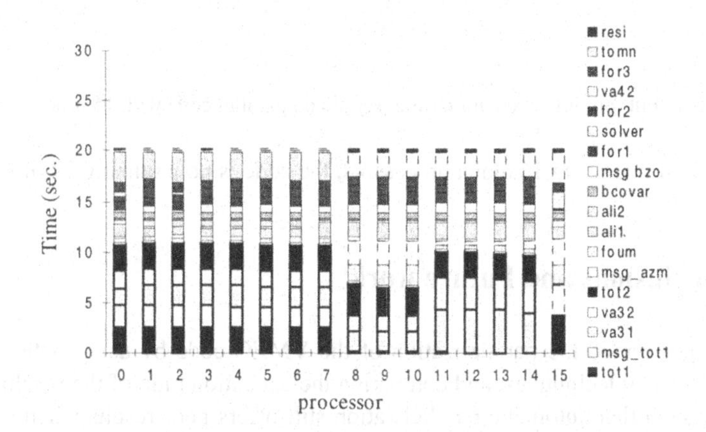

Fig. 6. CPU time of routines in *funct3d.* (16 processors, 50 iterations)

Comparing these results with those presented in Fig. 5, it can be observed how the new arrangement significantly increases the time in which the processors are busy. The new arrangement for the routines increases the speedup to almost 7 (using 8 processors), and to almost 11 (for 16 Alpha processors) (Fig. 7).

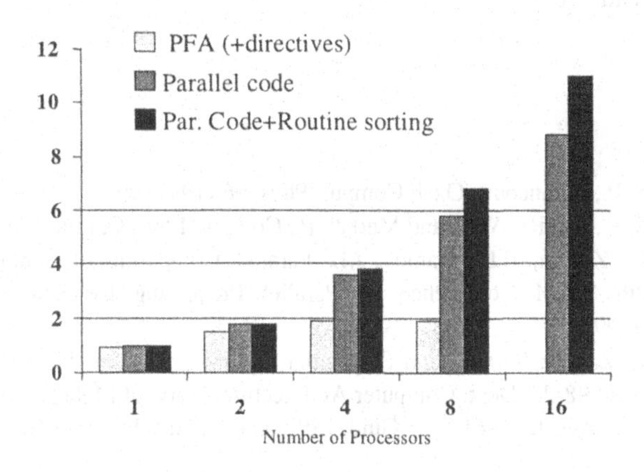

Fig. 7. Efficiency

Finally, for a typical set of parameter values, and also considering some optimizations for the sequential code [4], a significant reduction of the execution time can be observed:

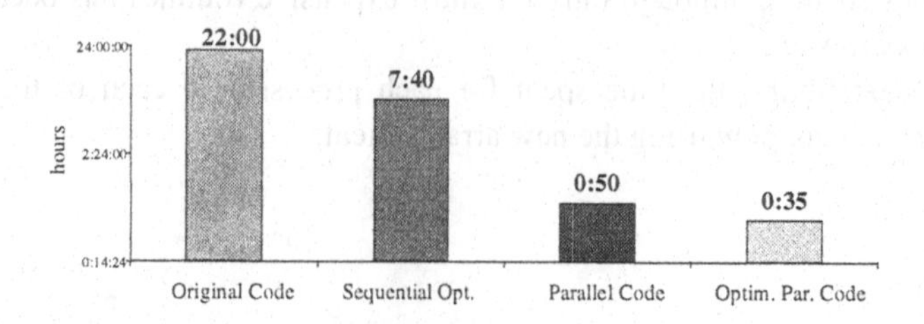

Fig. 8. Full program execution time (serial and parallel code with 16 processors)

It can be seen that using 16 alpha processors, the code is now running 37 times faster

5 Conclusions and Future work

We have performed the parallelization of the VMEC code by using both automatic tools and manual techniques, and comparing the execution time of the resulting codes we have found that automatic parallelization still offers poor results when deals with complex codes. In this work, we present a simple technique which can be used to extract the maximum parallelism to VMEC by combining data and task parallelism. The proposed algorithm significantly reduce the communication overhead and globally obtain a good load balance. It can also be extended to other applications.

In future work we will consider a dynamic scheduling of the loops by including the proposed technique in the own code. The inclusion in future parallelizing compilers will also be considered.

References

1 Hirschman, S. P., Betancourt, O.: J. Comput. Phys. 96 (1991) 99
2 Hirschman, S. P., van Rij, W. I. and Merkel, P., Comput. Phys. Comm. 39 (1986) 143-
3 Romero, L.F., Zapata, E.L., Ramos, J.I.: Parallel Computing of Semiconductor Laser Equations. 8th SIAM Conference on Parallel Processing for Scientific Computing. Minneapolis (1997)
4 Romero, L.F., Zapata: Parallel and Sequential Optimisations for the VMEC code. Tech. Rep. UMA-DAC-98-23, Dept. Computer Architecture, Univ. of Malaga (1998)
5 POWER Fortran Accelerator User's Guide. Silicon Graphics, Inc., 007-0711, (1997)

Rational Krylov Algorithms for Eigenvalue Computation and Model Reduction

Axel Ruhe[1] * and Daniel Skoogh[1]

Department of Mathematics,
Chalmers Institute of Technology and the University of Göteborg,
S-41296 Göteborg, Sweden
{ruhe,skoogh}@math.chalmers.se

Abstract. Rational Krylov is an extension of the Lanczos or Arnoldi eigenvalue algorithm where several shifts (matrix factorizations) are performed in one run. A variant has been developed, where these factorizations are performed in parallel.
It is shown how Rational Krylov can be used to find a reduced order model of a large linear dynamical system. In Electrical Engineering, it is important that the reduced model is accurate over a wide range of frequencies, and then Rational Krylov with several shifts comes to advantage.
Results for numerical examples coming from Electrical Engineering applications are demonstrated.

1 Introduction, Eigenvalues and Linear Dynamic Systems

In many scientific and engineering contexts one needs to solve an algebraic eigenvalue problem,

$$(A - \lambda B)x = 0\,, \tag{1}$$

for eigenvalues λ_k and eigenvectors x_k. Its solution is used as the first step on the way from a static to a dynamic description of a system of interacting entities. Think of a network of electric components or a mechanical construction of masses and springs! Small perturbations of a static, equilibrium, position will be formed as eigenvectors and move as the eigenvalues indicate.

The foremost mathematical modeling tool is the Laplace transform, where we map a description in time and space to one involving decay rates and oscillation frequences, these are the real and imaginary parts of the eigenvalues. In this realm we use eigenvalues to get the solution of a linear system of ordinary differential equations,

$$\dot{x} = Ax\,,\; x(0) = a\,,$$

* Partial support given by TFR, the Swedish Research Council for Engineering Sciences, Dnr 222, 96-555, and from the Royal Society of Arts and Sciences in Göteborg.

over time as a linear combination of eigensolutions,

$$x(t) = \Sigma_k \alpha_k x_k e^{\lambda_k t} .$$

Today we will concentrate on how to find reduced order models of large linear dynamic systems. Let us specialize, and consider a model of an electric RLC network, that is a network containing only passive components, resistances (R) inductances (L) and capacitances (C). Such a network follows the equations,

$$\begin{aligned} C\dot{x} &= -Gx + ru \\ y &= l^T x \end{aligned} , \tag{2}$$

where x is the state vector, u is the vector of inputs (controls) and y is the vector of outputs. G is the matrix of DC components (resistances) and C is the matrix of components with memory (inductances and capacitances). The external connections for input and output are given by r and l, which for the single input single output case are just column vectors.

For an RLC network, the state vector and matrices are partitioned into,

$$x = \begin{pmatrix} v \\ i \end{pmatrix} \quad C = \begin{pmatrix} C & 0 \\ 0 & L \end{pmatrix} \quad G = \begin{pmatrix} G & B \\ -B^T & 0 \end{pmatrix} , \tag{3}$$

where v is the vector of node voltages, i is the vector of currents in those branches that have inductors, C is the matrix of capacitances, L of inductances and G of conductances (reciprocal of resistance). The rectangular matrix B is just the incidence matrix of the inductor branches.

In an RLC network, the matrices C, L and G are all symmetric positive semidefinite. In realistic cases we may also safely assume that the pencil (G, C) is regular.

Make a Laplace transform of the system (2) and get,

$$\begin{aligned} sCX &= -GX + rU \\ Y &= l^T X \end{aligned} , \tag{4}$$

giving the input output relation,

$$\begin{aligned} Y &= l^T (G + sC)^{-1} rU \\ &= H(s)U \end{aligned} , \tag{5}$$

which defines the transfer function $H(s)$.

We see that the transfer function has poles in the complex plane at the eigenvalues s_k of the pencil

$$(G + sC)x = 0 . \tag{6}$$

Purely imaginary eigenvalues correspond to persistent oscillations, and we are often interested in the norm of $H(s)$ along the imaginary axis. An eigenvalue close to the imaginary axis shows up as a peak of this norm, and an excitation at this frequency will give rise to a slowly decaying oscillation. This is, of course, provided the system is *stable*, which for our choice of sign in the pencil (6) occurs when all eigenvalues are in the right half plane.

The task is now to find a new *reduced order model*, with a state space of a much smaller dimension, chosen so that the transfer function $\tilde{H}(s)$ of the reduced model keeps all the important properties of the original transfer function (5). Let us assume that we have found a subspace spanned by the first k columns of the nonsingular matrix S, and express an arbitrary vector x in the state space by,

$$x = S\tilde{x}\,.$$

The linear system (2) is now transformed into,

$$\begin{array}{rl} S^{-1}\boldsymbol{C}S\dot{\tilde{x}} = & -S^{-1}\boldsymbol{G}S\tilde{x} + S^{-1}\boldsymbol{r}u \\ y = & \boldsymbol{l}^T S\tilde{x} \end{array}, \tag{7}$$

and we get the system of the reduced order model by keeping the appropriate leading $k \times k$ submatrices of these.

Many different ways to choose a basis S have been tried. An evident choice is to take the eigenvectors that correspond to eigenvalues in the frequency range of interest, this is standard practice in Mechanical Engineering. Here one is most often interested in the lowest frequencies, or in frequencies close to a slow external load, like the hum of an engine or the shaking of an earthquake. Recently Krylov spaces, computed by Lanczos or Arnoldi algorithms applied to a shifted and inverted matrix pencil (8) have caught on, especially in Electrical Engineering, (Freund and Feldmann at Bell Laboratories [4, 5], Elfadel and Ling *et al* at IBM [2]) but also in Mechanical Engineering (E. Grimme *et al* [6]) and even in Quantum Chemistry, the tridiagonal matrix from Lanczos has been used to predict photoionization cross sections (H. Karlsson [7]). A Rational Krylov approach in the spirit we are going to discuss today has already been tried at several places, see [6] and [1]!

2 Rational Krylov

A standard practice to find eigenvalues of large matrix pencils (1) is to apply a Krylov space algorithm, Lanczos or Arnoldi, to a shift invert spectral transformation,

$$C = (A - \mu B)^{-1} B\,. \tag{8}$$

It will compute eigenvalues close to the shift point μ in the complex plane[3].

The Rational Krylov algorithm [9] is a further development of this approach, where an orthonormal basis V_j is built up one column at a time. In each step a shifted and inverted matrix operator (8) is applied, and the resulting vector is orthogonalized to the already computed part of the basis by means of Gram

Schmidt. Rational Krylov differs from shifted and inverted Arnoldi, in that it may use different shifts μ_j in different steps j.

Let us first give the algorithm step by step, then derive the basic recursion between the computed vectors, and last show how it is equivalent to a certain shifted and inverted Arnoldi iteration.

ALGORITHM RKS

---

1. Start with vector v_1, where $\|v_1\| = 1$. Set $t_1 = e_1$.
2. For $j = 1, 2, \ldots$ until *convergence*
   (a) Choose shift μ_j.
   (b) Set $r = V_j t_j$, *(continuation vector).*
   (c) Operate $r := (A - \mu_j B)^{-1} B r$
   (d) Orthogonalize $r := r - V_j h_j$ where $h_j = V_j^H r$, *(Gram Schmidt).*
   (e) Get new vector $v_{j+1} = r / h_{j+1,j}$, where $h_{j+1,j} = \|r\|$, *(normalize).*
   (f) Compute approximate solutions, $\lambda_i^{(j)}$ and $y_i^{(j)}$, and test for convergence.
   (g) Get t_{j+1}, *(continuation combination).*
3. Compute eigenvectors .

END.

Let us follow what happens during a typical step j of ALGORITHM RKS. Eliminate the intermediate vector r, used in steps 2b – 2e, and get

$$V_{j+1} h_j = (A - \mu_j B)^{-1} B V_j t_j ,$$

the vector h_j now has length $j + 1$. Multiply from the left by $(A - \mu_j B)$,

$$(A - \mu_j B) V_{j+1} h_j = B V_j t_j . \tag{9}$$

Separate terms with A to the left and B to the right,

$$A V_{j+1} h_j = B V_{j+1} (h_j \mu_j + t_j) ,$$

with a zero added to the bottom of the vector t_j to give it length $j + 1$.

This is the relation for the j th step, now put the corresponding vectors from the previous steps in front of this and get,

$$A V_{j+1} H_{j+1,j} = B V_{j+1} K_{j+1,j} , \tag{10}$$

with two $(j + 1) \times j$ Hessenberg matrices, $H_{j+1,j} = [h_1 h_2 \ldots h_j]$, containing the Gram Schmidt orthogonalization coefficients, and

$$K_{j+1,j} = H_{j+1,j} \operatorname{diag}(\mu_i) + T_{j+1,j} \tag{11}$$

with the upper triangular matrix $T_{j+1,j} = [t_1 t_2 \ldots t_j]$, built up from the continuation combinations used in step 2b.

If we choose all zero shifts, $\mu_j = 0$, and continue with the latest vector, $t_j = e_j$, in all steps j, we get back the shifted and inverted Arnoldi,

$$A^{-1}BV_j = V_{j+1}H_{j+1,j} \,.$$

In the general case, it is also possible to bring the recursion (10) into an Arnoldi recursion by careful application of orthogonal transformations to the Hessenberg matrices H and K, bringing one of them to triangular form. We find orthogonal matrices $Q_{j+1,j+1}$ and $Z_{j,j}$ so that the transformed matrices,

$$\begin{aligned} \tilde{K} &= Q^H_{j+1,j+1}K_{j+1,j}Z_{j,j} && \text{is Hessenberg,} \\ R &= Q^H_{j+1,j+1}H_{j+1,j}Z_{j,j} && \text{is upper triangular,} \\ W_{j+1} &= V_{j+1}Q_{j+1,j+1} && \text{is a new orthogonal basis.} \end{aligned}$$

This transformation is done one column at a time, it is easiest to understand if we follow one minor step in detail. Assume that $j = 4$ and that the first 2 columns are already in the desired form,

$$H = \begin{bmatrix} x & x & x & x \\ & x & x & x \\ & & x & x \\ & & y & x \\ & & & x \end{bmatrix}, K = \begin{bmatrix} x & x & x & x \\ x & x & x & x \\ & x & x & x \\ & & x & x \\ & & & x \end{bmatrix}.$$

Apply a rotation $R_{3,4}$ in the $(3,4)$ plane from the left to zero out the element y at $h_{4,3}$. However this rotation fills $k_{4,2}$, which has to be chased up left along the main diagonal. Let us write this in a table, where the arrows denote multiplication from left or right,

$$\begin{aligned} &R_{3,4} \rightarrow \text{ zeros } h_{4,3} && \text{fills } k_{4,2} \\ &S_{2,3} \leftarrow \text{ zeros } k_{4,2} && \text{fills } h_{3,2} \\ &R_{2,3} \rightarrow \text{ zeros } h_{3,2} && \text{fills } k_{3,1} \\ &S_{1,2} \leftarrow \text{ zeros } k_{3,1} && \text{fills } h_{2,1} \\ &R_{1,2} \rightarrow \text{ zeros } h_{2,1}. \end{aligned}$$

After this sweep both matrices have the same nonzero pattern as before, except that the element y at $h_{4,3}$ is put to zero. After we had made one such sweep for each column up to column j, we have the desired form,

$$AW_jR_{j,j} = BW_{j+1}\tilde{K}_{j+1,j} \,, \tag{12}$$

and can get an approximate eigensolution from

$$\begin{aligned} \tilde{K}_{j,j}y_i^{(j)} &= R_{j,j}y_i^{(j)}\theta_i^{(j)} && \text{approximate eigenvalue } \theta_i^{(j)} \,, \\ \tilde{x}_i &= W_jR_{j,j}y_i^{(j)} && \text{approximate eigenvector } \tilde{x}_i \,. \end{aligned}$$

Note that this pencil has the same nonzero pattern as the intermediate matrix in the QZ algorithm. Note also that this could have been done on a shifted pencil, and yield a more accurate recursion than we described in [8]. For an appropriate choice of shift, w_{j+1} will be a good continuation vector as described in [9].

Note, however also that even if (12) is an Arnoldi recursion, it does not start at the original starting vector v_1 but at w_1, which is a linear combination of all the vectors computed during the whole RKS algorithm.

3 Reduced Order Model

Now return to the linear dynamic system (2) and its response function $H(s)$ defined by (5). Let us specialize to the single input single output case where both matrices $\boldsymbol{l}$ and $\boldsymbol{r}$ are single vectors and we get,

$$H(s) = \boldsymbol{l}^T(\boldsymbol{G} + s\boldsymbol{C})^{-1}\boldsymbol{r}\,. \tag{13}$$

Our task is now to evaluate this for many values of the frequency parameter s. We see that for each s we need to solve the system

$$(\boldsymbol{G} + s\boldsymbol{C})x(s) = \boldsymbol{r} \tag{14}$$

for the vector $x(s)$ and then do a scalar multiplication with $\boldsymbol{l}$.

This can be done with few factorizations of shifted large matrices (14) in the following way.

Apply the Rational Krylov algorithm to the matrix pencil (6) i. e. set $A = \boldsymbol{G}$, $B = \boldsymbol{C}$ and shift with $\mu_j = -s_j$, some appropriate frequencies for which the matrices $\boldsymbol{G} + s_i\boldsymbol{C}$ are nonsingular, and chosen so that the frequency range of interest is covered. Start at the vector $v_1\beta_0 = (\boldsymbol{G} + s_1\boldsymbol{C})^{-1}\boldsymbol{r}$, and note that in this case we do not use a random start.

After j steps we may compute an approximate solution to $x(s)$ (14) as,

$$\tilde{x}(s) = V_{j+1}(K_{j+1,j} + s_1 H_{j+1,j})\,\tilde{y}(s)\,,$$

where,

$$\tilde{y}(s) = (K_{j,j} + sH_{j,j})^{-1}e_1\beta_0\,, \tag{15}$$

a solution to a $j \times j$ linear system with a coefficient matrix of Hessenberg form.

To see how this comes about, shift the basic Rational Krylov recursion (10) and get,

$$(\boldsymbol{G} + s\boldsymbol{C})V(K + s_i H) = (\boldsymbol{G} + s_i\boldsymbol{C})V(K + sH)\,,$$

and multiply with a shifted and inverted matrix to get

$$(\boldsymbol{G} + s_i\boldsymbol{C})^{-1}(\boldsymbol{G} + s\boldsymbol{C})V(K + s_i H) = V(K + sH)\,. \tag{16}$$

This is useful to get an expression for the premultiplied residual,

$$r(s) = (\boldsymbol{G} + s_i\boldsymbol{C})^{-1}\{(\boldsymbol{G} + s\boldsymbol{C})\,\tilde{x}(s) - \boldsymbol{r}\}\,. \tag{17}$$

The choice to start at $v_1\beta_0 = (\boldsymbol{G} + s_1\boldsymbol{C})^{-1}\boldsymbol{r}$ and the relation (16) will now for $i = 1$ imply that,

$$\begin{aligned} r(s) &= (\boldsymbol{G} + s_1)^{-1}(\boldsymbol{G} + s\boldsymbol{C})V_{j+1}(K_{j+1,j} + s_1 H_{j+1,j})\, \tilde{y}(s) - v_1\beta_0 \\ &= V_{j+1}\{(K_{j+1,j} + sH_{j+1,j})\, \tilde{y}(s) - e_1\beta_0\} \\ &= v_{j+1}(k_{j+1,j} + sh_{j+1,j})(\, \tilde{y}(s))_j \\ &= v_{j+1}h_{j+1,j}(s - s_j)(\, \tilde{y}(s))_j \,, \end{aligned} \tag{18}$$

a unit length vector v_{j+1} multiplied by the last Gram Schmidt orthogonalization coefficient $h_{j+1,j}$, the difference $s - s_j$ between the evaluation frequency s and the last shift s_j, and finally the last element $(\, \tilde{y}(s))_j$ of the solution to the $j \times j$ system (15). This last factor is likely to get small when we have used shifts s_i reasonably close to s during the Rational Krylov process.

Compare to the similar expressions one gets when applying Lanczos or Arnoldi to an eigenvalue problem or GMRES to a linear system of equations. Our way (15) to get the approximate solution actually corresponds to the FOM (Full Orthogonalization Method), which is closely related to GMRES.

One advantage of using FOM, instead of the more accurate GMRES, is that we can represent the reduced order model as,

$$\tilde{H}(s) = \tilde{\boldsymbol{l}}^T (\tilde{\boldsymbol{G}} + s\tilde{\boldsymbol{C}})^{-1}\tilde{\boldsymbol{r}}\,, \tag{19}$$

with

$$\begin{aligned} \tilde{\boldsymbol{l}}^T &= \boldsymbol{l}^T V_{j+1}(K_{j+1,j} + s_1 H_{j+1,j}) \\ \tilde{\boldsymbol{G}} &= K_{j,j} \\ \tilde{\boldsymbol{C}} &= H_{j,j} \\ \tilde{\boldsymbol{r}} &= e_1\beta_0\,. \end{aligned}$$

The error of the approximate solution $\tilde{x}(s)$ can be expressed as

$$\begin{aligned} \Delta\, \tilde{x}(s) &= x(s) - \tilde{x}(s) \\ &= (\boldsymbol{G} + s\boldsymbol{C})^{-1}(\boldsymbol{G} + s_i\boldsymbol{C})r(s) \\ &= (\boldsymbol{I} + (s_i - s)(\boldsymbol{G} + s\boldsymbol{C})^{-1}\boldsymbol{C})r(s), \end{aligned} \tag{20}$$

where $r(s)$ is the premultiplied residual (17). An upper bound for the error of the approximate transfer function $\tilde{H}(s)$ is

$$|\epsilon(s)| \leq \|\, l \,\|\, |\alpha(s)|(1 + |s_i - s|\, \|\, (\boldsymbol{G} + s\boldsymbol{C})^{-1}\boldsymbol{C}v_{j+1}\, \|). \tag{21}$$

where $\alpha(s) = h_{j+1,j}(s - s_j)(\, \tilde{y}(s))_j$, see (18).

In order to compute an efficient and reliable error estimate we need to estimate $\|\, (\boldsymbol{G} + s\boldsymbol{C})^{-1}\boldsymbol{C}v_{j+1}\, \|$. How this is done is described in [10].

4 Two Numerical Examples

Let us report tests on two models of medium size radio frequency circuits.

The first test example is provided by Zhaojun Bai and is of dimension $n = 256$. The rational Krylov algorithm is applied with the shifts $\mu_1 = -j10^8$, $\mu_2 = -j10^9$ and $\mu_3 = -j10^{10}$, where $j = \sqrt{-1}$. The shifts are applied 5, 7 and 7 times respectively. In figure 2 we show the approximate transfer function $\tilde{H}(s)$, and in figure 3 we show the error and an error estimate of $\tilde{H}(s)$. Note that the true error is below the error estimate. The eigenvalue distribution is given in figure 1. Compare the eigenvalue distribution with the plots for the transfer function and error estimate, and note that the transfer function has a top near the imaginary part of each eigenvalue.

We did another run with the same matrices but the shifts closer together at $\mu_1 = -j10^9$, $\mu_2 = -2j \times 10^9$ and $\mu_3 = -7j \times 10^9$, used 4, 4 and 8 times respectively. Look at figures 4 and 5 and note that now the errors are much smaller at the higher frequencies, where there are many poles. It is evident that an adaptive strategy where the shifts are chosen where we want small errors will be of advantage.

The second test example is provided by Peter Feldmann and is of dimension $n = 1841$. Now the rational Krylov algorithm is applied with the shifts $\mu_1 = -j10^8$, $\mu_2 = -j10^9$ and $\mu_3 = -4j \times 10^9$, applied 4, 8 and 13 times respectively. In figure 6 we show the approximate transfer function $\tilde{H}(s)$, and in figure 7 the error and an error estimate of $\tilde{H}(s)$. The eigenvalue distribution is given in figure 8.

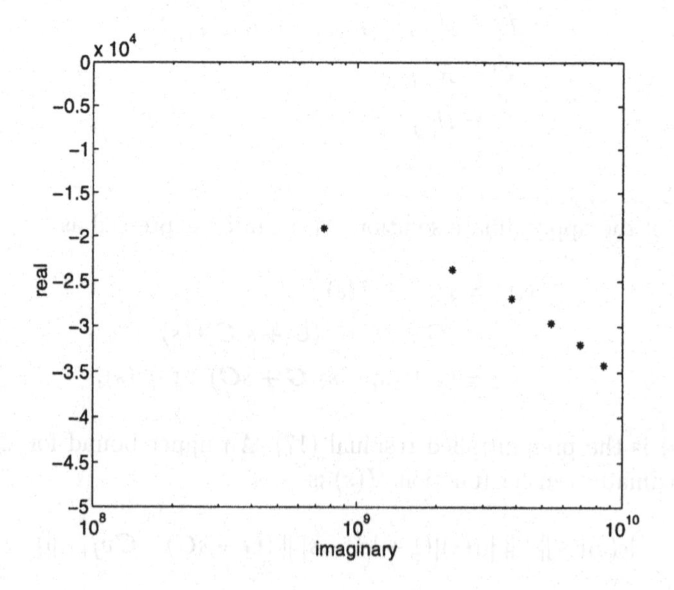

Fig. 1. The eigenvalue distribution of the first test example given by Bai, $n = 256$.

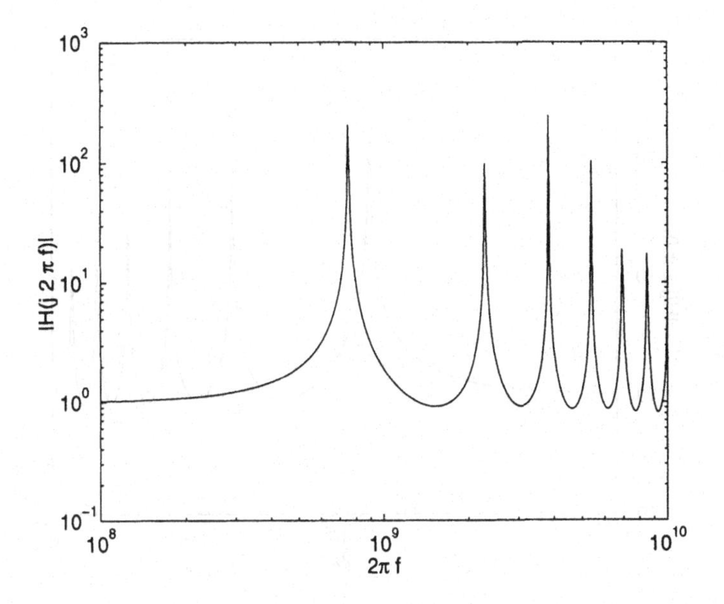

Fig. 2. The first test example of dimension $n = 256$. The figure shows the approximate transfer function $\tilde{H}(s)$. The rational Krylov algorithm is applied with the shifts $\mu_1 = -j10^8$, $\mu_2 = -j10^9$ and $\mu_3 = -j10^{10}$, where $j = \sqrt{-1}$. The shifts are applied 5, 7 and 7 times respectively.

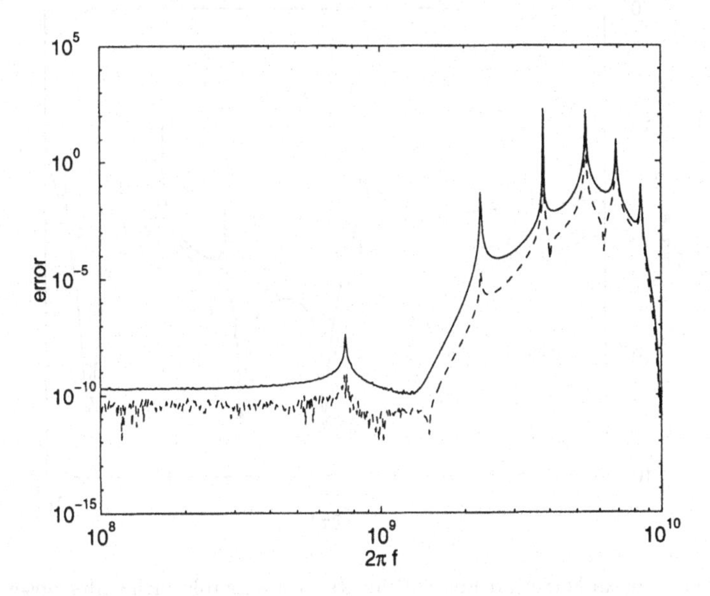

Fig. 3. The error and error estimate of the first test example. The error estimate is plotted with a solid line –, and the true error is plotted with a dashed line – –

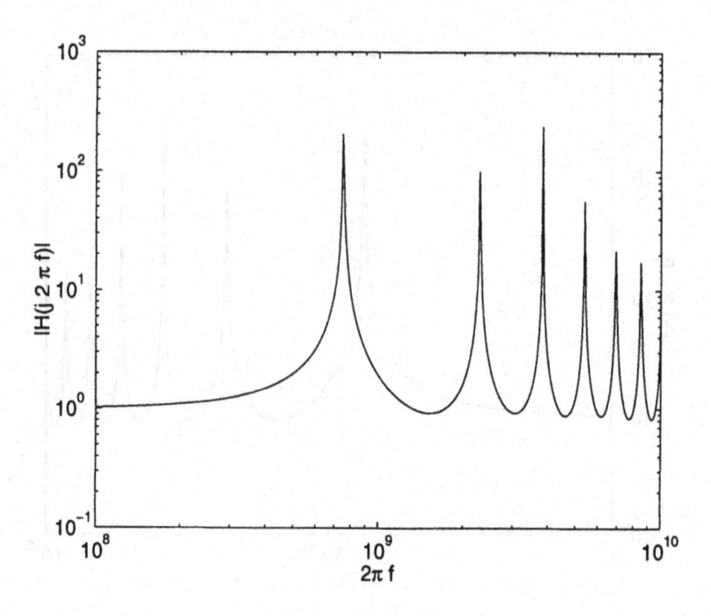

Fig. 4. First test example is of dimension $n = 256$. Approximate transfer function $\tilde{H}(s)$ for rational Krylov algorithm is applied with the shifts closer together at $\mu_1 = -j10^9$, $\mu_2 = -2j \times 10^9$ and $\mu_3 = -7j \times 10^9$, applied 4, 4 and 8 times respectively.

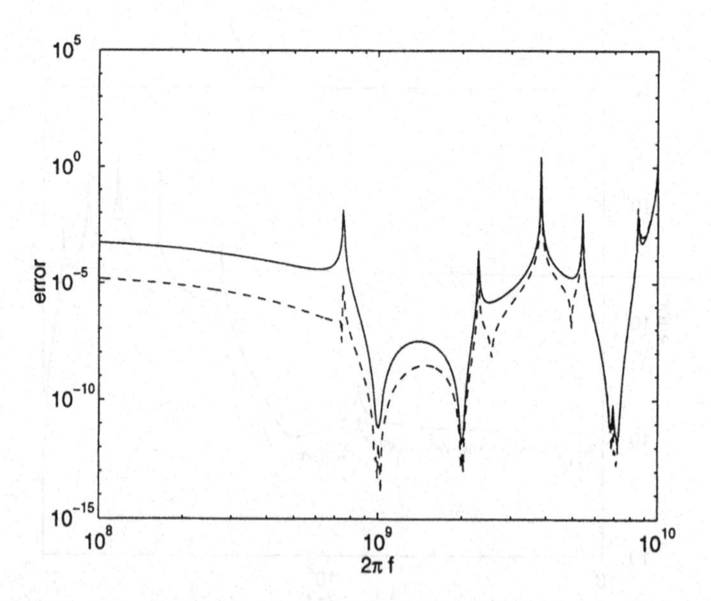

Fig. 5. The error and error estimate of the first test example with shifts closer together, corresponding to the figure 4. The error estimate is plotted with a solid line –, and the true error is plotted with a dashed line – –

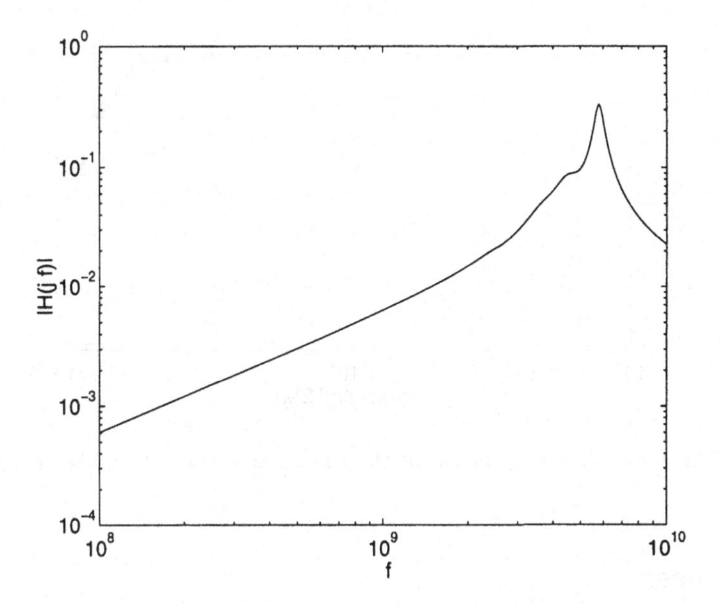

Fig. 6. Second test example, dimension $n = 1841$. Approximate transfer function $\tilde{H}(s)$ for the rational Krylov algorithm applied with shifts $\mu_1 = -j10^8$, $\mu_2 = -j10^9$ and $\mu_3 = -4j \times 10^9$, where $j = \sqrt{-1}$. The shifts are applied 4, 8 and 13 times respectively.

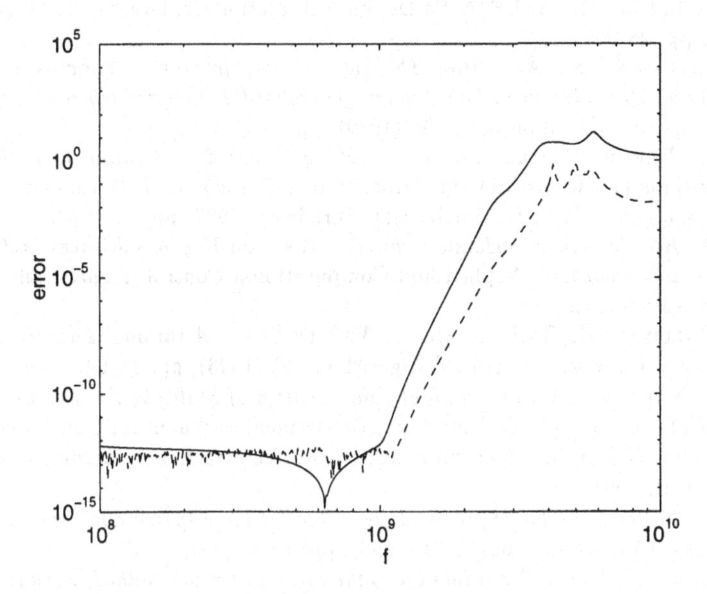

Fig. 7. The error and error estimate of the second test example. The error estimate is plotted with a solid line –, and the true error is plotted with a dashed line – –

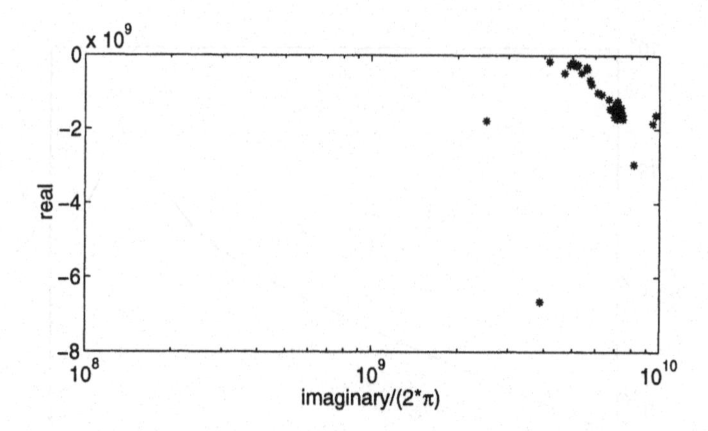

Fig. 8. The eigenvalue distribution of the second test example given by Feldmann, $n = 1841$.

References

1. I. M. Elfadel and D. D. Ling, *A block rational Arnoldi algorithm for multipoint passive model-order reduction of multiport RLC networks*, in Technical Digest of the 1997 IEEE/ACM International Conference on Computer-Aided Design, IEEE Computer Society Press, 1997, pp. 66–71.
2. ——, *Zeros and passivity of Arnoldi-reduced-order models for interconnect networks*, in Proc 34nd IEEE/ACM Design Automation Conference, ACM, New York, 1997, pp. 28–33.
3. T. Ericsson and A. Ruhe, *The spectral transformation Lanczos method for the numerical solution of large sparse generalized symmetric eigenvalue problems*, Mathematics of Computation, 35 (1980), pp. 1251–1268.
4. R. W. Freund, *Circuit simulation techniques based on Lanczos-type algorithms*, in Systems and Control in the Twenty-First Century, C. I. Byrnes, B. N. Datta, D. S. Gilliam, and C. F. Martin, eds., Birkäuser, 1997, pp. 171–184.
5. ——, *Reduced-order modeling techniques based on Krylov subspaces and their use in circuit simulation*, Applied and Computational Control, Signals, and Circuits, (1998, to appear).
6. K. Gallivan, E. Grimme, and P. Van Dooren, *A rational Lanczos algorithm for model reduction*, Numerical Algorithms, 12 (1996), pp. 33–64.
7. H. O. Karlsson, *Atomic and Molecular Density-of-States by Direct Lanczos Methods*, PhD thesis, Uppsala University, Department of Quantum Chemistry, 1994.
8. A. Ruhe, *Eigenvalue algorithms with several factorizations – a unified theory yet?* to appear, 1998.
9. ——, *Rational Krylov, a practical algorithm for large sparse nonsymmetric matrix pencils*, SIAM J. Sci. Comp., 19 (1998), pp. 1535–1551.
10. D. Skoogh, *Model order reduction by the rational Krylov method.* work in progress, 1998.

Solution of Distributed Sparse Linear Systems Using PSPARSLIB*

Yousef Saad[1] and Maria Sosonkina[2]

[1] Department of Computer Science and Engineering, University of Minnesota, 200 Union Street S.E., Minneapolis, MN 55455, USA,
saad@cs.umn.edu

[2] Department of Computer Science, University of Minnesota at Duluth, 320 Heller Hall, 10 University Drive, Duluth, MN 55812, USA,
masha@d.umn.edu

Abstract. In a parallel linear system solution, an efficient usage of a multiprocessor system is usually achieved by implementing algorithms with high degree of parallelism and good convergence properties as well as by tuning parallel codes to a particular system. Among the software tools that facilitate this development is PSPARSLIB, a suite of codes for solving sparse linear systems of equations. PSPARSLIB takes a modular approach to constructing a solution method and has logic-transparent computational kernels that can be adapted to the problem at hand. Here, we outline a few parallel solution methods incorporated recently in PSPARSLIB. We give a rationale for implementing these techniques and present several numerical experiments.

1 Distributed Sparse Matrices: Basic Concepts

On a distributed memory parallel computer, every linear system solution begins with an assignment of the equations to the processors. In these mappings it is typical to have each processor hold a subset of equations (rows of the sparse linear system) along with a vector of the variables associated with these rows. The collection of these different subsets for each processor can be determined by a graph partitioner or *ad hoc* from *a priori* knowledge of the problem. PSPARSLIB adopts these rather general ways of distributing a sparse linear system that are closely related to the physical viewpoint. Subsequently, the developed iterative techniques view the system as a *distributed object.* To create this distributed object, PSPARSLIB provides two routines that built a list of neighboring processors and data structures local to each processor. This preprocessing phase distinguishes among three types of unknowns (Fig. 1): (1) Interior variables, those coupled with local variables only; (2) Local interface variables, those coupled with non-local (external) variables as well as local variables; and (3) External interface variables, those variables in other processors that are coupled with local variables.

* This work was supported in part by NSF under grant CCR-9618827 and in part by the Minnesota Supercomputer Institute.

PSPARSLIB provides preprocessing routines, a number of simple matrix and communication kernels, and iterative solvers. The preprocessing module comprises, among other things, routines for setting the distributed data structure described above and multicoloring routines. The basic kernels comprise simple operations with matrices, such as distributed matrix-vector products, their block variants, and matrix-vector products with the transpose. The communication part includes various local data exchange routines used by the matrix-vector operation and by the preconditioners. Finally, one can find a number of preconditioners and accelerators. A typical way of using PSPARSLIB is to set up the equations independently on each processor, possibly repartition using a parallel version of Metis, and then invoke the preprocessing and solution routines in turn for solving the resulting distributed linear system. PSPARSLIB is an ongoing project, in that additional solvers and techniques developed get added regularly.

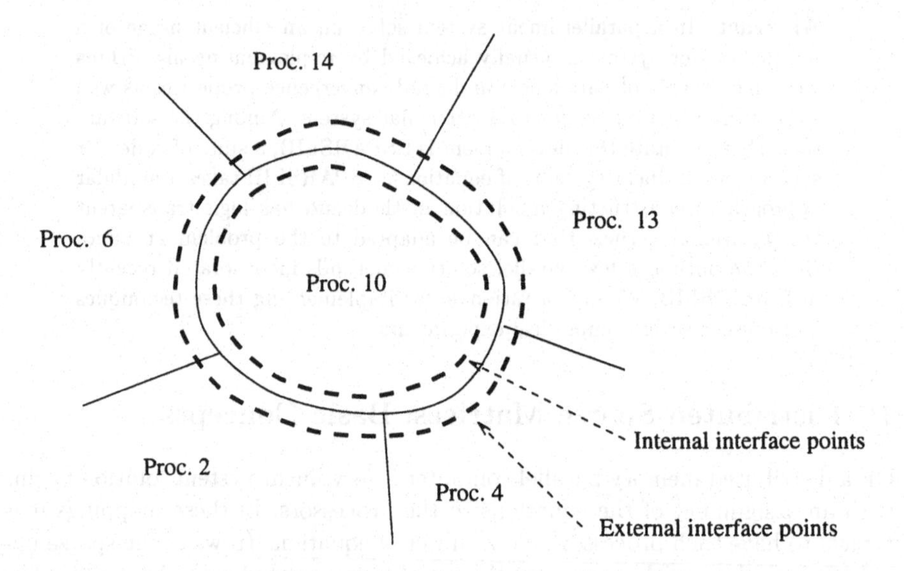

Fig. 1. A local view of a distributed sparse matrix.

2 Krylov subspace accelerators

Krylov subspace accelerators are fairly straightforward to implement in parallel. Aside from matrix-vector products and preconditioning, the main operations in an algorithm, such as GMRES, are the orthogonalization of vector sets and vector updates. If the components of all the vectors are distributed in accordance with the distribution of the vector of unknowns, vector updates are trivial to parallelize. In contrast, a global dot product requires a global sum of the separate dot products of the subvectors in each processor and may become a potential bottleneck in parallel implementations of a GMRES procedure.

We have experimented with distributed Flexible GMRES (FGMRES) [8] on various platforms: SGI cluster, IBM cluster, CRAY T3E, Intel Paragon, and SP2. (Message Passing Interface (MPI) [2] has been used as a communication library.) Table 1 illustrates the effect on the execution time of various strategies used in

Table 1. Execution times (sec.) for three orthogonalization strategies on the IBM SP2

Proc	MGSO	CGSO	SCGSO
1	66.4	65.2	65.1
4	37.6	26.9	21.1
8	32.1	15.4	11.7

the Gram-Schmidt orthogonalization. The matrix tested, called VENKAT01, originates from a fluid flow problem. It is an irregularly structured matrix of dimension $n = 62,424$ with $1,717,792$ nonzero elements. The system is solved using GMRES(30) without preconditioning, so that the iteration count is the same in each case. The standard classical Gram-Schmidt algorithm (CGSO) is used without reorthogonalization. Each orthonormalization requires two synchronization points, one to compute the inner products $h_{ij} = (w, v_i)$ needed to form the linear combination $w - \sum_{i=1}^{j} h_{ij} v_i$ and one to get the norm of the result in order to normalize it. This second synchronization point can be avoided by postponing the normalization to the next step (call the resulting algorithm the synchronized classical Gram-Schmidt (SCGSO) algorithm). The FGMRES time represents the time spent in the accelerator, excluding the time for the matrix-vector products. Table 1 shows that for 8 processors, Modified Gram-Schmidt (MGSO) has a speed-up close to 2, whereas SCGSO has a speed-up of about 6.

An experiment from [4] (Fig. 2) gives an idea on how the orthogonalization time compares with the rest of the computation. This experiment has been conducted on an SGI cluster with the same VENKAT01 test matrix. Overlapping block Jacobi (additive Schwarz) is now used for preconditioning. The relative tolerance is $\epsilon = 10^{-6}$ and the dimension of the Krylov subspace is $m = 15$. Fig. 2 shows that all contributions decrease as the number of processors increases except for the FGMRES time (dominated by classical Gram-Schmidt orthogonalization), which moves up slightly in going from 8 to 16 processors. Observe that for 16 processors this time becomes dominant. However, later experiments on the CRAY T3E indicated that even for a rather large number of processors, the impact of the cost of the Gram-Schmidt orthogonalization on the overall efficiency remains rather moderate. It is clear that the impact on efficiency will tend to become non-negligible with decrease in the size of each local matrix because the ratio of communication to computation increases in this case.

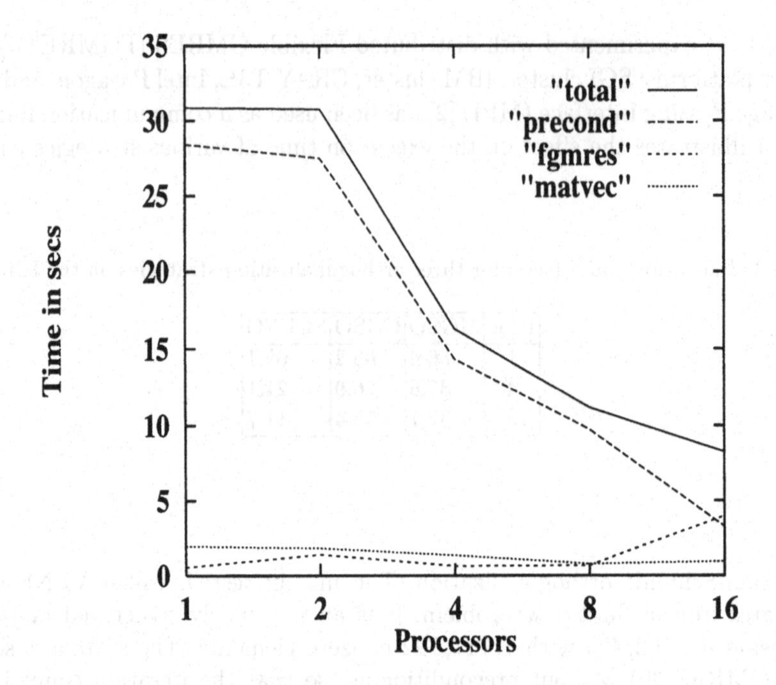

Fig. 2. Contributions to the total execution time for a distributed GMRES iteration preconditioned with overlapped additive Schwarz procedure

3 Preconditioning of Distributed Sparse Matrices

3.1 Distributed ILU

A distributed version of ILU(0) has also been developed by taking a similar local viewpoint [5]. The ILU(0) preconditioning approach works quite well for many problems. However, its parallel version, like other parallel preconditioners, may suffer from slow convergence because of its 'local nature'. A good way to circumvent convergence degradation is to use more accurate factorizations. A version of ILUT is available in SPARSKIT [6] and several variants are being used in industrial applications. Thus, it is important to develop similar algorithms for distributed sparse matrices. There are at least three different ways of obtaining parallel versions of ILUT.

1. Extensions of the Domain-Decomposition viewpoint mentioned above for ILU(0),
2. Extensions of ILUM [7, 10] to distributed sparse matrices and,
3. Schur complement-based techniques.

A natural way of developing an ILUT strategy is to consider again the two-level ordering idea employed for ILU(0) in [5]. The internal nodes are processed first by an ILUT procedure then ILU(0) is used on the interface nodes. This is a local process. In the second phase, the local interface nodes for each domain could be processed using a similar priority rule as with ILU(0) to determine the order in which processors must eliminate their interface nodes [3].

3.2 Variants of Schur Complement Techniques

Schur complement techniques exploit the observation that it is possible to obtain a smaller (dense) system involving equations for interface points alone.

Let the local matrix A_i residing in processor i be block-partitioned according to the internal variables followed by the (local) interface variables:

$$A_i = \begin{pmatrix} B_i & F_i \\ E_i & D_i \end{pmatrix} . \tag{1}$$

If we call x_i the vector of the internal variables, and y_i the vector of the interface variables, we can write all local equations as follows.

$$\begin{aligned} B_i x_i + F_i y_i &= f_i \\ E_i x_i + D_i y_i + \textstyle\sum_{j \in N_i} E_{ij} y_j &= g_i . \end{aligned} \tag{2}$$

The term $E_{ij} y_j$ is the contribution from the adjacent subdomain number j and N_i is the set of subdomains that are adjacent to subdomain i. The variable x_i can be eliminated from this system by extracting from the first equation $x_i = B_i^{-1}(f_i - F_i y_i)$ which yields,

$$S_i y_i + \sum_{j \in N_i} E_{ij} y_j = g_i - E_i B_i^{-1} f_i , \tag{3}$$

in which S_i is the 'local' Schur complement:

$$S_i = D_i - E_i B_i^{-1} F_i \ . \tag{4}$$

The equations (3) for all subdomains i altogether constitute a system of equations which involves only the interface points y_j, $j = 1, 2, \ldots, s$.

A number of techniques have already been developed in the Domain Decomposition framework for solving Schur complement systems, see [11] and references therein. In [4], a well-known relation [1] between a global Gauss-Seidel type iteration for the variables $\binom{x_i}{y_i}$ and a similar iteration for the variables y_i was exploited. Specifically, the y-part of the iterates associated with a Gauss-Seidel or Jacobi iteration for the system (2) is actually a Gauss-Seidel or Jacobi sequence for the Schur system (3). This sequence can now be accelerated with GMRES. This general process provides a simple mechanism for defining preconditioned Krylov subspace methods for the Schur complement system – without forming or even explicitly referring to this system. One general problem with Schur complement techniques is that the inner solves must be accurate. This is because the underlying coefficient matrix (4) involves B_i^{-1}.

It is also possible to exploit ILU factorizations in approximating the local Schur complements S_i. The main observation [8] is that if we compute the (exact)

LU factorization of A_i and consider the matrix $L_{y,i}$ and $U_{y,i}$ extracted from their lower right corners, i.e., related to the y variables, then we have $S_i = L_{y,i}U_{y,i}$. Therefore, we can form some sparse incomplete LU factorization of A_i and *retain only its lower principal parts* which can be used to precondition S_i. [9]. Thus, the construction and application of the preconditioning to S_i adds no cost to the global system solution. Since any type of ILU factorization can be used in these local approximations, it is preferable to choose one that yields an accurate local factorization. Consider, for example, ILUT. The number of fill-in elements, controlled by the parameter `lfil`, helps to reach the desired accuracy in preconditioning A_i. Table 2 shows the ratios of the average number of nonzero elements per row in the part of L and U factors used for preconditioning the Schur complement system over that required for the L and U factors for preconditioning A_i. This is done for several processor numbers (4,8,32,48,64) and two different values of the `lfil` parameter. Since the ratios of nonzeros are different in each processor, we show their average over all the processors. The matrix VENKAT01 has been used for testing. In Table 2, all the ratios are slightly greater than one. Thus, the amount of fill-in obtained in $L_{y,i}$ and $U_{y,i}$ is roughly the same as that obtained in the rest of the matrix A_i. This is because the nonzero elements in the L and U factors tend to concentrate near the main diagonal.

Table 2. Ratio between the average number of nonzero elements per row in S_i and in A_i

lfil	4	8	32	48	64
20	1.01	1.08	1.07	1.14	1.05
50	1.19	1.22	1.16	1.07	1.31

4 Conclusion

Because of the complexity in developing parallel solution methods, software packages and tools can be quite helpful, more so than in the sequential context. PSPARSLIB is a package devoted to the iterative solution of distributed sparse linear systems. A simple data structure stores the local equations. All solvers and tools are then built for distributed sparse matrices using this data structure.

Of the preconditioners developed so far for general sparse linear system, we found that the Schur complement-based approach performs best for a large number of processors. It is, in effect, a multilevel technique, which has some attributes of a global solver because it attempts to solve the global Schur complement system. Our experiments reveal that communication time is usually not damaging to performance when a good partitioning of the data is used and the local problem is large enough. Poor load balancing may have a more serious impact.

References

1. Chan T. F., Goovaerts D.: On the relationship between overlapping and nonoverlapping domain decomposition methods. SIAM J. on Matrix Analysis and Applications **13** (1992) 663–670
2. Gropp W, Lusk E, Skjellum A.: Using MPI: Portable Parallel Programming with the Message Passing Interface. MIT press Boston (1994)
3. Karypis G.: Graph Partitioning and its Applications to Scientific Computing. PhD thesis. Department of Computer Science, University of Minnesota, Minneapolis, MN (1996)
4. Lo G.-C., Saad Y.: Iterative solution of general sparse linear systems on clusters of workstations. Technical Report UMSI 96/117 & UM-IBM 96/24 Minnesota Supercomputer Institute, University of Minnesota, Minneapolis, MN (1996)
5. Ma S., Saad Y.: Distributed ILU(0) and SOR preconditioners for unstructured sparse linear systems. Technical Report 94–027, Army High Performance Computing Research Center, University of Minnesota, Minneapolis, MN (1994)
6. Saad Y.: SPARSKIT: A basic tool kit for sparse matrix computations. Technical Report 90-20, Research Institute for Advanced Computer Science, NASA Ames Research Center, Moffet Field, CA (1990)
7. Saad Y.: ILUM: a parallel multi-elimination ILU preconditioner for general sparse matrices. SIAM J. on Scientific Computing. **17** (1996) 830–847
8. Saad Y.: Iterative Methods for Sparse Linear Systems. PWS publishing, New York (1996)
9. Saad Y., Sosonkina M.: Distributed Schur Complement Techniques for General Sparse Linear Systems. Technical Report UMSI 97/159 Minnesota Supercomputer Institute, University of Minnesota, Minneapolis, MN (1997)
10. Saad Y., Zhang J.: BILUM: Block Versions of Multi-Elimination and Multi-Level ILU Preconditioner for General Sparse Linear Systems. Technical Report UMSI 97/126 Minnesota Supercomputer Institute, University of Minnesota, Minneapolis, MN (1997)
11. Smith B., Bjørstad P., Gropp W.: Domain decomposition: Parallel multilevel methods for elliptic partial differential equations. Cambridge University Press, New York (1996)

Parallelization of the DAO Atmospheric General Circulation Model

William Sawyer[1,2], Robert Lucchesi[1], Peter Lyster[1,2], Lawrence Takacs[1], Jay Larson[1,2], Andrea Molod[1], Sharon Nebuda[1,2], and Carlos Pabon-Ortiz[1]

[1] NASA Goddard Space Flight Center, Data Assimilation Office
Code 910.3, Greenbelt MD, 20771, USA
{Sawyer, Lucchesi, Lyster, Takacs, Larson, Molod, Nebuda, Pabon-Ortiz}@dao.gsfc.nasa.gov
http://dao.gsfc.nasa.gov

[2] Department of Meteorology, University of Maryland at College Park
College Park MD, 20742-2425, USA
{sawyer, lys, jlarson}@atmos.umd.edu

Abstract. In this paper we present a parallel Atmospheric General Circulation Model currently under development at the NASA Data Assimilation Office. The constituent simulation parameterizations of the AGCM are briefly described, and the parallelization strategy discussed. The key performance obstacles to attaining target benchmarks are treated, as well as our plans for code optimization. We also present the most current benchmarks arising from the this High Performance Computing and Communications Initiative Grand Challenge project.

1 Introduction

The goal of the Data Assimilation Office is to produce accurate gridded datasets of atmospheric fields by assimilating a range of observations along with physically consistent model forecasts. The DAO produces datasets that are used by the climate research community. The data come from conventional sources that are used for weather forecasts (e.g., radiosondes, earth-surface measurements, and satellite temperature retrievals), as well as new sources such as satellites that will be launched in the coming years. The end-to-end Goddard Earth Observing System (GEOS) Data Assimilation System (DAS) currently supports stratospheric flight missions and reanalysis projects for NASA.

The key components of GEOS DAS are the gridpoint-based atmospheric general circulation model (AGCM) [1] and a data analysis system [2]. These are currently being parallelized using MPI [3] in order to meet the requirement of performing thirty days of assimilation in one calendar day. The parallelization of the data analysis system and the resulting performance is discussed extensively in [4]. The parallelization of the AGCM in the framework of GEOS DAS is discussed in this paper.

2 Implementation of the Parallel GEOS AGCM

The AGCM uses a finite-difference algorithm to integrate the primitive equations. The model uses a geographic spherical coordinate grid with horizontal spacing of 2° latitude by 2.5^o longitude, with a 70-level vertical sigma coordinate. Physical processes are modeled using land-surface, moisture, turbulence, and longwave and shortwave radiative transfer schemes [1,5]. The dynamical core [6] of the AGCM computes the time tendencies due to advection of winds, temperatures, surface pressure and an arbitrary number of tracers.

The AGCM used in our work is similar to that described in [7]. However, GEOS DAS — which performs data assimilation as opposed to only simulation — requires numerous extensions to support stratospheric flight missions in a production environment. First, the underlying latitude-longitude grid contains additional vertical levels to encompass the stratosphere; the work mentioned here generally uses 70 levels with either $4^o \times 5^o$, $2^o \times 2.5^o$, or $1^o \times 1^o$ horizontal resolution. Secondly, prognostic and diagnostic data required in such simulations require a parallel concept to remove the potential I/O bottleneck. Thirdly, in the stratospheric model numerical instabilities occur near the poles which do not play a role in the tropospheric model. These instabilities require that the computational grid be transformed or "rotated" such that they disappear.

2.1 Parallelization of Physical and Dynamical Parameterizations

The physical parameterizations within the AGCM involve dependencies only in the vertical coordinate and are communication-free if the gridded data are distributed with a horizontal domain decomposition (checkerboard) [8]. The scalability of the Physics thus depends only on the load balance, cache performance, and the availability of sufficient work for each processing element (PE). Although the dynamical core has horizontal dependencies, these give rise only to nearest-neighbor communication, and thus considerable data locality can be exploited.

Table 1. Cray T3E-600 timing results of AGCM Physics and Dynamics components ($2^o \times 2.5^o \times 70$ resolution) for each call and for a simulation day.

Parameterization	Call frequency (simulation day)	Run times for given number of PEs (seconds)					
		32 PEs		64 PEs		128 PEs	
		per call	per day	per call	per day	per call	per day
Dynamics	480	0.54	259.20	0.32	153.60	0.19	91.20
Moisture	160	0.88	140.80	0.44	70.40	0.22	35.20
Turbulence	48	2.82	135.36	1.42	68.10	0.71	34.08
Shortwave	24	16.87	404.88	8.48	203.52	4.26	102.24
Longwave	8	26.85	214.80	13.40	107.20	6.70	53.60
TOTAL			1155.04		602.82		316.32

Table 1 indicates the time for a call of each portion of the AGCM, as well as the total time for a full day of simulation. For large numbers of PEs, the dynamical core becomes a bottleneck, as it does not scale as well as the Physics. The performance plateau of the dynamical core can be partially overcome by increasing the grid resolution. However, a Courant-Friedrich-Levy (CFL) condition requires that the time step be reduced for finer resolutions, increasing the frequency of the dynamics and making it by far the dominant computational portion. Therefore the dynamical core has been optimized with Cray SHMEM communication to alleviate this potential bottleneck.

2.2 Parallel I/O

The input and output of data to the GEOS Data Assimilation System (GEOS DAS) are the responsibility of the GEOS Parallel I/O Subsystem (GPIOS). Efficient parallel I/O is critical if a distributed-memory version of GEOS DAS is to achieve strategic performance goals. The assimilation of thirty days per day requires daily input of more than 1.5 gigabytes and produces 25 to 35 gigabytes of gridded output data in HDF format. I/O in the GEOS DAS is complicated by the need to translate data from the internal computational grid to a grid appropriate for users. These transformations can be computationally expensive and require significant communication when done in a distributed domain.

Table 2. The AGCM output timing and throughput results on the SGI Origin indicate that the effective I/O rate for constant output size can be nearly maintained even as the number of compute PEs rises substantially.

I/O PEs	3	3	3	3	3
Compute PEs	2	4	8	16	24
Data size (MB)	926	926	926	926	926
Computation Wait Time (s.)	29	29	29	32	35
Effective I/O Rate (MB/s.)	32	32	32	29	26

The simplest and perhaps the most common method of performing I/O from a distributed memory system is gathering and writing from the root PE. The first implementation of parallel output in GEOS-DAS is very similar to this method with two important differences. First, there is more than one PE designated for output. The number of output PEs can be easily increased at run-time to allow flexibility in the definition of output streams and memory management on machines with limited memory nodes. Second and most importantly, the output PEs are in an MPI group distinct from the other compute PEs and will be dedicated to I/O and related functions. They act as a cache for output data, allowing computation in the AGCM to continue once the MPI transfer of data from compute PEs to I/O PEs is complete. One advantage of such an approach is that it allows data transformations in preparation for output to occur locally on I/O PEs without slowing the forward progression of the assimilation system.

We are considering the integration of maturing parallel I/O libraries such as MPI-IO and parallel HDF to enable efficient and scalable raw I/O. When coupled with scalable transformation algorithms it is likely that I/O processing will scale well enough to be done in a truly parallel manner making dedicated I/O nodes unnecessary.

2.3 Pole Rotation for Latitude-longitude Grids

Due to the stratospheric modeling of GEOS DAS computational instabilities from finite difference schemes can arise in the polar regions when a strong cross-polar flow occurs. By rotating the computational grid to the geographic equator, however, the instability near the computational pole is removed due to the vanishing Coriolis term. A discussion of the pole rotation can be found in [5].

In order to "rotate a grid," the coordinates of the original lat-lon grid are mapped to a new lat-lon grid. Interpolation coefficients are determined by the proximity of rotated grid points of one grid to grid points on the other grid. Various interpolation schemes can be employed, e.g., bi-linear or bi-cubic, the latter being employed in GEOS DAS. Once the transformation matrix is defined, sets of grid values, e.g., individual levels or planes of atmospheric data, can be transformed ad infinitum.

If the data of both grids are decomposed in a "checkerboard" manner the problem has little data locality. Although both grids must have this decomposition to ensure good performance in other parts of the application, the data locality, and hence communication overhead, can be improved by permuting the PE ownership of one of the grids. This permutation ensures adequate local work to hide communication latencies for the non-local data. Performance results in Fig. 1 indicate that this problem can be solved efficiently and at the GFlop/s rates needed to solve high-resolution earth science models.

Grid rotation is one of a set of *grid transformations* which are employed in the GEOS DAS. Other transformations include the interpolation between grids of different resolution, or between a "stretched" grid (with non-constant intervals) and a fixed grid. Such grid transformations are supported in a parallel library called PILGRIM [9] currently under development at the DAO.

Acknowledgments

We would like to thank Max Suárez, and Dan Schaffer for their valuable suggestions. The work of Will Sawyer, Peter Lyster and Jay Larson was partially funded by the High Performance Computing and Communications Initiative (HPCC) Earth and Space Science (ESS) program.

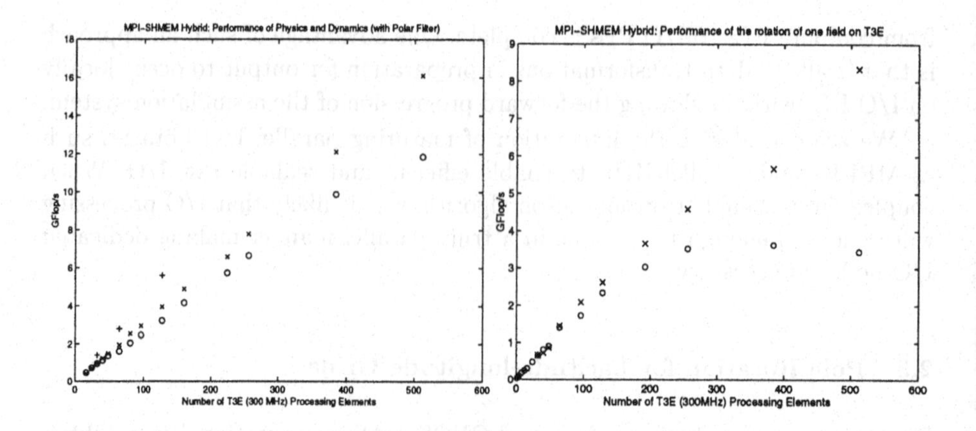

Fig. 1. On the left the performance of the MPI GEOS DAS Physics (+, the ensemble of turbulence, moisture, shortwave and longwave radiation) indicates linear scaling if the resolution (here $2^o \times 2.5^o \times 70$ levels) is sufficiently large to occupy the PEs. Dynamics, which performs nearest neighbor communication using SHMEM, scales only slightly less well for both $2^o \times 2.5^o \times 70$ (o) and $1^o \times 1^o \times 43$ (x) resolutions. On the right the performance of an optimized MPI-SHMEM hybrid implementation of the pole rotation for current resolution (o, $2^o \times 2.5^o$) as well as the future resolution (x, $1^o \times 1^o$) indicates scalability to large numbers of PEs, although the absolute performance is less than for the Physics and Dynamics.

References

1. L. L. Takacs, A. Molod, and T. Wang. Documentation of the Goddard Earth Observing System (GEOS) General Circulation Model — Version 1. Technical Memorandum 104606, NASA Code 910.3, 1994.
2. A. da Silva and J. Guo. Documentation of the Physical-space Statistical Analysis System (PSAS). DAO Office Note 96-02, NASA Code 910.3, 1996.
3. MPI Forum. MPI: A Message-Passing Interface Standard. *International Journal of Supercomputer Applications*, 8(3&4):157–416, 1994.
4. C. H. Q. Ding and R. D. Ferraro. Climate Data Assimilation on a Massively Parallel Supercomputer. In *Proceedings Supercomputing '96*. ACM-IEEE, 1996.
5. DAO staff. Algorithm Theoretical Basis Document. NASA Technical Memorandum, NASA Code 910.3, 1996.
6. M. J. Suárez and L. L. Takacs. Documentation of the ARIES/GEOS Dynamical Core: Version 2. NASA Technical Memorandum 104606, NASA Code 910.3, 1995.
7. D. S. Schaffer and M. J. Suárez. Next Stop: Teraflop; The Parallelization of an Atmospheric General Circulation Model. In A. Tentner, editor, *Proceedings of High Performance Computing 1998*. SCS, 1998.
8. A. Mirin, W. Dannevik, P. Eltgroth, M. Wehner J. Brown, and B. Chan. Performance of a Portable, Parallel Atmospheric General Circulation Model. In R. Sincovec, editor, *Proceedings of the Sixth SIAM Conference on Parallel Processing for Scientific Computing*. SIAM, 1992.
9. W. Sawyer, A. da Silva, P. M. Lyster, and L. L. Takacs. Parallel Grid Manipulations in Earth Science Calculations. In *VECPAR 1998 Proceedings*. Faculdade de Engenharia da Universidade do Porto, 1998.

Dynamic Performance Callstack Sampling: Merging TAU and DAQV

Sameer Shende, Allen D. Malony, and Steven T. Hackstadt

Computational Science Institute, Department of Computer and Information Science, University of Oregon, Eugene OR 97403
{sameer, malony, hacks}@cs.uoregon.edu

Abstract. Observing the performance of an application at runtime requires economy in what performance data is measured and accessed, and flexibility in changing the focus of performance interest. This paper describes the performance callstack as an efficient performance view of a running program which can be retrieved and controlled by external analysis tools. The performance measurement support is provided by the TAU profiling library whereas tool-program interaction support is available through the DAQV framework. How these systems are merged to provide dynamic performance callstack sampling is discussed.

1 Introduction

There are several motivations for wanting to observe the performance of a parallel application during its execution (e.g., to terminate a long running job or to steer performance variables). The downside in doing so is the deleterious effect on performance that may result. This trade-off forces consideration of a means to capture just enough performance information to make possible essential performance "views." Additionally, there is the issue of how external performance analysis tools access the performance data resident in the program's execution state, and what overhead is incurred as a result.

In this paper, we describe the dynamic performance callstack tool that we are developing as part of the TAU profiling package [1] for parallel, multi-threaded C++ programs. We also discuss how the performance callstack information is made available to analysis and visualization tools running in the program's computational environment. This is done using the DAQV-II tool interaction framework [4].

2 Performance Callstacks

A *callstack* at a point in execution shows the current execution location(s) and the sequence of procedure calls that led to it. A *performance callstack* is a view on a program's performance execution profile at runtime. The execution profile shows where time is being spent in a program's code (mainly with respect to routines) for each thread of execution. A *performance callstack profile* at a point in time is defined as the profile of the application with respect to the active profiled blocks on the callstack if the application had terminated at that time.

Selective profiling allows a user to profile a group of profiled blocks. A profiled block is said to be active if its instrumentation is enabled at runtime. If a profiled block is not active, it does not appear on the callstack and no statistics are maintained for it. A profiled block can be a routine (function) or a user-defined statement-level timer that has both inclusive and exclusive profiled quantities associated with it. A profiled quantity could be time or the value of a hardware performance counter such as the number of secondary data cache misses.

The performance callstack view is defined only for those routines in the calling stack of a thread at the point where the callstack is sampled. Looking at a single sample, the performance callstack shows, for each profiled block in the calling stack, its current execution profile statistics. These statistics include the number of invocations or calls, the number of profiled subroutines called by it, the aggregate exclusive and inclusive time spent in the routine, and the instance exclusive and inclusive time spent in the routine since the start of each instance of the profiled block on the callstack.

3 TAU Portable Profiling Library

The TAU portable profiling library [8] is used to build the performance callstack view. The library features the ability to capture performance data for C++ function, method, basic block, and statement execution, as well as template instantiation. It also supports the definition of profiling groups for organizing and controlling instrumentation. The performance callstack contains the TAU profiling data for those functions in the calling stack. From the profiling data collected, TAU's profile analysis procedures can then generate a wealth of performance information for the user. It can show the exclusive and inclusive time spent in each function with nanosecond resolution. For templated entities, it shows the breakup of time spent for each instantiation. Other data includes the number of times each function was called, the number of profiled functions each function invoked, and the mean inclusive time per call. Time information can also be displayed relative to nodes, contexts, and threads [3]. All of this analysis is also available to the user of the performance callstack view.

4 Runtime Access to Performance Callstack View

The performance callstack information provides a snapshot on a program's performance during execution. The ideal solution for accessing this information at runtime would not impact the performance of the application significantly, yet would allow callstack data to be accessed in a distributed environment with a simple interface for tools that obviates concern for where the individual parts of the callstack (node, context, and thread parts) are located in the application's execution environment. The application should not be bothered by the number and location of tools.

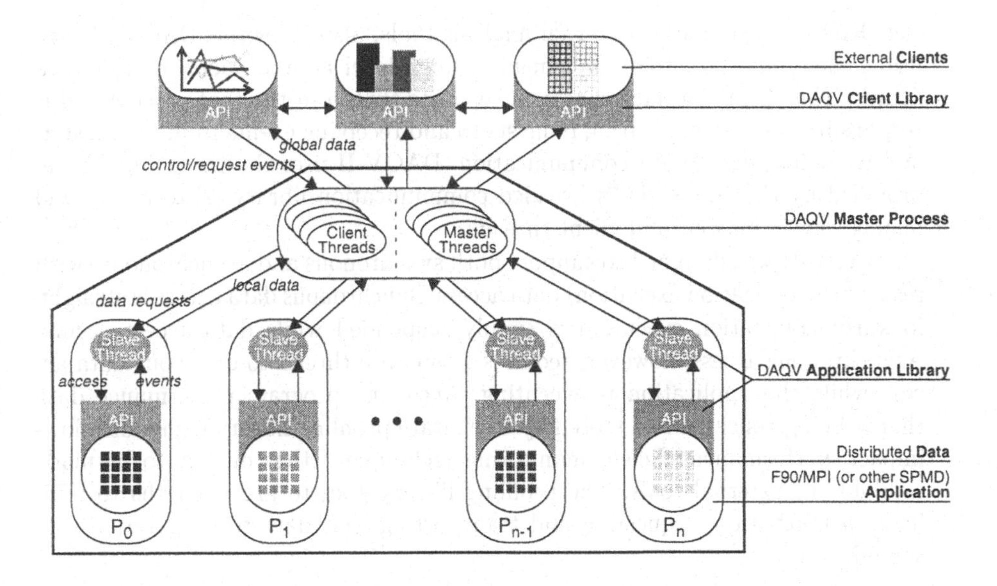

Fig. 1. DAQV-II Framework

The common approaches of writing data to trace files for post-execution analysis and building runtime interfaces into applications to access data over a network both require that complicated functionality be built into analysis tools to deal with multiple files or network connections and to coordinate callstack data access. In addition to the communications functionality, there are issues concerning application-tool synchronization, callstack data consistency, and performance perturbation.

Thus, the "glue" between the tool and application interfaces must implement the mapping of a high-level, callstack data view to its individual parts and runtime location while servicing callstack access requests from multiple, easily-developed analysis tools. We decided to use the DAQV framework to implement the interfaces and glue for runtime access to performance callstack views.

5 The DAQV Program Interaction Framework

The DAQV-II [4, 6] framework for program interoperability provides external tools with a view of distributed data as a logical global array that can be accessed selectively via a high-level array reference; see Figure 1.

Each application process is linked with a DAQV library that provides a simple API for describing how data is distributed and to indicate places in the code where it may be accessed. The execution of each application process is shadowed by a *slave* thread, which maintains information about available data and performs accesses to that data when requested. The individual slave threads are coordinated by a separate process called the DAQV *master*, which acts as the "single point of contact" for the set of application processes and analysis tools in the DAQV environment. External analysis tools, also known as DAQV *clients*, use a client interface that implements a high-level access model to registered arrays. The client interface runs as a separate thread in the tool process and is responsible for communicating requests to and receiving events from the master, and participating in data communication. DAQV-II uses the threading system provided by the Nexus multithreaded communication library [2] to create and manage slave, master, and client threads.

DAQV-II was designed to support both synchronous and asynchronous (with respect to application execution) data access. Synchronous data access is straightforward; application execution is simply suspended while data access occurs. Asynchronous access, however, requires a separate thread to carry out data access while the application is executing. Accessing program performance data that is being collected and stored by a separate profiling library demands a low-impact, performance-efficient monitoring technique if that data is to be made available to external tools. The profiling library goes to great lengths to minimize perturbation; requesting and transporting that data to an external tool should, too.

6 Integration of TAU and DAQV

Integrating online monitoring capabilities with a profiling library requires several considerations. The details of managing multiple network connections to each of the processes in a parallel application greatly complicates client development. Furthermore, excessive synchronization, like that which can occur in traditional client/server architectures can deteriotate application performance.

In our first implementation of the DAQV system [5], our primary objective was to simplify tool interaction by allowing clients to view a multi-process, parallel application as a single entity and make logical requests for data from the global array structure defined by the collection of individual program arrays. Our most recent work, DAQV-II, extends this by addressing the limitations of synchronous data access by using the *probe* and *mutate* metaphors. The asynchronous data access supported by DAQV-II allows program execution to continue while a separate thread reads or writes application data.

The model supported by DAQV-II is particularly suitable for callstack monitoring. First, it allows asynchronous access and minimizes the synchronization overhead experienced by the application. Second, it supports a simple abstraction for interacting with parallel applications and eases the tool development process. Third, DAQV allows multiple client tools to access the "global" performance callstack simultaneously. We use the DAQV-II framework to access the performance callstack and deliver the profile data to external analysis tools.

Figure 2 shows how we have merged the DAQV-II framework with the TAU performance callstack measurements. The performance callstacks for the parallel threads are distributed across the processing nodes in the parallel execution. DAQV-II allows this distributed callstack data to be described as a single global callstack array that can be requested by clients. The callstack data is collected

Fig. 2. TAU-DAQV integration

in each thread when the TAU_MONITOR() routine is executed. This callstack data snapshot is then registered with the DAQV-II system, which informs DAQV of the location and size of the data so that it may fill subsequent requests for it. The parallel program does not need to perform a barrier operation during callstack access, and it can continue to execute with minimal intrusion. The synchronization operations are off-loaded to the DAQV slave threads, which are responsible for communication with the master process, further reducing the intrusion in the parallel program.

Fig. 3. Callstack on node 0 of a POOMA 2D Diffusion equation simulation

Figure 3 shows the callstack view of a two dimensional diffusion equation which tracks the progression of the diffusion of a heat source on a mesh with respect to time. It was implemented using the POOMA [7] object oriented scientific computing framework.

7 Conclusions

The merging of the TAU and DAQV-II systems described above has been implemented for performance callstack sampling. Callstack analysis and visualization tools have also been constructed. The TAU portable profiling library captures performance data in parallel C++ programs for functions, methods, basic blocks, statements, and template instatiations. Access to this information at runtime, in the form of the TAU performance callstack, can yield a wide range of useful performance information to the user. DAQV-II provides an interoperability and data exchange infrastructure appropriate for program and performance monitoring. Asynchronous data access minimizes overhead and perturbation, and DAQV's model simplifies the development of client tools. The system we have built by merging the TAU portable profiling library with the DAQV-II interaction framework facilitates efficient and convenient access to performance callstack information by external tools.

Acknowledgments

This work was supported in part by the DOE ASCI (#C70660017-3) and DOE 2000 (#DEFC0398ER259986) programs. We would like to thank the Los Alamos Natl. Laboratory, and Steve Karmesin and Pete Beckman in particular, for their support. Ariya Lohavanichbutr, Chad Busche, and Michael Kaufman at the Univ. of Oregon, contributed to the implementation of the system.

References

1. Advanced Computing Laboratory (LANL): TAU Portable Profiling URL:http://www.acl.lanl.gov/tau. (1998)
2. Foster, I., Kesselman, C., Tuecke, S.: The Nexus Approach to Integrating Multithreading and Communication, Jour. of Parallel and Distributed Computing. Vol. 37 (1). Aug (1996) pp. 70–82.
3. Gannon, D., Beckman, P., Johnson, E., Green, T., Levine, M.: HPC++ and the HPC++LIB Toolkit, Technical Report Department of Computer Science, Indiana University (1998).
4. Hackstadt, S., Harrop, C., Malony, A.: A Framework for Interacting with Distributed Programs and Data, In: Proc. of the Seventh Int'l Symp. on High Performance Distributed Computing 1998 (HPDC-7). IEEE, July (1998).
5. Hackstadt, S., Malony, A.: DAQV: Distributed Array Query and Visualization Framework, Journal of Theoretical Computer Science, special issue on Parallel Computing Vol. 196, No. 1-2, April (1998) pp. 289–317.
6. Malony, A. D., Hackstadt, S.: Performance of a System for Interacting with Parallel Applications, Intl. Jour. of Parallel and Distributed Systems and Networks. (1998)
7. Reynders, J. et. al.: Pooma: A Framework for Scientific Simulation on Parallel Architectures, In: Wilson, G., Lu, P. (Eds.): Parallel Programming using C++, M.I.T. Press (1996) pp. 553–594.
8. Shende, S., Malony, A. D., Cuny, J., Lindlan, K., Beckman, P., Karmesin, S.: Portable Profiling and Tracing for Parallel, Scientific Applications using C++, Proc. of ACM SIGMETRICS Symp. on Parallel and Distributed Tools. Aug (1998) pp. 134–145.

A Parallel Rational Krylov Algorithm for Eigenvalue Computations

Daniel Skoogh

Department of Mathematics, Chalmers University of Technology and the University of Göteborg, S-41296 Göteborg, Sweden
skoogh@math.chalmers.se

Abstract. An implementation of a parallel rational Krylov method for the generalised matrix eigenvalue problem is discussed. The implementation has been done on a MIMD computer and on a cluster of workstations. The Rational Krylov algorithm is an extension of the shifted and inverted Arnoldi method where several shifts are used to compute basis vectors for one subspace. In this parallel implementation, the different shifted matrices are factorised each on one processor and then the iteration vectors are generated in parallel.

1 Introduction and Purpose

Arnoldi [2] and Lanczos [2], the currently most successful basic iterative eigenvalue algorithms, are both based on Krylov subspaces. They start with an arbitrary starting vector and build up a basis one vector at a time, letting the matrix operate on the most recent basis vector and including this new direction into the subspace. Any vector in the Krylov subspace can be expressed as a polynomial of the matrix applied to the starting vector.

The rational Krylov method [4–7] is a generalisation of the shift and invert Arnoldi method. In the shift and invert Arnoldi method we choose one shift μ in the complex plane where we want the eigenvalues to converge fast, while in the rational Krylov method we choose several shifts $\mu_1, \ldots, \mu_{\bar{\imath}}$.

Consider the generalised eigenproblem

$$Au = \lambda Bu. \tag{1}$$

The space $\mathcal{S}_{\bar{k}+1}$ in the rational Krylov method starting at the vector v_1 is

$$\begin{aligned} \mathcal{S}_{\bar{k}} \equiv span\{&v_1, (A-\mu_1 B)^{-1}Bv_1, \ldots, ((A-\mu_1 B)^{-1}B)^{\bar{\jmath}_1}v_1, \\ &(A-\mu_2 B)^{-1}Bv_1, \ldots, ((A-\mu_2 B)^{-1}B)^{\bar{\jmath}_2}v_1, \\ &\ldots \\ &(A-\mu_{\bar{\imath}} B)^{-1}Bv_1, \ldots, ((A-\mu_{\bar{\imath}} B)^{-1}B)^{\bar{\jmath}_{\bar{\imath}}}v_1\}. \end{aligned} \tag{2}$$

where $\bar{k} = \bar{\jmath}_1 + \bar{\jmath}_2 + \ldots + \bar{\jmath}_{\bar{\imath}}$ and $\mathrm{Dim}(\mathcal{S}_{\bar{k}+1}) \le \bar{k}+1$. We use over bar $\bar{\imath}$, $\bar{\jmath}$ and $\bar{k}$ to denote the last (largest) element in a sequence of integers i, j, k. In the rational Krylov method every vector $x \in S_{\bar{k}}$ can be written as

$$x = r(B^{-1}A)v_1,$$

where r is a rational function.

In theory it is possible to generate the basis vectors in parallel by operating with shifted and inverted matrices with different shifts one on each processor. The purpose of this work is to investigate if this can be done in practice [8].

2 Parallel Rational Krylov Algorithm

In the parallel algorithm, the basis vectors are generated in parallel. The matrix operations themselves are not parallelised.

The key to the parallel algorithm is that it does not matter in exact arithmetic in which order the operators $(A - \mu_i B)^{-1}B$, $i = 1, \dots, \bar{\imath}$ are applied in building a basis for the subspace $S_{\bar{k}+1}$ (2).

There are several ways to implement a parallel rational Krylov algorithm. In our approach we use $\bar{p}$ different processors to compute $r_p = (A - \mu_p B)^{-1}Br_p$, $p = 1, \dots, \bar{p}$, and then we let each processor orthogonalise its own vector (Same Program Multiple Data).

The algorithm uses a total number of $\bar{\imath}$ shifts, $\mu_1, \dots, \mu_{\bar{\imath}}$. Each shift μ_i, $1 \le i \le \bar{\imath}$, is used $\bar{\jmath}$ times in the matrix operator $(A - \mu_i B)^{-1}B$. At a given moment at a processor p, $1 \le p \le \bar{p}$, the letter k stands for the total number of basis vectors. The vector v_{k+1} is the basis vector being calculated in the current iteration, and v_{k_p} is the vector that the matrix $(A - \mu_i B)^{-1}B$ is applied to. When the algorithm has run to completion, the total number of basis vectors is $\bar{k} + 1$, where $\bar{k} = \bar{\imath}\bar{\jmath} = \bar{m}\bar{p}\bar{\jmath}$ and each processor takes $\bar{m}$ shifts.

Parallel Rational Krylov Algorithm

1 Choose v_1 such that $\| v_1 \| = 1$ (The same vector on all processors)
2 **start program on** $\bar{p}$ processors, $p = 1 : \bar{p}$
3 **for** $m = 1 : \bar{m}$
4 $\quad i = (m-1)\bar{p} + p$
5 $\quad LU = A - \mu_i B$, (factorise)
6 $\quad$ **for** $j = 1 : \bar{\jmath}$
7 $\quad\quad k = (m-1)\bar{\jmath}\bar{p} + (j-1)\bar{p} + p$
8 $\quad\quad k_p = \max(1, k - \bar{p} + 1)$
9 $\quad\quad r_{k+1} = U^{-1}L^{-1}Bv_{k_p}$, (Operate)
10 $\quad\quad H(1 : k_p, k) = V_{k_p}^H r_{k+1}$, (Gram Schmidt, old vectors)
11 $\quad\quad r_{k+1} = r_{k+1} - V_{k_p}H(1 : k_p, k)$
12 $\quad\quad$ **receive** v_l **and** h_{l-1} **from their processors**, $l = k_p + 1 : k$
13 $\quad\quad H(k_p + 1 : k, k) = V_{k_p+1:k}^H r_{k+1}$, (Gram Schmidt, new vectors)
14 $\quad\quad r_{k+1} = r_{k+1} - V_{k_p+1:k}H(k_p + 1 : k, k)$
15 $\quad\quad h_{k+1,k} = \| r_{k+1} \|$

16 $\quad\quad v_{k+1} = r_{k+1}/h_{k+1,k}$, (deliver new vector)
17 $\quad\quad$ (v_{k+1} will be the start vector v_{k_p} in the next iteration)
18 $\quad\quad$ **send** v_{k+1} and h_k to the other processors
19 $\quad$ **end**
20 **end**
21 **receive** v_l and h_{l-1} from their processors, $l = k+2 : k + (\bar{p} - p) + 1$
(from processor $p+1, \ldots, \bar{p}$)

In our experimental implementation each processor (or node) keeps a copy of all basis vectors that are generated. After the processor p has carried out the operation on line 9, $r_{k+1} = U^{-1}L^{-1}Bv_{k_p}$, the vector r_{k+1} is first orthogonalised against the basis vectors located at the local processor. The last basis vectors are not yet available because the other processors are working on them. After the first part of the orthogonalisation is finished, the processor is now ready to receive the last basis vectors and orthogonalise r_{k+1} against them and to send the resulting vector v_{k+1} to the other processors. For a description of how the approximate eigenvalue problem is calculated, see [5–7].

3 Implementation Details

3.1 Programs

The first parallel program is a straightforward implementation of parallel rational Krylov algorithm described in section 2. It uses a blocking `receive` and a non-blocking `send`, see [8].

For the case when only a blocking `send` and `receive` is available we have made a special version, see [8].

We have also made a Master Slave implementation, see [8].

3.2 Implementation Details

For the shifted and inverted operations we have used direct band solvers from LAPACK [1]. General sparse solvers could also be used, but then the analysing phase needs to be done in parallel on all processors, and later be distributed to the different processors.

Whenever appropriate we have used LAPACK (Linear Algebra PACKage) and BLAS (Basic Linear Algebra Subroutines) routines, see [1].

To ensure that the basis is orthonormal to working precision every vector r_k is orthogonalised twice against the basis.

The programs are implemented in Fortran and uses PVM (Parallel Virtual Machine) [3] for communications between the different processors. We have also implemented a semi parallel MATLAB code for convergence tests.

4 Tests

4.1 A Test of Speedup

The programs have been tested on IBM-SP2, and on a cluster of SUN ELC workstations. All tests are taken from [8].

The test matrices consists of a finite difference discretisation of a convection diffusion problem. On the IBM-SP2 the dimension of the test matrix is 10 000 and the bandwidth is 100. On the SUN ELC configuration the dimension of the test matrix is 1000 and the bandwidth is 100. The larger test problem would not fit on the SUN ELC configuration. The bandwidth is chosen in such a way that the same time ratio for orthogonalisation and solving the equation is obtained for both test sites.

In figure 2 and 1 we show speedup for the SUN ELC and IBM-SP2 respectively for the parallel programs. The total number of basis vectors generated was 150 and the total number of shifts was 6. The programs run on 6, 3, 2 and 1 processors. We did not manage to get the non-blocking variant to work on the IBM-SP2, probably due to a bug in the PVM implementation (PVMe). For tests on a Master Slave implementation, see [8].

4.2 A Test of Linear Independence

It has appeared during our tests, that the matrices $\boldsymbol{H}$ and $\boldsymbol{K}$ get nearly rank deficient, when the number of columns k increases.

Let us demonstrate how this comes about by choosing $\boldsymbol{A} = \text{diag}(1:500)$, $\boldsymbol{B} = \boldsymbol{I}$, a total of 60 iterations, and two shifts $\mu_1 = 100.5$ and $\mu_2 = 110.5$.

In figure 3 we show the normalised singular values for the matrices $\boldsymbol{H}$ and $\boldsymbol{K}$ as the number of iterations k grows for the parallel test. The pluses $+$ are the singular values of the matrix $\boldsymbol{H}$ and the circles $\circ$ are the singular values of the matrix $\boldsymbol{K}$. For the corresponding sequential test, the singular values of the $\boldsymbol{H}$ and $\boldsymbol{K}$ matrices are at normal values.

The test shows that the matrices $\boldsymbol{H}$ and $\boldsymbol{K}$ get a null space of dimension one in the parallel test for $\bar{p} = 2$ processors. A singular value decomposition of the matrices shows that they get a common null space in this case.

If we project away the null space when the approximate eigenvalues are calculated, we get the same rate of convergence for the parallel as for the sequential case, see [8]. However the null space could be a source of inaccuracy, and in general fewer eigenvalues will converge when one uses the same sequence of shifts as in the corresponding sequential algorithm.

If we use $\bar{p} > 2$ processors, we have observed that we get a null space of dimension $\bar{p} - 1$. The null space occurs gradually.

References

1. E. Anderson, Z. Bai, C. Bischof, J. Demmel, J. Dongarra, J. Du Croz, A. Greenbaum, S. Hammarling, A. McKenney, S. Ostrouchov, and D. Sorensen. *LAPACK Users' Guide, Release 2.0.* SIAM, Philadelphia, 1995. 324 pages.

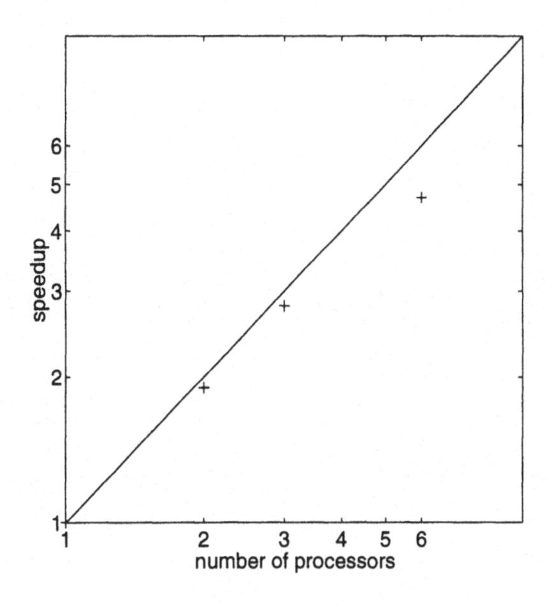

Fig. 1. Speedup graph for the IBM-SP2 configuration. The pluses + are the speedup for the parallel program with a blocking **send**

2. W.E. Arnoldi. The principle of minimized iteration in the solution of the matrix eigenvalue problem. *Quart. Appl. Math.*, 9:17–29, 1951.
3. A. Geist, A. Beguelin, J. Dongarra, W. Jiang, R. Manchek, and V. Sunderam. *PVM: Parallel Virtual Machine A Users' Guide and Tutorial for Networked Parallel Computing.* MIT Press, 1994.
4. Axel Ruhe. Rational Krylov sequence methods for eigenvalue computation. *Lin. Alg. Appl.*, 58:391–405, 1984.
5. Axel Ruhe. Rational Krylov algorithms for nonsymmetric eigenvalue problems, II: Matrix pairs. *Lin. Alg. Appl.*, 197/198:283–296, 1994.
6. Axel Ruhe. The Rational Krylov algorithm for nonsymmetric eigenvalue problems. III: Complex shifts for real matrices. *BIT*, 34:165–176, 1994.
7. Axel Ruhe. Rational Krylov, a practical algorithm for large sparse nonsymmetric matrix pencils. *SIAM J. Sci. Comp.*, 19:1535–1551, 1998.
8. Daniel Skoogh. An implementation of a parallel rational Krylov algorithm. Licentiate Thesis, Chalmers University of Technology, Göteborg, Sweden, 1996.

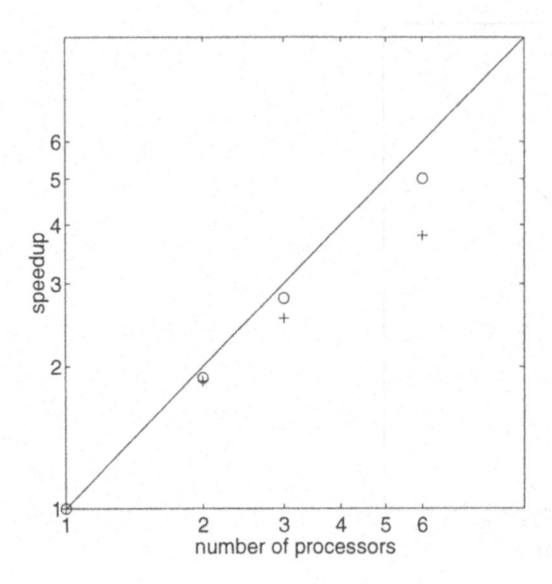

Fig. 2. Speedup graph for the SUN configuration. The pluses + are the speedup for the parallel program with a blocking **send** and the circles o are the speedup for the parallel program with a non-blocking **send**

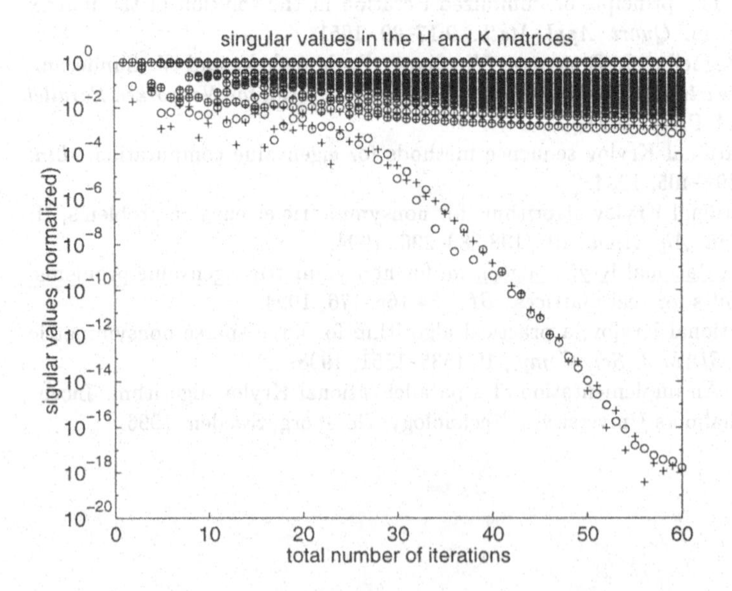

Fig. 3. Normalised singular values for the matrices $\boldsymbol{H}$ and $\boldsymbol{K}$ for the parallel test with 2 shifts and 2 processors. The pluses + are the singular values of the matrix $\boldsymbol{H}$ and the circles o are the singular values of the matrix $\boldsymbol{K}$

Portable Implementation of Real-Time Signal Processing Benchmarks on HPC Platforms*

Jinwoo Suh and Viktor K. Prasanna

Department of EE-Systems, EEB-200C
University of Southern California
Los Angeles, CA 90089-2562
prasanna@usc.edu
http://ceng.usc.edu/~prasanna.html

Abstract. For the evaluation of HPC systems for real-time signal processing, real-time benchmarks have recently been proposed by the US DoD signal processing and HPC communities. For the implementation of real-time benchmarks, we have developed efficient communication algorithms for M-to-N K-block-cyclic communication. Using our algorithms, we have implemented real-time 2D-FFT and Corner Turn benchmarks that have been defined by MITRE and Rome Lab on SP2 and T3E. Our results show that the number of processors required to perform the 2D-FFT benchmark was reduced by up to 53% compared with earlier implementation. The total corner turn time was also reduced by up to 78%. The performance of IBM SP2 and SGI/Cray T3E are compared using these benchmarks.

1 Introduction

Many embedded applications require high performance computing to achieve real-time performance. These include Space-Time Adaptive Processing (STAP), Synthetic Aperture Radar(SAR), Sonar systems, Automatic Target Recognition/Tracking systems, and Vision applications, among others [7,8,11,12]. Implementation of these applications on High Performance Computing (HPC) platforms is being considered by many researchers[2,3,11].

To evaluate HPC systems for real-time applications, several benchmarks have been proposed. These include Hartstone[5], Rhealstone[6], TPC[10], C^3I[1], and MITRE benchmarks[4].

In these benchmark suites, one of the essential operations is communication between processors. Because many processors are employed, efficient communication algorithms are needed to obtain high performance and scalable implementation.

* Effort sponsored by the Dod High Performance Computing Modernization Office, and Rome Laboratory, Air Force Materiel Command, USAF, under agreement number F30602-97-2-0016. The U.S. Government is authorized to reproduce and distribute reprints for Governmental purposes notwithstanding any copyright annotation thereon.

One of the most important communication primitives in real-time signal processing is M-to-N K-block-cyclic communication. In the M-to-N K-block-cyclic communication, M source processors send data to N destination processors using the K-block-cyclic distribution pattern (See Section 3) [7].

This paper presents a set of communication algorithms for M-to-N K-block-cyclic communication. The number of communication steps in the proposed communication algorithms is reduced to as small as $\lg(M+N)$.

Our implementation on the IBM SP2 and the SGI/Cray T3E performed using C and the Message Passing Interface (MPI) standard[9] is portable. For the sake of comparison, a serial communication algorithm[3] was also implemented. Our results show: 1) reduction in the total communication time, 2) reduction in the minimum number of processors needed to process a specific size data, and 3) increase in the maximum input data size for 2D-FFT that can be processed by a given number of processors.

2 MITRE Benchmarks

The MITRE benchmarks[4] provide a methodology to assess the performance of HPC platforms for real-time embedded applications. These benchmarks provide both timing and functional specifications. The minimum size of a machine that can process the given problem is determined. In this section, the 2D Real-Time FFT and the Corner Turn benchmarks are briefly described.

The input data for the **2D real-time FFT benchmark** are complex vectors of size $n \times n$, where n is a positive integer. The real-time FFT consists of two steps: Row FFT and Column FFT.

The timing specification consists of the following two cases.

Case 1: Period = latency = 1 second.

Case 2: Period = 1 second, no restriction on latency.

The period is the time interval between successive input matrices to the machine. The latency is the elapsed time between the data input to the machine and its corresponding output. For Case 1 and 2, the minimum size of the machine in terms of the number of processing nodes is to be determined.

The **corner turn operation** is the transpose of a matrix distributed on more than one processor. This consists of three steps: local data rearrangement, data redistribution, and unpacking of received messages. In local data rearrangement, the data stored in row major order is rearranged by columns. In the second step (data redistribution), the data is transferred among the processors for data remapping. In the third step, in each destination processor, the received data is rearranged by columns. To analyze the scalability, the number of processors (n) is increased, while the size of the input matrix is kept fixed. The values of n to be considered are 256, 512, 1K, ..., 16K.

P_0	P_1	P_2	P_3
0	0	0	1
1	1	2	2
2	3	3	3

Fig. 1. An example of 3 × 4 *dpt*

3 Proposed Communication Algorithm

Block-cyclic data redistribution occurs in many High Performance Computing (HPC) applications such as radar, sonar, and automatic target recognition. ScaLAPACK, a mathematical software for dense linear algebra computations, also uses block-cyclic distribution for good load balance and computing efficiency.

In this section, an efficient algorithm for M-to-N K-block-cyclic communication is shown. We assume that N is a multiple of M. In our approach, node contention is eliminated by reorganizing the communication steps. Start up cost is reduced by reorganizing the communication pattern. The M-to-N K-block-cyclic communication problem is defined using the following definition of destination processor table.

Definition: An $m \times n$ *destination processor table* (*dpt*) is defined as a table where the jth column $(0 \leq j \leq n-1)$ consists of m distinct indices of the destination processors of the data blocks that processor P_j has to send.

An example of *dpt* is shown in Figure 1. In this example, column 0 indicates that the P_0 needs to send its first block of data to P_0, the second block to P_1, and the third block to P_2.

In M-to-N K-block-cyclic communication, an array data is first divided into KM blocks each of size x. The blocks are distributed to the M source processors in round robin fashion as shown in Figure 3 (a). Then, K consecutive blocks are collected to the same destination processor as shown in Figure 3 (b). For example, the first three blocks (0, 1, and 2) are sent to P_8, and the next three consecutive blocks (3 ,4 , and 5) are sent to P_9. For this example, the *dpt* is shown in Figure 3 (c), and the final *dpt* is shown in Figure 3 (d).

The initial *dpt* can be transformed to the final *dpt* using row and column transformations. A column transformation in a *dpt* corresponds to the reorganization of elements in a processor's memory, and a row transformation corresponds to interprocessor communication.

Our solution uses the notion of *generalized circulant matrix form.*

Definition: An $m \times n$ $(m \leq n)$ matrix is a *circulant matrix*, if row K = (row 0 which is right (left) circular-shifted K times), $0 \leq K \leq m-1$.

Definition: Given a $p \times q$ matrix, suppose the matrix can be partitioned into blocks of size $s \times t$, where $p = m \cdot s$ and $q = n \cdot t$ for some $s (\geq 1)$ and $t (\geq 1)$. Then, the matrix is said to be a *generalized circulant matrix*, if row block k = (row block 0 which is right (left) circular-shifted k) $(0 \leq k \leq m-1)$ times and each block is either a circulant matrix or a generalized circulant matrix.

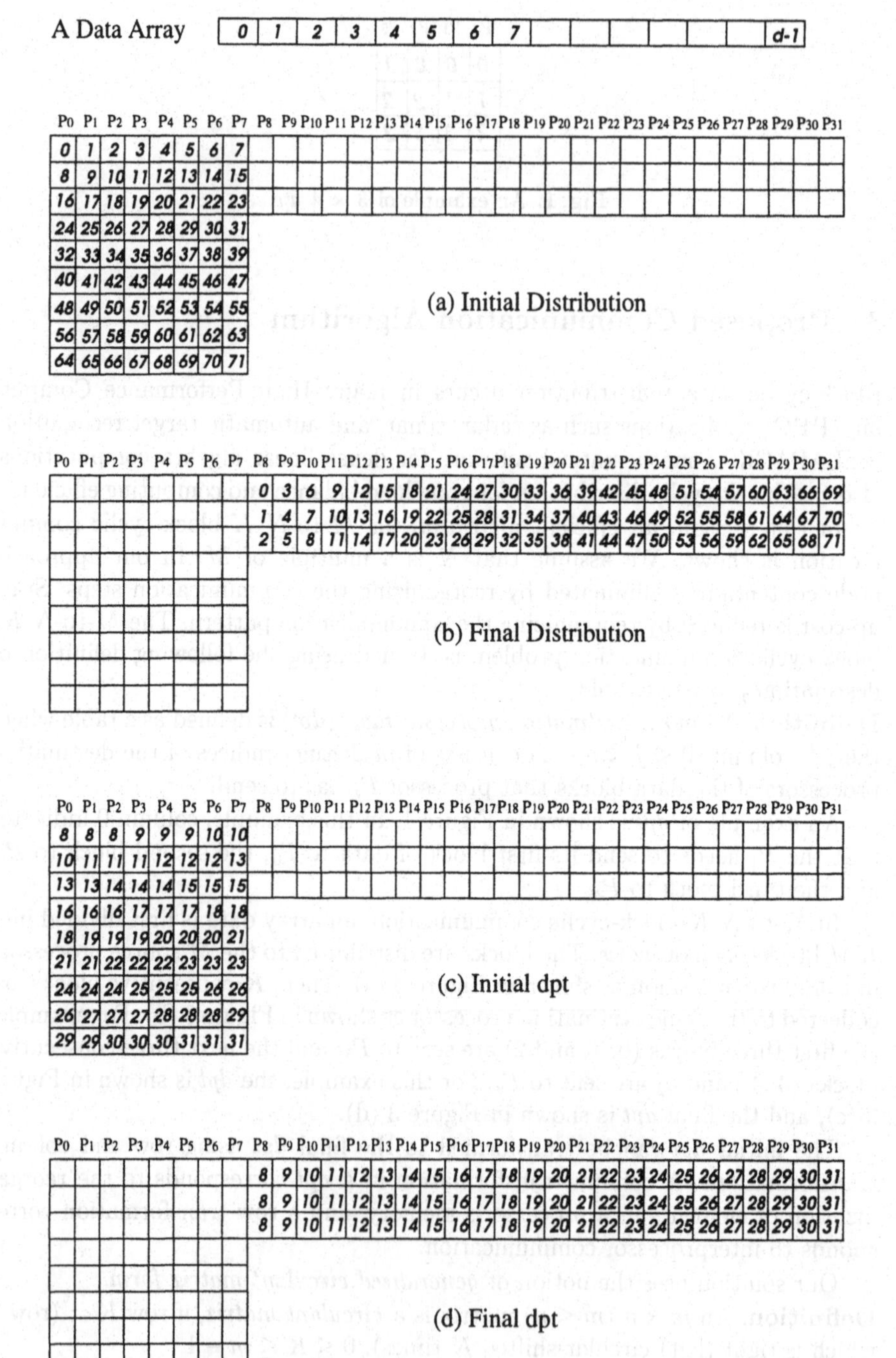

Fig. 2. M-to-N K-block-cyclic communication (M=8, N=24, and K=3, each block in the diagram denotes an x byte data block.)

An example of $K \times N$ *generalized circulant matrix* is shown in Figure 3.

$$\mathbf{C} = \begin{bmatrix} C_0 & C_1 & \cdots & C_{N'-1} \\ C_1 & C_2 & \cdots & C_0 \\ C_2 & C_3 & \cdots & C_1 \\ \vdots & \vdots & & \vdots \\ C_{K'-1} & C_{K'} & \cdots & C_{K'-2} \end{bmatrix} \qquad C_j = \begin{bmatrix} c_{j,0} & c_{j,1} & \cdots & c_{j,G-1} \\ c_{j,1} & c_{j,2} & \cdots & c_{j,0} \\ \vdots & \vdots & & \vdots \\ c_{j,G-1} & c_{j,0} & \cdots & c_{j,G-2} \end{bmatrix}$$

Fig. 3. Examples of circulant and generalized circulant matrix.

Also, our solution uses the concepts of the direct, indirect, and hybrid schedules. In direct schedule, the data is sent from the source processor to the final destination processor directly. In indirect schedule, the data is sent to the final destination processor through one or more intermediate processors. Hybrid schedule uses some steps of indirect schedule and one step of direct schedule.

Our approach divides the M-to-N K-block-cyclic communication into two subproblems. By dividing the problem and rearranging the data deliberately, we can convert the M-to-N K-block-cyclic communication into a set of data scattering problems and a set of K-block-cyclic communication problems among the same set of processors as follows. Due to space limitations, the proofs have been omitted.

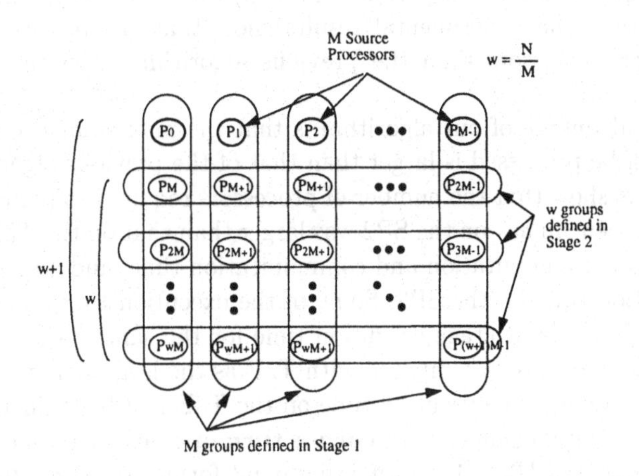

Fig. 4. Partitioning of the processors in the first and the second stages.

Stage 1 : The destination processors are partitioned into M groups as in Figure 3. The initial *dpt* is transformed to a generalized circulant matrix. This stage can be performed in $\lceil \lg(N/M + 1) \rceil$ steps.

Stage 2 : In this stage, a new processor partition is defined. The processors included in the $i^{th} (0 \le i \le N/M - 1)$ group are $P_j (iM \le j \le (i+1)M - 1)$ (See Figure 3). In each group, redistribution from $cyclic(x, M)$ to $cyclic(Kx, M)$ is performed. The communication can be performed in (i) K communication steps, (ii) $\lceil \lg K' \rceil + \lceil \lg G \rceil + 1$ communication steps using an indirect schedule, where $G = \gcd(K, G)$ and $K' = K/G$, or (iii) $d + \lceil \frac{K}{2^d} \rceil$ communication steps using a hybrid schedule with d degree of indirection.

Thus, the M-to-N K-block-cyclic communication can be performed in (i) $\lceil \lg(N/M+1) \rceil + K$ communication steps, (ii) $\lceil \lg(N/M+1) \rceil + \lceil \lg K' \rceil + \lceil \lg G \rceil + 1$ communication steps using an indirect schedule, or (iii) $\lceil \lg(N/M+1) \rceil + d + \lceil \frac{K}{2^d} \rceil$ communication steps using a hybrid schedule with d degree of indirection.

4 Experimental Results

The Real-time 2D-FFT and the corner turn benchmarks were implemented on the IBM SP2 at the Maui High Performance Computing Center (MHPCC) and on the SGI/Cray T3E at the San Diego Supercomputer Center(SDSC). For the sake of comparison, the 2D-FFT and the corner turn were also implemented using the previous M-to-N communication algorithm[3]. The results of the **real-time 2D-FFT** implementation are shown in Figure 5 and Figure 6.

The minimum numbers of processors required to process the 2D-FFT for various data sizes are shown in Figure 5. The experiments were performed for two cases: i) Case 1: latency = period = 1 sec, ii) Case 2: latency = no constraint, period = 1 sec. The experimental results show that the proposed algorithm requires fewer processors than the previous algorithm when the data size is large.

Another advantage of our algorithm is that the maximum size of the input data that can be processed is larger than that of the previous algorithm.

The results show that the number of processors required to perform the real-time 2D-FFT benchmark on the SP2 was larger than that on the T3E. The main reasons are larger computation and communication time, and larger fluctuation of the execution time on the SP2. Some of the execution times on the SP2 are much longer than the average execution time for the same parameter settings. The SP2 results are worse compared with T3E as the largest time is used in the evaluation. However, the execution time on the T3E had little fluctuation.

The experimental results for the **corner turn operation** are shown in Figure 7. The data size was 1K $\times$ 1K, in floating point format (4 bytes/data).

The communication time of the previous serial algorithm increases as the number of processors increases. However, the communication time of the proposed algorithm monotonically decreases as the number of the processors increases. As the number of source processors increases, there is more available

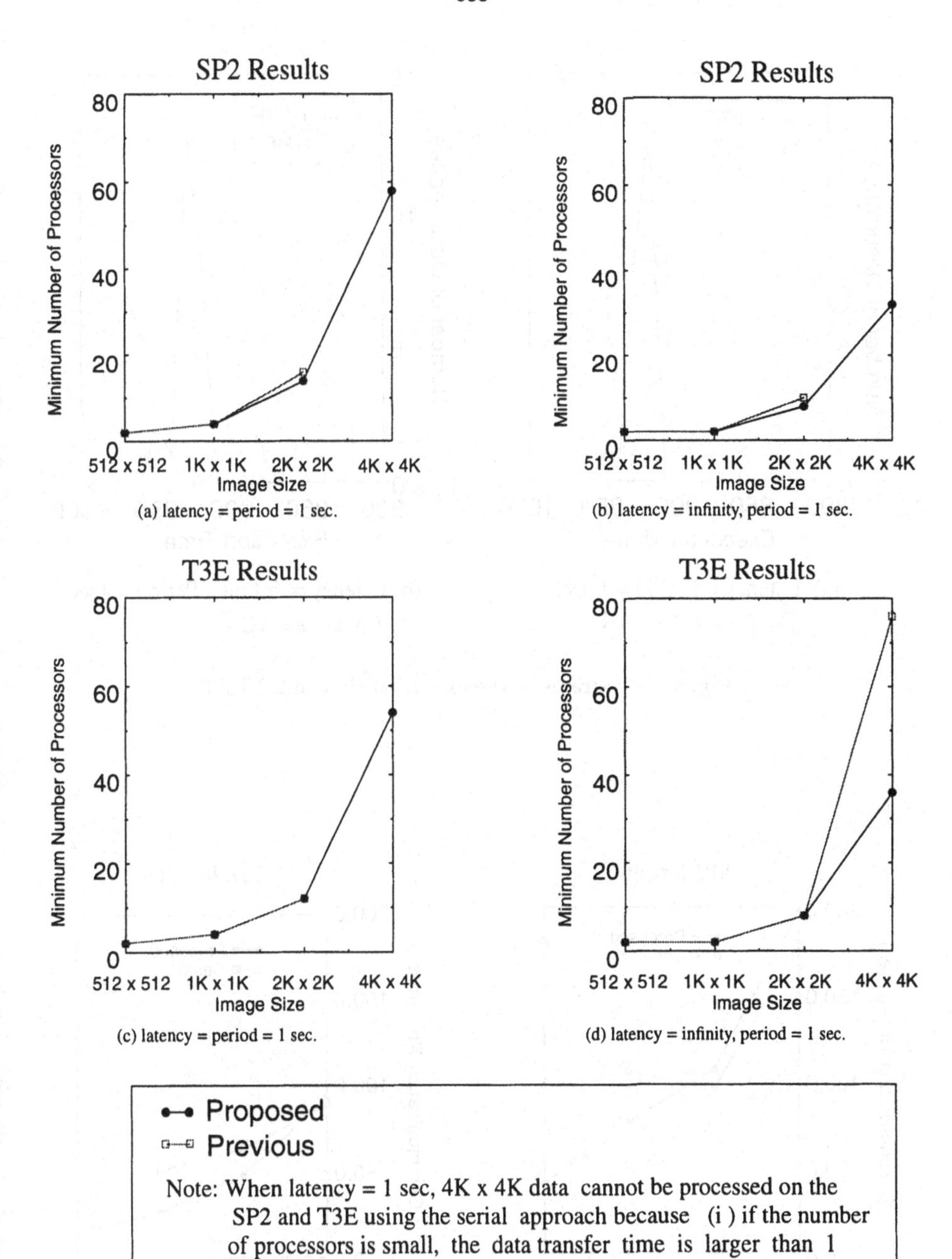

Fig. 5. 2D-FFT benchmark results

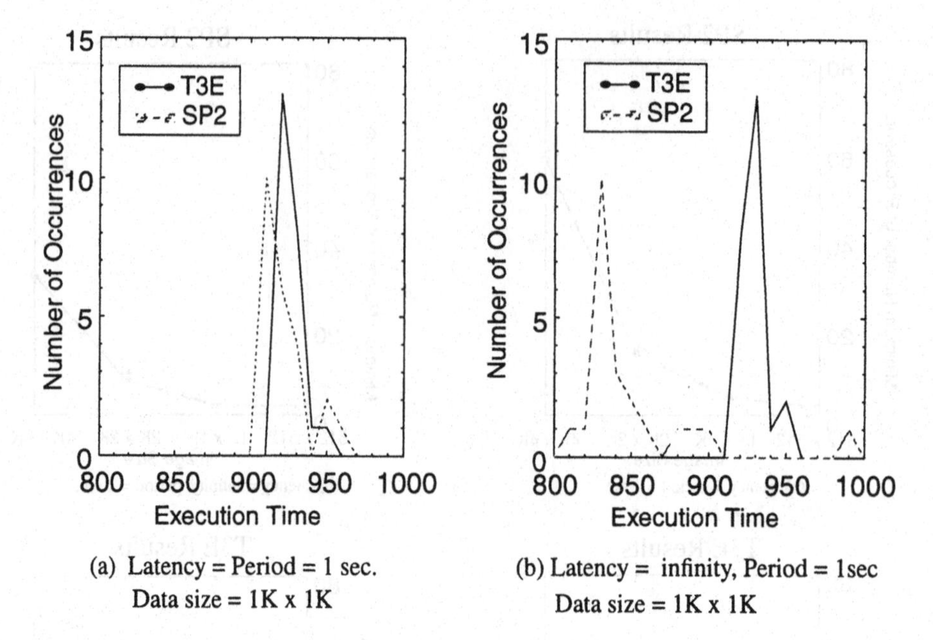

(a) Latency = Period = 1 sec.
Data size = 1K x 1K

(b) Latency = infinity, Period = 1sec
Data size = 1K x 1K

Fig. 6. Histograms of the execution time of 2D-FFT

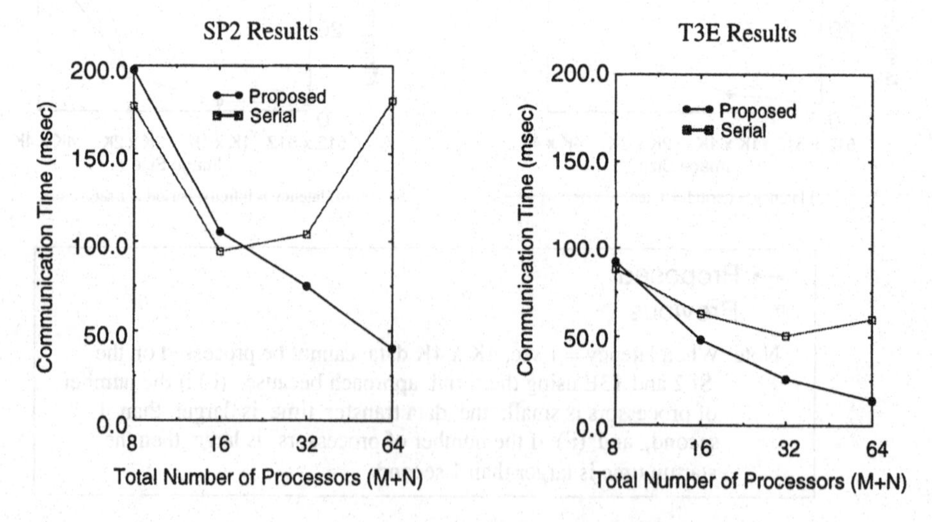

Fig. 7. Corner turn benchmark implementation results

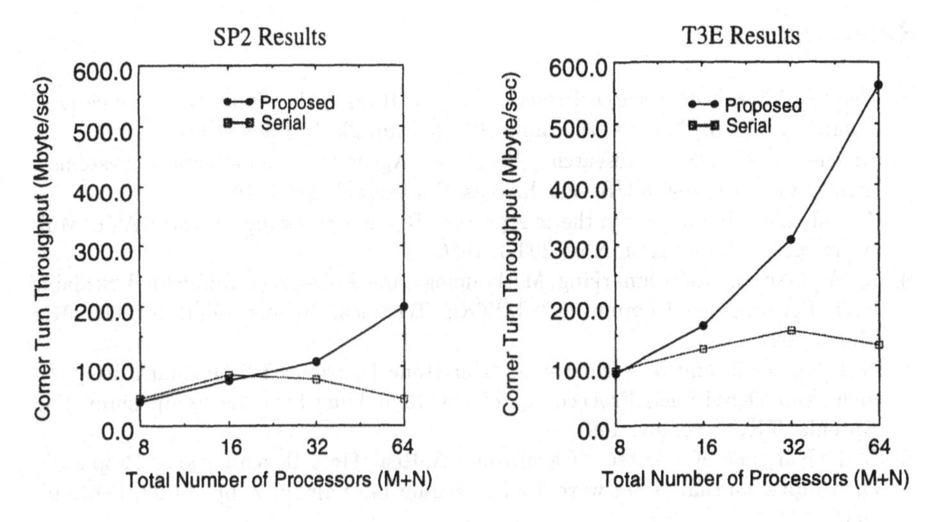

Fig. 8. Corner turn throughput performance (Data size = 1K × 1K)

bandwidth which is exploited by the proposed algorithm. Figure 8 shows the corner turn throughput which is the sustained bandwidth for the corner turn operation. The proposed algorithm shows monotonic increase in bandwidth; however, the previous algorithm shows decrease as the number of processors is increased beyond 16 (32) on the SP2 (T3E). This is due to the fact that our algorithm exploits the available bandwidth as the number of processors increases.

5 Conclusion

In this paper, we implemented the MITRE benchmarks on the SP2 and the T3E. For efficient implementation of the real-time benchmarks, we proposed portable M-to-N communication primitive.

By using the proposed algorithms, the number of processors required to process the real-time 2D-FFT was reduced significantly. Also, the maximum data size that can be processed was increased.

Using the implemented benchmarks, we compared the performance of SP2 and T3E. In general, T3E showed higher performance in performing real-time benchmarks (using the previous as well as the proposed algorithms).

The proposed M-to-N communication primitive can be used for parallelizing multi-stage applications where each stage has different computational requirement. Typical examples of such applications are Space-Time Adaptive Processing (STAP), Synthetic Aperture Radar(SAR), Sonar systems, Automatic Target Recognition/Tracking systems, and Vision applications.

References

1. Defense Advanced Research Projects Agency/Rome Lab, "Real-Time Benchmark Summary," http://www.se.rl.af.mil:8001/benchmarks/ c3ipbs_rt.html, 1997.
2. Defense Advanced Research Projects Agency, "Embeddable Systems," http://www.ito.arpa.mil/ResearchAreas/Embeddable.html, 1996.
3. T. Einstein, "Realtime Synthetic Aperture Radar Processing on the RACE Multicomputer," Application Note 203.0, 1996.
4. R. A. Games, "Benchmarking Methodology for Real-Time Embedded Scalable High Performance Computing," MITRE Technical Report MTR 96B0000010, March 1996.
5. N. I. Kamenoff and N. Weiderman, "Hartstone Distributed Benchmark: Requirement and Definitions, Proceedings of the Real-Time Systems Symposium, San Antonio, TX, November 1991.
6. R. P. Kar, and K. Porter, "Rhealstone, A Real-Time Benchmarking Proposal", Dr. Dobb's Journal of Software Tools, Volume 14, Number 2, pp. 14-22, February 1989.
7. Y. W. Lim, P. B. Bhat, and V. K. Prasanna, "Efficient Algorithms for Block-Cyclic Redistribution of Arrays," Eighth IEEE Symposium on Parallel and Distributed Processing, New Orleans, Louisiana, October 1996.
8. Myungho Lee, Wenheng Liu, and Viktor K. Prasanna, "A Mapping Methodology for Designing Software Task Pipelines for Embedded Signal Processing," 3rd International Workshop on Embedded HPC Systems and Applications (EHPC'98) at 12th International Parallel Processing Symposium (IPPS'98), and 9th Symposium on Parallel and Distributed Processing (SPDP'98), Orlando, Florida, April 1998.
9. Message Passing Interface Forum, MPI: A Message-Passing Interface Standard, Technical Report CS-94-230, University of Tennessee, Knoxville, May 1994.
10. TPC, "TPC Benchmark Descriptions," http://www.tpc.org/ home.page.html, 1996.
11. C.-L. Wang, "High Performance Computing for Vision on Distributed Memory Machines," Ph.D. Thesis, University of Southern California, August 1995.
12. C. R. Yerkes, and E. Webster, "Implementation of ω-k synthetic aperture radar imaging algorithm on a massively parallel supercomputer," SPIE Proceedings Vol. 2230, pp. 171-178, Orlando, FL, April 1994.

Large Scale Active Networks Simulation*

Karthik Swaminathan, Radharamanan Radhakrishnan,
Philip A. Wilsey, and Perry Alexander

Department of ECECS, PO Box 210030,
University of Cincinnati, Cincinnati, Ohio 45221–0030, U.S.A.
{skarthik, ramanan, pawi}@ececs.uc.edu

Abstract. With the addition of user-controllable computing capabilities to conventional data networks, the network can no longer be perceived as a passive conveyor of bits. It transforms into a generic computational engine or an "active network". While such capabilities open up many exciting possibilities for network designers, simulation of such large scale high-performance networks is exacerbated by a number of issues. Issues such as scalability and expressibility, will be addressed by the simulator designer as simulation requires the designer to "implement" models of the active networks systems. This paper proposes the use of a parallel simulation (using relaxed synchronization strategies, such as Time Warp) environment to accurately simulate the behavior and profile the performance of large scale, active networks.

1 Objective

Active networks modify present day network architectures by allowing the network to change from a passive carrier of bits to a more generic computational engine. Information injected into the system may be modified, stored or redirected while being transported. With the current trend of decreasing computing power costs, it becomes increasingly worthwhile to embed computational power into the components of the network to process the information flowing through the network. Consequently, in an active network, the routers or nodes of the network perform customized computations on the datagrams flowing through them. These datagrams are referred to as *smart packets* since the packets often carry control information for the active routers. A more detailed description of the internal structure and organization of active networks is available in the literature [1]. Examining the current active nets architectures developed by researchers at MIT (ANTS) [11], Pennsylvania (SwitchWare) [5], and Columbia [12], reveals a common, emergent structure for active nodes based on *services* and *policies*. In this model, services represent methods and resources available to an executing packet while policies govern and protect these services. Exploiting this common node structure permits the development of simulation models for network nodes and network architectures in an implementation independent manner.

* Support for this work was provided in part by the Advanced Research Projects Agency under contracts J–FBI–93–116 and DABT63–96–C–0055.

Simulation models of policies and architectures can be developed and evaluated independent of the implementation. Although formal analysis of models of policies and architectures is being carried out to validate active network architecture, well known problems of scalability and expressibility still exist. In large networks (millions of nodes), it will prove impossible to provide formal proofs of correctness using traditional theorem proving approaches. Simulation based techniques will be needed to test the feasibility and correctness of these large scale network designs. In addition, simulation of such network models will improve our knowledge of active networks and can potentially motivate the design and development of applications that exploit the processing power of active networks.

This paper proposes the use of an active networks simulation environment (ANSE) for the design and analysis of large scale active network systems. Simulation models for active nodes, smart packets and network topology have been developed such that large scale simulations involving all three basic components of an active network can be performed. To achieve throughput necessary for a reasonable study of scalability and performance, the WARPED Time Warp parallel simulation engine [7] is being specialized to simulate active networking systems. Libraries of active network models have been developed to allow designers to simulate large (1,000,000+ node) systems. The simulation environment allows the user to specify different internetwork topologies and simulate them using various active node models. The remainder of the paper is organized as follows. Section 2 introduces the Time Warp mechanism and gives a brief description of the WARPED simulation kernel. A description of the constituents of an active network is presented in Section 3. Section 4 presents some preliminary scalability and performance results. Finally, Section 5 contains some concluding remarks.

2 Time Warp Simulation

Time Warp is an optimistic synchronization strategy for parallel discrete-event simulation. To ensure simulation correctness in parallel simulation environments, events must be executed in causal order across all simulation processes. Process synchronization assures this through either *conservative* [8] or *optimistic* [3] methods. The Time Warp [6] synchronization protocol is an instance of optimistic synchronization. In Time Warp, global adherence to causality is not enforced and may occasionally be violated. Violations of causality are repaired by a rollback recovery mechanism.

In the WARPED [7] simulation kernel, simulation objects are grouped into entities called *logical processes*, or LPs. Processor parallelism occurs at the LP level and each LP is responsible for global time management, communication management, and scheduling for the simulation objects that it contains. In addition, communication between simulation objects within the same LP is performed by direct insertion into the input queue of the receiving object. Off-processor communication is achieved through MPI [4] calls. A more detailed description of the internal structure and organization of the WARPED kernel is available on the www at `http://www.ececs.uc.edu/ paw/warped`.

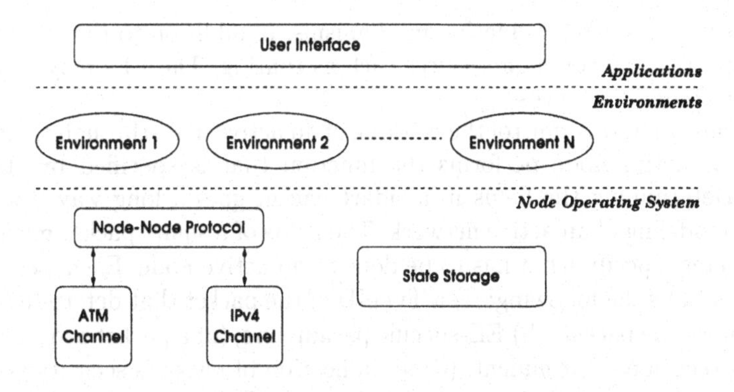

Fig. 1. Components of an Active Node

3 Active Networks Simulation

Large scale active network simulation is carried out with the help of the Active Networks Simulation Environment (ANSE). The ANSE consists of a simulation workbench that allows the user to model the components of an active network and simulate models that contains a large number of active nodes and a large population of smart packets. In order to provide the user with this facility, we need to provide realistic and scalable models of active nodes and smart packets. We model the active node and the smart packet based on the current thinking in this area [10]. The ANSE also provides the means to generate the network topology using Georgia Tech's Internetwork Topology Models (GT-ITM) [2]. The ANSE also provides the user with an analysis framework that allows the user to analyze the results of the simulation. This section deals exclusively with the modeling of the active node and the smart packet. A more detailed description of the ANSE is available in the literature [9]. The rest of this section discusses the active nodes and the smart packets respectively.

The active node is responsible for the processing of smart packets. The two major components of the active node are the *node operating system* and the *execution environment* (EE). The general relationship between these components is shown in Figure 1. An EE provides an interface through which the user accesses the resources and capabilities of an active node. These capabilities are provided for by the node operating system. Our model allows for multiple EEs to be supported at the node. The system is also designed in such a way that not every available EE needs to be supported at every node. As soon as a packet arrives at an active node its passed to the node OS. The node OS first validates the packet after looking into the various security parameters provided in the packet. Once the packet passes the validation phase, the node OS determines which EE gets the packet. It also allocates resources as required by the packet before passing it on to the EE. Once the processing is complete, the node OS is responsible for transmitting the packet. Thus the node OS provides for the following: (a) a node-node protocol or packet transmission strategy; (b) demultiplexing of packets to determine which EE has been requested; (c) resource allocation strategies; and

(d) security and packet validation mechanisms. In addition to this, the node OS also supports some common services such as routing. The EEs may or may not use these common services.

The smart packets control the working of an active node through their various fields. The active node performs the function that is specified by the smart packet. Determining the fields in a smart packet goes a long way towards the realistic modeling of an active network. The fields of a smart packet must be able to completely specify what has to be done at an active node. Each smart packet ideally contains the following: (i) a TypeID of the packet that determines the EE that handles the packet; (b) EE-specific parameters; (c) a packet program for the specific execution environment; (d) an indication of packet's security status; (e) resource requirements for processing the packet; (f) an indication of the channel on which the packet was received; (g) the packet's source and destination ids; (h) an indication if the packet is to be forwarded or not in case the requested EE was not available at the node; and (i) an indicator of the energy available in the packet to avoid congestion of the network by packets that never reach their destination. Modeling these characteristics has allowed us to construct realistic models of active networks and the following section presents our initial simulation results.

4 Preliminary Simulation Results

Table 1. Performance values for 50 simulation time units

Number of Nodes	Number of Packets	Execution Time (seconds)			
		LP0	LP1	LP2	LP3
1000	4000	1.21	1.21	1.24	1.25
5000	20,000	9.45	9.42	9.42	9.66
10,000	40,000	32.81	32.77	33.29	33.34
50,000	200,000	1,218.50	1,219.40	1,218.50	1,218.30
100,000	400,000	7,870.50	8,027.60	8,312.40	8,342.80
150,000	600,000	24,252.70	24,663.50	25,753.30	25,246.70

The preliminary performance studies have been conducted using a network of SUN workstations and the data provided in this paper has been collected by running the simulations on a 4 processor Sun SparcCenter 1000 (running Solaris 2.6). The simulations were tested with fixed configurations of 4 LPs (one for each of the processor in the multiprocessor). Table 1 illustrates some preliminary performance results. For all the test cases, an inter-arrival time of 10 simulation time units was used. The network to be simulated is dynamically generated, thereby cutting down on the compile time. At runtime, the *setup time* includes the time to construct the connectivity of the network and the time taken to instantiate the network nodes. This setup time is currently in the order of minutes for the large simulations. It is seen that as the number of nodes scale, the execution time of the simulation also scales accordingly. For the larger cases (100,000 and 150,000 nodes), the execution time is disproportionately large. This is due to the higher memory consumption of the simulation. Currently, research is ongoing to reduce this memory footprint. The preliminary results are commensurate with our prediction that large scale active network simulation is feasible.

5 Conclusions

Active networks are currently being designed to address the mismatch between the rate at which user requirements change and the pace at which network infrastructure can be deployed. This is accomplished by increasing the degree and sophistication of network based computation. As the sophistication of these networks increase, it becomes increasingly important to test the correctness of these complex large scale systems. Simulation of these large networks will need to be performed to validate the designs.

This paper demonstrates the use of an active networks simulation environment to successfully simulate large scale active networks. Current research indicates that network simulation can be achieved in an optimistic environment with little or no violations of causality. Communications in networks is not synchronous, allowing parallel simulation processes to run largely without rollback. Further, ongoing research is developing means for: (i) optimizing communication; (ii) optimizing rollback; and (iii) avoiding rollback when causality is violated. By avoiding rollback, simulation throughput is dramatically enhanced.

References

1. Jonathan M. Smith David L. Tennenhouse, W. David Sincoskie, David J. Wetherall, and Gary J. Minden. A survey of active network research. *IEEE Communications Magazine*, 35(1):80–86, January 1997.
2. K. Calvert E. Zegura and S. Bhattacharjee. How to model an internetwork. In *Proceedings of IEEE INFOCOM*, April 1996.
3. R. Fujimoto. Parallel discrete event simulation. *Communications of the ACM*, 33(10):30–53, October 1990.
4. W. Gropp, E. Lusk, and A. Skjellum. *Using MPI: Portable Parallel Programming with the Message-Passing Interface.* MIT Press, Cambridge, MA, 1994.
5. C. Gunter, S. Nettles, and J. Smith. The switchware active network architecture. Available at http://www.cis.upenn.edu/~switchware/, Nov 1997.
6. D. Jefferson. Virtual time. *ACM Transactions on Programming Languages and Systems*, 7(3):405–425, July 1985.
7. D. E. Martin, T. J. McBrayer, and P. A. Wilsey. WARPED: A Time Warp simulation kernel for analysis and application development. In H. El-Rewini and B. D. Shriver, editors, *29th Hawaii International Conference on System Sciences (HICSS-29)*, volume Volume I, pages 383–386, January 1996.
8. J. Misra. Distributed discrete-event simulation. *Computing Surveys*, 18(1):39–65, March 1986.

9. D. M. Rao, K. Swaminathan, R. Radhakrishnan, P. A. Wilsey, and P. Alexander. ANSE: An Active Networks Simulation Environment. In *Proceedings of the Austrian-Hungarian Workshop on Distributed and Parallel Systems (DAPSYS'98)*, September 1998. (forthcoming).
10. Active Networks research group. Architectural framework for active networks. *Active Nets'98 Workshop*, March 1998.
11. D. Wetherall, J. Guttag, and D. Tennenhouse. Ants: A toolkit for building and dynamically deploying network protocols. In *Proceedings of IEEE OPENARCH*, April 1998.
12. Y. Yemini and S. da Silva. Towards programmable networks. In *Proceedings of the FIP/IEEE International Workshop on Distributed Systems*, Oct 1996.

Forward Dependence Folding as a Method of Communication Optimization in SPMD Programs

Zdzislaw Szczerbinski

Polish Academy of Sciences, Institute for Theoretical and Applied Computer Science,
Baltycka 5, 44–100 Gliwice, Poland
zdzich@iitis.gliwice.pl
http://www.iitis.gliwice.pl/~zdzich.zor

Abstract. In the paper, a method is proposed for optimizing communication in SPMD programs executed in distributed-memory environments. The programs in question result from parallelizing single loops whose dependence graphs are acyclic. Upon introduction to the basics of data dependence theory, the idea of forward dependence folding is presented. Next, it is shown how dependence folding may be coupled with message aggregation as a method of reducing the number of time-costly interprocessor message transfers. Theoretical considerations are accompanied by experimental results from applying the method to programs executed in a network of workstations.

1 Introduction

In this paper we describe a method of communication optimization in SPMD programs executed in a distributed-memory environment. The programs result from parallelizing single (not nested) loops whose dependence graphs are acyclic.

As a natural consequence of the flow of data in a program, a (*data*) *dependence* exists between two statements S' (the *source*) and S'' (the *sink*, executed after S') if they both access the same variable and at least one access results in assigning a new value to the variable. This paper is concerned with *true dependences* only i.e. dependences in which the value is assigned in the source only. A dependence inside a loop is *loop-independent* if the accesses of S' and S'' to the variable are in the same iteration; otherwise it is *loop-carried.* The *distance* of a dependence in a single (not nested) loop is the number of iterations between its source and sink. A loop-carried dependence is *forward* if the source lexically precedes the sink, and *backward* if the sink precedes the source or they are the same statement ($S' = S''$)[1].

A parallel loop is represented as a *dependence graph* whose vertices are the loop body's statements and edges symbolize dependences. This paper is concerned with loops whose dependence graphs are *acyclic.* If the graph is acyclic

[1] The loop is assumed to be a result of structured programming i.e. it contains no backward branches and the lexical order of statements in its body corresponds to the order of execution in a single iteration.

then, by sorting it topologically, the loop may be restructured into a form in which all dependences are forward.

In a distributed-memory environment, a loop is parallelized so that individual iterations or groups of successive iterations are assigned to separate processors. This model of parallel program execution is called *Single Program Multiple Data* (*SPMD*). In the SPMD model, all processors execute an identical code, parametrized only by the processor identifier. In the paper, *cyclic* scheduling of iterations and decomposition of arrays [1] are assumed as well as the *owner computes* rule [9].

If a loop-carried dependence occurs in a parallelized loop, the execution of the loop's iterations must be synchronized so that the source of the dependence is executed before its sink: the *dependence* is *synchronized.* In a distributed-memory environment the synchronization is ensured through interprocessor communication in the form of a *message transfer*. Since transmitting a message between processors adds a substantial overhead, as little communication as possible is desired in a SPMD program.

The paper is concerned with the issue of reducing the number of message transfers responsible for synchronization of loop-carried dependences. The objective was to find a way of optimally aggregating messages so that as many individual data items as possible could be transmitted together rather than as individual messages.

2 Forward dependence folding

In a shared-memory system, where all synchronization is through access to common locations in memory, synchronizing a dependence D_k may imply that, for another dependence D_j, its source is executed before its sink even though no explicit synchronization has been provided to ensure this. The synchronization of dependence D_j may then be eliminated.

Two forward dependences may be *disjoint, adjacent, overlapping* or *nested* (one in the other) (Fig. 1). According to the synchronization elimination rule for two forward dependences in a shared-memory program [5], the synchronization of forward dependence D_k eliminates the synchronization of another forward dependence D_j of the same distance iff D_k is nested in D_j. If dependences D_j and D_k overlap (Fig. 1c) then, by ensuring that the source of D_k is executed before the sink of D_j, an imaginary dependence D_l is synchronized (Fig. 2a). Since D_l is nested in both D_j and D_k, its synchronization eliminates the synchronizations of D_j and D_k. Instead of two, there is now only one synchronization required. The above process is called *dependence folding* [5]. In Fig. 2b a larger group of overlapping dependences of the same distance are folded. In general, the lexical order of dependences' sinks need not correspond to the order of their sources, as is the case in Fig. 2b. Some dependences may be nested in others rather than overlapping with them. In any case, the source of the condensation is the (lexically) last source, and its sink — the first sink in the group of overlapping/nested dependences.

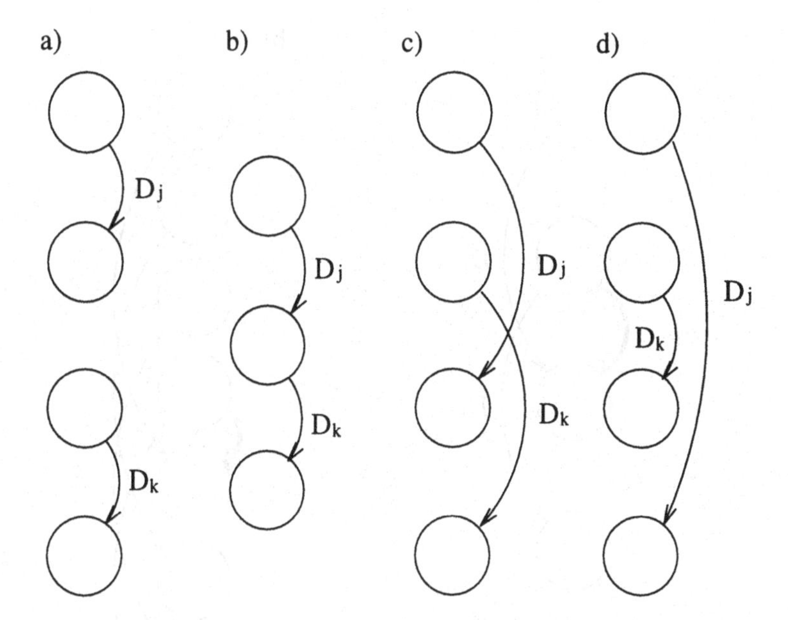

Fig. 1. Two forward dependences: a) disjoint, b) adjacent, c) overlapping, d) one nested in the other

Dependence folding as described above pertains to a given, defined sequence of statements in the loop. By applying *statement reordering* [10] additional elimination of synchronizations may be achieved. A method for obtaining the optimal order of statements in a loop for the folding of forward dependences of the same distance has been described in [5, 6]. The developed algorithm ELMAX sorts the loop's acyclic dependence graph topologically in a specific manner. It has been proved in [5] that in the loop restructured with ELMAX the maximum possible number of dependence synchronizations may be eliminated.

Virtually all elements of data dependence theory are common to programs executed on both shared- and distributed-memory systems. However, any optimization concept for dependence synchronization in a shared-memory program is valid for a distributed-memory program only if it is ensured that, for every dependence, the value assigned in the dependence's source is actually *transmitted* to the processor executing the dependence's sink.

3 Optimizing communication/synchronization in SPMD programs by dependence folding and message aggregation

Optimal dependence folding may be coupled with message aggregation to form an effective communication optimization method. The idea of *message aggregation* [7] is to group into one message a number of data items sent to the same

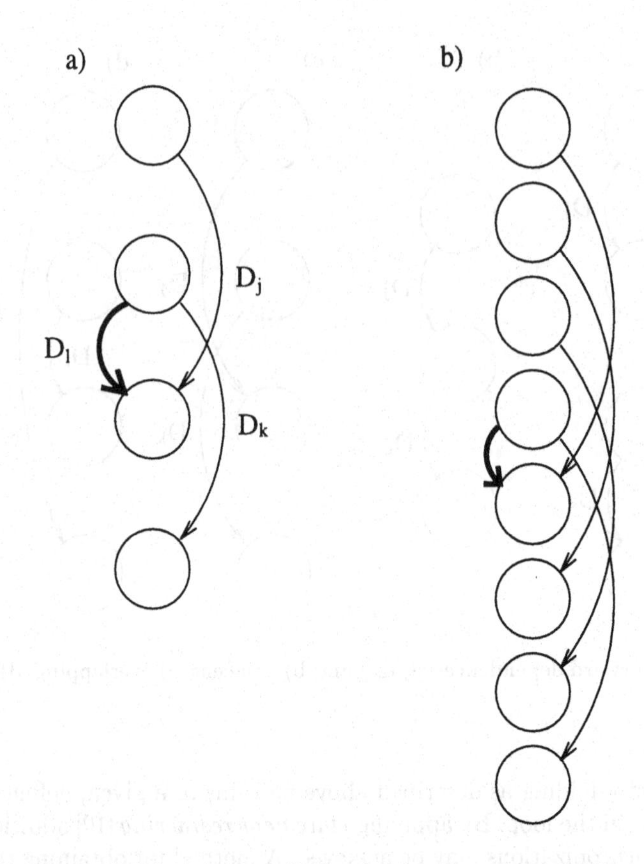

Fig. 2. Synchronization elimination by dependence folding

processor even if the items are unrelated. The rationale is that transmitting a message between processors is far more costly than extra buffering. If we treat D_j and D_k in Fig. 2a as loop-carried dependences in a distributed-memory program, synchronizing the dependences separately involves two message transfers: one after executing the source of D_j (value v_j sent), the other after executing the source of D_k (value v_k sent). However, D_j and D_k may be folded into D_l. It now suffices to send v_j and v_k as one message upon execution of the source of D_l. Both D_j and D_k are now synchronized with one message transfer. Moreover, the values assigned in the dependences' sources are physically transmitted to the processor executing their sinks so that the values may be used in the sinks. Dependence folding in conjunction with message aggregation is thus valid as a method of synchronization elimination in distributed-memory programs.

If there are a number of dependences of the same distance in a loop and the loop's dependence graph is acyclic then the algorithm ELMAX may be employed. As a result, the maximum number of messages are aggregated in the same way as the maximum number of dependence synchronizations are eliminated in the corresponding shared-memory program.

As an illustration, a trivial example of dependence folding and message aggregation is shown below. The original sequential Fortran loop has the form (all arrays are of the type 4-byte integer):

```
     do i = 2,n
S1:    y(i) = y(i) + i
S2:    w(i) = w(i) + i
S3:    x(i) = y(i-1) + 2
S4:    z(i) = w(i-1) + 5
     enddo
```

In the loop there are two overlapping forward dependences of distance 1: $S_1 \rightarrow S_3$ and $S_2 \rightarrow S_4$. The SPMD code for the parallelized loop is as follows.

```
     do i = lb,ub
S1:    y(i) = y(i) + i
       if (i.lt.n) call send(label1,y(i),4,mod((i+1),nproc))
S2:    w(i) = w(i) + i
       if (i.lt.n) call send(label2,w(i),4,mod((i+1),nproc))
       if (i.gt.2) call receive(label1,temp1,4)
S3:    x(i) = temp1 + 2
       if (i.gt.2) call receive(label2,temp2,4)
S4:    z(i) = temp2 + 5
     enddo
```

lb and *ub* are the loop's local bounds (parametrized by the processor identifier), 4 is the number of transmitted bytes and *nproc* is the number of processors. After dependence folding and message aggregation the number of message transfers is reduced:

```
     do i = lb,ub
S1:    y(i) = y(i) + i
       buffer(1) = y(i)
S2:    w(i) = w(i) + i
       buffer(2) = w(i)
       if (i.lt.n) call send(label,buffer(1),8,mod((i+1),nproc))
       if (i.gt.2) call receive(label,buffer(1),8)
S3:    x(i) = buffer(1) + 2
S4:    z(i) = buffer(2) + 5
     enddo
```

In a more sophisticated example, the algorithm ELMAX ensures optimal message aggregation. The loop's acyclic dependence graph is given in Fig. 3a. The solid and dashed line edges represent loop-carried and loop-independent dependences, respectively. All loop-carried dependences are assumed to be of the same distance. Figure 3b shows the dependence graph sorted topologically with ELMAX (for picture clarity, loop-independent dependences are not visualized). The bold arrows symbolize aggregated message transfers. Figure 3c presents an example sequence of statements resulting from topologically sorting the dependence graph in a manner different from the one adopted in ELMAX. Again, dependence folding is possible. However, the number of message transfers is now four compared to two if ELMAX is performed (and seven if dependences are not folded).

4 Experimental results

The developed optimization method has been implemented in practice. Several SPMD programs, both with and without message aggregation, were run on five Sun Ultra workstations working in a local area network. The distributed-memory model of parallel computation was provided by `mpich` [3], a public-domain implementation of the MPI message-passing standard [4]. The analyzed programs were written in High Performance Fortran (HPF) and transformed into SPMD executables by ADAPTOR, a public-domain HPF compilation system [2]. The SPMD program form was then edited — buffers for aggregated messages were introduced and individual calls to communication procedures replaced by adequate calls with the buffers as parameters. Also, in a number of cases, the order of statements in the loop was changed as prompted by ELMAX. The restructured program was next compiled and linked together with ADAPTOR's runtime system, realized by using MPI routines. In all test runs, significant decrease of computation time was observed. Table 1 shows timing results for the simple loop with two dependences in section 3.

Table 1. Execution times, in seconds, of an example SPMD program

n	100	1000	10000	100000
without message aggregation	0.05	0.58	5.7	60.0
with message aggregation	0.03	0.26	3.0	28.5

5 Conclusion

A method has been proposed for optimizing communication in SPMD programs resulting from parallelizing single loops whose dependence graphs are acyclic. Theoretical considerations have been verified by applying the developed method to example SPMD programs executed in a real-world distributed-memory environment. The timing results confirm the method's effectiveness and applicability to practical parallel programming.

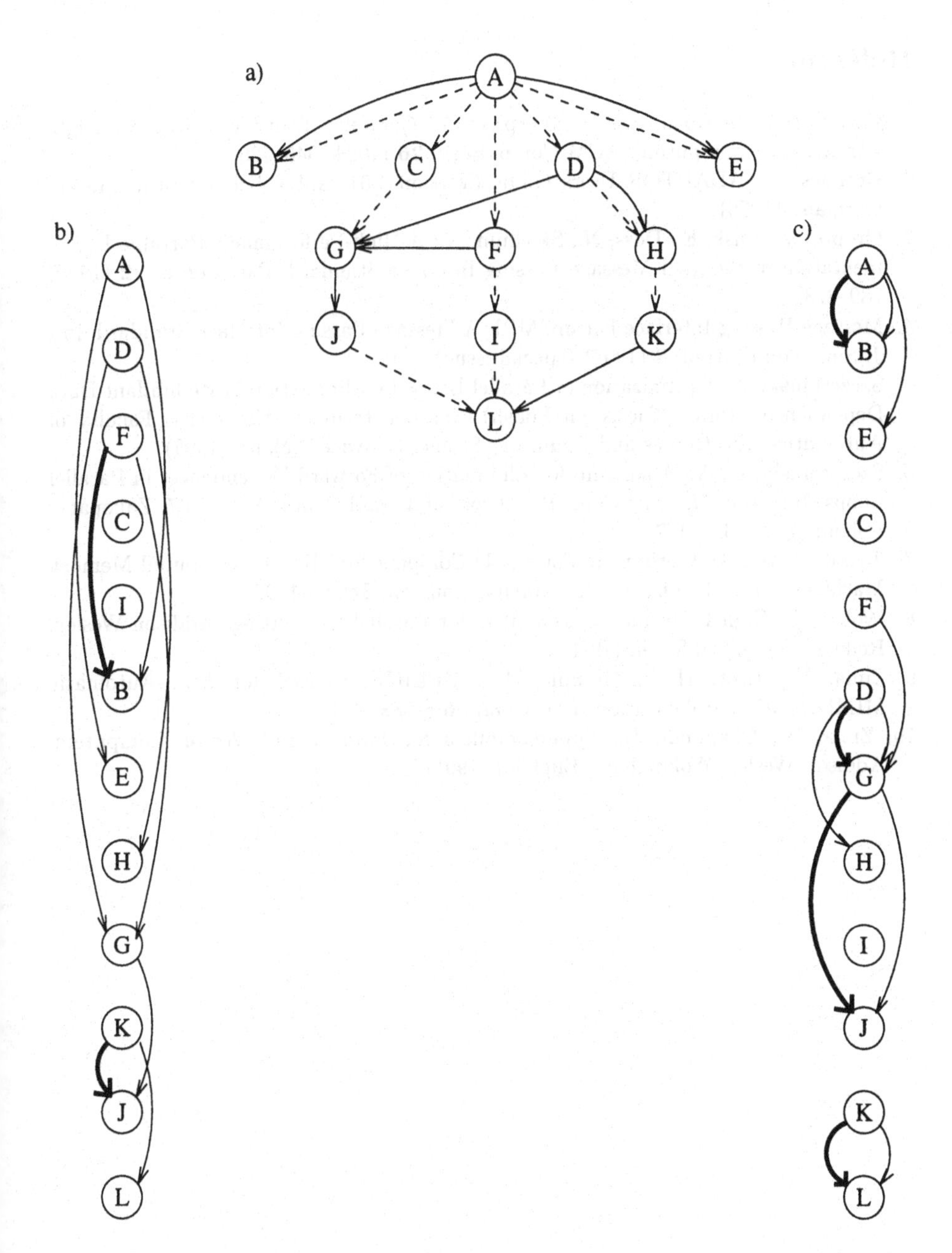

Fig. 3. Dependence folding and message aggregation with ELMAX: a) the loop's dependence graph, b) the graph sorted with ELMAX, c) the graph sorted arbitrarily

Suggested future research in this area includes relaxing the owner computes rule (e.g. by adopting the scheme where the owner of the most operands does the computation [8]) and exploring *block* and *block-cyclic* [1] array decompositions and iteration schedulings.

References

1. Bacon, D. F., Graham, S. L., Sharp, O. J.: Compiler Transformations for High-Performance Computing. ACM Comp. Surv. **26** (1994) 345–420
2. Brandes, T.: ADAPTOR Users Guide (Version 4.0). GMD, Schloss Birlinghoven, Germany (1996)
3. Gropp, W., Lusk, E., Doss, N., Skjellum, A.: A High-Performance, Portable Implementation of the MPI Message Passing Interface Standard. Par. Comp. **22** (1996) 789–828
4. Message Passing Interface Forum: MPI: A Message-Passing Interface Standard. Int. J. Supercomp. Appl. **8** (1994) (special issue)
5. Szczerbinski, Z.: Optimization of Parallel Loops by Elimination of Redundant Data Dependences. Ph.D. Thesis (in Polish), Silesian Technical University, Faculty of Automatics, Electronics and Computer Science, Gliwice, Poland (1995)
6. Szczerbinski, Z.: An Algorithm for Elimination of Forward Dependences in Parallel Loops. In: Proc. 2nd Int. Conf. Par. Proc. and Appl. Math. PPAM'97, Zakopane, Poland (1997) 398–407
7. Tseng, C.-W.: An Optimizing Fortran D Compiler for MIMD Distributed-Memory Machines. Ph.D Thesis, Rice University, Houston, Texas (1993)
8. Wolfe, M.: High Performance Compilers for Parallel Computing. Addison-Wesley, Redwood City, California (1996)
9. Zima, H., Bast, H.-J., Gerndt, M.: SUPERB: A Tool for Semi-Automatic MIMD/SIMD Parallelization. Par. Comp. **6** (1988) 1–18
10. Zima, H., Chapman, B.: Supercompilers for Parallel and Vector Computers. Addison-Wesley, Wokingham, England (1991)

A Parallel Genetic Clustering for Inverse Problems

Telega H[1], Schaefer R[1], Cabib E[2]

[1]Institute of Computer Science, Jagiellonian University, Poland
{telega, schaefer}@ii.uj.edu.pl
[2]Department of Civil Engineering, University of Udine, Italy
cabib@hydrus.cc.uniud.it

Abstract. A parallel global optimization strategy for an ill posed inverse problem in computational mechanics is proposed. It contains a rough recognition of basins of attraction of local minima (clustering) with the use of genetic algorithms. Two levels of parallelism are involved. Basic asymptotic properties of the proposed genetic clustering will be proved. The computational example of the optimal pretractions design in a network structure (hanging roof) will be also shortly described.

Many inverse problems in computational mechanics, e.g. parameters identification, constitute optimization tasks with a multimodal objective function. Almost all local minima are often needed to be found. Rough information about their basins of attraction can be also desirable.

We will focus on a description of a new global optimization method that belongs to the class of clustering methods [4]. We propose to use genetic algorithms (GA) [3] as a part of the clustering strategy. We expect, that such method will speed up the process of finding local minima and their basins of attraction. It will find all local minima that have adequately large basins with a sufficiently large objective variability. The method is twofold parallel: firstly by finding clusters concurrently (implicit genetic parallelism), secondly due to its parallel and distributed implementation. We also intend to show an asymptotic feature of the proposed method.

1 Outline of the proposed algorithm

The aim of the algorithm is to find a satisfactory set of isolated local minima and to get rough information about their basins of attraction. Additionally, the algorithm deals with many local minima of the fitness function f (the objective) as well as with large differences between values of local minima and large plateaus.

Basins of attraction will be approximated by sets of connected hypercubes that have the relative Lebesgue measure equal to θ. We allow omitting minima with the relative volume of the basin less than 2θ and the local variation of the objective function less than δ.

Hereafter, by clusters we will mean approximated basins of attraction. Parts of the basins that are recognized in one step of the algorithm will be called subclusters.
The outline of the algorithm is as follows:

1. Divide the feasible set D into p subdomains.
2. Set all subdomains to be "active".
3. $C \leftarrow \varnothing$ (set of recognized clusters), $MAX \leftarrow 0$ (current estimated maximum value of f on D), $SUM \leftarrow 0$ (total fitness).
4. `REPEAT`

|| in "active" subdomains:

- 4.1 Generate initial population (from uniform distribution).
- 4.2 Evaluate fitness function f outside recognized clusters.
- 4.3 Modify fitness function ($f \leftarrow MAX$ on clusters). Update total fitness SUM.
- 4.4 Steps of GA - evaluating new generations until the complex stop criterion is satisfied: a) clusters can be recognized, or b) GA recognizes plateau outside known clusters (then the subdomain is set to be "passive").
- 4.5 Clusters recognition. (output: information about new subclusters and new "passive" subdomains).

Update set of recognized clusters (join subclusters).
`UNTIL` all subdomains are "passive" `OR` satisfactory set of clusters is found.

The fitness modification (point 4.3) results in repelling individuals from clusters (subclusters) that are already known. Steps of GA will be described in more detail in the next section. Below, we give the idea of the clusters recognition. A single basin of attraction can be recognized in one or several steps of the loop (4). Each subdomain is divided into hypercubes of the volume δ. After the GA is stopped, new subclusters can be detected by the density analysis of individuals in the hypercubes. The hypercube that contains the best individual is selected as the seed of a new subcluster. Neighbor hypercubes, with the density of individuals $\rho > \rho_t$ (an arbitrary constant), are attached to the cluster. A rough local optimization method is started in each new subcluster and the result of this optimization is retained. Subclusters are joined (the last point in (4)). If the local method that starts from a new subcluster ends in the already recognized cluster, then the subcluster is attached to the cluster.

2 Genetic kernel of clustering

We have chosen the Simple Genetic Algorithm model (SGA) [3,5] to be implemented in our method, because some theoretical results concerning SGA were proved [5]. This enables us to derive some properties of our algorithm.

2.1 Mathematical background

SGA with a finite population of size n can be understood as a Markov chain with M possible states - populations. Vose in [5] gives the formula to evaluate the transition matrix Q. Each population can be identified as a point from the unit simplex $\Lambda = \left\{ x \in \Re^r : x \geq 0, \sum_{j=0}^{r-1} x_j = 1 \right\}$, where r denotes the number of possible individuals, the j-th coordinate of x is equal to the frequency of the j-th individual in a population identified by x. If $n < \infty$, then there exists a weak limit of probability distributions of population occurrences $\pi_n = \lim_{k \to \infty} \pi_n^k = \lim_{k \to \infty} \pi_n^0 Q^k$ for any π_n^0 (the initial distribution), where k stands for the generation number. When $n \to \infty$, the set X_n of points corresponding to possible populations becomes dense in Λ. SGA with the infinite population defines the unique operator $\Gamma : \Lambda \to \Lambda$ (see [5]). Let F be the set of fixed points of Γ. If for $x \in \Lambda$ $\lim_{t \to \infty} \Gamma^t(x) \in F$ (Γ^t denotes the composition of Γ) then $\pi^*(F) = 1$, where $\pi^* \overset{w}{=} \lim_{n \to \infty} \pi_n^k$ [5]. π^* is a probabilistic measure defined on Λ, also π_n^k, $k = 1, \ldots, \infty$ can be treated as discrete probabilistic measures on Λ, such that $\pi_n^k(\Lambda \setminus X_n) = 0$. We assume, that F is a finite set of fixed points of Γ. Each point $\$ \in F$ corresponds to some local minima of f. It means that individuals which constitute the population described by $\$$ are encoded local minimizers of f. Moreover, all minima are represented in F. The above formalism allows us to move from the source clustering problem to the problem of finding measures sufficiently concentrated on attractors of fixed points of Γ and their proper estimators.

2.2 Asymptotics

If a population is large enough and a sufficient number of generations has been performed, then the resulting probabilistic measure π_n^k is arbitrary concentrated on a metric neighborhood of fixed points of Γ. This can be stated as the lemma:

Lemma. Let $F_\varepsilon = \left\{ x \in \Lambda : d(x, y) < \varepsilon, \forall y \in F \right\}$, d is a distance in $\Re^r$. Then $\forall \varepsilon > 0$ $\forall \eta > 0$ $\exists N$ $\exists K$ so that $\forall n > N$ and $\forall k > K$ $\pi_n^k(F_\varepsilon) > 1 - \eta$.

Proof (outline): For any integer n and $\forall \varepsilon > 0$ $\exists \$ \leq \varepsilon$ such that $F_\$$ is π_n-continuous. The set $F_\$$ is also π^*-continuous. The weak convergence $\pi_n \Rightarrow \pi^*$ implies $\pi_n(F_\$) \to \pi^*(F_\$)$ (see [1]), so $\forall \eta$ $\exists N : \forall n > N$ $\left| \pi_n(F_\$) - \pi^*(F_\$) \right| < \frac{\eta}{2}$. We have also $\pi^*(F_\$ \setminus F) = 0$ (see [5]), then $\pi_n(F_\$) > 1 - \frac{\eta}{2}$. Similarly, $\pi_n^k \Rightarrow \pi_n$ implies that $\exists K$ such that $\forall k > K$ $\left| \pi_n(F_\$) - \pi_n^k(F_\$) \right| < \frac{\eta}{2}$. Hence, $\pi_n^k(F_\$) > 1 - \eta$. From monotonicity of π_n^k we have $\pi_n^k(F_\varepsilon) > 1 - \eta$ (since $F_\$ \subset F_\varepsilon$).

2.3 SGA local and global stop criteria

The local stop criterion should distinguish two basic kinds of SGA behavior. The first one is that SGA finds clusters after few generations and the second is that SGA converges to the uniform distribution of individuals. This corresponds to the recognition of a plateau outside of the already known clusters. Other cases are treated as the situation when SGA does not fit to the particular problem and a refinement of SGA parameters is suggested.
The stopping strategy is as follows:

- Check if an arbitrary number of hypercubes has the density of individuals below an arbitrary threshold ρ_t. If so, then begin clustering procedure, otherwise, go to the next step.
- `If` the Cauchy-like criterion (convergence in the space of populations Λ) is satisfied `then` check the χ^2 test for uniform distribution of individuals.

    `If` the distribution is uniform `then` set the subdomain to be "passive"
    `endif`

  `else` refinement of SGA parameter is suggested. Break computations `endif`.

The recognition which subdomain is "passive" is essential for the global stop criterion. This decision is taken basing on the given above asymptotics and the estimation of the probabilistic measure taken from SGA. The uniform distribution of individuals corresponds to the only fixed point of Γ for infinite populations (that is the point situated in the center of Λ). For finite populations it corresponds to points near the center of Λ.

3 Case study

The considered algorithm will be applied to the optimal pretractions design of a network structure made of elastic unconnected fibers fastened at their ends to a square, rigid frame (hanging roof). The linearized and homogenized governing equation

$$-\mathrm{div}(\sigma Dw) = q \text{ in } \Omega, \quad w = 0 \text{ on } \partial\Omega \tag{1}$$

delivers the relationship between the compliance $w(x)$, pretraction tensor $\sigma(x) = diag(\sigma_1(x_2), \sigma_2(x_1))$, and the transversal loading density $q(x)$ on the frame area Ω. Given q, we try to find $\sigma^* \in \left(L^\infty(\Omega)\right)^2$ such that $\mathsf{F}(\sigma^*) \le \mathsf{F}(\sigma)$ for all $\sigma \in \left(L^\infty(\Omega)\right)^2$, and $w_{\sigma^*} \in H_0^1(\Omega)$ satisfies the state equation (1). The cost functional $\mathsf{F}(\sigma) = \mathsf{E}(\sigma) + \mathsf{P}(\sigma)$, where $\mathsf{E}(\sigma) = \int_\Omega \sigma |Dw_\sigma|^2 dx$ is the stored energy of the network, and $\mathsf{P}(\sigma)$ denotes a small penalty which forces pretractions suitable for an available assortment of fibers. $\mathsf{P}(\sigma)$ is a multimodal, nonnegative function which reaches zero for many admissible pretractions. Moreover, we define the constraints: $0 < \lambda \leq \sigma_1, \sigma_2 \leq \Lambda$ in Ω, $\int_\Omega (\sigma_1 + \sigma_2) dx = S$ with $S \in [2\lambda, 2\Lambda]$.

We will consider the case of a balanced loading $\int_\Omega q dx = 0$. Cabib et al. [2] proved that under above assumptions there is more than one minimizer σ^*.

4 Implementation and first tests

Computations were carried out on a cluster of workstations. Point (4.3) from the algorithm and the subclusters merging were processed by a master agent, the other points were performed by local agents invoked dynamically. PVM library was used. We tested the problem with 10 fibers in each direction. The dimension of the optimization problem was equal to 5 due to the symmetry. Although not all tests were finished yet (in particular some work is currently being done concerning GA parameters and a dynamically changing raster), the first results are encouraging. With the constraints parameters: S=14800, λ_1=10, λ_2=1000, we obtained 15 local minima with 6 different values of the objective function. The population size was 2000. After 4 generations (8000 solver calls) we obtained such density of individuals in basins of attraction of 4 best local minima that would require more than 360 000 solver calls by using uniform random sample. The overhead for additional operations in genetic clustering was strongly dominated by the objective evaluation. The attempt to recognize basins of attraction more precisely reduces the advantage of genetic clustering, but in cases of large plateaus it is still significant.

5 Conclusions

- A new global optimization clustering method has been proposed. Its optimistic complexity is better than Monte Carlo methods. The effectiveness of SGA as a part of the clustering, results from unconditional convergence in the space of probabilistic measures. The measure estimator (population) exhibits clusters after few generations.
- Basic asymptotic properties have been proved.
- The proposed method can recognize all minimizers that satisfy assumed conditions regardless of the objective values.
- The feature mentioned above can be useful in polyoptimization in which the objective expresses a scalarization of multiple optimization criteria.
- In particular, the algorithm has been successfully used in tests concerning the optimal pretraction design of hanging roofs.

References

1. Billingsley, P.:Probability and measure. John Wiley & Sons, New York (1979)
2. Cabib, E., Davini, C., Chong-Quing Ru: A Problem in the Optimal Design of Networks under Transverse Loading. Quarterly of Appl. Mathematics, Vol. XLVIII, (1990) 252-263
3. Goldberg D.: Genetic Algorithms in Search, Optimization and Machine Learning. Addison-Wesley (1989)
4. Horst, Panos M. Pardalos: Handbook of Global Optimization. Kluwer Ac. Publisher (1995)
5. Vose, M.: Modeling Genetic Algorithms with Markov Chains. Annals of Mathematics and Artificial Intelligence, no 5 (1992) 79-88

A Parallel Hierarchical Solver for Finite Element Applications[1]

Clemens-August Thole[1], Alexander Supalov[1], Stefan Mayer[2]

[1]GMD-SCAI.WR, Schloß Birlinghoven, D-53754 Sankt Augustin
{Clemens-August.Thole, Alexander.Supalov}@gmd.de
[2]MacNeal Schwendler GmbH, Innsbrucker Ring 15, D-81612 München
Stefan.Mayer@macsch.com

Abstract. Subject of the European Esprit project PARASOL is the development of fast parallel direct solvers and test of parallel iterative solvers on their applicability and robustness in an industrial framework. Target application codes are the CFD-code POLYFLOW, the structural analysis codes MSC/NASTRAN and DNV SESAM as well as the deep drawing code INDEED and the composite modelling code ARC3D. P-elements in MSC/NASTRAN linear static analysis of solid structures allow the polynomial degree of the base functions to be specified either globally or for each element. Discretisations with lower p-level can therefore be used as coarser grids for a multilevel iterative method. This paper presents a parallel version of such an iterative solver.

1 Objectives

As part of the European Esprit project EUROPORT, 38 commercial and industrial simulation codes were parallelized for distributed memory architectures. During the project, sparse matrix solvers turned out to be a major obstacle for high scalability of the parallel version of several codes. The European Commission therefore launched the PARASOL project[1] to bring together developers of industrial finite element codes and experts in the area of parallel sparse matrix solvers in order to test innovative solvers in an industrial environment for performance and robustness. Direct and iterative parallel sparse matrix solvers will be developed and integrated into the PARASOL library. Based on the Rutherford-Boeing sparse matrix file format specification[2], an interface specification was developed suited for parallel machines[3]. This parallel interface is also able to transport geometry and other information which might be needed by iterative solvers. Based on this interface, parallel direct solvers as well as iterative sparse matrix solvers are being developed by CERFACS, GMD, ONERA, RAL and University Bergen. The iterative solvers use

[1] This work was supported by the European Commission as part of the Esprit project PARASOL (No. 20160).

domain decomposition methods and hierarchical approaches. Industrial simulation codes are modified in such a way that they support the parallel interface and will evaluate the different solvers on industrial test cases. In addition to the CFD-code POLYFLOW, PARASOL involves the structural analysis codes MSC/NASTRAN and DNV SESAM as well as the metal forming code INDEED from INPRO and ARC3D from APEX (simulation of composite rubber metal part deformations).

Typical problem sizes for MSC/NASTRAN and DNV SESAM are in the order of 10^6-10^7 degrees of freedom. Metal forming and CFD codes require the solution of a sparse matrix in each time step. This restricts the problem size to about 10^5-10^6 in these cases.

2 Methods

2.1 Hierarchical P-Solver

MSC/NASTRAN is a commercial structural analysis code used for the optimisation of the static and dynamic behaviour of structures. This includes a variety of products such as automobiles, aeroplanes, satellites, ships and buildings. An overview of the technical basis of MSC/NASTRAN is given in [4].

In order to support the user in achieving the desired accuracy with minimal effort, MSC has extended the MSC/NASTRAN element basis by p-elements. For these elements, an error estimator is provided and in several steps the order of each base function is adapted until the desired accuracy is reached. Neighbouring elements might arrive at different order (even in different directions). Solid elements, which may be used as p-elements, are the standard solid elements (see also [4]).

For p-elements the stiffness matrix $\lfloor K \rfloor$ and vectors $\lfloor u \rfloor$ for p-elements are organised as follows:

$$[K] = \begin{bmatrix} K_{gg} & K_{gh} \\ K_{hg} & K_{hh} \end{bmatrix}; \; [u] = \begin{bmatrix} u_g \\ u_h \end{bmatrix} \tag{1}$$

K_{gg} and u_g correspond to the degrees of freedom of linear base functions (geometric degrees of freedom), K_{hh} and u_h to higher order base functions (hierarchical degrees of freedom). For the solution of

$$[K][u] = [f] \tag{2}$$

an algorithm similar to those discussed by Axelsson[5] was implemented to improve the approximation $\lfloor u^{v} \rfloor$of the solution $\lfloor u \rfloor$:

Method R(n):

$$\left[u^{\nu,0}\right]=\left[u^{\nu}\right]$$

for i = 1 to n do:

$$\left[u_g^{\nu,i}\right]=\text{RELAX}\left(\left[K_{gg}\right],\left(\left[f_g\right]-\left[K_{gh}\right]\left[u_h^{\nu,i-1}\right]\right),\left[u_h^{\nu,i-1}\right]\right)$$

$$\left[u_h^{\nu,i}\right]=\text{RELAX}\left(\left[K_{hh}\right],\left(\left[f_h\right]-\left[K_{hg}\right]\left[u_g^{\nu,i}\right]\right),\left[u_h^{\nu,i-1}\right]\right) \quad (3)$$

$$\left[u_g^{\nu,n+1}\right]=\left[K_{gg}\right]^{-1}\left(\left[f_g\right]-\left[K_{gh}\right]\left[u_h^{\nu,n}\right]\right),\ \left[u_h^{\nu,n+1}\right]=\left[u_h^{\nu,n}\right]$$

for i = n+2 to 2n+1 do:

$$\left[u_h^{\nu,i}\right]=\text{RELAX}\left(\left[K_{hh}\right],\left(\left[f_h\right]-\left[K_{hg}\right]\left[u_g^{\nu,i-1}\right]\right),\left[u_h^{\nu,i-1}\right]\right)$$

$$\left[u_g^{\nu,i}\right]=\text{RELAX}\left(\left[K_{gg}\right],\left(\left[f_g\right]-\left[K_{gh}\right]\left[u_h^{\nu,i}\right]\right),\left[u_g^{\nu,i-1}\right]\right)$$

$$\left[u^{\nu+1}\right]=\left[u^{\nu,2n+1}\right]$$

$\text{RELAX}\left(\left[K\right],\left[f\right],\left[\overline{u}\right]\right)$ performs a smoothing step on the problem $\left[K\right]\left[u\right]=\left[f\right]$ with $\left[\overline{u}\right]$as initial approximation to the solution. For the numerical experiments, pointwise Gauß-Seidel iterations have been used. R(n)-CG denotes the conjugate gradient algorithm, which uses the method R(n) as preconditioner.

In order to be able to treat very huge problems, the sequential p-element solver was directly integrated into MSC/NASTRAN for evaluation purposes. The individual components are optimised either by calling computational kernels from MSC/NASTRAN or by using BLAS I routines. The module fully supports out-of-core calculations such that the problem size in only limited by available disk space. **Table 1** summarises performance results for the R(1)-CG algorithm.

Table 1. Elapsed times and Iteration counts for a number of test cases (Measurements by MSC)

Test case	dofs	Non-zeros	Elapsed time for direct solver (IBM workstation model)	Elapsed time for p-solver solver (IBM workstation model)	Number of iterations
Knuckle-P	10.827	1.233.457	13 s (39h)	46 s (39h)	84
Tetra-P	32.871	2.573.533	216 s (39h)	49 s (390)	20
Crancksseg-P	52.804	10.614.210	879 s (39h)	226 s (590)	62
Audicss-P	314.973	77.651.641		24.308 s (591)	404

For all Tetra-P and Knuckle-P the p-solver is much faster than the direct solvers. Due to huge fill-in, the factor for the Audicss-P is much larger than 10 GByte and only special workstations will be able to handle this application (with unacceptable elapsed times). The coming MSC/NASTRAN version 70.5 will contain a new iterative solver (BIC) based on a modified incomplete Cholesky preconditioner. This solver was faster than the p-solver for the Crankseg-P and the Audicss-P test cases. (However, incomplete Cholesky preconditioners are even harder to parallelise than direct solvers.) A more detailed discussion on the numerical properties of the hierarchical p-solver is contained in [6].

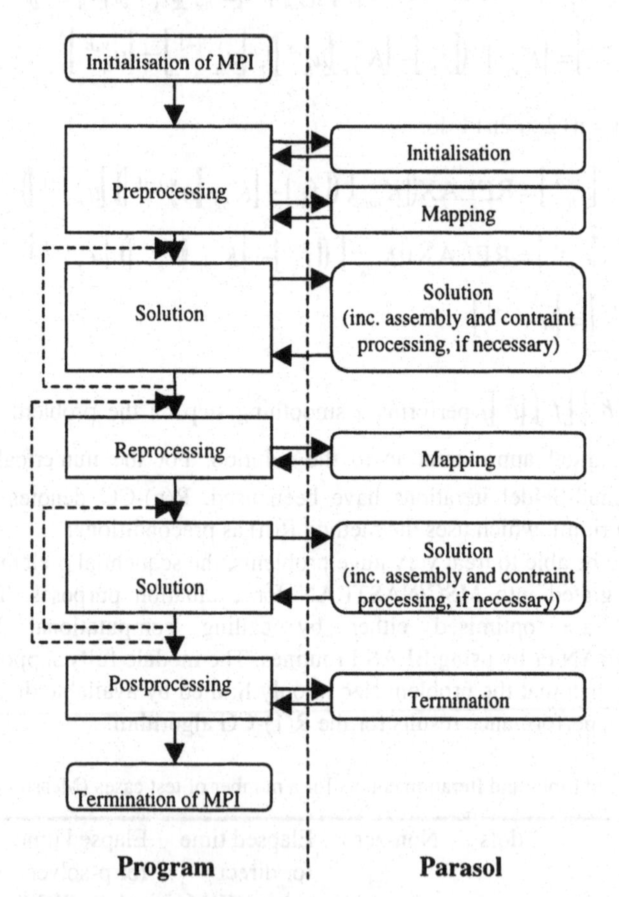

Fig. 1. Structure of the PARASOL library interface

2.2 Parallelisation of the hierarchical p-solver

As part of the PARASOL project a parallel version of the p-solver has been developed and has been integrated with the PARASOL library interface. **Fig. 1** shows the basic structure of the PARASOL library interface[3]. For each of the solvers a solver specific mapping module and a solution module has to be developed.

2.3 Mapping

The mapping of the degrees of freedom to processors is based on grid partitioning by the METIS package (Version 3.0.5)[7]. The partitioning is either based on the connectivity graph of either the finite element grid, the geometric degrees of freedom, or all degrees of freedom. In the first two cases the connectivity graph is much smaller and therefore less time is spent for grid partitioning. Currently each hierarchical degree of freedom is assigned according to the mapping of its geometric neighbours and the processors with the lowest id is selected. This generates load imbalance but minimises the number of synchronisation steps for the multicoloured relaxation (see next section).

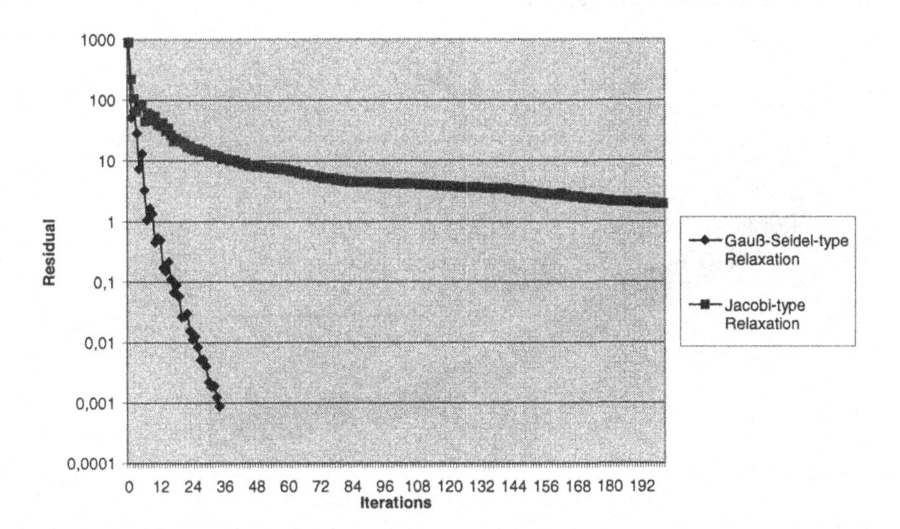

Fig. 2. Convergence history for the parallel p-element method of 32 nodes: Jacobi-type relaxation vs. Gauß Seidel - type relaxation

2.4 Parallelisation of the relaxation

The original relaxation for the p-element solver was sequential and had to be algorithmically modified in order to run efficiently in parallel. Experiments showed that the Gauß-Seidel nature of the relaxation has to be preserved. An independent relaxation of the degrees of freedom related to each processor resulted in an unacceptable increase of the iteration numbers (c.f. **Fig 2**). Therefore the degrees of freedom at processor boundaries are coloured and each relaxation is performed in several steps according to the colouring of the degrees of freedom. The interior degrees of freedom are distributed over the colouring steps in order to support load balancing.

Although the ordering of the dofs is changed substantially, only little variation of the convergence behaviour was observed, as long as the global Gauß-Seidel nature of the relaxation was preserved.

2.5 Parallel direct solver

In each iteration the treatment of the geometric degrees on freedom requires the direct solution of a sparse matrix problem. The parallel sparse matrix solver WSSMP of Gupta et. al. [8] has been used for this purpose.

2.6 Parallel simulation results

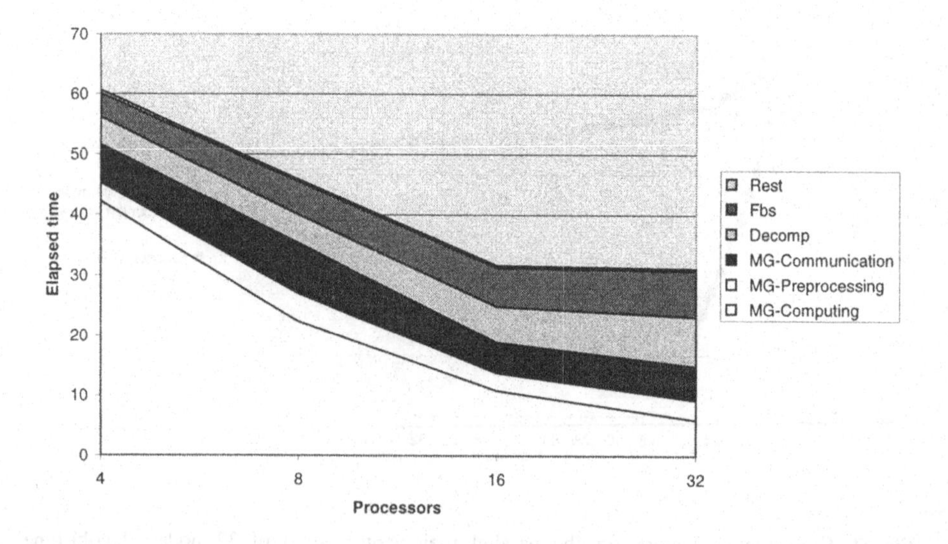

Fig. 3. Timing analysis for the parallel p-solver on the cranckseg-P test case using IBM SP-2 nodes

Figure 3 shows a performance analysis of the parallel p-solver on the Cranckseg-P test case using up to 32 SP-2 processors. The Cranckseg-P test case consists of 2nd order elements resulting in 52.804 dofs and10.614.210 non-zeros. 16 processors can be used efficiently for this test case. The parallel performance of the direct solver and the time spent in communication for the MG-components (relaxation and conjugate gradient algorithm) limit the scalability.

The time spent in communication for the MG-components is caused by load imbalance and not by actual data transfer. This is evident from the output of the VAMPIR trace tool from Pallas shown in **Figure 4**. Colour/dark portions of the horizontal bars for each node indicate the time spent in communication routines during one relaxation step. Here node 1 uses nearly all its time for computing, while others are waiting on intermediate results.

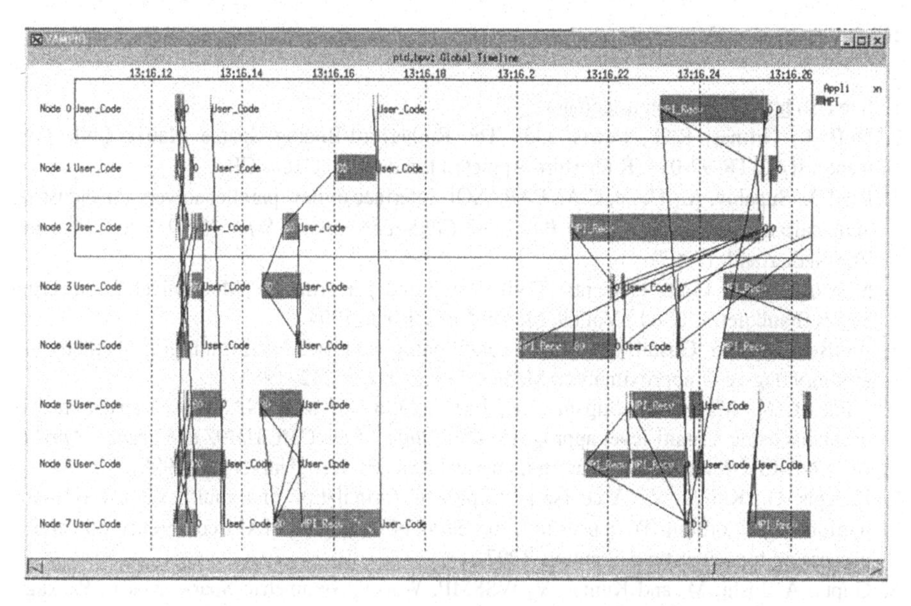

Fig. 4. VAMPIR output for p-solver of 8 processors for one relaxation step

The elapsed time for the parallel hierarchical p-solver took about 30 sec on 16 processors. More processors are expected to be used efficiently on larger problems.

3 Conclusions

A parallel iterative hierarchical p-solver has been implemented and tested with medium size industrial test cases. Up to 16 processors could be efficiently used for these test cases, which is a substantial improvement compared to results achieved within EUROPORT-1[9,10].

In order to preserve the numerical properties of the algorithm, the global Gauß-Seidel nature of the relaxation has to be kept unchanged, however, the actual numbering does not influence the numerical performance.

An improvement of the scalability is expected on larger test cases.

Acknowledgements

The authors would like to thank Anshul Gupta from IBM for the permission to use his solver package for numerical experiments.

References

1. See also http://www.genias.de/parasol
2. Duff, I.; Grimes, R.G.; Lewis, J.G.: The Rutherford-Boeing Sparse Matrix Collection. Report RAL-TR-97-031, Rutherford Appleton Laboratory, Oxon, GB.
3. Reid, J., Supalov, A., Thole, C.A.: PARASOL Interface to new parallel solvers for industrial applications, *Proceedings of the ParCo'97 Conference,.Bonn, Sept. 16-19, 1997*, Elsevier, The Netherlands (1997).
4. MacNeal, R.H.: Finite elements: Their design and performance. Mechanical Engineering Series (Faulkner, L.L. ed.) Vol 89, Marcel Dekker Inc., 1994.
5. Axelsson, O.and Gustafsson, I.: Preconditioning and two-level multigrid methods for arbitrary degree of approximation. Math. Comp., 40: 219-242 (1983).
6. Thole, C.A., Mayer, S., Supalov, A., Fast solution of MSC/NASTRAN sparse matrix problems using a multi-level approach, *Proceedings of the CMCMM97 conference, April 6-11, 1997.* Electronic transactions on numerical analysis, Kent University (1998).
7. Karypis, G.; Kumar, V.: A coarse grain parallel formulation of a multilevel k-way graph partitioning algorithm. To appear in: Proceeding of the eighth SIAM conference on parallel processing for scientific computing, 1997.
8. Gupta, A, Joshi, M, and Kumar, V, WSSMP: Watson Symmetric Sparse Matrix Package, IBM Research Report RC 20923 (92669), IBM Research Devision, Computer Science/Mathematics, Watson (1997).
9. Stüben, K.; Mierendorff, H.; Thole, C.A.; Thomas, O.: Industial parallel computing with real codes. Parallel Computing 22 (1996), pp. 725-737.
10. Stüben, K.; Mierendorff, H.; Thole, C.A.; Thomas, O.: EUROPORT: Parallel CFD for industrial applications. In: Schiano, P.; Ecer, A.; Periaux, J.; Satofuka, N.: Parallel Computational Fluid Dynamics, Elsevier Amsterdam, 1997.

Parallel Computation and Visualization of 3D, Time-Dependent, Thermal Convective Flows

Ping Wang and Peggy Li

Jet Propulsion Laboratory,
California Institute of Technology,
MS 168-522, 4800 Oak Grove Drive,
Pasadena CA 91109-8099, U.S.A.
wangp@rockymt.jpl.nasa.gov

Abstract. A high-resolution numerical study on parallel systems is reported on three-dimensional ($3D$), time-dependent, thermal convective flows. Numerical results are obtained for Rayleigh numbers up to 5×10^7 and for a Prandtl number 0.733 equivalent to that of air, in a cubical enclosure, which is heated differentially at two vertical sidewalls. A parallel implementation of the finite volume method with a multigrid scheme is discussed, and a parallel and distributed visualization system is developed for visualizing the flow. The details of the 3D, time-dependent flow are described. Separations of the flow near the horizontal walls occur, and y-variations of the flow are strong. Periodical flow patterns appear. The 3D solutions for such high Rayleigh number 5×10^7 become strong convective, time-dependent, and periodical.

1 Introduction

Thermal convection driven by imposed horizontal density gradients finds many applications in engineering and in nature: reactor cooling systems, crystal growth procedures, and solar-energy collectors. The most numerically studied form of this problem is the case of a rectangular cavity with differentially heated sidewalls. To date, the two dimensional version of this problem has received considerable attention[1] [2] [3] [4], but for the 3D case, very few results have been obtained. The first numerical solution of the 3D thermal cavity problem was reported by Mallinson and de Vahl Davis [5], and a numerical study of 3D natural convection for air in the cubic enclosure was reported in [6]. For large Rayleigh number 3D flows, there are few numerical results available. The existing 3D numerical simulations are still in a rudimentary stage. Most existing numerical studies have suffered from insufficient resolution, and the prominent characteristics of complicated 3D flows have not been discussed in sufficient depth because the accuracy of the solution is not good enough. In particular, at large Rayleigh numbers, greatly enhanced numerical capabilities are essential in order to catch the significant dynamic features in thin boundary layers. The main reason of these was due to the limitation of computing power and memory size.

Recent advances in computing hardware have dramatically affected the prospect of modeling 3D time-dependent flows. The significant computational resources of massively parallel supercomputers promise to make such studies feasible. In addition to using advanced hardware, designing and implementing a well-optimized parallel thermal convection code will significantly improve the computational performance and reduce the total research time to complete these studies. Based on the finite volume method, a numerical study for the 3D problems is recently investigated by using massively parallel computing systems [7], where 3D numerical solutions of steady state for various Rayleigh numbers up to 10^7 are reported. But for the 3D time-dependent flows, more work needs to be done.

In the present paper, a numerical study for 3D, time-dependent thermal convective flows is investigated. An efficient numerical scheme with a general parallel implementation on massively parallel supercomputers is presented. Parallel visualization of numerical results is also discussed. Visualization of large 3D, time dependent scientific datasets can not be effectively performed by use of workstation-based visualization systems due to limits in computing power and memory. Therefore, a parallel rendering system on massively parallel supercomputer is desirable for visualizing this type of dataset. A distributed visualization system, called ParVox, has been designed and implemented on the MPP to visualize multiple time-step, 3D scientific datasets in a structured grid.

The paper is organized as follows: the mathematical formulation of the 3D time dependent thermal cavity flows is given in Section 2. The finite volume method with a multigrid scheme and a parallel implementation are described in Section 3. In section 4, the ParVox system and its parallel implementation is presented. The 3D time-dependent numerical solutions for a high Rayleigh number 5×10^7 are discussed in Section 5. In the same Section, how to use ParVox to visualize the results is also discussed. The summary of the present work is given in Section 6.

2 Mathematical Formulation

The flow domain is a rectilinear box of $0 < x < l, 0 < y < d$, and $0 < z < h$. The appropriate governing equations, subject to the Boussinesq approximation, can be written in non-dimensional form as

$$\frac{\partial u}{\partial x} + \frac{\partial v}{\partial y} + \frac{\partial w}{\partial z} = 0, \tag{1}$$

$$\frac{1}{\sigma}\left(\frac{\partial u}{\partial t} + u\frac{\partial u}{\partial x} + v\frac{\partial u}{\partial y} + w\frac{\partial u}{\partial z}\right) = -\frac{\partial p}{\partial x} + \nabla^2 u, \tag{2}$$

$$\frac{1}{\sigma}\left(\frac{\partial v}{\partial t} + u\frac{\partial v}{\partial x} + v\frac{\partial v}{\partial y} + w\frac{\partial v}{\partial z}\right) = -\frac{\partial p}{\partial y} + \nabla^2 v, \tag{3}$$

$$\frac{1}{\sigma}\left(\frac{\partial w}{\partial t} + u\frac{\partial w}{\partial x} + v\frac{\partial w}{\partial y} + w\frac{\partial w}{\partial z}\right) = -\frac{\partial p}{\partial z} + \nabla^2 w + RT, \tag{4}$$

$$\frac{\partial T}{\partial t} + u\frac{\partial T}{\partial x} + v\frac{\partial T}{\partial y} + w\frac{\partial T}{\partial z} = \nabla^2 T. \tag{5}$$

The dimensionless variables in these equations are the fluid velocities u, v, w , the temperature T, and the pressure p, where $\sigma = \nu/\kappa$ is the Prandtl number and $R = g\beta\Delta T h^3/\kappa\nu$ is the Rayleigh number. Here ν is the kinematic viscosity, κ is the thermal diffusivity, β is the coefficient of thermal expansion, and g is the acceleration due to gravity. The rigid end walls on x=0, l are maintained at constant temperatures T_0 and $T_0 + \Delta T$, respectively, while the other boundaries are assumed to be insulating. So the boundary conditions on the rigid walls of the cavity are

$$u = v = w = 0 \quad \text{on} \quad x = [0, l], \quad y = [0, d], \quad z = [0, h], \tag{6}$$

$$T = 0 \quad \text{on} \quad x = 0, \tag{7}$$

$$T = 1 \quad \text{on} \quad x = l, \tag{8}$$

$$\frac{\partial T}{\partial z} = 0 \quad \text{on} \quad z = 0, h, \tag{9}$$

$$\frac{\partial T}{\partial y} = 0 \quad \text{on} \quad y = 0, d. \tag{10}$$

In general the motion is controlled by the parameters σ, R and the flow domain.

3 Numerical Approach and Parallel Implementation

The numerical approach is based on the widely used finite volume method [8] with an efficient and fast elliptic multigrid solver for predicting incompressible fluid flows. This approach has been proven to be a remarkably successful implicit method. A normal, staggered grid configuration is used and the conservation equations are integrated over a macro control volume. Local, flow-oriented, upwind interpolation functions are used in the scheme to prevent the possibility of unrealistic oscillatory solutions at high Rayleigh numbers. The discretized equations derived from the scheme, including a pressure equation which consumes most of the computation time, are solved by using a parallel multigrid method. This method acts as a convergence accelerator and reduces the CPU time significantly for the entire computation.

The main idea of the multigrid approach is to use the solution on a coarse grid to revise the required solution on a fine grid. In view of this, a hierarchy of grids of different mesh sizes is used to solve the fine grid problem. It has been proven theoretically and practically that the multigrid method has a better rate of convergence [9]. In the present computation, a V-Cycle scheme with a flexible number of grid levels is implemented with the Successive Over-Relaxation as the smoother. Injection and linear interpolation are used as restriction operators and interpolation operators, respectively.

The finite volume method with multigrid scheme for 3D, time-dependent, thermal convective flows is implemented on distributed memory systems [7]. A flexible parallel code for such flow problems is designed by using domain decomposition techniques and the MPI (Massage Passing Interface) communication

API (Application Programming Interface). In order to achieve load balancing and to exploit maximal parallelism as much as possible, a general and portable parallel structure based on the domain decomposition techniques is designed for the 3D flow domain, which has 1D, 2D and 3D partition features and can be chosen according to different geometries. MPI software is used for the internal communication which is encountered when each subdomain on each processor needs its neighbor's boundary data information. The implementation is carried out on the distributed memory systems, and the code currently runs on the Intel Paragan, the Cray T3D, the Cray T3E, the IBM SP2, the Beowulf system, and the HP/Convex SPP2000, which can be easily ported to other parallel systems.

4 Parallel Visualization

A distributed visualization system, called ParVox, is used to visualize the multiple timestep 3D thermal convective flow on the MPP (Massively Parallel Processor). ParVox is a parallel volume rendering system designed for visualizing large volumes of time dependent, 3D scientific datasets. It is a parallel implementation of the forward-feeding splatting algorithm [10] utilizing both object-space decomposition and image-space decomposition. Currently, it runs on the Cray T3D and the Cray T3E using Cray's shmem library for interprocessor communication, and also on the HP/Convex SPP2000 system using the one-sided communication API defined in the MPI2.0 standard. The asynchronous one-sided communication allows the ParVox system effectively overlapping a rendering process and message passing which is required to resort the data from the rendering stage to the compositing stage.

ParVox is capable of visualizing 3D volume data as a translucent volume with adjustable opacity for each different physical value, or as multiple iso-surfaces at different thresholds and different opacities. It can also slice through the 3D volume and only view a set of slices in either of the three major orthogonal axes. Moreover, it is capable of animating multiple time-step 3D datasets at any selected viewpoint. ParVox is designed to serve as either an interactive visualization tool for post-processing, or a rendering API to be linked with any application program. As a distributed visualization system, ParVox provides four functional modules: 1. an X Window based GUI program for display and viewing control, 2. a parallel input library for reading 4D volume datasets in NetCDF format, 3. a network interface program that interfaces with the GUI running on a remote workstation, 4. a parallel wavelet image compression library capable of supporting both lossless and lossy compression. The detailed description of the ParVox system and its parallel implementation can be found in [11].

5 Results and Discussion

Various numerical tests have been performed on the 3D code. The numerical scheme is robust and efficient, and the general parallel structure allows us to use different partitions to suit various physical domains. Numerical results are obtained for a wide range of Rayleigh number up to 5×10^7. The steady-state solutions for Rayleigh number up to 10^7 can be found in [7]. Here 3D time-dependent numerical solutions for a Rayleigh number 5×10^7 with a Prandtl number 0.733 in $0 \leq x, y, z \leq 1$ are presented, and the solution for $R = 10^7$ was used as the initial state. A grid size $128 \times 128 \times 128$ with a 3D partition was used for the computation. The results are displayed in terms of velocity and temperature. In all the figures the hot wall is on the right. All numerical results and visualization have been carried out on the Cray T3E.

Figure 1(a-f) illustrates the velocity field for Rayleigh number 5×10^7 in air at different time. At the beginning in Figure 1(a), it is easy to see that the flow rises from the hot side, travels horizontally, and sinks on the cold side. With the increase of time in Figure 1(b), the flow structure becomes very complicated. Multiple eddies are generated in the entire flow domain, and separations of flow occur on the bottom near the left corner and on the top near the right corner. When $t = 0.3$ in Figure 1(c), two big eddies appear in the middle area with some corner eddies near the lower left corner and the upper right corner. Once $t = 0.4$ in Figure 1(d), a large eddy, which slightly tilts, is formed in the center of the domain. It becomes multiple flows again with two large eddies in Figure 1(e). The flow continuously changes, and forms a very sophisticated flow pattern at $t = 0.6$ in Figure 1(f). It is obvious to notice that three large eddies are in the middle area-one is in the center and one each close to its sidewall. Corner eddies are strong, and the reverse flows appear on the bottom near the left corner and on the top near the right corner. Figure 1(f) has a similar flow pattern as the one in Figure 1(b). From the time sequence, the solutions for such high Rayleigh number become strong convective, time-dependent, and periodical.

In order to have more detailed information about the flow structure, an enlarged plot for the velocity field near the lower corner is displayed in Figure 2. The separation of flow occurs at about $x = 0.3$, and the corner eddy is formed in $0 < x < 0.2, 0 < z < 0.2$ along the entire y direction. An eddy near the cold wall in $0 < x < 0.2, 0.2 < z < 0.8$ is also noticed. Figure 3 shows the detailed profile of the velocity component (v) on $x - y$ plane at $z = 0.9$. Here the y-variations of the flow are appreciable, especially at the corners of the box the velocities vary rapidly and strong flows (v) are generated over there. But the variations at the middle region are much less than those at the corner areas. Those flows affect the main streams gradually through the corners. In conclusion the flow is no longer a two dimensional motion, and the y-variations are not negligible. The temperature field at $t = 0.6$ is plotted in Figure 4, which shows very thin boundary layers on the two sidewalls and the near-linear temperature stratification in the interior.

Because of the complexity of the flow structure, visualizing such a data set in detail is a challenge. The main problem is how to capture the dynamics in the velocity field, such as the multiple eddies and flow separations, and high speed flow near the walls. Those features cover a wide range of velocity values. In order to display the global behavior of vector fields, using a vector magnitude is considered. After some experiments, it is fund that the most effectively way to represent the velocity volume is to use direct volume rendering with a condensed spectrum color map and an opacity map with mostly low opacity except the values at the lower end and the higher end. The condensed colormap is used to distinguish velocity difference in a small value range. The selected opacity map highlights only the high velocity and the low velocity; the less interesting mid-range velocities are mostly transparent, thus revealing the internal low-speed flow. Figure 5 is a snapshot of the ParVox GUI showing the colormap and the opacity map described above and the rendered image based on this specific classification.

6 Conclusions

In this paper high Rayleigh number 3D time-dependent numerical solutions have been obtained by using massively parallel computers. The results show that the numerical scheme is robust and efficient. The 3D numerical results for Rayleigh number 5×10^7 with time series show many interesting flow structures. They give a complete description of the 3D flow. The flow field gradually changes to a multiple roll structure from an initial single flow circulation as the time increases. The flow structures become very complicated, and regions of reverse flow on the horizontal walls are present. Very thin thermal boundary layers are formed on the two sidewalls. The y-variations are very strong near the corners, and eventually affect the main flow so the overall motion of the flow becomes three-dimensional. The solutions for such high Rayleigh number become strong convective, time-dependent, and periodical.

The ParVox system allows users interactively visualize the time-dependent, high-resolution, 3-D thermal convective flow on a remote supercomputer. The color and opacity editing capability equipped in ParVox can effectively change the volume classification on-the-fly, thus selectively revealing different physics of the volume data. The ParVox system can deliver about 2 frames/second end-to-end rendering rate from a remote supercomputer to user's desktop workstations via a low-speed network. It is proven to be a powerful tool for exploration and analysis of large volumes of 3D, time-dependent volume datasets.

In spite of the difficulties associated with large Rayleigh number simulation, our results presented here clearly demonstrate the great potential for applying this approach to solve high resolution, large Rayleigh number flow in realistic, 3D geometries using parallel systems. Much higher Rayleigh numbers computations of thermal convection in 3D for various applications are under investigation, which include the deep, rapidly rotating atmospheres of the outer planets, such as the Jupiter and the Saturn in planetary atmospheres science, and the 3D thermal convection in ocean science.

References

1. A. Bejan and C. L. Tien. Laminar natural convection heat transfer in a horizontal cavity with different end temperatures. *Trans. A.S.M.E. : J. Heat Trans*, 100:641–647, 1978.

2. J.E. Drummond and S.A. Korpela. Natural convection in a shallow cavity. *J.Fluid Mech.*, 182:543–564, 1987.
3. P. Wang and P.G.Daniels. Numerical solutions for the flow near the end of a shallow laterally heated cavity. *J.Eng.Math.*, 28:211–226, 1994.
4. P. G. Daniels and P.Wang. On the evolution of thermally-driven shallow cavity flows. *J. Fluid Mech.*, 259:107–124, 1994.
5. G.D. Mallinson and de Vahl Davis. Three-dimensional natural convection in a box. *J. Fluid Mech.*, 83:1–31, 1977.
6. T Fusegi, J.M. Hyun, K.Kuwahara, and B. Farouk. A numerical study of three-dimensional nature convection in a differentially heated cubical enclosure. *Int. J. Heat Mass Transfer*, 34:1543–1557, 1991.
7. P. Wang. Massively parallel finite volume computation of three-dimensional thermal convective flows. *Advances in Engineering Software*, 29:451–461, 1998.
8. S.V. Patankar. *Numerical Heat Transfer and Fluid Flow.* Hemisphere, New York, 1980.
9. S.F. McCormick. *Multilevel Adaptive Methods for Partial Differential Equations.* Frontiers on Applied Mathematics, SIAM, Philadelphia, 1989.
10. L. Westover. Footprint evaluation for volume rendering. *Computer Graphics, Proceedings of Siggraph*, 24, 1990.
11. P. Li and et. al. Parvox – a parallel splatting volume rendering system for distributed visualization. *Proceedings of the 1997 IEEE Parallel Rendering Symposium*, pages 7–14, 1997.

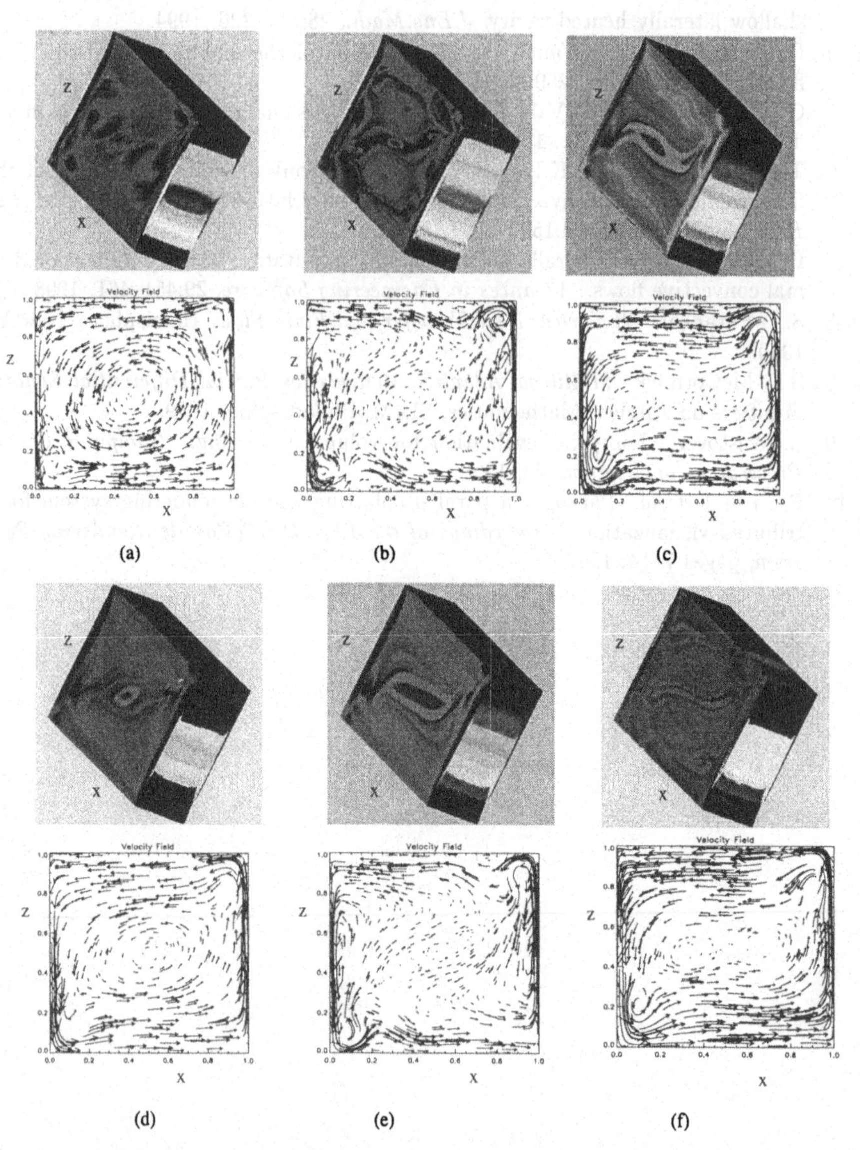

Figure 1. The velocity field for R=5x10^7 with Pr=0.733 on the entire flow domain (colour) and y=0.5 (black and white) at (a) t=0.1, (b) t=0.2, (c) t=0.3, (d) t=0.4, (e) t=0.5, (f) t=0.6.

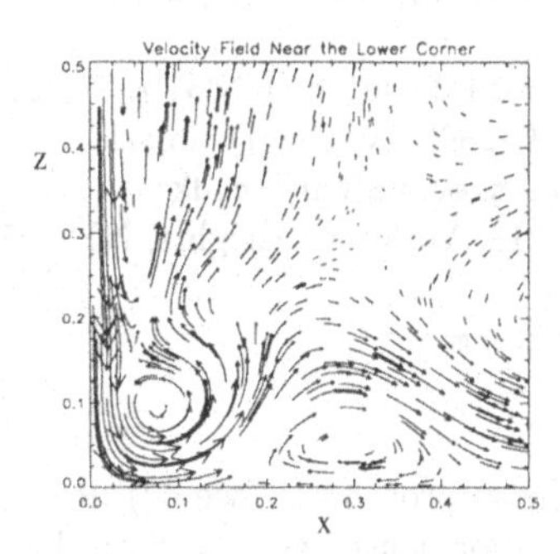

Figure 2. The enlarged velocity field near the lower corner at t=0.6.

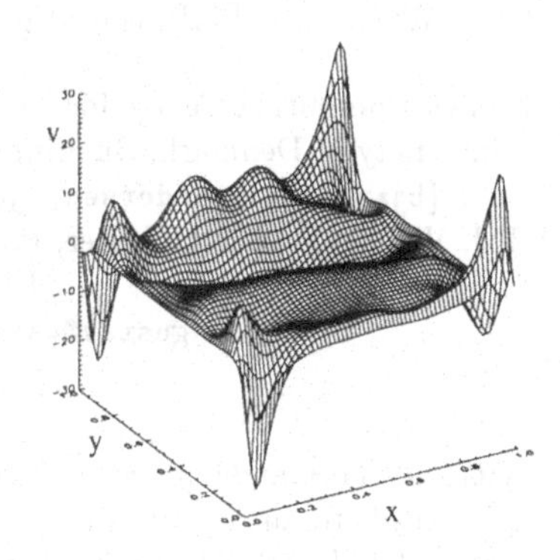

Figure 3. The variations of the velocity component (v) on the x-y plane at z=0.9.

Figure 4. The temperature field on the entire domain for R=5x10^7 with Pr=0.733.

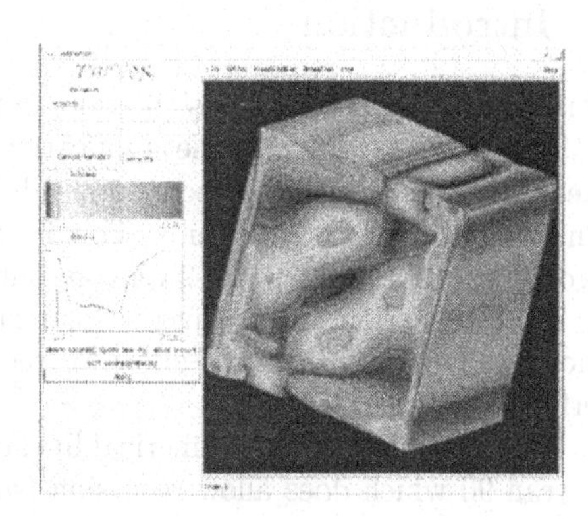

Figure 5. The velocity field for R=5x10^7 with Pr=0.733 generated by the ParVox syste.

Recursive Formulation of Cholesky Algorithm in Fortran 90

Jerzy Waśniewski[1], Bjarne Stig Andersen[1] and Fred Gustavson[2]

[1] The Danish Computing Centre for Research and Education (UNI•C), Technical University of Denmark, Building 304, DK-2800 Lyngby, Denmark.
{bjarne.stig.andersen, jerzy.wasniewski}@uni-c.dk
[2] IBM T.J. Watson Research Center, P.P. Box 218, Yorktown Heights, NY 10598. USA.
gustav@watson.ibm.com

Abstract. Fortran 90 allows writing recursive procedures (see [7]). Recursion leads to automatic variable blocking for dense linear-algebra algorithms (see [5, 8]). The recursive way of programming algorithms eliminate the use of BLAS level 2 in the factorization steps. For this and other reasons recursion usually speed up the algorithms.
The formulation of the Cholesky factorization algorithm using recursion in Fortran 90 is presented in this paper.

1 Introduction

This work is a continuation of the work of Gustavson and Toledo described in [5, 8]. These papers describe the application of recursion to the numerical dense linear algebra algorithms. Recursion leads to automatic variable blocking for the dense linear-algebra algorithms. This is some modification of the LAPACK [1] algorithms. LAPACK's level-2 version routines are transformed into level-3 codes by using recursion. In the paper [5] recursion is emulated in FORTRAN 77 since this language does not allow for recursion. The paper [5] shows very nice performance results.

Here we present these numerical linear algebra algorithms using recursion in Fortran 90 which does allow recursion (see [7]). The programs are simpler and the recursion part is automatic as it is handled by the compiler. The intermediate subroutines have Fortran 90 standard too (see [3]).

Algorithm of the Cholesky Factorization

$$A = L\,L^T = U^T U$$

written in Fortran 90 is attached in the appendix of this paper.

2 Cholesky Decomposition

We would like to compute the solution to a system of linear equations $AX = B$, where A is real, symmetric and positive definite. X and B are rectangular matrices or vectors.

The Cholesky decomposition should be used to factor A. The Cholesky algorithm requires that only the upper or the lower triangular part of matrix A is stored.

$$A = U^T U, \quad \text{if upper triangular part of } A \text{ is given}$$

or

$$A = L L^T, \quad \text{if lower triangular part of } A \text{ is given}$$

U is an upper triangular matrix and L is a lower triangular ($L = U^T$). The factored form of A is then used to solve the system of equations $A X = B$.

2.1 Lower Triangular Case

Given an $n \times n$ matrix A, compute a lower triangular matrix L.

The Cholesky decomposition algorithm can be written in this case (see [4] or [2]):

Algorithm 2.11

$$\begin{aligned} &For\ j = 1, 2, \ldots n;\ l_{jj} = (a_{jj} - \textstyle\sum_{k=1}^{j-1} l_{jk}^2)^{1/2} \\ &\quad For\ i = j+1, \ldots n;\ l_{ij} = (a_{ij} - \textstyle\sum_{k=1}^{j-1} l_{ik} l_{jk})/l_{jj} \end{aligned}$$

where a_{ij} and l_{ij} for $i = 1, 2, \ldots n$ and $j = 1, 2, \ldots i$ are elements of the matrices A and L respectively.

This algorithm requires $n^3/3$ flop operations. If A is not positive definite, then this algorithm will fail by attempting to compute the square root of a negative number or by dividing by zero. This is the cheapest way to test if a symmetric matrix is a positive definite.

The matrices A and L can be partitioned to smaller submatrices:

$$A = \begin{pmatrix} A_{11} & \\ A_{21} & A_{22} \end{pmatrix} \quad \text{and} \quad L = \begin{pmatrix} L_{11} & \\ L_{21} & L_{22} \end{pmatrix} \tag{1}$$

The sizes of the submatrices are: for A_{11} and L_{11} is $(p \times p)$, for A_{21} and L_{21} is $((n-p) \times p)$, and for A_{22} and L_{22} is $((n-p) \times (n-p))$, where $p = n/2$. Matrices A_{11}, L_{11}, A_{22} and L_{22} are lower triangular. Matrices A_{21} and L_{21} are rectangular.

The formulae in the algorithm 2.11 can be modified so that the computations can be done on small matrices of (1):

1. For A_{11}, $i = 1, 2, \ldots p$ and $j = 1, 2, \ldots i$.
2. For A_{21}, $i = p+1, \ldots n$ and $j = 1, 2, \ldots p$.
3. For A_{22}, $i = p+1, \ldots n$ and $j = p+1, \ldots i$.

Algorithm 2.12 *Computation on submatrix A_{11}.*

$$\begin{aligned} &For\ j = 1, 2, \ldots p;\ l_{jj} = (a_{jj} - \textstyle\sum_{k=1}^{j-1} l_{jk}^2)^{1/2} \\ &\quad For\ i = j+1, \ldots p;\ l_{ij} = (a_{ij} - \textstyle\sum_{k=1}^{j-1} l_{ik} l_{jk})/l_{jj} \end{aligned}$$

This is just the Cholesky algorithm for submatrix A_{11}.

Algorithm 2.13 *Computation on submatrix A_{21}.*

For $j = 1, 2, \dots p$
For $i = p+1, \dots n$; $l_{ij} = (a_{ij} - \sum_{k=1}^{j-1} l_{ik}l_{jk})/l_{jj}$

Multiplying by l_{jj} ($l_{jj} \neq 0$), ordering and simplifying the formulae:

$$\sum_{k=1}^{j} l_{ik}l_{jk} = a_{ij} \quad \text{for} \quad i = p+1, \dots n \quad \text{and} \quad j = 1, 2, \dots p$$

This is just linear triangular system of equations with multiple RHS

$$L_{21}L_{11}^T = A_{21},$$

where L_{21} is unknown, L_{11}^T coefficient matrix and A_{21} RHS matrix. Such a system can be solved by the **BLAS** *subroutine* **TRSM**.

Algorithm 2.14 *Computation on submatrix A_{22}.*

for $j = p+1, \dots n$: $l_{jj} = (a_{jj} - \sum_{k=1}^{j-1} l_{jk}^2)^{1/2}$
for $i = j+1, \dots n$: $l_{ij} = (a_{ij} - \sum_{k=1}^{j-1} l_{ik}l_{jk})/l_{jj}$

By simple modification:

for $j = p+1, \dots n$: $l_{jj} = (\hat{a}_{jj} - \sum_{k=p+1}^{j-1} l_{jk}^2)^{1/2}$
for $i = j+1, \dots n$: $l_{ij} = (\hat{a}_{ij} - \sum_{k=p+1}^{j-1} l_{ik}l_{jk})/l_{jj}$

where $\hat{a}_{jj} = a_{jj} - \sum_{k=1}^{p} l_{jk}^2$ and $\hat{a}_{ij} = a_{ij} - \sum_{k=1}^{p} l_{ik}l_{jk}$ are elements of a modified submatrix

$$\hat{A}_{22} = A_{22} - L_{21}L_{21}^T, \quad \textbf{BLAS} \textit{ subroutine } \textbf{SYRK}.$$

We should apply the Cholesky algorithm for submatrix $\hat{A}_{22}$.

We can formulate the Cholesky recursive algorithm now:

Algorithm 2.15 *Cholesky recursive algorithm if lower triangular part of A is given:*

- *if $n > 1$ then*
  - L_{11} := *cholesky of* A_{11}
  - $L_{21}L_{11}^T = A_{21}$ (**TRSM**) *and* • $\hat{A}_{22} := A_{22} - L_{21}L_{21}^T$ (**SYRK**)
  - L_{22} := *cholesky of* $\hat{A}_{22}$
- *otherwise* • $L := \sqrt{A}$

2.2 Upper Triangular Case

The matrices A and U are in this case:

$$A = \begin{pmatrix} A_{11} & A_{12} \\ & A_{22} \end{pmatrix} \text{ and } L = \begin{pmatrix} U_{11} & U_{12} \\ & U_{22} \end{pmatrix}$$

Algorithm 2.21 *Cholesky Decomposition of matrix A.*

For $j = 1, 2, \ldots n;\ l_{jj} = (a_{jj} - \sum_{k=1}^{j-1} l_{jk}^2)^{1/2}$

For $i = j+1, \ldots n;\ l_{ij} = (a_{ij} - \sum_{k=1}^{j-1} l_{ik} l_{jk})/l_{jj}$

Algorithm 2.22 *Cholesky recursive algorithm if upper triangular part of A is given:*

- *if* $n > 1$ *then*
  - $U_{11} :=$ *cholesky of* A_{11}
  - $U_{11}^T U_{12} = A_{12}$ *(* **TRSM** *) and* • $\hat{A}_{22} := A_{22} - U_{12}^T U_{12}$ *(* **SYRK** *)*
  - $U_{22} :=$ *cholesky of* $\hat{A}_{22}$
- *otherwise* • $U := \sqrt{A}$

3 Performance Issues and Further Development

The project on "Recursive Numerical Linear Algebra" at UNI•C just began. The Recursive Cholesky and LU with partial pivoting factorization Algorithms are already programmed in Fortran 90. More information will be given in the next publications. Some performance results can be fund at present in [5] and [6]. The **TRSM** and **SYRK** subroutines can be made recursive too.

References

1. E. Anderson, Z. Bai, C. H. Bischof, J. Demmel, J. J. Dongarra, J. Du Croz, A. Greenbaum, S. Hammarling, A. McKenney, S. Ostrouchov and D. C. Sorensen. *LAPACK Users' Guide Release 2.0.* SIAM, Philadelphia, 1995.
2. J.W. Demmel. *Applied Numerical Linear Algebra.* SIAM, Philadelphia, 1997.
3. J. Dongarra, and J. Waśniewski. *High Performance Linear Algebra Package – LAPACK90.* Lawn number 134: http://www.netlib.org/lapack/lawns/lawn134.ps Report UNIC-98-01, UNI•C, Lyngby, Denmark, 1998. Report ut-cs-98-384, University of Tennessee, Computer Science Department, Knoxville, April, 1998.
4. G.H. Golub and C.F. Van Loan. *Matrix Computations.* Johns Hopkins University Press, Baltimore, MD, Any ed. from 1983.
5. F.G. Gustavson. *Recursive Leads to Automatic Variable Blocking for Dense Linear-Algebra Algorithms.* IBM Journal of Research and Development, Volume 41, Number 6, November 1997.
6. F.G. Gustavson, A. Henriksson, I. Jonsson, B. Kågström and P. Ling. *Superscalar GEMM-based Level 3 BLAS – The On-going Evolution of a Portable and High-Performance Library. This Proceedings*, Springer Verlag, 1998.
7. S. Metcalf and J. Reid. *Fortran 90/95 Explained.* Oxford, New York, Tokyo, Oxford University Press, 1996.
8. S. Toledo. *Locality of Reference in LU Decomposition with Partial Pivoting.* SIAM Journal on Matrix Analysis and Applications, vol. 18, No. 4, 1997.

A Fortran 90 Subroutine of Cholesky Recursive Algorithm

```
RECURSIVE SUBROUTINE RPOTRF( A, UPLO, INFO )
   USE LA_PRECISION, ONLY: WP => DP
   USE LA_AUXMOD, ONLY: ERINFO, LSAME
   USE F90_RCF, ONLY: RCF => RPOTRF, RTRSM, RSYRK
   IMPLICIT NONE
   CHARACTER(LEN=1), OPTIONAL, INTENT(IN) :: UPLO
   INTEGER, OPTIONAL, INTENT(OUT) :: INFO
   REAL(WP), INTENT(INOUT) :: A(:,:)

   CHARACTER(LEN=*), PARAMETER :: SRNAME = 'RPOTRF'
   REAL(WP), PARAMETER :: ONE = 1.0_WP
   CHARACTER(LEN=1) :: LUPLO
   INTEGER :: N, P, LINFO

   N = SIZE(A,1); LINFO = 0
   IF( PRESENT(UPLO) )THEN; LUPLO = UPLO; ELSE; &
                                 LUPLO = 'U'; ENDIF

   IF( N < 0 .OR. N /= SIZE(A,2) )THEN; LINFO = -1
   ELSE IF( .NOT. (LSAME(LUPLO,'U').OR.LSAME(LUPLO,'L')) ) &
                                          THEN; LINFO = -2

   ELSE IF (N == 1) THEN; A(1,1) = SQRT(A(1,1))
   ELSE IF( N > 0 )THEN; P=N/2
     IF( LSAME(LUPLO,'L') )THEN
       CALL RCF( A(1:P,1:P), LUPLO, LINFO )
       CALL RTRSM( A(1:P,1:P), A(P+1:N,1:P), UPLO=LUPLO, &
                                   SIDE='R', TRANSA='T' )
       CALL RSYRK( A(P+1:N,1:P), A(P+1:N,P+1:N), ALPHA=-ONE, &
                                              UPLOC=LUPLO )
       CALL RCF( A(P+1:N,P+1:N), LUPLO, LINFO )
     ELSE
       CALL RCF( A(1:P,1:P), LUPLO, LINFO )
       CALL RTRSM( A(1:P,1:P), A(1:P,P+1:N), TRANSA='T' )
       CALL RSYRK( A(1:P,P+1:N), A(P+1:N,P+1:N), ALPHA=-ONE, &
                                               TRANSA='T' )
       CALL RCF( A(P+1:N,P+1:N), LUPLO, LINFO )
     ENDIF
   ENDIF

   CALL ERINFO( LINFO, SRNAME, INFO )
END SUBROUTINE RPOTRF
```

High Performance Linear Algebra Package for FORTRAN 90

Jerzy Waśniewski[1] and Jack Dongarra[2]

[1] The Danish Computing Centre for Research and Education (UNI•C), Technical University of Denmark, Building 304, DK-2800 Lyngby, Denmark.
jerzy.wasniewski@uni-c.dk
[2] Department of Computer Science, University of Tennessee, 107 Ayres Hall, Knoxville, TN 37996-1301, USA and Mathematical Sciences Section, Oak Ridge National Laboratory, P.O.Box 2008, Bldg. 6012, Oak Ridge, TN 37831-6367, USA.
dongarra@cs.utk.edu

Abstract. LAPACK90 is a set of FORTRAN90 subroutines which interfaces FORTRAN90 with LAPACK. All LAPACK driver subroutines (including expert drivers) and some LAPACK computationals have both generic LAPACK90 interfaces and generic LAPACK77 interfaces. The remaining computationals have only generic LAPACK77 interfaces. In both types of interfaces no distinction is made between single and double precision or between real and complex data types.

1 LAPACK

LAPACK is a library of FORTRAN 77 subroutines for solving the most commonly occurring problems in numerical linear algebra. It has been designed to be efficient on a wide range of modern high-performance computers. The name LAPACK is an acronym for Linear Algebra PACKage.

LAPACK provides routines for solving systems of simultaneous linear equations, least-squares solutions of linear systems of equations, eigenvalue problems, and singular value problems. The associated matrix factorizations (LU, Cholesky, QR, SVD, Schur, generalized Schur) are also provided, as are related computations such as reordering of the Schur factorizations and estimating condition numbers. Dense and banded matrices are handled, but not general sparse matrices. In all areas, similar functionality is provided for real and complex matrices, in both single and double precision.

The original goal of the LAPACK project was to make the widely used EISPACK and LINPACK libraries run efficiently on shared-memory vector and parallel processors. On these machines, LINPACK and EISPACK are inefficient because their memory access patterns disregard the multi-layered memory hierarchies of the machines, thereby spending too much time moving data instead of doing useful floating-point operations. LAPACK addresses this problem by reorganizing the algorithms to use block matrix operations, such as matrix multiplication, in the innermost loops. These block operations can be optimized for each architecture to account for the memory hierarchy, and so provide a transportable way to achieve high efficiency on diverse modern machines. LAPACK requires that highly optimized block matrix operations be already implemented on each machine.

LAPACK routines are written so that as much as possible of the computation is performed by calls to the Basic Linear Algebra Subprograms (BLAS). While LINPACK and EISPACK are based on the vector operation kernels of the Level 1 BLAS. LAPACK is designed at the outset to exploit the Level 3 BLAS – a set of specifications for FORTRAN subprograms that do various types of matrix multiplication and the solution of triangular systems with multiple right-hand sides. Because of the coarse granularity of the Level 3 BLAS operations, their use promotes high efficiency on many high-performance computers, particularly if specially coded implementations are provided by the manufacturer.

Highly efficient machine-specific implementations of the BLAS are available for many modern high-performance computers. The BLAS enable LAPACK routines to achieve high performance with transportable software. Although a model FORTRAN implementation of the BLAS is available from netlib in the BLAS library [9]. It is not expected to perform as well as a specially tuned implementation on most high-performance computers. On some machines it may give much worse performance. But it allows users to run LAPACK software on machines that do not offer any other implementation of the BLAS.

For more information on LAPACK and references on BLAS, LINPACK and EISPACK see [1].

2 ScaLAPACK

ScaLAPACK is a library of high-performance linear algebra routines for distributed memory message-passing MIMD (Multiple Instruction Multiple Data) computers and networks of workstations supporting PVM (Parallel Virtual Machine) and/or MPI (Message Passing Interface). ScaLAPACK is a continuation of the LAPACK project (see section 1). Both libraries (LAPACK and ScaLAPACK) contain routines for solving systems of linear equations, least squares problems, and eigenvalue problems. The goals of both projects are efficiency (to run as fast as possible), scalability (as the problem size and number of processors grow), reliability (including error bounds), portability (across all important parallel machines), flexibility (so users can construct new routines from well-designed parts), and ease of use (by making the interface to LAPACK and ScaLAPACK look as similar as possible). Many of these goals, particularly portability, are aided by developing and promoting standards , especially for low-level communication and computation routines. ScaLAPACK has been successful in attaining these goals, limiting most machine dependencies to two standard libraries called the BLAS (Basic Linear Algebra Subprograms) and BLACS (Basic Linear Algebra Communication Subprograms) [10]. LAPACK runs on any machine where the BLAS are available, and ScaLAPACK runs on any machine where both the BLAS and the BLACS are available.

The library is currently written in FORTRAN 77 (with the exception of a few symmetric eigenproblem auxiliary routines written in C to exploit IEEE arithmetic) in a Single Program Multiple Data (SPMD) style using explicit message passing for interprocessor communication. The name ScaLAPACK is an acronym for Scalable Linear Algebra PACKage, or Scalable LAPACK

For more information on ScaLAPACK and references on BLAS, BLACS, PBLAS, PVM and MPI see [2].

3 FORTRAN 90

FORTRAN has always been the principal language used in the fields of scientific, numerical, and engineering. A series of revisions to the standard defining successive versions of the language has progressively enhanced its power and kept it competitive with several generations of rivals. The present FORTRAN standard is 90/95. A summary of the new features is:

- Array operations.
- Pointers.
- Improved facilities for numerical computations including a set of numerical inquiry functions.
- Parameterization of the intrinsic types, to permit processors to support short integers, very large character sets, more than two precisions for real and complex, and packed logicals.
- User-defined derived data types composed of arbitrary data structures and operations upon those structures.
- Facilities for defining collections called "modules", useful for global data definitions and for procedure libraries. These support a safe method of encapsulating derived data types.
- Requirements on a compiler to detect the use of constructs that do not conform to syntax of the language or are obsolescent.
- A few source form, more appropriate to use at a terminal
- New control constructs such as the SELECT CASE construct and a new form of the DO.
- The ability to write internal procedures and recursive procedures, and to call procedures with optional and keyword arguments.
- Dynamic storage (automatic arrays, allocatable arrays, and pointers).
- Improvements to the input-output facilities, including handling partial records and a standardized NAMELIST facility.
- Many new intrinsic procedures.

Taken together, the new features contained in FORTRAN 90/95 ensure that the FORTRAN language will continue to be used successfully for a long time to come. The fact that it contains the whole of FORTRAN 77 as a subset means that conversion to FORTRAN 90/95 is as simple as conversion to another FORTRAN 77 processor. For more information on FORTRAN 90/95 see [7].

4 High Performance FORTRAN (HPF)

FORTRAN is reaching its limitations on the latest generations of high perforce computers. FORTRAN was originally developed for serial machines with linear memory architectures. In the past several years it has become increasingly apparent that a language design relying on this architectural features creates difficulties when executing on parallel machines. One symptom of this is the proliferation of parallel FORTRAN dialects, each specialized to the machine where it was first implemented. As the number of competing parallel machines on the market increases, the lack of a standard parallel FORTRAN is becoming increasingly serious. HPF solves this problem. The overriding goal of HPF was therefore to produce a dialect of FORTRAN that could be used on variety of parallel machines. HPF is an extension of FORTRAN 90/95. The array calculation and dynamic storage allocation features of FORTRAN 90, and the **FORALL** statement, the **PURE** and **EXTRINSIC** attributes of FORTRAN 95, make it natural base for HPF. The new HPF language futures fall into four categories with respect to FORTRAN 90/95:

- New directives.
- New language syntax.
- Library routines.
- Language restrictions.

For more information on HPF see [5].

5 LAPACK90

5.1 LAPACK for FORTRAN 90

All LAPACK driver subroutines (including expert drivers) and some LAPACK computationals have both generic LAPACK90 interfaces and generic LAPACK77 interfaces. The remaining computationals have only generic LAPACK77 interfaces. In both types of interfaces no distinction is made between single and double precision or between real and complex data types. The use of the LAPACK90 (LAPACK77) interface requires the user to specify the F90_LAPACK (F77_LAPACK) module.

For example, the LA_GESV driver subroutine, which solves a general system of linear equations, can be called in the following ways:

- CALL LA_GESV(A, B, IPIV=ipiv, INFO=info)
  Module F90_LAPACK is needed in this case.
- CALL LA_GESV(N, NRHS, A, LDA, IPIV, B, LDB, INFO)
  Module F77_LAPACK is needed in this case.

The documentation for LAPACK90 see in [3,4]. The LAPACK90 library and documentation is successively updated. The LAPACK 90 Users' Guide is in progress. The present implementation of the LAPACK90 can be summarized in the following titles:

- Driver Routines for Linear Equations.
- Expert Driver Routines for Linear Equations.
- Driver Routines for Linear Least Squares Problems.
- Driver Routines for generalized Linear Least Squares Problems.
- Driver Routines for Standard Eigenvalue and Singular Value Problems.
- Divide and Conquer Driver Routines for Standard Eigenvalue Problems.
- Expert Driver Routines for Standard Eigenvalue Problems.
- Driver Routines for Generalized Eigenvalue and Singular Value Problems.
- Some Computational Routines for Linear Equations and Eigenproblems.

5.2 Future Work: ScaLAPACK for HPF

Work on an HPF interface for ScaLAPACK has been started by a number of groups: such as the University of Tennessee (see [6, 8]), and at UNI•C. The plan is to develop an HPF interface for several of the more heavily used ScaLAPACK subroutines and test programs.

References

1. E. Anderson, Z. Bai, C. H. Bischof, J. Demmel, J. J. Dongarra, J. Du Croz, A. Greenbaum, S. Hammarling, A. McKenney, S. Ostrouchov and D. C. Sorensen. *LAPACK Users' Guide Release 2.0.* SIAM, Philadelphia, 1995.
2. L.S. Blackford, J. Choi, A. Ceary, E. D'Azevedo, J. Demmel, I. Dhilon, J. Dongarra, S. Hammarling, G. Henry, A. Petitet, K. Stanley, D. Walker, and R.C. Whaley. *ScaLAPACK Users' Guide.* SIAM, Philadelphia, 1997.
3. L.S. Blackford, J.J. Dongarra, J. Du Croz, S. Hammarling, and J. Waśniewski. *LAPACK90 - FORTRAN90 version of LAPACK.* Website: http://www.netlib.org/lapack90/ (1997)
4. J. Dongarra, and J. Waśniewski. *High Performance Linear Algebra Package – LAPACK90.* Lawn number 134: http://www.netlib.org/lapack/lawns/lawn134.ps Report UNIC-98-01, UNI•C, Lyngby, Denmark, 1998. Report ut-cs-98-384, University of Tennessee, Computer Science Department, Knoxville, April, 1998.
5. C.H. Koelbel, D.B. Lovemann, R.S. Schreiber, G.L. Steele Jr., and M.E. Zosel. *The High Performance FORTRAN Handbook.* The MIT Press Cambridge, Massachusetts, London, England, 1994.
6. *P.A.R. Lorenzo, A. Mü ller, Y. Murakami, and B.J.N. Wylie.* High Performance FORTRAN Interfacing to ScaLAPACK. In J. Waśniewski, J. Dongarra, K. Madsen, and D. Olesen (Eds.), Applied Parallel Computing, Industrial Computation and Optimization, Third International Workshop, PARA'96, Lyngby, Denmark, August 1996, Proceedings, Lecture Notes in Computer Science No. 1184, Springer-Verlag, 1996, pp. 457-466
7. M. Metcalf and J. Reid. *FORTRAN 90/95 Explained.* Oxford, New York, Tokyo, Oxford University Press, 1996.
8. *R.C. Whaley. HPF Interface to ScaLAPACK.* Website: http://www.netlib.org/scalapack/prototype/ (1997).
9. *BLAS (Basic Linear Algebra Subprograms).* Website: http://www.netlib.org/blas/
10. *BLACS (Basic Linear Algebra Communication Subprograms).* Website: http://www.cs.utk.edu/~rwhaley/Blacs.html

Author Index

Springer

Lecture Notes in Computer Science

For information about Vols. 1–1458
please contact your bookseller or Springer-Verlag

Vol. 1459: D.G. Feitelson, L. Rudolph (Eds.), Job Scheduling Strategies for Parallel Processing. Proceedings, 1998. VII, 257 pages. 1998.

Vol. 1460: G. Quirchmayr, E. Schweighofer, T.J.M. Bench-Capon (Eds.), Database and Expert Systems Applications. Proceedings, 1998. XVI, 905 pages. 1998.

Vol. 1461: G. Bilardi, G.F. Italiano, A. Pietracaprina, G. Pucci (Eds.), Algorithms – ESA'98. Proceedings, 1998. XII, 516 pages. 1998.

Vol. 1462: H. Krawczyk (Ed.), Advances in Cryptology - CRYPTO '98. Proceedings, 1998. XII, 519 pages. 1998.

Vol. 1463: N.E. Fuchs (Ed.), Logic Program Synthesis and Transformation. Proceedings, 1997. X, 343 pages. 1998.

Vol. 1464: H.H.S. Ip, A.W.M. Smeulders (Eds.), Multimedia Information Analysis and Retrieval. Proceedings, 1998. VIII, 264 pages. 1998.

Vol. 1465: R. Hirschfeld (Ed.), Financial Cryptography. Proceedings, 1998. VIII, 311 pages. 1998.

Vol. 1466: D. Sangiorgi, R. de Simone (Eds.), CONCUR'98: Concurrency Theory. Proceedings, 1998. XI, 657 pages. 1998.

Vol. 1467: C. Clack, K. Hammond, T. Davie (Eds.), Implementation of Functional Languages. Proceedings, 1997. X, 375 pages. 1998.

Vol. 1468: P. Husbands, J.-A. Meyer (Eds.), Evolutionary Robotics. Proceedings, 1998. VIII, 247 pages. 1998.

Vol. 1469: R. Puigjaner, N.N. Savino, B. Serra (Eds.), Computer Performance Evaluation. Proceedings, 1998. XIII, 376 pages. 1998.

Vol. 1470: D. Pritchard, J. Reeve (Eds.), Euro-Par'98: Parallel Processing. Proceedings, 1998. XXII, 1157 pages. 1998.

Vol. 1471: J. Dix, L. Moniz Pereira, T.C. Przymusinski (Eds.), Logic Programming and Knowledge Representation. Proceedings, 1997. IX, 246 pages. 1998. (Subseries LNAI).

Vol. 1472: B. Freitag, H. Decker, M. Kifer, A. Voronkov (Eds.), Transactions and Change in Logic Databases. Proceedings, 1996, 1997. X, 396 pages. 1998.

Vol. 1473: X. Leroy, A. Ohori (Eds.), Types in Compilation. Proceedings, 1998. VIII, 299 pages. 1998.

Vol. 1474: F. Mueller, A. Bestavros (Eds.), Languages, Compilers, and Tools for Embedded Systems. Proceedings, 1998. XIV, 261 pages. 1998.

Vol. 1475: W. Litwin, T. Morzy, G. Vossen (Eds.), Advances in Databases and Information Systems. Proceedings, 1998. XIV, 369 pages. 1998.

Vol. 1476: J. Calmet, J. Plaza (Eds.), Artificial Intelligence and Symbolic Computation. Proceedings, 1998. XI, 309 pages. 1998. (Subseries LNAI).

Vol. 1477: K. Rothermel, F. Hohl (Eds.), Mobile Agents. Proceedings, 1998. VIII, 285 pages. 1998.

Vol. 1478: M. Sipper, D. Mange, A. Pérez-Uribe (Eds.), Evolvable Systems: From Biology to Hardware. Proceedings, 1998. IX, 382 pages. 1998.

Vol. 1479: J. Grundy, M. Newey (Eds.), Theorem Proving in Higher Order Logics. Proceedings, 1998. VIII, 497 pages. 1998.

Vol. 1480: F. Giunchiglia (Ed.), Artificial Intelligence: Methodology, Systems, and Applications. Proceedings, 1998. IX, 502 pages. 1998. (Subseries LNAI).

Vol. 1481: E.V. Munson, C. Nicholas, D. Wood (Eds.), Principles of Digital Document Processing. Proceedings, 1998. VII, 152 pages. 1998.

Vol. 1482: R.W. Hartenstein, A. Keevallik (Eds.), Field-Programmable Logic and Applications. Proceedings, 1998. XI, 533 pages. 1998.

Vol. 1483: T. Plagemann, V. Goebel (Eds.), Interactive Distributed Multimedia Systems and Telecommunication Services. Proceedings, 1998. XV, 326 pages. 1998.

Vol. 1484: H. Coelho (Ed.), Progress in Artificial Intelligence – IBERAMIA 98. Proceedings, 1998. XIII, 421 pages. 1998. (Subseries LNAI).

Vol. 1485: J.-J. Quisquater, Y. Deswarte, C. Meadows, D. Gollmann (Eds.), Computer Security – ESORICS 98. Proceedings, 1998. X, 377 pages. 1998.

Vol. 1486: A.P. Ravn, H. Rischel (Eds.), Formal Techniques in Real-Time and Fault-Tolerant Systems. Proceedings, 1998. VIII, 339 pages. 1998.

Vol. 1487: V. Gruhn (Ed.), Software Process Technology. Proceedings, 1998. VIII, 157 pages. 1998.

Vol. 1488: B. Smyth, P. Cunningham (Eds.), Advances in Case-Based Reasoning. Proceedings, 1998. XI, 482 pages. 1998. (Subseries LNAI).

Vol. 1489: J. Dix, L. Fariñas del Cerro, U. Furbach (Eds.), Logics in Artificial Intelligence. Proceedings, 1998. X, 391 pages. 1998. (Subseries LNAI).

Vol. 1490: C. Palamidessi, H. Glaser, K. Meinke (Eds.), Principles of Declarative Programming. Proceedings, 1998. XI, 497 pages. 1998.

Vol. 1491: W. Reisig, G. Rozenberg (Eds.), Lectures on Petri Nets I: Basic Models. XII, 683 pages. 1998.

Vol. 1492: W. Reisig, G. Rozenberg (Eds.), Lectures on Petri Nets II: Applications. XII, 479 pages. 1998.

Vol. 1493: J.P. Bowen, A. Fett, M.G. Hinchey (Eds.), ZUM '98: The Z Formal Specification Notation. Proceedings, 1998. XV, 417 pages. 1998.

Vol. 1497: V. Alexandrov, J. Dongarra (Eds.), Recent Advances in Parallel Virtual Machine and Message Passing Interface. Proceedings, 1998. XII, 412 pages. 1998.

Vol. 1498: A.E. Eiben, T. Bäck, M. Schoenauer, H.-P. Schwefel (Eds.), Parallel Problem Solving from Nature – PPSN V. Proceedings, 1998. XXIII, 1041 pages. 1998.

Vol. 1499: S. Kutten (Ed.), Distributed Computing. Proceedings, 1998. XII, 419 pages. 1998.

Vol. 1501: M.M. Richter, C.H. Smith, R. Wiehagen, T. Zeugmann (Eds.), Algorithmic Learning Theory. Proceedings, 1998. XI, 439 pages. 1998. (Subseries LNAI).

Vol. 1502: G. Antoniou, J. Slaney (Eds.), Advanced Topics in Artificial Intelligence. Proceedings, 1998. XI, 333 pages. 1998. (Subseries LNAI).

Vol. 1503: G. Levi (Ed.), Static Analysis. Proceedings, 1998. IX, 383 pages. 1998.

Vol. 1504: O. Herzog, A. Günter (Eds.), KI-98: Advances in Artificial Intelligence. Proceedings, 1998. XI, 355 pages. 1998. (Subseries LNAI).

Vol. 1505: D. Caromel, R.R. Oldehoeft, M. Tholburn (Eds.), Computing in Object-Oriented Parallel Environments. Proceedings, 1998. XI, 243 pages. 1998.

Vol. 1506: R. Koch, L. Van Gool (Eds.), 3D Structure from Multiple Images of Large-Scale Environments. Proceedings, 1998. VIII, 347 pages. 1998.

Vol. 1507: T.W. Ling, S. Ram, M.L. Lee (Eds.), Conceptual Modeling – ER '98. Proceedings, 1998. XVI, 482 pages. 1998.

Vol. 1508: S. Jajodia, M.T. Özsu, A. Dogac (Eds.), Advances in Multimedia Information Systems. Proceedings, 1998. VIII, 207 pages. 1998.

Vol. 1510: J.M. Zytkow, M. Quafafou (Eds.), Principles of Data Mining and Knowledge Discovery. Proceedings, 1998. XI, 482 pages. 1998. (Subseries LNAI).

Vol. 1511: D. O'Hallaron (Ed.), Languages, Compilers, and Run-Time Systems for Scalable Computers. Proceedings, 1998. IX, 412 pages. 1998.

Vol. 1512: E. Giménez, C. Paulin-Mohring (Eds.), Types for Proofs and Programs. Proceedings, 1996. VIII, 373 pages. 1998.

Vol. 1513: C. Nikolaou, C. Stephanidis (Eds.), Research and Advanced Technology for Digital Libraries. Proceedings, 1998. XV, 912 pages. 1998.

Vol. 1514: K. Ohta, D. Pei (Eds.), Advances in Cryptology – ASIACRYPT'98. Proceedings, 1998. XII, 436 pages. 1998.

Vol. 1515: F. Moreira de Oliveira (Ed.), Advances in Artificial Intelligence. Proceedings, 1998. X, 259 pages. 1998. (Subseries LNAI).

Vol. 1516: W. Ehrenberger (Ed.), Computer Safety, Reliability and Security. Proceedings, 1998. XVI, 392 pages. 1998.

Vol. 1517: J. Hromkovič, O. Sýkora (Eds.), Graph-Theoretic Concepts in Computer Science. Proceedings, 1998. X, 385 pages. 1998.

Vol. 1518: M. Luby, J. Rolim, M. Serna (Eds.), Randomization and Approximation Techniques in Computer Science. Proceedings, 1998. IX, 385 pages. 1998.

Vol. 1520: M. Maher, J.-F. Puget (Eds.), Principles and Practice of Constraint Programming - CP98. Proceedings, 1998. XI, 482 pages. 1998.

Vol. 1521: B. Rovan (Ed.), SOFSEM'98: Theory and Practice of Informatics. Proceedings, 1998. XI, 453 pages. 1998.

Vol. 1522: G. Gopalakrishnan, P. Windley (Eds.), Formal Methods in Computer-Aided Design. Proceedings, 1998. IX, 529 pages. 1998.

Vol. 1524: G.B. Orr, K.-R. Müller (Eds.), Neural Networks: Tricks of the Trade. VI, 432 pages. 1998.

Vol. 1525: D. Aucsmith (Ed.), Information Hiding. Proceedings, 1998. IX, 369 pages. 1998.

Vol. 1526: M. Broy, B. Rumpe (Eds.), Requirements Targeting Software and Systems Engineering. Proceedings, 1997. VIII, 357 pages. 1998.

Vol. 1528: B. Preneel, V. Rijmen (Eds.), State of the Art in Applied Cryptography. Revised Lectures, 1997. VIII, 395 pages. 1998.

Vol. 1529: D. Farwell, L. Gerber, E. Hovy (Eds.), Machine Translation and the Information Soup. Proceedings, 1998. XIX, 532 pages. 1998. (Subseries LNAI).

Vol. 1530: V. Arvind, R. Ramanujam (Eds.), Foundations of Software Technology and Theoretical Computer Science. XII, 369 pages. 1998.

Vol. 1531: H.-Y. Lee, H. Motoda (Eds.), PRICAI'98: Topics in Artificial Intelligence. XIX, 646 pages. 1998. (Subseries LNAI).

Vol. 1096: T. Schael, Workflow Management Systems for Process Organisations. Second Edition. XII, 229 pages. 1998.

Vol. 1532: S. Arikawa, H. Motoda (Eds.), Discovery Science. Proceedings, 1998. XI, 456 pages. 1998. (Subseries LNAI).

Vol. 1533: K.-Y. Chwa, O.H. Ibarra (Eds.), Algorithms and Computation. Proceedings, 1998. XIII, 478 pages. 1998.

Vol. 1538: J. Hsiang, A. Ohori (Eds.), Advances in Computing Science – ASIAN'98. Proceedings, 1998. X, 305 pages. 1998.

Vol. 1540: C. Beeri, P. Buneman (Eds.), Database Theory – ICDT'99. Proceedings, 1999. XI, 489 pages. 1999.

Vol. 1541: B. Kågström, J. Dongarra, E. Elmroth, J. Waśniewski (Eds.), Applied Parallel Computing. Proceedings, 1998. XIV, 586 pages. 1998.

Vol. 1542: H.I. Christensen (Ed.), Computer Vision Systems. Proceedings, 1999. XI, 554 pages. 1999.

Vol. 1543: S. Demeyer, J. Bosch (Eds.), Object-Oriented Technology ECOOP'98 Workshop Reader. 1998. XXI, 571 pages. 1998.

Vol. 1544: C. Zhang, D. Lukose (Eds.), Multi-Agent Systems. Proceedings, 1998. VII, 195 pages. 1998. (Subseries LNAI).

Vol. 1546: B. Möller, J.V. Tucker (Eds.), Prospects for Hardware Foundations. Survey Chapters, 1998. X, 468 pages. 1998.

Vol. 1548: A.M. Haeberer (Ed.), Algebraic Methodology and Software Technology. Proceedings, 1999. XI, 531 pages. 1999.